Control Systems

Third Edition

Control Systems

Third Edition

Dr. N.C. Jagan
B.E., M.E., Ph.D., MISTE, FIE

Professor in Electrical Engineering (Retd.)

University College of Engineering
Osmania University
Hyderabad.

BSP **BS Publications**
A unit of **BSP Books Pvt., Ltd.**

4-4-309/316, Giriraj Lane, Sultan Bazar,
Hyderabad - 500 095
Phone : 040 - 23445605, 23445688

Control Systems, ***Third Edition by Dr. N.C. Jagan***

Published by :

BSP **BS Publications**
A unit of **BSP Books Pvt., Ltd.**

4-4-309/316, Giriraj Lane, Sultan Bazar,
Hyderabad - 500 095
Phone : 040 - 23445605, 23445688
e-mail : info@bspbooks.net

ISBN : 978-93-52300-80-8 (HB)

Preface to Third Edition

I am glad to bring the third edition of this book. In the previous edition, the text was revised to suit the needs of syllabi of various universities to be covered in one semester.

Now in this edition, the Questions from Gate Examination in the years 2003 to 2014 in EEE and ECE, pertaining to control systems, in the new pattern of examination are added yearwise.

This would go a long way in helping the students to prepare well for the Gate Examination.

I am thankful to the publishers for bringing this third edition.

- Author

Preface to First Edition

The primary purpose of this book is to provide a good textbook on control systems in simple language so that students with urban as well as rural background can easily understand. As many engineering colleges are situated in rural areas, where competent faculty is difficult to find, it is hoped that this book will enable the students to study the subject by themselves. The principles of feedback control systems are explained in detail and illustrated with suitable examples. Specifically the contents are selected to suit the syllabus framed by J.N. Technological University and some other Technological Universities.

As a first step in studying the Control Systems, a mathematical model has to be developed for the system to be controlled. Different models for physical systems are developed in chapter 2. Block diagram representation and signal flow graph representation are developed. Different aspects of open loop and closed loop Control Systems are brought out clearly. The components used in a Control Systems are explained in detail and mathematical models to represent them are also developed. The good and bad effects of feedback are explained clearly. Sensitivity analysis is also included in chapter 2.

The time response specifications of a Control System are developed in chapter 3. Stability of Control Systems is introduced in chapter 4 and algebraic criterion for determining the absolute stability of systems is also discussed. The locus of closed loop poles as one of the parameter changes, is developed as root locus technique in chapter 5.

Frequency response analysis is described in chapter 6. The correlation between time and frequency domain specifications is also established. Chapter 7. describes the frequency domain technique of determining the stability of the systems. The concept of gain margin and phase margin are introduced as measures of relative stability.

The ultimate aim in studying the control theory is to enable the student to design a suitable controller to obtain the desired response from a given plant. The design of a controller based on frequency domain technique is discussed in chapter 8.

With the advent of computers, state variable representation of Control Systems has been employed invariably, in developing various techniques to design modern Control Systems. An introduction to the state variable representation and the methods of analysis of these systems is given in chapter 9.

MATLAB, a software tool is invariably used today to solve all the problems in Control Systems. Time domain response, frequency domain response, root locus plots, Bode plots, Nyquist plots etc., can be conveniently obtained by using this mathematical tool. Design also can be carried out in a simple way using MATLAB. As present day students are all computer savy no attempt is made to present programs in MATLAB to perform various tasks, in this text. The students are, however encouraged to solve all the problems in the text, using this software.

This book, as already mentioned, is intended as a text book for one semester course in Control Systems. The author has taught this subject several times and it is felt that this book will help not only the students but also the teachers offering this course.

The author wishes to express his sincere thanks to M/S BS publications for their encouragement and help in bringing out this book. The author also wishes to offer his appreciation to the efforts put in by Mr. Naresh, Ms. Kalpana, who are responsible for processing the book on the computer.

I also wish to thank all my family members who have supported me throughout the preparation of this book.

-Author

Contents

1 Introduction

1.1 Why Automatic Control ?

Automatic control of many day to day tasks relieves the human beings from performing repetitive manual operations. Automatic control allows optimal performance of dynamic systems, increases productivity enormously, removes drudgery of performing same task again and again. Imagine manual control of a simple room heating system. If the room temperature is to be maintained at a desired temperature T^0C, by controlling the current in an electrical heating system, the current may be adjusted by moving the variable arm in a rheostat. The temperature of the room depends on a host of factors : number of persons in the room, the opening and closing of doors due to persons moving in and out, fluctuation in the supply voltage etc. A human operator has to continuously monitor the temperature indicated by a thermometer and keep on adjusting the rheostat to maintain the temperature all the twenty four hours. The operator should be continuously alert and relentlessly perform a simple job of moving the arm of the rheostat. Any mistake on his part may result in great discomfiture to the persons in the room.

Now, imagine the same operation of measuring the temperature, estimating the error between the desired temperature and the actual temperature, moving the arm of the rheostat accurately by an automatic controller. Since error between the actual temperature and the desired temperature is continuously obtained and used to activate the controller, any disturbances caused due to movements of persons occupying the room, supply variations etc. will be automatically taken care of. How much of a relief it is ! This is only a simple task, but many complex industrial processes, space craft systems, missile guidance systems, robotic systems, numerical control of machine tools employ automatic control systems. There is no field in engineering where automatic control is not employed. Even human system is a very complex automatic feedback control system. The modern engineers and scientists must, therefore, have a thorough knowledge of the principles of automatic control systems.

The first automatic control was invented by James Watt. He employed a centrifugal or fly ball governor for the speed control of a steam engine in 1770. But much of the advances had to wait for more than a hundred years, until *Minorsky*, *Hazen* and *Nyquist* contributed significantly in the development of control system theory. Hazen coined the word "*servo mechanisms*" to describe feedback control systems in which the variable to be controlled is a mechanical position, velocity or acceleration of a given object. During 1940s, frequency response methods and root locus techniques were developed to design linear, stable, closed loop control systems with given performance measures. In later part of 1950s, much emphasis was given in designing systems, which not only satisfied given performance measures, but also provided optimum design in a given sense. As the systems became more and more complex with more number of inputs and outputs and with the advent of digital computers, modern control theory reverted back to methods based on time domain analysis and synthesis using state variable representations.

In the period between 1960 and 1980, to cope up with the complexity and stringent requirements on accuracy, speed and cost adaptive control was developed. Both deterministic and stochastic systems were considered and controllers were designed which were optimal, adaptive and robust. The principles developed in automatic control theory were not only used in engineering applications, but also in non engineering systems like economic, socio economic systems and biological systems.

1.2 Open Loop and Closed Loop Control Systems

Open Loop Control Systems : A system in which the output has no effect on the control action is known as an open loop control system. For a given input the system produces a certain output. If there are any disturbances, the out put changes and there is no adjustment of the input to bring back the output to the original value. A perfect calibration is required to get good accuracy and the system should be free from any external disturbances. No measurements are made at the output.

A traffic control system is a good example of an open loop system. The signals change according to a preset time and are not affected by the density of traffic on any road. A washing machine is another example of an open loop control system. The quality of wash is not measured; every cycle like wash, rinse and dry cycle goes according to a preset timing.

Closed Loop Control Systems : These are also known as feedback control systems. A system which maintains a prescribed relationship between the controlled variable and the reference input, and uses the difference between them as a signal to activate the control, is known as a feedback control system. The output or the controlled variable is measured and compared with the reference input and an error signal is generated. This is the activating signal to the controller which, by its action, tries to reduce the error. Thus the controlled variable is continuously fedback and compared with the input signal. If the error is reduced to zero, the output is the desired output and is equal to the reference input signal.

1.3 Open Loop Vs Closed Loop Control Systems

The open loop systems are simple and easier to build. Stability, which will be discussed in later chapters, is not a problem. Open loop systems are cheaper and they should be preferred whenever there is a fixed relationship between the input and the output and there are no disturbances. Accuracy is not critical in such systems.

Closed loop systems are more complex, use more number of elements to build and are costly. The stability is a major concern for closed loop systems. We have to ensure that the system is stable and will not cause undesirable oscillations in the output. The major advantage of closed loop system is that it is insensitive to external disturbances and variations in parameters. Comparatively cheaper components can be used to build these systems, as accuracy and tolerance do not affect the performance. Maintenance of closed loop systems is more difficult than open loop systems. Overall gain of the system is also reduced.

Open Loop Systems

Advantages

1. They are simple and easy to build.
2. They are cheaper, as they use less number of components to build.
3. They are usually stable.
4. Maintenance is easy.

Disadvantages

1. They are less accurate.
2. If external disturbances are present, output differs significantly from the desired value.
3. If there are variations in the parameters of the system, the output changes.

Closed Loop Systems

Advantages

1. They are more accurate.
2. The effect of external disturbance signals can be made very small.
3. The variations in parameters of the system do not affect the output of the system i.e. the output may be made less sensitive to variation is parameters. Hence forward path components can be of less precision. This reduces the cost of the system.
4. Speed of the response can be greatly increased.

Disadvantages

1. They are more complex and expensive
2. They require higher forward path gains.
3. The systems are prone to instability. Oscillations in the output many occur.
4. Cost of maintenance is high.

1.4 Feedback Control Systems

Fig. 1.1 represents a feedback control system.

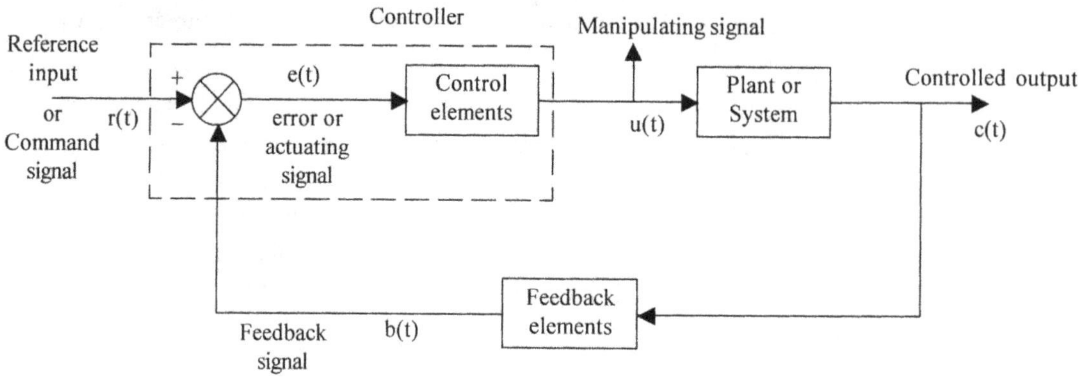

Fig. 1.1 A feedback control system

A feedback control system is represented as an interconnection of blocks characterized by an input output relation. This method of representing a control system is known as a block diagram representation. While other methods are also used to represent the control system, this is more popular. The input to the entire system is called as a reference input or a command input, r(t). An error detector senses the difference between the reference input and the feedback signal equal to or proportional to the controlled output. The feedback elements measure the controlled output and convert or transform it to a suitable value so that it can be compared with the reference input. If the feedback signal, b(t), is equal to the controlled output, c(t), the feedback system is called as unity feedback system.

The difference between the reference input and the feedback signal is known as the error signal or actuating signal e(t), This signal is the input to the control elements which produce a signal known as manipulated variable, u(t). This signal manipulates the system or plant dynamics so that the desired output is obtained. The controller acts until the error between the output variable and the reference input is zero. If the feedback path is absent, the system becomes an open loop control system and is represented in Fig. 1.2.

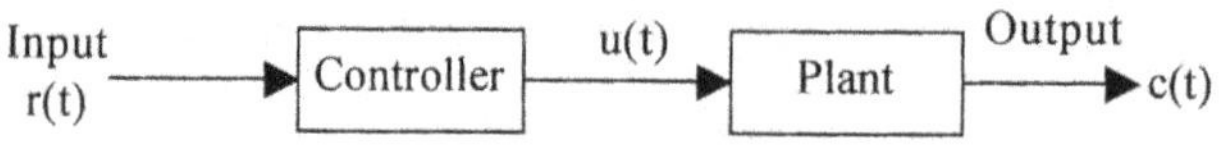

Fig. 1.2 Open loop control system

1.5 Classification of Control Systems

Depending on the type of signals present at the various parts of a feedback control system, the system may be classified as a (i) continuous time feedback control system or a (ii) discrete time feedback control system.

1.5.1 Continuous Time Feedback Control Systems

If the signals in all parts of a control system are continuous functions of time, the system is classified as continuous time feedback control system. Typically all control signals are of low frequency and if these signals are unmodulated, the system is known as a d.c. *control system*. These systems use potentiometers as error detectors, d.c amplifiers to amplify the error signal, d.c. servo motor as actuating device and d.c tachometers or potentiometers as feedback elements. If the control signal is modulated by an a.c carrier wave, the resulting system is usually referred to as an a.c control system. These systems frequently use synchros as error detectors and modulators of error signal, a.c amplifiers to amplify the error signal and a.c servo motors as actuators. These motors also serve as demodulators and produce an unmodulated output signal.

1.5.2 Discrete Data Feedback Control Systems

Discrete data control systems are those systems in which at one or more parts of the feedback control system, the signal is in the form of pulses. Usually, the error in such system is sampled at uniform rate and the resulting pulses are fed to the control system. In most sampled data control systems, the signal is reconstructed as a continuous signal, using a device called 'hold device'. Holds of different orders are employed, but the most common hold device is a zero order hold. It holds the signal value constant, at a value equal to the amplitude of the input time function at that sampling instant, until the next sampling instant. A typical discrete data control system is shown in Fig. 1.3 which uses a sampler and a data hold.

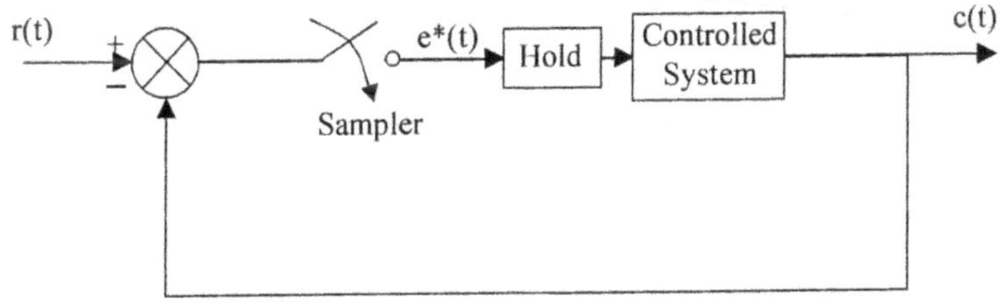

Fig. 1.3 Discrete data control system

These systems are also known as sampled data control systems.

Discreet data control systems, in which a digital computer is used as one of the elements, are known as digital control systems. The input and output to the digital computer must be binary numbers and hence these systems require the use of digital to analog and analog to digital converters. A typical digital control system is shown in Fig. 1.4.

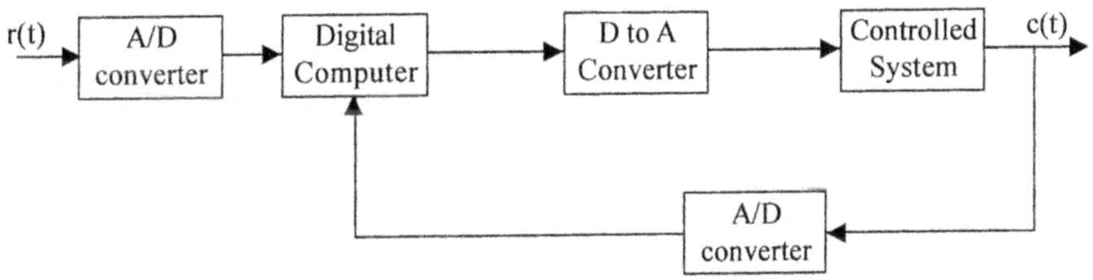

Fig. 1.4 Digital feedback control system

Digital devices may be employed in the feedback circuits as measuring elements.

A further classification of control systems can be made depending on the nature of the systems, namely,

1. Linear control systems
2. Non-linear control systems

1.5.3 Linear Control Systems

If a system obeys superposition principle, the system is said to be a linear system. Let $x_1(t)$ and $x_2(t)$ be two inputs to a system and $y_1(t)$ and $y_2(t)$ be the corresponding outputs. For arbitrary real constants k_1 and k_2, and for input $k_1 x_1(t) + k_2 x_2(t)$, if the output of the system is given by $k_1 y_1(t) + k_2 y_2(t)$, then the system is said to be a linear system. There are several simple techniques available for the analysis and design of linear control systems.

1.5.4 Non-Linear Control Systems

Any system which does not obey superposition principle is said to be a non-linear system. Physical systems are in general non-lienar and analysis of such systems is very complicated. Hence these systems are usually linearlised and well known linear techniques are used to analyse them.

These systems can be further classified depending on whether the parameters of the system are constants, or varying with respect to time. When the input to a system is delayed by T seconds, if the output is also delayed by the same time T, the system is said to be a time invariant system. Thus

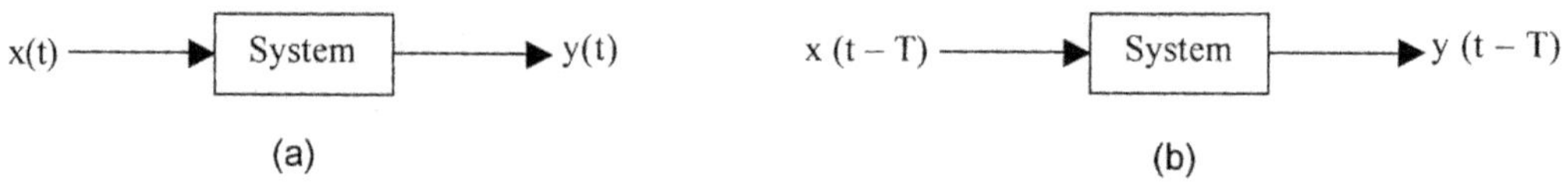

Fig. 1.5 Time invariant system.

On the other hand, if the output is dependent on the time of application of the input, the system is said to be a time varying system. Like non-linear systems, time varying systems also are more complicated for analysis. In this text we will be dealing with linear time invariant continuous systems only.

The layout of this book is as follows :

The mathematical modelling of processes suitable for analysis and design of controllers is discussed in Chapter 2. Typical examples from electrical, mechanical, pneumatic and hydraulic systems are given. The transfer function of the overall system is obtained by block diagram and signal flow graph representation of the systems. The effects of feedback on the performance of the system are also discussed.

In chapter 3, time domain specifications of the control system are defined with respect to the response of a typical second order system, for unit step input. Steady state errors are defined and the use of PID controllers is discussed to satisfy the design specifications of a control system. In chapter 4, the stability aspects of the system are discussed and algebraic criteria for obtaining the stability of the system are developed.

The roots of the characteristic equation of the system determine the behaviour and stability of a control system. The roots change as the parameters are changed. The concept of the locus of these roots as one of the parameters, usually the gain of the amplifier, is changed from o to ∞ is discussed in chapter 5. The design of a control system is rendered very easy by considering the response of the system to sinusoidal signals. Frequency domain analysis and development of frequency domain specifications are discussed in chapter 6.

Relative stability aspects are considered in chapter 7. Nyquist stability criterion is developed and measures of relative stability, viz, gain margin on phase margin are defined.

In chapter 8, design of compensating RC networks to satisfy the design specifications of a control system in frequency domain is discussed. In chapter 9, state space representation of control systems is developed, which enables modern techniques to be used in the design of control systems.

2 Mathematical Modelling of Physical Systems

2.1 Introduction

Whenever a task is to be performed, a set of physical objects are connected together and a suitable input is given to them, to obtain the desired output. This group of objects is usually termed as the '*system*'. The system may consist of physical objects and it may contain components, biological economical or managerial in nature. In order to analyse, design or synthesise a complex system, a physical model has to be obtained. This physical model may be a simplified version of the more complex system. Certain assumptions are made to describe the nature of the system. Usually all physical systems in the world are nonlinear in nature. But under certain conditions these systems may be approximated by linear systems. Hence for certain purposes, a linear model may be adequate. But if stringent accuracy conditions are to be satisfied, linear model may not be suitable. Similarly, the parameters of the system may be functions of time. But if they are varying very slowly, they may be assumed to be constant. In many engineering systems the elements are considered to be lumped and their behaviour is described by considering the effect at its end points called *terminals*. Long lines transmitting electrical signals, may not be adequately represented by lumped elements. A distributed parameter representation may be called for in this case. Hence depending on the requirements of a given job, suitable assumptions have to be made and a '*physical model*' has to be first defined. The behaviour of this physical model is then described in terms of a mathematical model so that known techniques of mathematical analysis can be applied to the given system.

2.2 Mathematical Models of Physical Systems

The system may be considered to be consisting of an inter connection of smaller components or elements, whose behaviour can be described by means of mathematical equations or relationships. In this book, we will be considering systems made up of elements which are linear, lumped and time invariant. An element is said to be linear if it obeys the principle of super position and homogeneity. If the responses of the element for inputs $x_1(t)$ and $x_2(t)$ are $y_1(t)$ and $y_2(t)$ respectively, the element is linear if the response to the input, $k_1\, x_1(t) + k_2\, x_2(t)$ is $k_1\, y_1(t) + k_2\, y_2(t)$ as shown in Fig. 2.1.

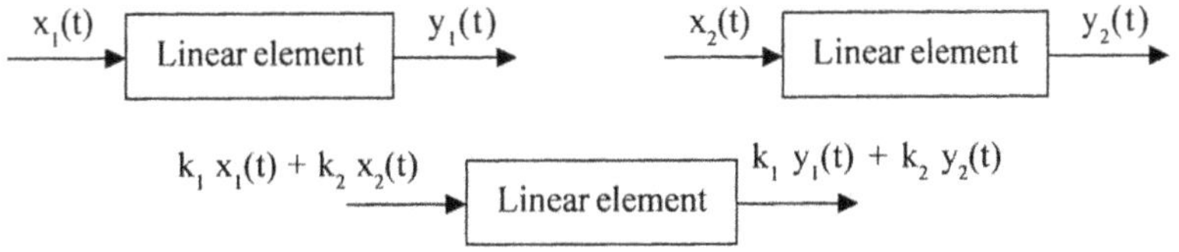

Fig. 2.1 Definition of linear element

An element is said to be '*lumped*' if the physical dimensions or spacial distribution of the element does not alter the signal passing through it. The behaviour of such elements are adequately represented by the effects at the end points called *terminals*. The temperature of a body may be treated as same, at all points of the body under certain conditions. Similarly the mass of a body may be considered as concentrated at a point. A rotating shaft may be considered as rigid. An electrical resistor may be represented by a lumped element, since the current entering at one terminal leaves the other terminal without undergoing any change. The voltage distribution in the physical body of the resistor is not considered. Only the voltage across its terminals is taken for analysis. These are some examples of lumped elements.

If the parameters representing the elements are not changing with respect to time, the element is said to be time invariant. Thus if a system is composed of linear, lumped and time invariant elements, its behaviour can be modelled by either linear algebraic equations or linear differential equations with constant coefficients. If the input output relations are algebraic, the system is said to be a static system. On the other hand, if the relations are described by differential equations, the system is said to be a dynamic system. We are mostly concerned with dynamic response of the systems and therefore, one of the ways by which a system is mathematically modelled is by differential equations. Another most useful and common mathematical model is the '*Transfer function*' of the system. It is defined as the ratio of Laplace transform of the output to the Laplace transform of the input. This is illustrated in Fig. 2.2.

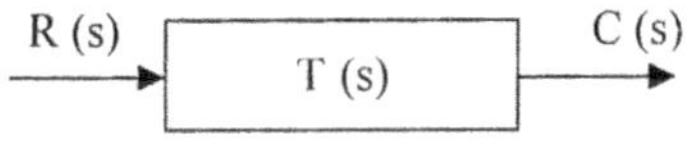

Fig. 2.2 Transfer function of a system

In Fig. 2.2, the transfer function is,

$$T(s) = \frac{C(s)}{R(s)} \quad .....(2.1)$$

In defining the transfer function, it is assumed that all initial conditions in the system are zero.

Having defined the two common ways of describing linear systems, let us obtain the mathematical models of some commonly occuring Electrical, Mechanical, Thermal, Fluid, Hydraulic systems etc.

2.2.1 Electrical Systems

Most of the electrical systems can be modelled by three basic elements : Resistor, inductor, and capacitor. Circuits consisting of these three elements are analysed by using Kirchhoff's Voltage law and Current law.

(a) *Resistor :* The circuit model of resistor is shown in Fig. 2.3 (a)

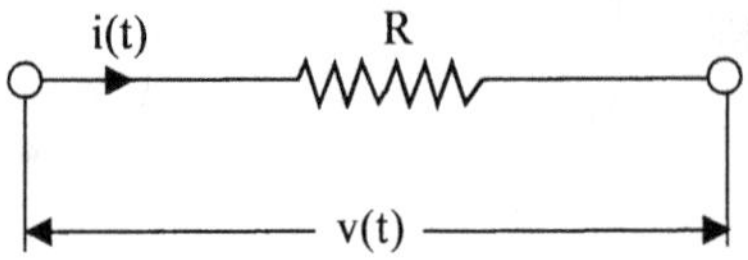

Fig. 2.3(a) Circuit model of resistor

The mathematical model is given by the Ohm's law relationship,

$$v(t) = i(t)\,R; \qquad i(t) = \frac{v(t)}{R} \qquad .....(2.2)$$

(b) *Inductor :* The circuit representation is shown in Fig. 2.3 (b)

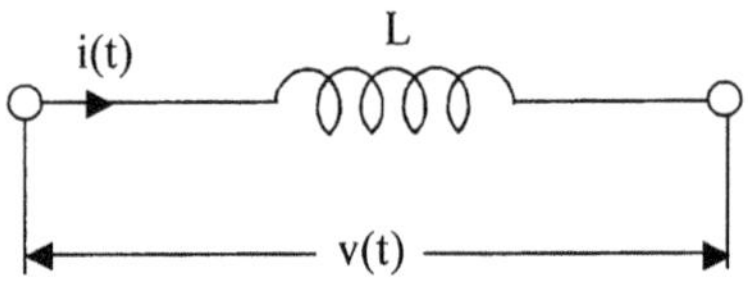

Fig. 2.3(b) Circuit model of inductor

The input output relations are given by Faraday's law,

$$v(t) = L\frac{di(t)}{dt} \qquad .....(2.3)$$

or

$$i(t) = \frac{1}{L}\int v\,dt \qquad .....(2.4)$$

where $\int v\,dt$ is known as the flux linkages $\Psi(t)$. Thus

$$i(t) = \frac{\Psi(t)}{L} \qquad .....(2.5)$$

If only a single coil is considered, the inductance is known as self inductance. If a voltage is induced in a as second coil due to a current in the first coil, the voltage is said to be due to mutual inductance, as shown in Fig. 2.3(c).

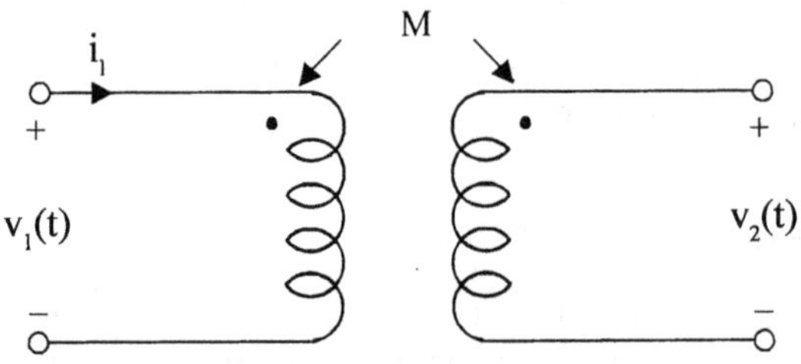

Fig. 2.3(c) Mutual inductance

In Fig. 2.3(c), $$v_2(t) = M \frac{di_1}{dt} \qquad .....(2.6)$$

(c) *Capacitor :* The circuit symbol of a capacitor is given in Fig. 2.3 (d).

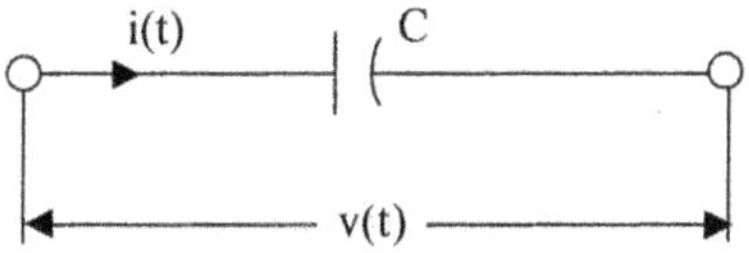

Fig. 2.3(d) Circuit symbol of a capacitor

$$v(t) = \frac{1}{C} \int idt \qquad .....(2.7)$$

or $$i(t) = C \frac{dv}{dt} \qquad .....(2.8)$$

In eqn. (2.7), $\int idt$ is known as the charge on the capacitor and is denoted by '*q*'. Thus

$$q = \int idt \qquad .....(2.9)$$

and $$v(t) = \frac{q(t)}{C} \qquad .....(2.10)$$

Another useful element, frequently used in electrical circuits, is the ideal transformer indicated in Fig. 2.4.

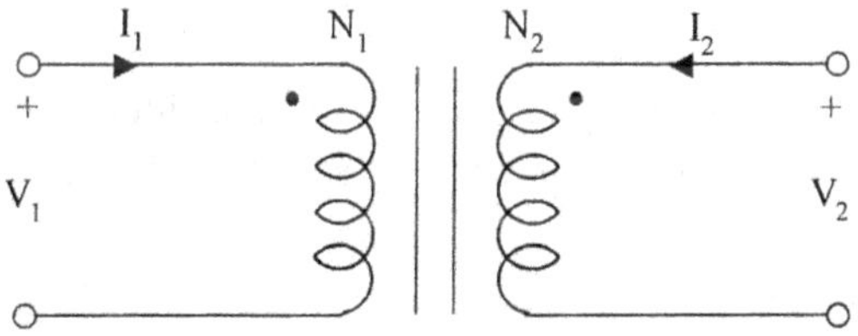

Fig. 2.4 Model of a transformer

The mathematical model of a transformer is given by,

$$\frac{V_2}{V_1} = \frac{N_2}{N_1} = \frac{I_1}{I_2} \qquad ....(2.11)$$

Electrical networks consisting of the above elements are analysed using Kirchhoff's laws.

Example 2.1

Consider the network in Fig. 2.5. Obtain the relation between the applied voltage and the current in the form of (a) Differential equation (b) Transfer function

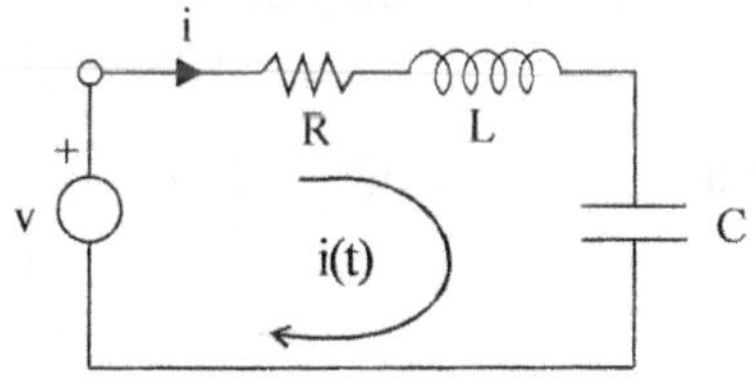

Fig. 2.5 An R, L, C series circuit excited by a voltage source.

Solution :

(a) *Writing down the Kirchhoff's voltage law equation for the loop, we have*

$$R\,i(t) + L\,\frac{d\,i(t)}{dt} + \frac{1}{C}\int_{-\infty}^{t} i(t)\,dt = v \qquad .....(2.12)$$

Denoting $\int_{-\infty}^{t} i(t)\,d(t)$ by q(t), we can also write eqn. (2.8) as

$$L\,\frac{d^2q(t)}{dt^2} + R\,\frac{dq(t)}{dt} + \frac{q(t)}{C} = v \qquad .....(2.13)$$

This is a 2nd order linear differential equation with constant coefficients.

(b) *Transfer function*

Taking Laplace transform of eqn. (2.13) with all initial conditions assumed to be zero, we have

$$Ls^2\,Q(s) + Rs\,Q(s) + \frac{1}{C}Q(s) = V(s) \qquad .....(2.14)$$

$$\frac{Q(s)}{V(s)} = \frac{1}{Ls + Rs + \frac{1}{C}} = \frac{C}{LCs^2 + RCs + 1} \qquad ....(2.15)$$

This is the transfer function of the system, if *q(t)* is considered as output. Instead, if *i(t)* is considered as the output, taking Laplace transform of eqn. (2.12), we have,

$$R\,I(s) + Ls\,I(s) + \frac{I(s)}{Cs} = V(s) \qquad .....(2.16)$$

$$\therefore \quad \frac{I(s)}{V(s)} = \frac{1}{Ls + R + \frac{1}{Cs}} = \frac{Cs}{LCs^2 + RCs + 1} \qquad .....(2.17)$$

Example 2.2

Consider the parallel RLC network excited by a current source (Fig. 2.6). Find the (a) differential equation representation and (b) transfer function representation of the system.

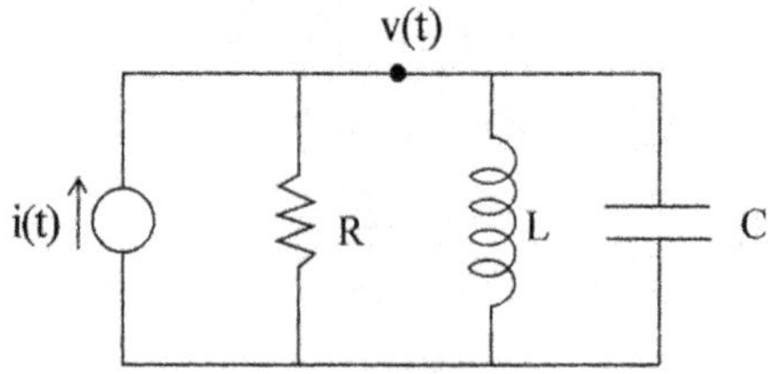

Fig. 2.6 Parallel RLC circuit excited by a current source

Solution :

(a) *Applying Kirchhoff's current law at the node,*

$$\frac{v(t)}{R}+\frac{Cdv(t)}{dt}+\frac{1}{L}\int v\,dt = i(t) \qquad .....(2.18)$$

Replacing $\int v\,dt$ by $\Psi(t)$, the flux linkages, we have,

$$C\frac{d^2\Psi(t)}{dt}+\frac{1}{R}\frac{d\psi(t)}{dt}+\frac{\psi(t)}{L} = i(t) \qquad .....(2.19)$$

(b) *Taking Laplace transform of eqn. (2.19), we have,*

$$Cs^2\,\Psi(s)+\frac{1}{R}s\,\Psi(s)+\frac{\Psi(t)}{L} = I(s)$$

$$\frac{\Psi(s)}{I(s)} = \frac{1}{Cs^2+\dfrac{s}{R}+\dfrac{1}{L}}$$

$$= \frac{L}{LCs^2+LGs+1} \qquad \left(\because \frac{1}{R}=G\right) \qquad .....(2.20)$$

If the voltage is taken as the output, taking Laplace transform of eqn. (2.18), we get

$$GV(s)+CsV(s)+\frac{1}{Ls}V(s) = I(s)$$

$$\therefore \qquad \frac{V(s)}{I(s)} = \frac{1}{Cs+G+\dfrac{1}{Ls}} = \frac{Ls}{LCs^2+LGs+1} \qquad .....(2.21)$$

Example 2.3

Obtain the transfer function I(s)/V(s) in the network of Fig. 2.7.

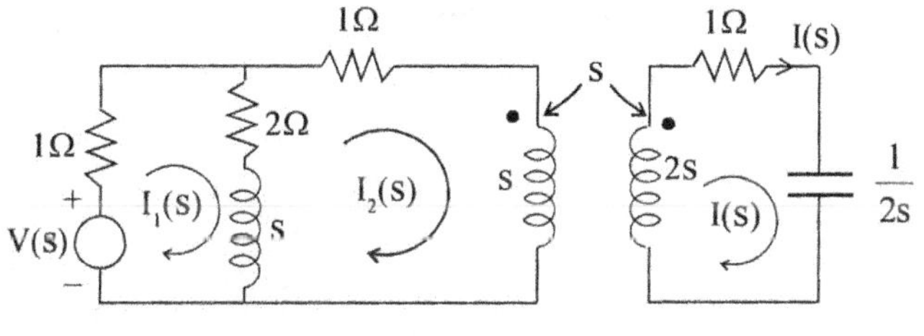

Fig. 2.7 Network for the Example 2.3

Solution :

Writing the loop equations

$$(3 + s)\, I_1(s) \quad - (s + 2)\, I_2(s) \quad = V(s) \qquad .....(1)$$

$$-(2 + s)\, I_1(s) \quad + (3 + 2s)\, I_2(s) - s.I(s) \quad = 0 \qquad .....(2)$$

$$-s\, I_2(s) + \left(1 + 2s + \frac{1}{2s}\right) I(s) = 0 \qquad .....(3)$$

Solving for I(s), we have

$$I(s) = \frac{\begin{vmatrix} 3+s & -(s+2) & V(s) \\ -(2+s) & (3+2s) & 0 \\ 0 & -s & 0 \end{vmatrix}}{\begin{vmatrix} 3+s & -(s+2) & 0 \\ -(2+s) & (3+2s) & -s \\ 0 & -s & \left(1+2s+\frac{1}{2s}\right) \end{vmatrix}}$$

$$= \frac{V(s).[s(s+2)]}{(3+s)\left[(3+2s)\left(1+2s+\frac{1}{2s}\right) - s^2\right] + (s+2)\left[-(s+2)\left(1+2s+\frac{1}{2s}\right)\right]}$$

$$I(s) = \frac{s(s+2)\,V(s).2s}{-(s+3)\,(4s^4 + 12s^3 + 9s^2 + 4s + 1)}$$

$$\frac{I(s)}{V(s)} = -\frac{2s^2(s+2)}{(s+3)\,(4s^4 + 12s^3 + 9s^2 + 4s + 1)}$$

2.2.2 Dual Networks

Consider the two networks shown in Fig. 2.5 and Fig. 2.6 and eqns. (2.13) and (2.19). In eqn. (2.13) if the variables q and v and circuit constant RLC are replaced by their dual quantities as per Table 2.1 eqn. (2.19) results.

Table 2.1 Dual Quantities

v(t)	↔	i(t)
i(t)	↔	v(t)
R	↔	$\frac{1}{R} = G$
C	↔	L
L	↔	C
$q = \int i\,dt$	↔	$\Psi = \int v\,dt$

The two networks, are entirely dissimilar topologically, but their equations are identical. If the solution of one equation is known, the solution of the other is also known.

Thus the two networks in Fig. 2.5 and 2.6 are known as *dual networks*. Given any planar network, a dual network can always be obtained using Table 2.1.

2.2.3 Mechanical Systems

Mechanical systems can be divided into two basic systems.

(a) Translational systems and (b) Rotational systems

We will consider these two systems separately and describe these systems in terms of three fundamental linear elements.

(a) *Translational systems :*

1. *Mass :* This represents an element which resists the motion due to inertia. According to Newton's second law of motion, the inertia force is equal to mass times acceleration.

$$f_M = Ma = M.\frac{dv}{dt} = M\frac{d^2x}{dt^2} \quad .....(2.22)$$

Where a, v and x denote acceleration, velocity and displacement of the body respectively. Symbolically, this element is represented by a block as shown in Fig. 2.8(a).

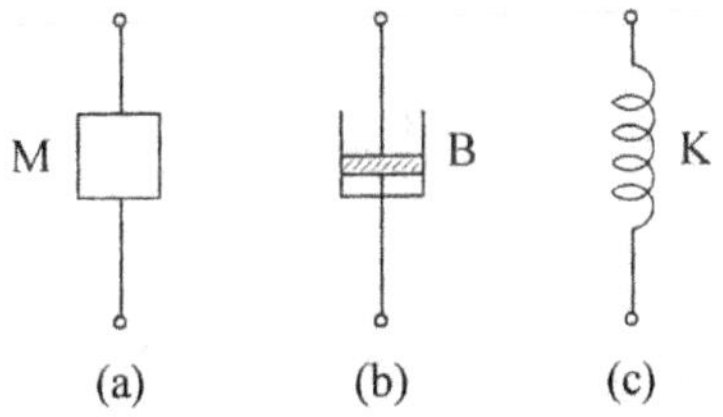

Fig. 2.8 Passive linear elements of translational motion (a) Mass (b) Dash pot (c) Spring.

2. *Dash pot :* This is an element which opposes motion due to friction. If the friction is viscous friction, the frictional force is proportional to *velocity*. This force is also known as dampling force. Thus we can write

$$f_B = Bv = B\frac{dx}{dt} \quad .....(2.23)$$

Where B is the damping coefficient. This element is called as dash pot and is symbolically represented as in Fig. 2.8(b).

3. *Spring :* The third element which opposes motion is the spring. The restoring force of a spring is proportional to the displacement. Thus

$$f_K = K\,x \quad .....(2.24)$$

Where K is known as the stiffness of the spring or simply spring constant. The symbol used for this element is shown in Fig. 2.8(c).

(b) ***Rotational systems :*** Corresponding to the three basic elements of translation systems, there are three basic elements representing rotational systems.

1. *Moment of Inertia :* This element opposes the rotational motion due to Moment of Inertia. The opposing inertia torque is given by,

$$T_I = Ja = J\frac{d\omega}{dt} = J\frac{d^2\theta}{dt^2} \qquad .....(2.25)$$

Where α, ω and θ are the angular acceleration, angular velocity and angular displacement respectively. J is known as the moment of inertia of the body.

2. *Friction :* The damping or frictional torque which opposes the rotational motion is given by,

$$T_B = B\omega = B\frac{d\theta}{dt} \qquad .....(2.26)$$

Where B is the rotational frictional coefficient.

3. *Spring :* The restoring torque of a spring is proportional to the angular displacement θ and is given by,

$$T_K = K\theta \qquad .....(2.27)$$

Where K is the torsimal stiffness of the spring. The three elements defined above are shown in Fig. 2.9.

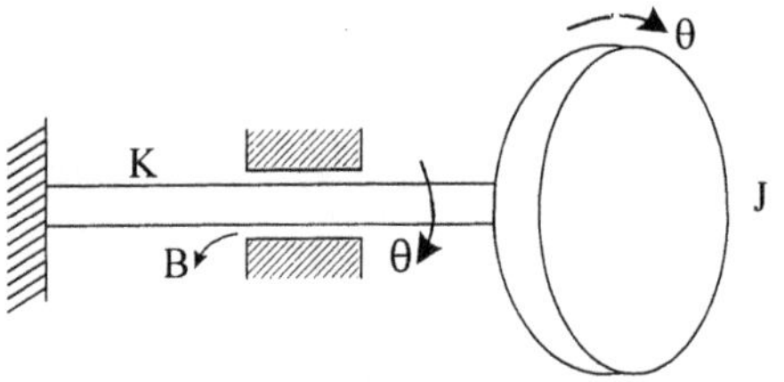

Fig. 2.9 Rotational elements

Since the three elements of rotational systems are similar in nature to those of translational systems no separate symbols are necessary to represent these elements.

Having defined the basic elements of mechanical systems, we must now be able to write differential equations for the system when these mechanical systems are subjected to external forces. This is done by using the D' Alembert's principle which is similar to the Kirchhoff's laws in Electrical Networks. Also, this principle is a modified version of Newton's second law of motion. The D' Alembert's principle states that,

"*For any body, the algebraic sum of externally applied forces and the forces opposing the motion in any given direction is zero*".

To apply this principle to any body, a reference direction of motion is first chosen. All forces acting in this direction are taken positive and those against this direction are taken as negative. Let us apply this principle to a mechanical translation system shown in Fig. 2.10.

A mass M is fixed to a wall with a spring K and the mass moves on the floor with a viscous friction. An external force f is applied to the mass. Let us obtain the differential equation governing the motion of the body.

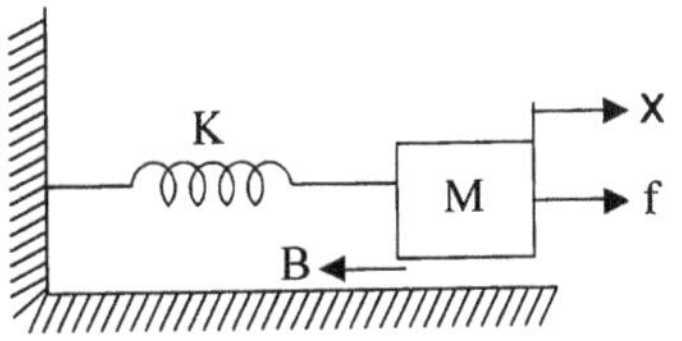

Fig. 2.10 A mechanical translational system

Let us take a reference direction of motion of the body from left to right. Let the displacement of the mass be x. We assume that the mass is a rigid body, ie, every particle in the body has the same displacement, x. Let us enumerate the forces acting on the body.

(a) external force = f

(b) resisting forces :

(i) Inertia force, $f_M = -M\frac{d^2x}{dt^2}$

(ii) Damping force, $f_B = -B\frac{dx}{dt}$

(iii) Spring force, $f_K = -Kx$

Resisting forces are taken to be negative because they act in a direction opposite to the chosen reference direction. Thus, using D' Alemberts principle we have,

$$f - M\frac{d^2x}{dt^2} - B\frac{dx}{dt} - Kx = 0$$

or
$$M\frac{d^2x}{dt^2} + B\frac{dx}{dt} + Kx = f \qquad .....(2.28)$$

This is the differential equation governing the motion of the mechanical translation system. The transfer function can be easily obtained by taking Laplace transform of eqn (2.28). Thus,

$$\frac{X(s)}{F(s)} = \frac{1}{Ms^2 + Bs + K}$$

If velocity is chosen as the output variable, we can write eqn. (2.28) as

$$M\frac{du}{dt} + Bu + K\int u\, dt = f \qquad .....(2.29)$$

Similarly, the differential equation governing the motion of rotational system can also be obtained. For the system in Fig. 2.11, we have

$$J\frac{d^2\theta}{dt^2} + B\frac{d\theta}{dt} + K\theta = T \qquad .....(2.30)$$

The transfer function of this system is given by

$$\frac{\theta(s)}{T(s)} = \frac{1}{Js^2 + Bs + K}$$

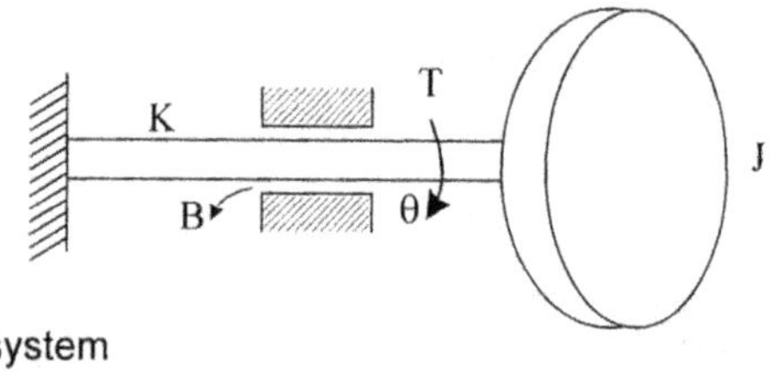

Fig. 2.11 Mechanical rotatinal system

Since eqn. (2.28) and eqn. (2.30) are similar, if the solution of one system is known, the solution for the other system can be easily obtained. Such systems are known as analogous systems. Further, the eqns. (2.15) and (2.19) of the electrical systems shown in Fig. 2.5 and 2.6 are similar to eqns. (2.28) and (2.30) of mechanical systems of Figures 2.10 and 2.11. Hence the electrical systems in Figures 2.5 and 2.6 are also analogous to mechanical systems in Figures 2.10 and 2.11.

2.2.4 Analogous Systems

Analogous systems have the same type of equations even though they have different physical appearance. Mechanical systems, fluid systems, temperature systems etc. may be governed by the same types of equations as that of electrical circuits. In such cases we call these systems as analogous systems. A set of convenient symbols are already developed in electrical engineering which permits a complex system to be represented by a circuit diagram. The equations can be written down easily for these circuits and the behaviour of these circuits obtained. Thus if an analogous electrical circuit is visualised for a given mechanical system, it is easy to predict the behaviour of the system using the well developed mathematical tools of electrical engineering. Designing and constructing a model is easier in electrical systems. The system can be built up with cheap elements, the values of the elements can be changed with ease and experimentation is easy with electrical circuits. Once a circuit is designed with the required characteristics, it can be readily translated into a mechanical system. It is not only true for mechanical systems but also several other systems like acoustical, thermal, fluid and even economic systems.

The analogous electrical and mechanical quantities are tabulated in Table 2.2.

Table 2.2 Analogous quantities based on force voltage analogy

Electrical system		Mechanical system			
		Translational		Rotational	
Voltage	V	Force	f	Torque	T
Current	i	Velocity	u	angular velocity	ω
Charge	q	Displacement	x	angular displacement	θ
Inductance	L	Mass	M	Moment of Inertia	J
Capacitance	C	Compliance	$\frac{1}{K}$	Compliance	$\frac{1}{K}$
Resistance	R	Damping coefficient	B	Damping coefficient	B

Comparing the eqn. (2.15) of the electrical system and eqn. (2.28) and eqn. (2.30) of mechanical systems it is easy to see the correspondence of the various quantities in Table 2.2. Since the external force applied to mechanical system, f_1, is made analogous to the voltage applied to the electrical circuit, this is called as *Force - Voltage analogy.*

If the mechanical systems are compared with the electrical circuit in Fig. 2.6, in which the source is a current source, force applied in mechanical system is analogous to the current. The analogous quantities based on force-current analogy are given in Table 2.3

The Table 2.3 can be easily understood by comparing eqn. (2.19) of the electrical system with eqns. (2.28) and (2.30) of mechanical systems.

Table 2.3 Analogous quantities based on Force - Current analogy

Electrical system	Mechanical system	
	Translational	Rotational
Current i	Force f	Torque T
Voltage v	Velocity u	angular velocity ω
Flux linkages Ψ	Displacement x	angular displacement θ
Capacitance C	Mass M	Moment of Inertia J
Conductance G	Damping coefficient B	Rotational Damping coefficient B
Inductance L	Compliance $\frac{1}{K}$	Compliance $\frac{1}{K}$

Now let us consider some examples of mechanical systems and construct their mathematical models.

Example 2.4

Write the equations describing the motion of the mechanical system shown in Fig. 2.12. Also find the transfer function $X_1(s)/F(s)$.

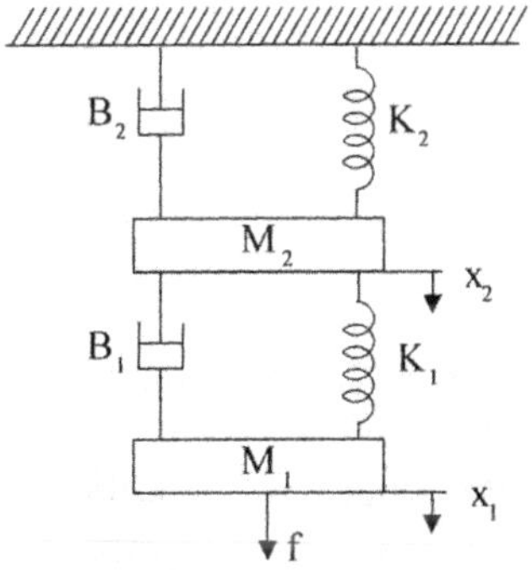

Fig. 2.12 A Mechanical system for example 2.4

Solution :

The first step is to identify the displacements of masses M_1 and M_2 as x_1 and x_2 in the direction of the applied external force *f*. Next we write the equilibrium equation for each of the masses by identifying the forces acting on them. Let us first find out the forces acting on mass M_1.

External force = f

Restoring forces :

(i) Inertia force, $-M_1 \frac{d^2x_1}{dt^2}$

(ii) Damping force, $-B_1 \frac{d^2(x_1 - x_2)}{dt^2}$

(iii) Spring force, $-K_1(x_1 - x_2)$

(If $x_1 > x_2$ the spring expands and restoring force is upwards)

(*Note :* Since top end of dash pot and spring are rigidly connected to mass M_2, their displacement is x_2 in the downward direction. Similarly, the bottom ends of dash pot spring are rigidly connected the mass M_1, they move downward by x_1. The relative displacement of the lower end upper end is $(x_1 - x_2)$ in the downward direction. Hence restoring forces are $-B_1 \frac{d(x_1 - x_2)}{dt}$ and $-K_1(x_1 - x_2)$ respectively due to the dash pot and spring).

Hence the equation of motion is

$$M_1 \frac{d^2x_1}{dt^2} + B_1 \frac{d(x_1 - x_2)}{dt} + K_1(x_1 - x_2) = f \qquad .....(1)$$

Now for mass M_2

External force = Zero

Restoring forces :

(i) Inertia force, $-M_2 \frac{d^2x_2}{dt^2}$

(ii) Damping forces,

(a) $-B_2 \frac{dx_2}{dt}$ (since one end is fixed and the other is connected to M_1 rigidly)

(b) $-B_1 \frac{d(x_2 - x_1)}{dt}$ (If $\frac{dx_2}{dt} > \frac{dx_1}{dt}$, the motion is in the downward direction and the frictional force is in the upward direction)

(iii) Spring forces :

(a) $-K_2 x_2$

(b) $-K_1(x_2 - x_1)$ (If $x_2 > x_1$, the spring is compressed and the restoring force is upward)

Hence the equation of motion for M_2 is,

$$M_2 \frac{d^2 x_2}{dt} + B_2 \frac{dx_2}{dt} + B_1 \frac{d(x_2 - x_1)}{dt} + K_2 x_2 + K_1 (x_2 - x_1) = 0 \qquad ....(2)$$

From eqns. (1) & (2), force voltage analogous electrical circuits can be drawn as shown in Fig. 2.13 (a).

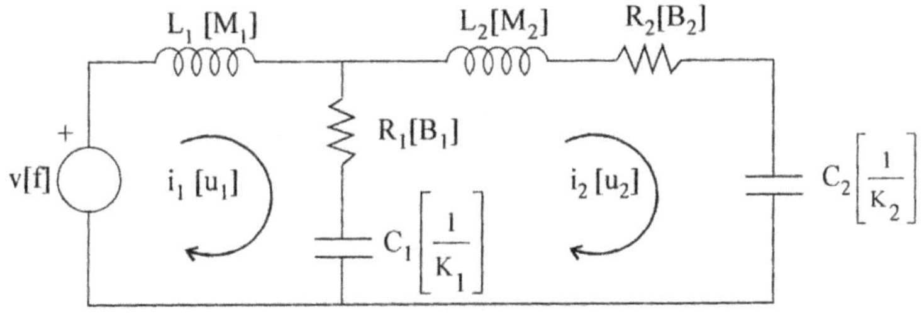

Fig. 2.13 (a) Force - Voltage analogous circuit for mechanical system of Fig. 2.12. Mechanical quantities are shown in parenthesis

Force - current analogous circuit can also be developed for the given mechanical system. It is given in Fig. 2.13 (b). Note that since the mass is represented by a capacitance and voltage is analogous to velocity, one end of the capacitor must always be grounded so that its velocity is always referred with respect to the earth.

$$\text{Transfer function} = \frac{X_1(s)}{F(s)}$$

Taking Laplace transform of eqns. (1) and (2) and assuming zero initial conditions, we have

$$M_1 s^2 X_1(s) + B_1 s [X_1(s) - X_2(s)] + K_1 [X_1(s) - X_2(s)] = F(s)$$

or $$(M_1 s^2 + B_1 s + K_1) X_1(s) - (B_1 s + K_1) X_2(s) = F(s) \qquad .....(3)$$

and $$M_2 s^2 X_2(s) + B_2 s X_2(s) + B_1 s [X_2(s) - X_1(s)] + K_2 X_2(s) + K_1 [(X_2(s) - X_1(s)] = 0$$

or $$-(B_1 s + K_1) X_1(s) + [M_2 s^2 + (B_2 + B_1)s + (K_2 + K_1) X_2(s) = 0 \qquad .....(4)$$

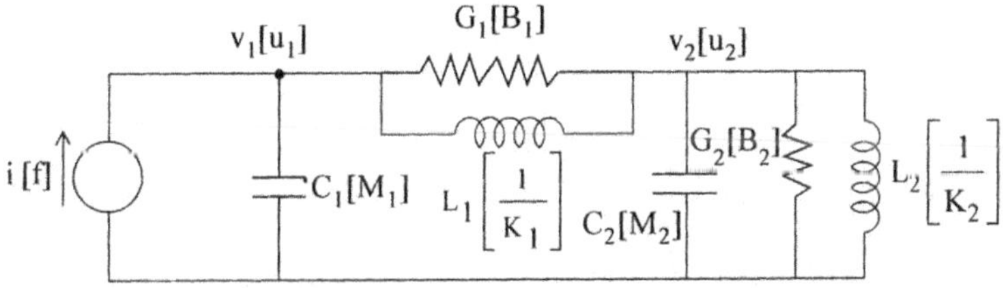

Fig. 2.13 (b) Force current analogous circuit for the mechanical system of Fig. 2.12

Solving for $X_2(s)$ in eqn (4), we get

$$X_2(s) = \frac{B_1 s + K_1}{M_2 s^2 + (B_1 + B_2)s + (K_1 + K_2)} X_1(s)$$

Substituting for $X_2(s)$ in eqn. (3)

$$(M_1 s^2 + B_1 s + K_1) X_1(s) - \frac{(B_1 s + K_1).(B_1 s + K_1)}{M_2 s^2 + (B_1 + B_2)s + K_1 + K_2} X_1(s) = F(s)$$

$$\therefore \quad X_1(s) = \frac{(M_2 s^2 + (B_1 + B_2)s + K_1 + K_2)}{(M_1 s^2 + B_1 s + K_1)[M_2 s^2 + (B_1 + B_2)s + (K_1 + K_2)] - (B_1 s + K_1)^2} F(s)$$

Thus $$T(s) = \frac{X_1(s)}{F(s)} = \frac{M_2 s^2 + (B_1 + B_2)s + K_1 + K_2}{(M_1 s^2 + B_1 s + K_1)[M_2 s^2 + (B_1 + B_2)s + (K_1 + K_2)] - (B_1 s + K_1)^2}$$

Which is the desired transfer function.

Example 2.5

Obtain the *f-v* and *f-i* analogous circuits for the mechanical system shown in Fig. 2.14. Also write down the equilibrium equations.

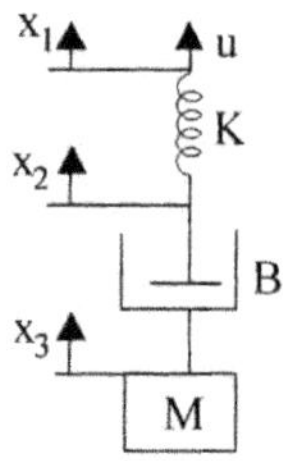

Fig. 2.14 Mechanical system for Ex. 2.5

Solution :

Let f be the force acting on the spring instead of velocity u. The displacements are indicated in the figure.

The equilibrium equations are :

$$K(x_1 - x_2) = f \qquad .....(1)$$

$$B(\dot{x}_2 - \dot{x}_3) + K(x_2 - x_1) = 0 \qquad .....(2)$$

$$B(\dot{x}_3 - \dot{x}_2) + M\ddot{x}_3 = 0 \qquad .....(3)$$

From eqs (1) and (2), we have

$$B(\dot{x}_2 - \dot{x}_3) = f \qquad .....(4)$$

From eqs (3) and (4), we have

$$M\ddot{x}_3 = f \qquad .....(5)$$

Force Current Analogous Circuit

Replacing the electrical quantities in equations (1), (4), and (5) by their force-current analogous quantities using Table 2.3, we have

$$\frac{1}{L}(\psi_1 - \psi_2) = i$$

or $$\frac{1}{L}\int (v_1 - v_2)dt = i \qquad .....(6)$$

$$G(\dot{\psi}_2 - \dot{\psi}_3) = i$$

or $$G(v_2 - v_3) = i \qquad .....(7)$$

$$C\,\ddot{\psi}_3 = i$$

or $$C\frac{dv_3}{dt} = i \qquad .....(8)$$

If i is produced by a voltage source *v*, we have the electrical circuit based on *f-i* analogy in Fig. 2.14 (a).

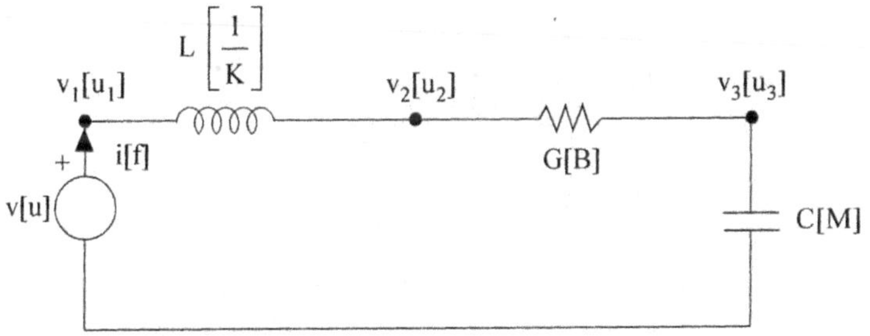

Fig. 2.14 (a) F-i analogous circuit for mechanical system in Ex. 2.5

Force Voltage Analogous Circuit

Using force voltage analogy, the quantities in eqs (1), (4) and (5) are replaced by the mechanical quantities to get,

$$\frac{1}{C}(q_1 - q_2) = v$$

or $$\frac{1}{C}\int (i_1 - i_2)\, dt = v \qquad .....(9)$$

$$R(\dot{q}_2 - \dot{q}_3) = v$$

or $$R(i_2 - i_3) = v \qquad .....(10)$$

$$L\frac{d^2q_3}{dt^2} = v$$

or $$L\frac{di_3}{dt} = v \qquad .....(11)$$

If the voltage is due to a current source i, we have the force voltage analogous circuit is shown in Fig. 2.14 (b)

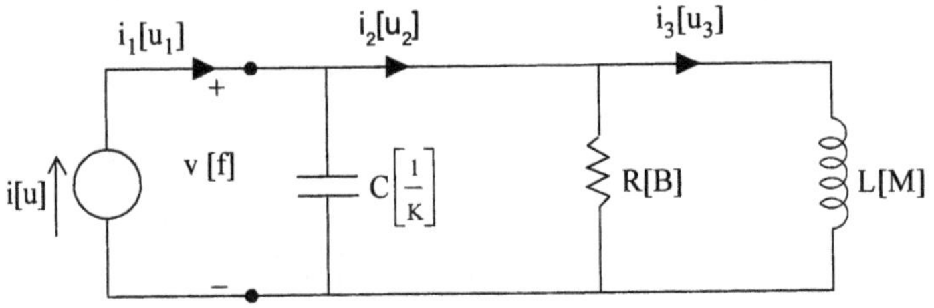

Fig. 2.14 (b) Force voltage analogous circuit for the mechanical system of Ex. 2.5

2.2.5 Gears

Gears are mechanical coupling devices used for speed reduction or magnification of torque. These are analogous to transformers in Electrical systems. Consider two gears shown in Fig. 2.15. The first gear, to which torque T_1 is applied, is known as the *primary gear* and has N_1 teeth on it. The second gear, which is coupled to this gear and is driving a load, is known as the *secondary gear* and has N_2 teeth on it.

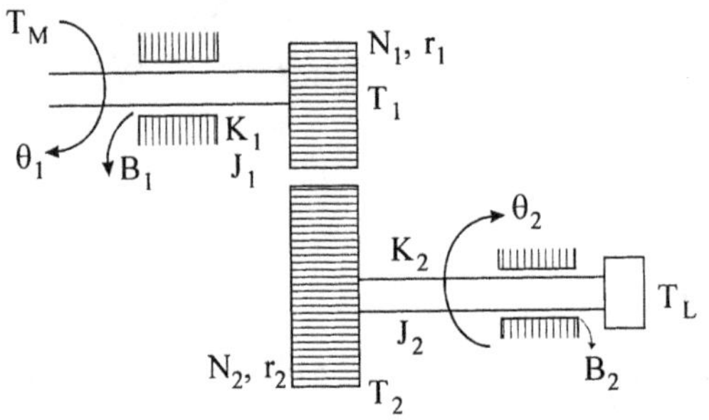

Fig. 2.15 Gear train

The relationships between primary and secondary quantities are based on the following principles.

1. The number of teeth on the gear is proportional to the radius of the gear

$$\frac{r_1}{r_2} = \frac{N_1}{N_2} \qquad .....(2.31)$$

2. The linear distance traversed along the surface of each gear is same.

   If the angular displacements of the two gears are θ_1 and θ_2 respectively, then

$$r_1\,\theta_1 = r_2\,\theta_2$$

$$\frac{r_1}{r_2} = \frac{\theta_2}{\theta_1} \qquad .....(2.32)$$

3. The work done by one gear is same as that of the other.

   If T_1 and T_2 are the torques on the two gears,

$$T_1\,\theta_1 = T_2\,\theta_2 \qquad .....(2.33)$$

   Using eqs (2.31), (2.32) and (2.33), we have,

$$\frac{\theta_1}{\theta_2} = \frac{T_2}{T_1} = \frac{N_2}{N_1} \qquad .....(2.34)$$

   Recall that, for an ideal transformer of primary turns N_1 and secondary turns N_2,

$$\frac{V_1}{V_2} = \frac{I_2}{I_1} = \frac{N_1}{N_2} \qquad .....(2.35)$$

   Eqn. (2.35) is similar to eqn. (2.34) and hence a gear train in mechanical system is analogous to an ideal transformer in the electrical system.

   Writing the equilibrium equations for the mechanical system of Fig. (2.14), we have

$$J_1\,\ddot{\theta}_1 + B_1\,\dot{\theta}_1 + K_1\,\theta_1 + T_1 = T_M \qquad .....(2.36)$$

$$J_2\,\ddot{\theta}_2 + B_2\,\dot{\theta}_2 + K_2\,\theta_2 + T_L = T_2 \qquad .....(2.37)$$

   Where T_M is the motor torque driving the shaft of the gear, T_1 is the torque available for the primary gear, T_2 is the torque at the secondary gear, T_L is the load torque and J_2 is the moment of inertia of secondary gear and the load.

   Using eqn. (2.34) in eqn. (2.37), we have

$$J_2\,\ddot{\theta}_2\,p + B_2\,\dot{\theta}_2 + K_2\,\theta_2 + T_L = \frac{N_2}{N_1}.T_1 \qquad .....(2.38)$$

   Substituting for T_1 in eqn. (2.36), using eqn. (2.38),

$$J_1\,\ddot{\theta}_1 + B_1\,\dot{\theta}_1 + K_1\,\theta_1 + \frac{N_1}{N_2}.(J_2\,\ddot{\theta}_2 + B_2\,\dot{\theta}_2 + K_2\,\theta_2 + T_L) = T_M \qquad .....(2.39)$$

   But, $\theta_2 = \frac{N_1}{N_2}\theta_1$

   $\therefore$ $\dot{\theta}_2 = \frac{N_1}{N_2}\dot{\theta}_1$

   and $\ddot{\theta}_2 = \frac{N_1}{N_2}\,\ddot{\theta}_1$

Using these relations in eqn. (2.39), we get

$$\left[J_1 + J_2\left(\frac{N_1}{N_2}\right)^2\right]\ddot{\theta}_1 + \left[B_1 + B_2\left(\frac{N_1}{N_2}\right)^2\right]\dot{\theta}_1 + \left[K_1 + K_2\left(\frac{N_1}{N_2}\right)^2\right]\theta_1 + \frac{N_1}{N_2}.T_L = T_M \qquad .....(2.40)$$

Replacing $J_1 + J_2\left(\frac{N_1}{N_2}\right)^2 = J_{1\,eq}$

$$B_1 + B_2\left(\frac{N_1}{N_2}\right)^2 = B_{1\,eq}$$

$$K_1 + K_2\left(\frac{N_1}{N_2}\right)^2 = K_{1\,eq}$$

and $\frac{N_1}{N_2}\,T_2 = T_{L\,eq}$

We have,

$$J_{1\,eq}\,\ddot{\theta}_1 + B_1\,\dot{\theta}_1 + K_{1eq}\,\theta_1 + T_{L\,eq} = T_M \qquad .....(2.41)$$

Thus the original system in Fig. 2.15 can be replaced by an equivalent system referred to primary side as shown in Fig. 2.16.

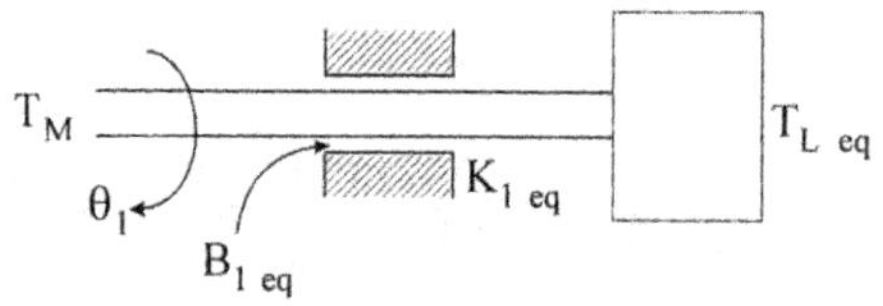

Fig. 2.16 Equivalent system refered to primary side

The moment of inertia J_2, frictional coefficient B_2 and torsional spring constant K_2 of secondary side are represented by their equivalents referred to primary side by,

$$J_{12} = J_2\left(\frac{N_1}{N_2}\right)^2$$

$$B_{12} = B_2\left(\frac{N_1}{N_2}\right)^2$$

$$K_{12} = K_2\left(\frac{N_1}{N_2}\right)^2$$

The load torque referred to primary side is given by,

$$T_{L\,eq} = T_L \cdot \frac{N_1}{N_2}$$

These equations can be easily seen to be similar to the corresponding electrical quantities in the ideal transformer.

All the quantities can also be referred to the secondary side of the gear. The relevant equations are

$$J_{2\,eq}\ddot{\theta}_2 + B_{2\,eq}\dot{\theta}_2 + K_{2\,eq}\theta_2 + T_L = T_{M\,eq} \quad .....(2.42)$$

where
$$J_{2\,eq} = J_2 + J_1\left(\frac{N_2}{N_1}\right)^2 ;\; B_{2\,eq} = B_2 + B_1\left(\frac{N_2}{N_1}\right)^2$$

$$K_{2\,eq} = K_2 + K_1\left(\frac{N_2}{N_1}\right)^2$$

and
$$T_{M\,eq} = \frac{N_2}{N_1} \cdot T_M$$

2.2.6 Thermal Systems

Thermal systems are those systems in which heat transfer takes place from one substance to another. They can be characterised by thermal resistance and capacitance, analogous to electrical resistance and capacitance. Thermal system is usually a non linear system and since the temperature of a substance is not uniform throughout the body, it is a distributed system. But for simplicity of analysis, the system is assumed to be linear and is represented by lumped parameters.

(a) *Thermal resistance*

There are two types of heat flow through conductors : Conduction or convection and radiation.

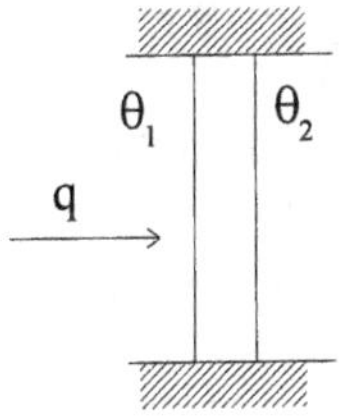

Fig. 2.17 Thermal resistance

For conduction of heat flow through a specific conductor, according to Fourier law,

$$q = \frac{KA}{\Delta X}(\theta_1 - \theta_2) \quad ......(2.42)$$

where,

q = Heat flow, Joules/Sec
K = Thermal conductivity, J/sec/m/deg k
A = Area normal to heat flow, m^2
ΔX = Thickness of conductor, m
θ = Temperature in ^{0}K

For convection heat transfer,

$$q = HA(\theta_1 - \theta_2) \quad .....(2.43)$$

where H = Convection coefficient, J/m^2/sec/deg k

The thermal resistance is defined by,

$$R = \frac{d\theta}{dq} = \frac{\Delta X}{KA} \quad \text{(Conduction)} \quad .....(2.44)$$

$$= \frac{1}{HA} \quad \text{(Convection)} \quad .....(2.45)$$

The unit of R is deg sec/J

For radiation heat transfer, the heat flow is governed by Stefan-Boltzmann law for a surface receiving heat radiation from a black body :

$$q = KAE(\theta^4{}_1 - \theta^4{}_2)$$
$$= A\sigma(\theta^4{}_1 - \theta^4{}_2) \quad .....(2.46)$$

where,

σ is a constant, 5.6697×10^{-8} J/sec/m^2/K^4
K is a constant
E is emissivity
A is surface in m^2

The radiation resistance is given by

$$R = \frac{d\theta}{dq} = \frac{1}{4A\sigma\theta_a^3} \text{ deg sec/J} \quad .....(2.47)$$

where θ_a is the average temperature of radiator and receiver. Since eqn. (2.46) is highly nonlinear, it can be used only for small range of temperatures.

(b) *Thermal Capacitance*

Thermal capacitance is the ability to store thermal energy. If heat is supplied to a body, its internal energy raises. For the system shown in Fig. 2.18,

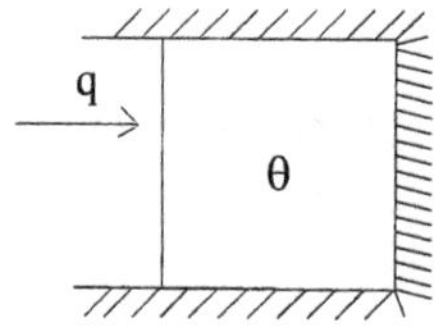

Fig. 2.18 Thermal Capacitance

$$C\frac{d\theta}{dt} = q \quad .....(2.48)$$

where C is the thermal capacitance

$$C = WC_P \quad .....(2.49)$$

where $W \rightarrow$ weight of block in kg

$C_P \rightarrow$ specific heat at constant pressure in J/deg/kg

2.2.7 Fluid Systems

Fluid systems are those systems in which liquid or gas filled tanks are connected through pipes, tubes, orifices, valves and other flow restricting devices. Compressibility of a fluid is an important property which influences the performance of fluid systems. If the velocity of sound in fluids is very high, compared to the fluid velocity, the compressibility can be disregarded. Hence compressibility effects are neglected in liquid systems. However compressibility plays an important role in gas systems.

The type of fluid flow, laminar or turbulent, is another important parameter in fluid systems. If the Reynolds number is greater than 4000, the flow is said to be turbulent and if the Reynolds number is less than 2000, it is said to be laminar flow.

For turbulent flow through pipes, orifices, valves and other flow restricting devices, the flow is found from Bernoulli's law and is given by

$$q = KA\sqrt{2g(h_1 - h_2)} \quad .....(2.50)$$

where $q \rightarrow$ liquid flow rate, m^3/sec

$K \rightarrow$ a flow constant

$A \rightarrow$ area of restriction, m^2

g is acceleration due to gravity, m/sec^2

h head of liquid, m

The turbulent resistance is found from

$$R = \frac{dh}{dq} = \frac{q}{gK^2A^2} = \frac{2(h_1 - h_2)}{q} \quad .....(2.51)$$

It can be seen that the flow resistance depends on *h* and *q* and therefore it is non linear. It has to be linearised around the operating point and used over a small range around this point.

The laminar flow resistance is found from the Poisseuille - Hagen law :

$$h_1 - h_2 = \frac{128\mu L}{\pi\gamma D^4} q \quad .....(2.52)$$

where $h \rightarrow$ head, m

$L \rightarrow$ length of the tube, m

$D \rightarrow$ inside diameter of the pipe, m

q $\rightarrow$ liquid flow rate, m^3/sec

m $\rightarrow$ absolute viscosity, kg-sec/m^2

γ $\rightarrow$ fluid density kg/m^3

Since the flow is laminar, head is directly proportional to the flow rate and hence, laminar flow resistance is given by

$$R = \frac{128\,\mu L}{\pi\gamma D^4}\ \text{sec/m}^2 \qquad .....(2.53)$$

Liquid storage tanks are characterised by the capacitance and is defined by,

$$C = \frac{dv}{dh}\ \text{m}^2 \qquad .....(2.54)$$

where v $\rightarrow$ volume of the liquid tank in m^3. Hence the capacitance of a tank is given by its area of cross section at a given liquid surface.

Gas systems consisting of pressure vessels, connecting pipes, valves etc. may be analysed by using the fundamental law of flow of compressible gases. Again, we have to consider two types of flow : turbulent and laminar flow. For turbulent flow through pipes, orifices, valves etc., we have

$$\omega = KA\gamma\sqrt{2g\,(p_1 - p_2)^{\gamma}} \qquad .....(2.55)$$

where ω $\rightarrow$ flow rate, kg/sec

K $\rightarrow$ flow constant

A $\rightarrow$ area of restriction, m^2

γ $\rightarrow$ gas density, kg/m^3

p $\rightarrow$ pressure in kg/m^2

Turbulent gas flow resistance is therefore given by

$$R = \frac{dp}{dw}\ \text{sec/m}^2 \qquad .....(2.56)$$

This is not easy to determine since the rational expansion factor depends on pressure. Usually the resistance is determined from a plot of pressure against flow rate for a given device.

The laminar gas flow resistance is obtained using eqn. (2.50).

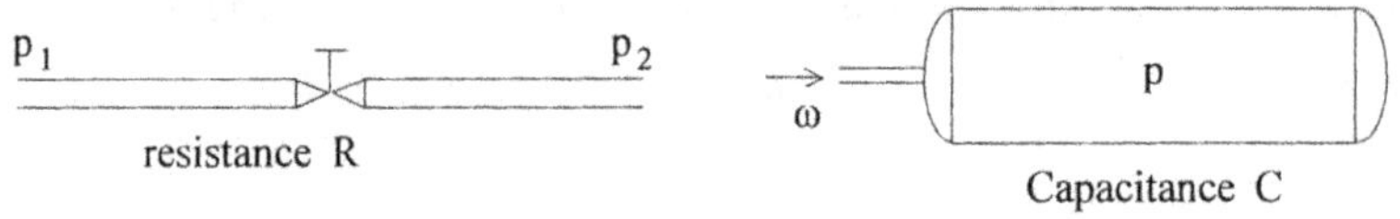

Fig. 2.19 Gas resistance and capacitance

The capacitance parameter for pressure vessels is defined as

$$C = \frac{dv}{dp} \qquad .....(2.57)$$

where $v \rightarrow$ weight of gas in vessel, kg
$p \rightarrow$ pressure kg/m^2

The flow rate and the gas pressure are related by the continuity law :

$$C\frac{dp}{dt} = \omega \qquad .....(2.58)$$

Where ω is the flow rate in kg/sec.

Let us now consider some thermal and fluid systems and obtain their transfer functions.

Example 2.6

Find the transfer function $\frac{C(s)}{V(s)}$ of the thermal system shown in Fig. 2.20. Heat is supplied by convection to a copper rod of diameter D.

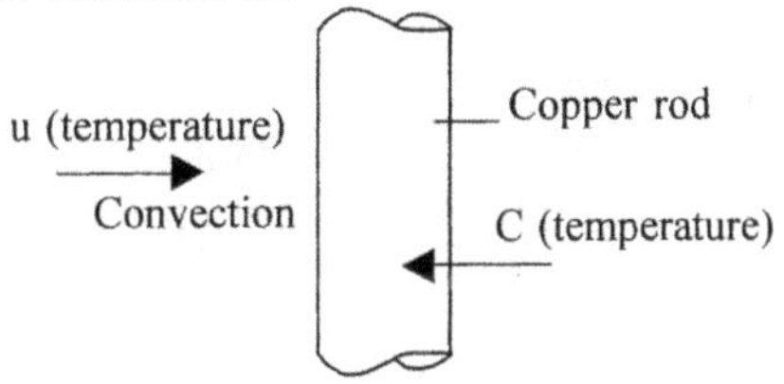

Fig. 2.20 Thermal System

Solution :

The Thermal resistance of the copper rod, from eqn. (2.45), is;

$$R = \frac{1}{HA}$$

Here A is the surface area of the rod.

Hence $A = \pi DL$

where L is the length of the rod.

$$\therefore \qquad R = \frac{1}{\pi DL} \text{ deg sec/J}$$

The thermal capacitance of the rod, from eqn. (2.49), is given by :

$$C = WC_p$$

$$= \frac{\pi D^2 L}{4} \rho C_p$$

where, C_p is the specific heat of copper

and ρ is the density of copper.

From eqn. (2.43), we have, $q = HA(u - c)$ and from eqn. (2.48),

$$C\frac{dc}{dt} = q$$

Combining these two equations,

$$C\frac{dc}{dt} = \frac{1}{R}(u - c)$$

$$C\frac{dc}{dt}+\frac{c}{R}=\frac{u}{R}$$

$$RC\frac{dc}{dt}+c=u$$

where RC = T is the time constant of the system.

$$\therefore \qquad T\frac{dc}{dt}+c=u$$

But $$T=RC=\frac{1}{\pi DLH}\times\frac{\pi D^2L}{4}\rho C_p$$

$$=\frac{D\rho C_p}{4H}$$

Thus the transfer function of the system is,

$$\frac{C(s)}{U(s)}=\frac{1}{Ts+1} \quad \text{where} \quad T=\frac{D\rho C_p}{4H}$$

Example 2.7

Obtain the transfer function $\frac{C(s)}{U(s)}$ for the system shown in Fig. 2.21. c is the displacement of the piston with mass M.

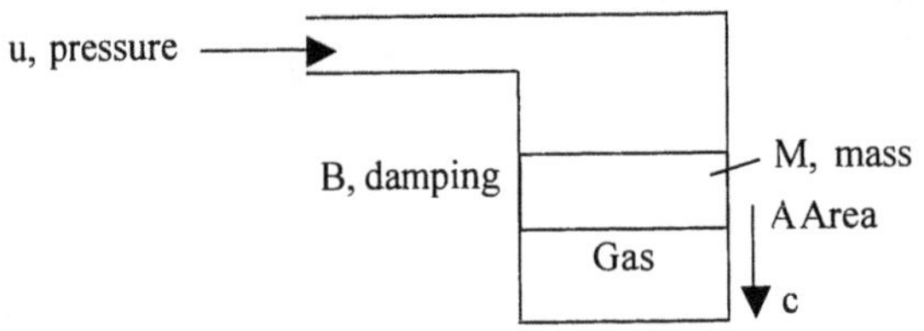

Fig. 2.21 A fluid system

Solution : The system is a combination of mechanical system with mass *M*, damping *B* and a gas system subjected to pressure.

The equilibrium equation is

$$M\ddot{c}+B\dot{c}=A\,[u-P_g] \qquad .....(2.1)$$

Where P_g is the upward pressure exerted by the compressed gas. For a small change in displacement of mass, the pressure exerted is equal to,

$$P_g=\frac{P}{V}Ac$$

where, *P* is the pressure exerted by the gas with a volume of gas under the piston to be *V*

But $\qquad PV = WRT$

where, *R* is the gas constant.

and T is the temperature of the gas.

$$\therefore \qquad P = \frac{WRT}{V}$$

$$\therefore \qquad P_g = \frac{WRT}{V^2} A\,c \qquad \text{......(2)}$$

Substituting eqn. (2) in eqn. (1) we have,

$$M\ddot{c} + B\,\dot{c} + \frac{WRT}{V^2} A^2 c = A\,u$$

The transfer function is,

$$\frac{C(s)}{U(s)} = \frac{A}{Ms^2 + Bs + K} \qquad \text{where } K = \frac{WRT}{V^2} A^2$$

Example 2.8

Obtain the transfer function for the liquid level system shown in Fig. 2.22.

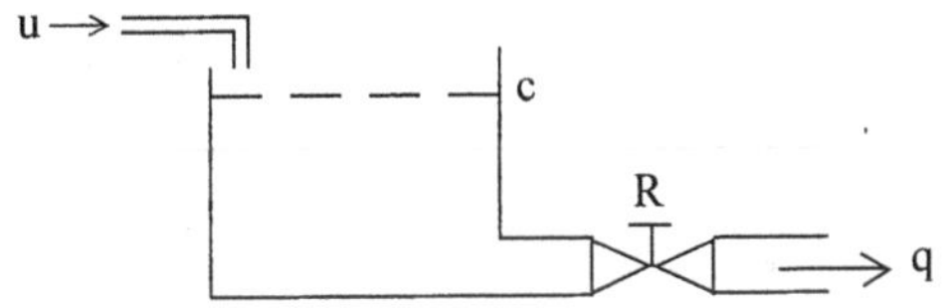

Fig. 2.22 Liquid level system

Solution : The capacitance of the vessel is C = A

where A is the area of cross section of the vessel.

The outflow q is equal to,

$$q = \frac{c}{R}$$

where R is the laminar flow resistance of the valve.

$$A\,\dot{c} = (u - q) = u - \frac{c}{R}$$

$$A\,\dot{c} + \frac{c}{R} = u$$

The transfer function is

$$\frac{C(s)}{u(s)} = \frac{R}{RAs + 1}$$

2.3 Block Diagrams of Closed Loop Systems

The transfer function of any system can be represented by a block diagram as shown in Fig. 2.23.

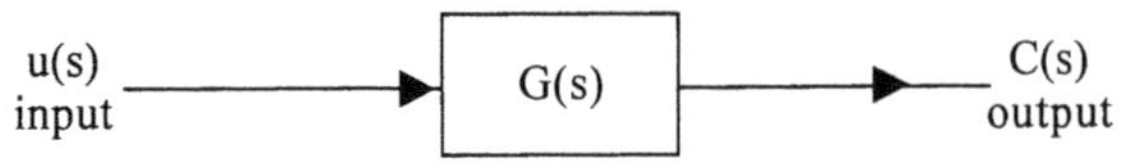

Fig. 2.23 Block diagram of a system

A complex system can be modelled as interconnection of number of such blocks. Inputs to certain blocks may be derived by summing or substracting two or more signals. The sum or difference of signals is indicated in the diagram by a symbol called summer, as shown in Fig. 2.24 (a).

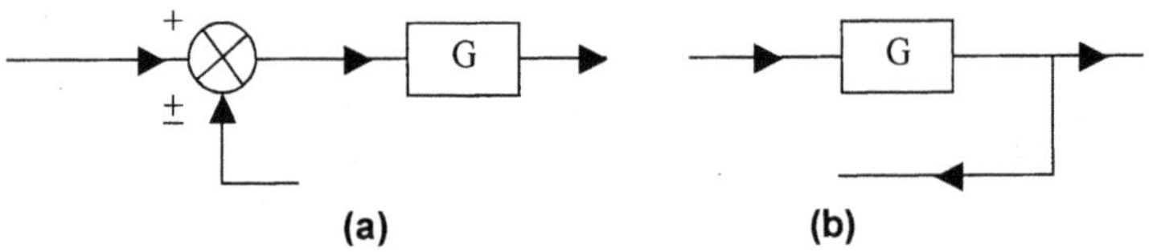

Fig. 2.24 (a) Symbol of a summer (b) Pick off point

On the other hand, a signal may be taken from the output of a block and given to another block. This point from which the signal is tapped is known as pick off point and is shown in Fig. 2.24 (b). A simple feedback system with the associated nomenclature of signals is shown in Fig. 2.25.

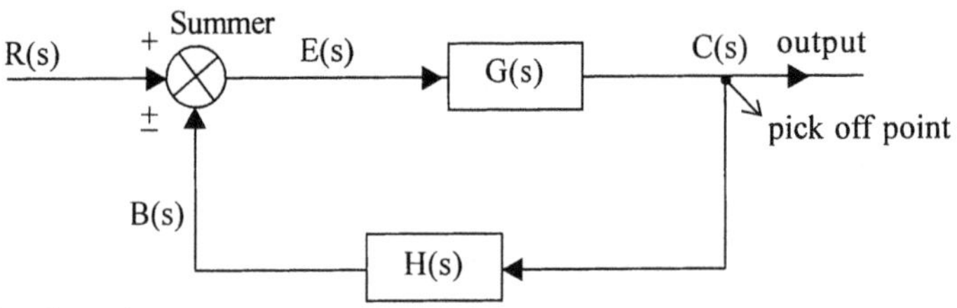

Fig. 2.25 Simple feedback system

$G(s) = \frac{C(s)}{E(s)}$; Forward path transfer function, or plant transfer function

$H(s) = \frac{B(s)}{E(s)}$; Transfer function of the feedback elements

where, R(s); Reference input or desired output

C(s); Output or controlled variable

B(s); Feedback signal

E(s); Error signal

G(s) H(s); Loop transfer function

This closed loop system can be replaced by a single block by finding the transfer function $\frac{C(s)}{R(s)}$.

$$C(s) = G(s)\,E(s) \qquad \text{.....(2.59)}$$

But
$$E(s) = R(s) \pm B(s)$$

$$= R(s) \pm H(s)\,C(s) \qquad \text{.....(2.60)}$$

$\therefore$
$$C(s) = G(s)\,[R(s) \pm H(s)\,C(s)] \qquad \text{.....(2.61)}$$

$$C(s)\,[1 \mp G(s)\,H(s)] = G(s)\,R(s) \qquad \text{.....(2.62)}$$

$$\frac{C(s)}{R(s)} = \frac{G(s)}{1 \mp G(s)\,H(s)} \qquad \text{.....(2.63)}$$

This transfer function can be represented by the single block shown in Fig. 2.26.

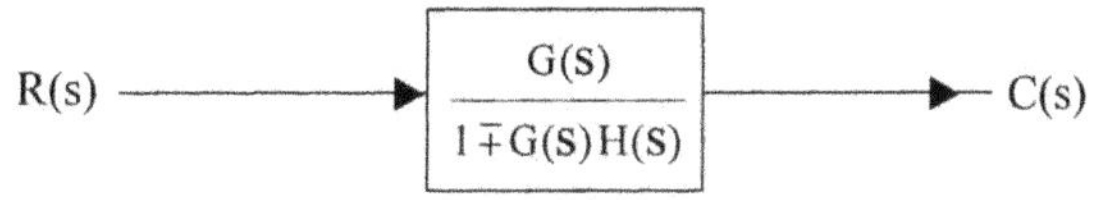

Fig. 2.26 Transfer function of the closed loop system

(**Note :** If the feedback signal is added to the reference input the feedback is said to be positive feedback. On the other hand if the feedback signal is subtracted from the reference input, the feedback is said to be negative feedback).

For the most common case of negative feedback,

$$\frac{C(s)}{R(s)} = \frac{G(s)}{1 + G(s)\,H(s)} \qquad \text{.....(2.64)}$$

The block diagram of a complex system can be constructed by finding the transfer functions of simple subsystems. The overall transfer function can be found by reducing the block diagram, using block diagram reduction techniques, discussed in the next section.

2.3.1 Block Diagram Reduction Techniques

A block diagram with several summers and pick off points can be reduced to a single block, by using block diagram algebra, consisting of the following rules.

Rule 1 *:* Two blocks G_1 and G_2 in cascade can be replaced by a single block as shown in Fig. 2.27.

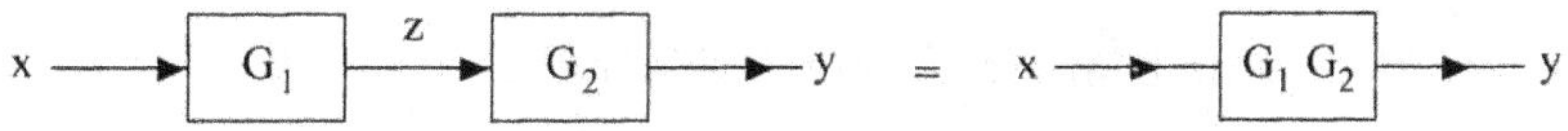

Fig. 2.27 Cascade of two blocks

Here
$$\frac{z}{x} = G_1, \quad \frac{y}{z} = G_2$$

$\therefore$
$$\frac{z}{x} \cdot \frac{y}{z} = \frac{y}{x} = G_1\,G_2$$

Rule 2 : A summing point can be moved from right side of the block to left side of the block as shown in Fig. 2.28.

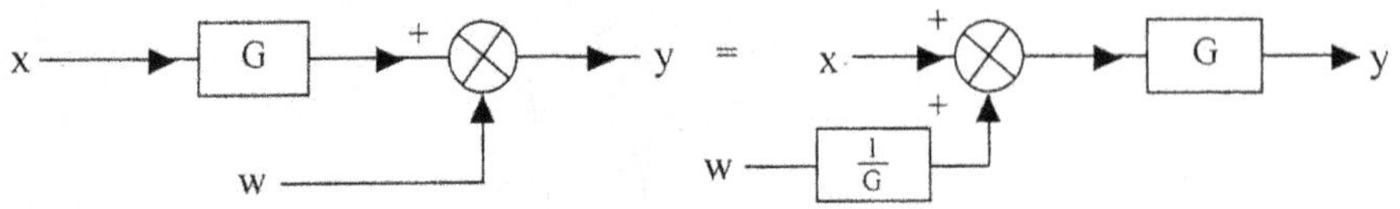

Fig. 2.28 Moving a summing point to left of the block

For the two systems to be equivalent the input and output of the system should be the same. For the left hand side system, we have,

$$y = G\,[x] + w$$

For the system on right side also,

$$y = G\left[x + \frac{1}{G}[w]\right] = G\,[x] + w$$

Rule 3 : A summing point can be moved from left side of the block to right side of the block as shown in Fig. 2.29.

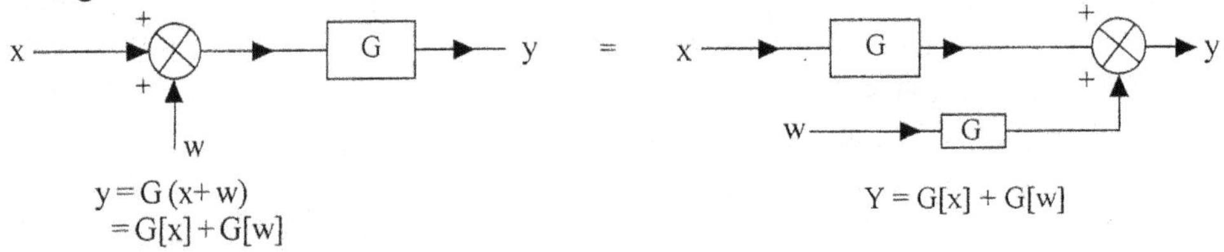

Fig. 2.29 Moving a summer to the right side of the block

Rule 4 : A Pick off point can be moved from the right side of the block to left side of the block as shown in Fig. 2.30.

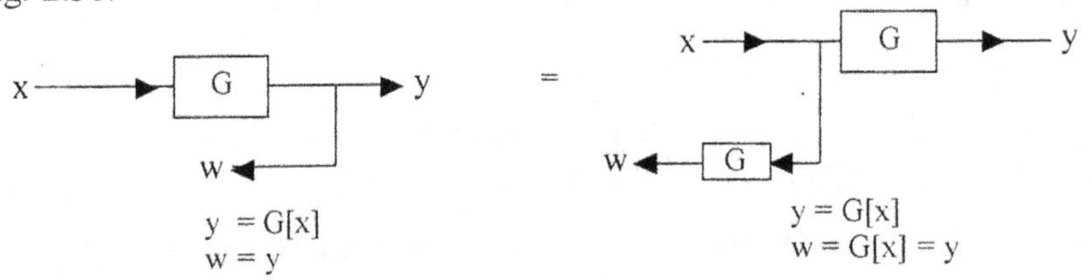

Fig. 2.30 Moving a pick off point to left side of the block

Rule 5 : A Pick off point can be moved from left side of the block to the right side of the block as shown in Fig. 2.31.

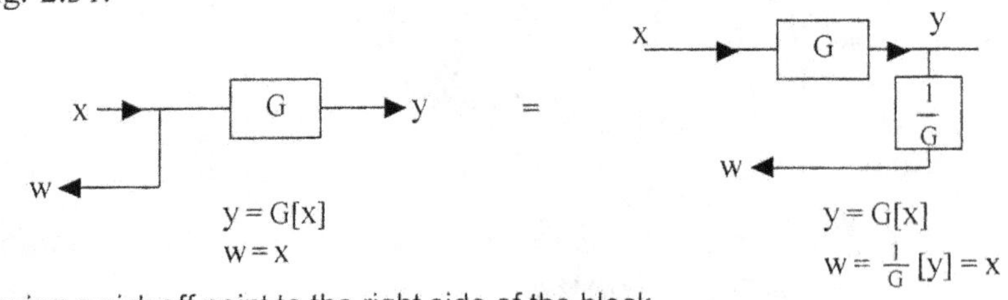

Fig. 2.31 Moving a pick off point to the right side of the block

Rule 6 : A feedback loop can be replaced by a single block as shown in Fig 2.32.

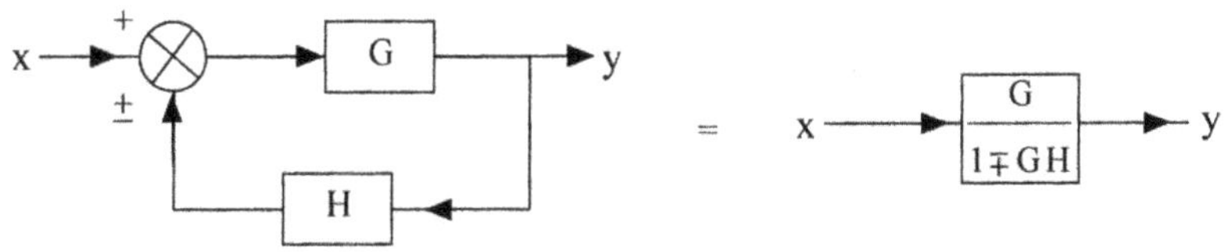

Fig. 2.32 Feedback loop is replaced by a single block

Using these six rules, which are summerised in Table 2.4, any complex block diagram can be simplified to a single block with an input and output. The summing points and pick off points are moved to the left or right of the blocks so that we have either two blocks in cascade, which can be combined, or a simple feedback loop results, which can be replaced by a single block.

This is illustrated in the following examples.

Table 2.4 Rules of Block diagram algebra

Rule no.	Given system	Equivalent system
1. Blocks in cascade	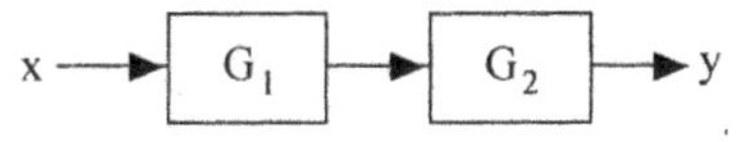	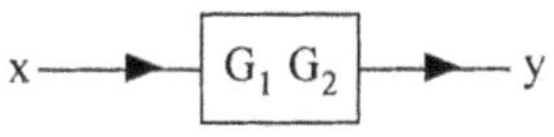
2. Moving summing point to left	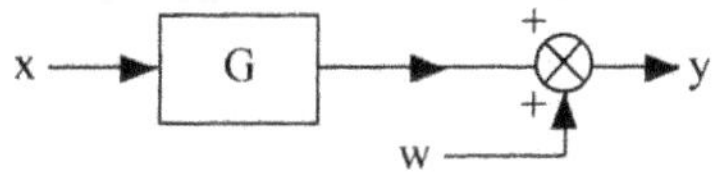	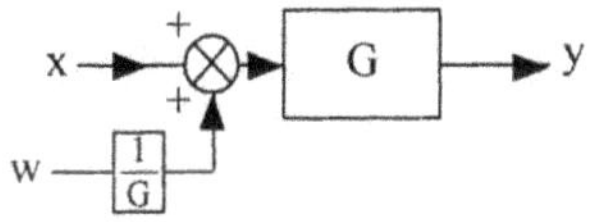
3. Moving summing point to right	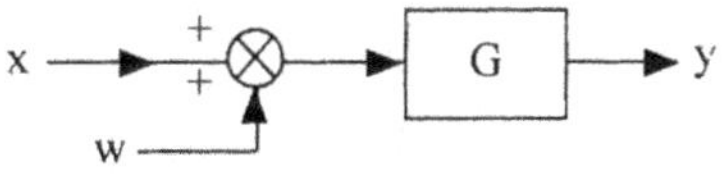	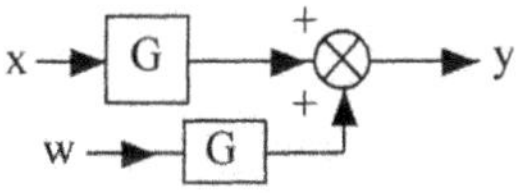
4. Moving pick off point to left	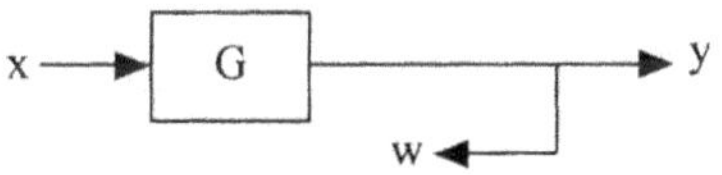	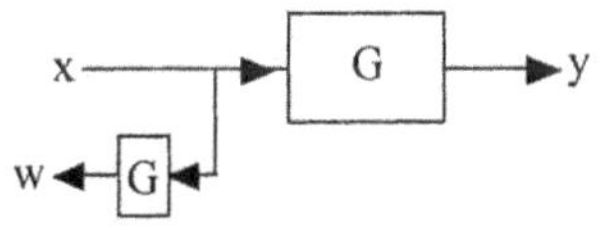
5. Moving pick off point to right	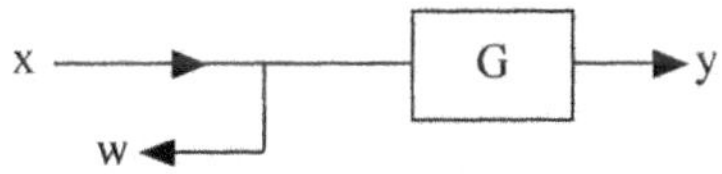	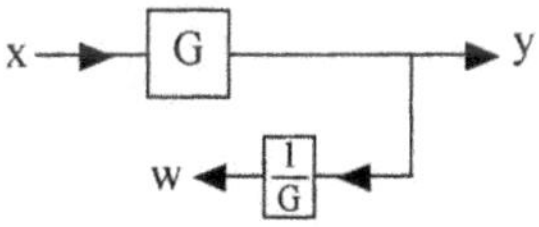
6. Absorbing a loop	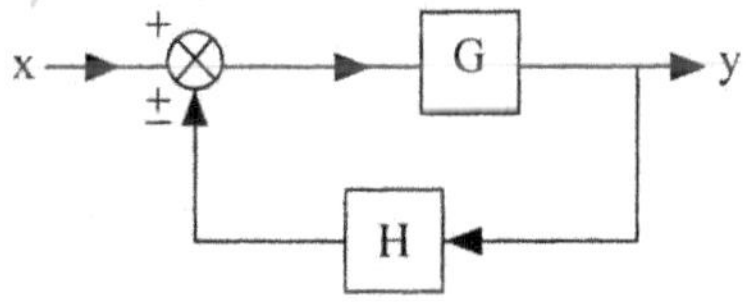	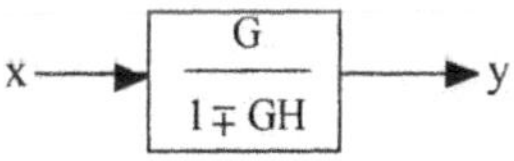

Example 2.9

Find the overall transfer function of the system shown in Fig. 2.33 using block diagram reduction technique.

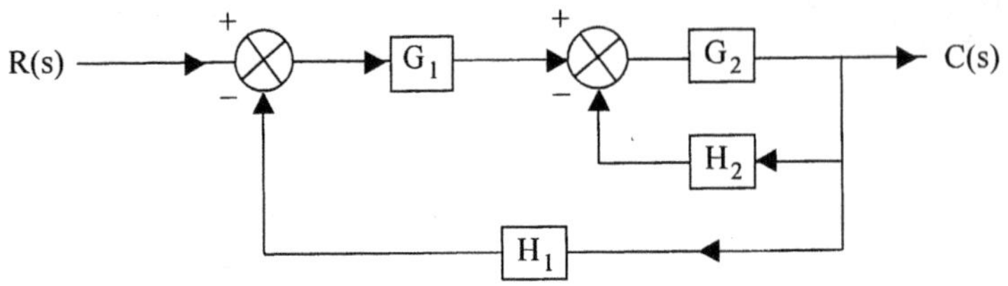

Fig. 2.33 Block diagram of a system

Solution :

Step 1 : Using rule 6, the feedback loop with G_2 and H_2 is replaced by a single block as shown in Fig. 2.33 (a).

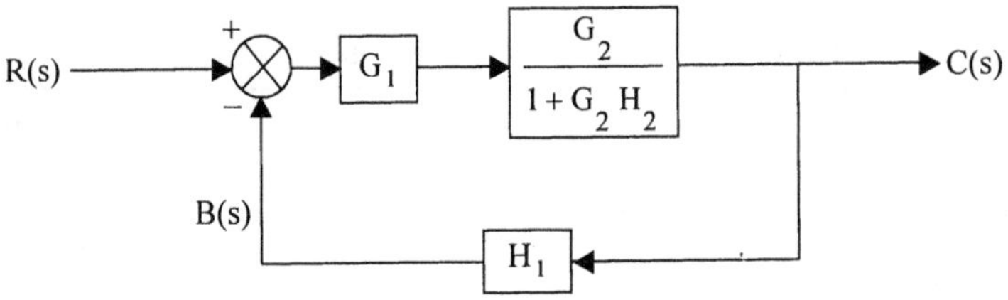

Fig. 2.33 (a)

Step 2 : The two blocks in the forward path are in cascade and can be replaced using rule 1 as shown in Fig. 2.33 (b).

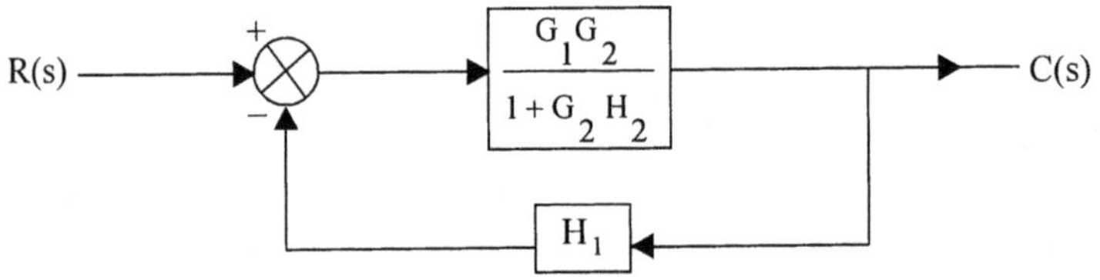

Fig. 2.33 (b)

Step 3 : Finally, the loop in Fig. 2.33 (b) is replaced by a single block using rule 6.

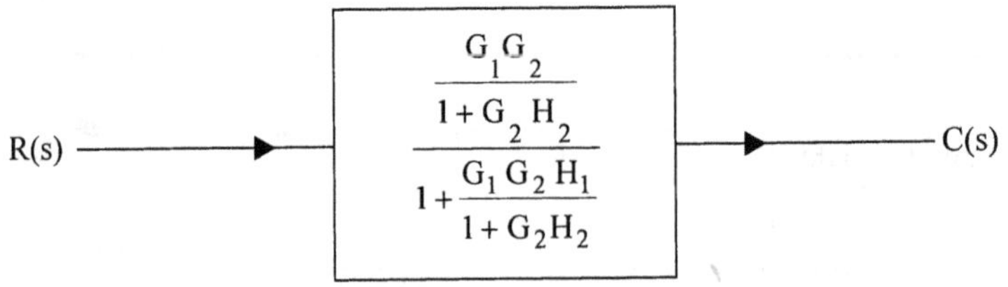

Fig. 2.33 (c)

The transfer function can be simplified as,

$$\frac{C(s)}{R(s)} = \frac{G_1G_2}{1 + G_2H_2 + G_1G_2H_1}$$

Example : 2.10

Obtain the overall transfer function of the system shown in Fig. 2.34 (a).

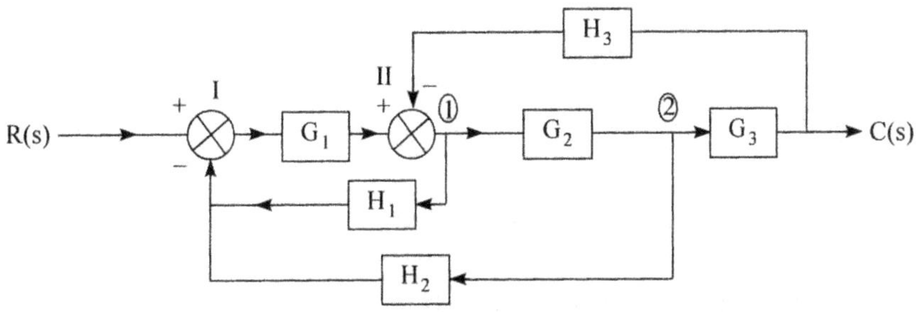

Fig. 2.34 (a) Block diagram of a system for Example 2.10

Solution :

Step 1 : Moving the pick off point (2) to the right of block G_3 and combining blocks G_2 and G_3 in cascade, we have,

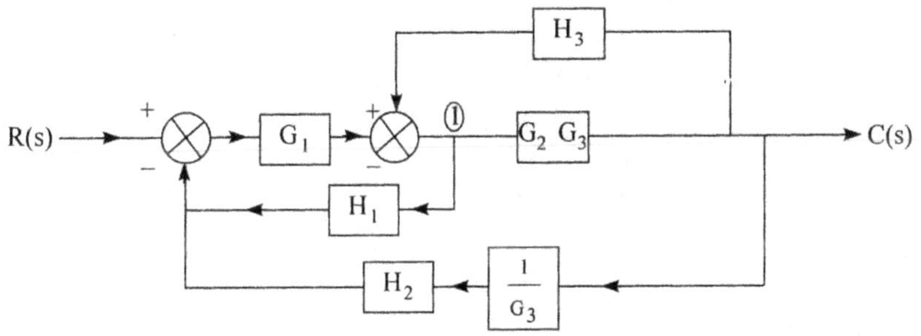

Fig. 2.34 (b) Pick off point (2) moved to right of block G_3

Step 2 : Moving the pick off point (1) to the right of block G_2 G_3

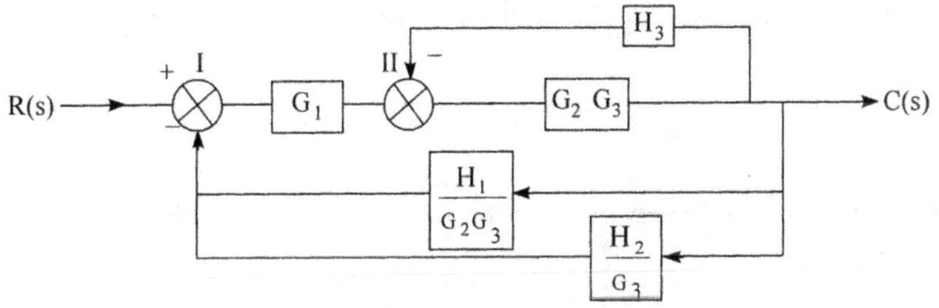

Fig. 2.34 (c) Moving pick off point (1) to right of G_2 G_3

Step 3 : Absorbing the loop with G_2 G_3 and H_3, and combining it with block G_1 in cascade, we have,

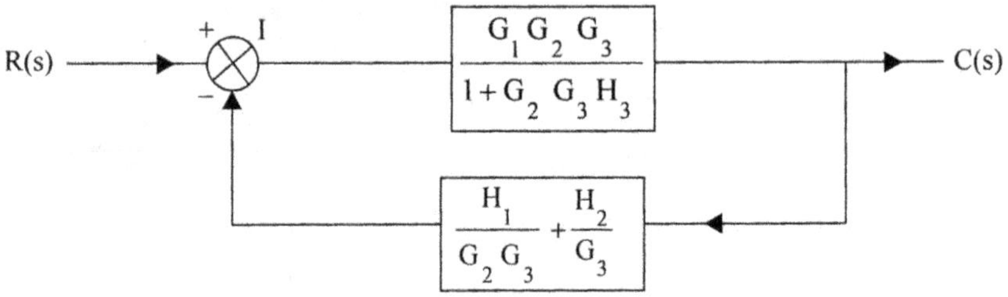

Fig. 2.34 (d) Absorbing the loop and combining it with G_1

The feedback paths $\frac{H_1}{G_2G_3}$ and $\frac{H_2}{G_3}$ have the same inputs and both are subtracted from R(s) at the summer I. Hence they can be added and represented by a single block as shown in Fig. 2.34 (d).

Step 4 : Finally the closed loop in Fig. 2.34(e) is absorbed and the simplified block is given by

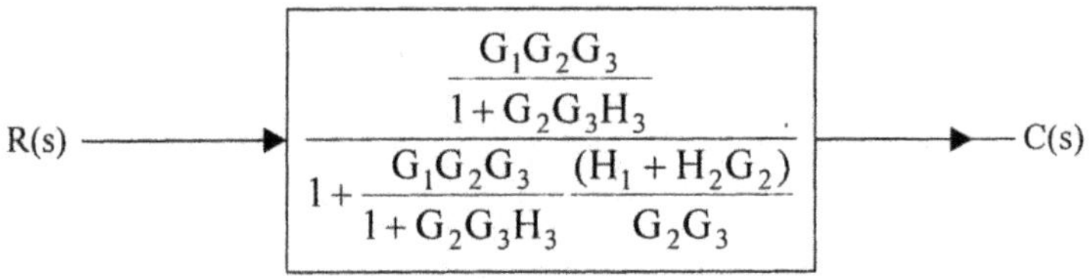

Fig. 2.34 (e) Final step in the block diagram reduction

The transfer function of the system is,

$$\frac{C(s)}{R(s)}=\frac{G_1\,G_2\,G_3}{1+G_2\,G_3\,H_3+G_1\,H_1+G_1\,G_2\,H_2}$$

Example 2.11

Reduce the block diagram and obtain $\frac{C(s)}{R(s)}$ in Fig. 2.35 (a).

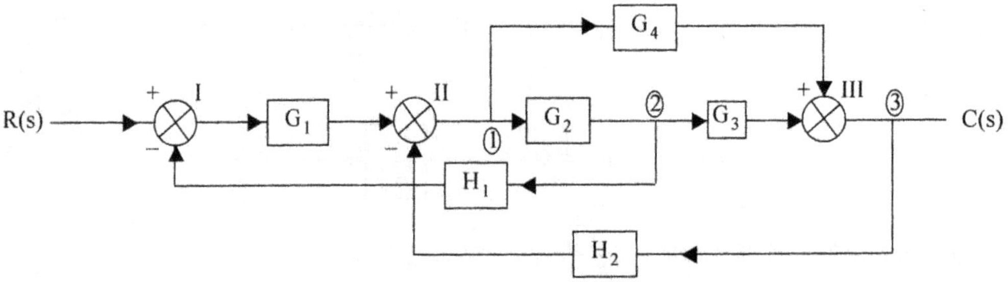

Fig. 2.35 (a) Block diagram of a system for Ex. 2.11

Solution :

Step 1 : Move pick off point (2) to left of G_2 and combine $G_2\ G_3$ in cascade. Further $G_2\ G_3$ and G_4 have same inputs and the outputs are added at summer III. Hence they can be added and represented by a single block as shown in Fig. 2.35 (b).

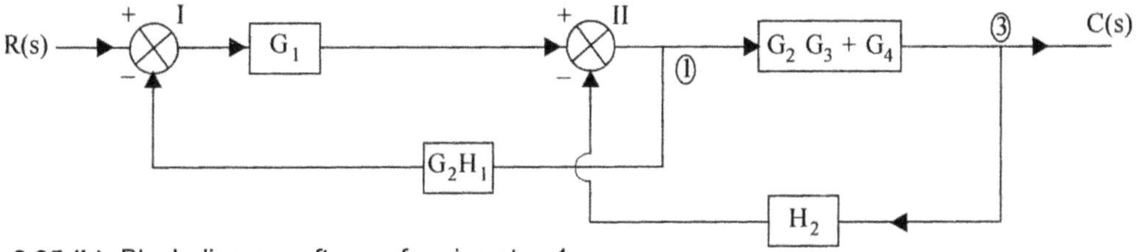

Fig. 2.35 (b) Block diagram after performing step 1

Step 2 : Moving the pick off point (1) to right of block ($G_2\ G_3 + G_4$), we have

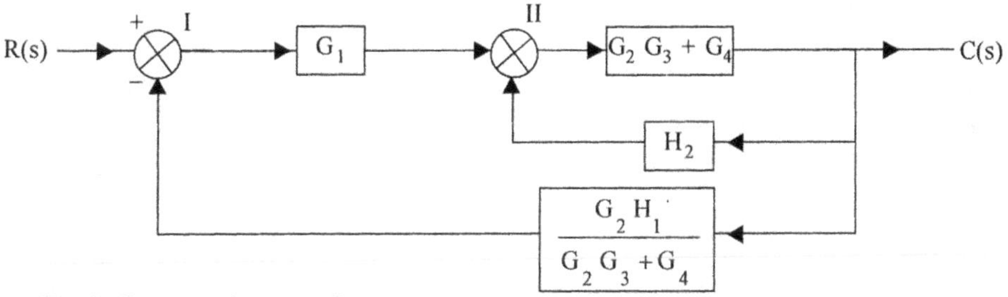

Fig. 2.35 (c) Block diagram after step 2

Step 3 : Absorbing loop with ($G_2\ G_3 + G_4$) and H_2

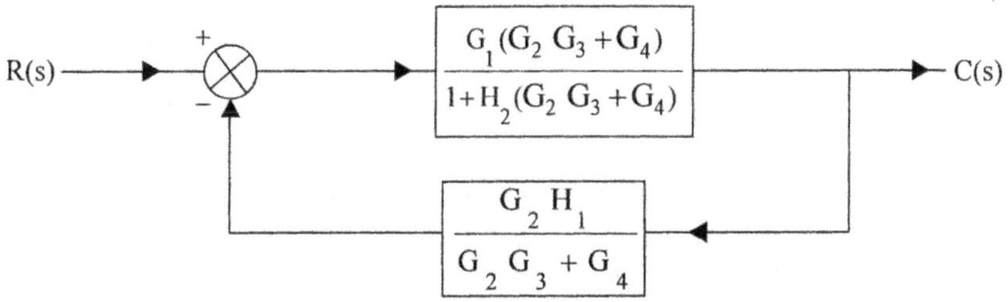

Fig. 2.35 (d) Block diagram after step 3

The transfer function of the system is

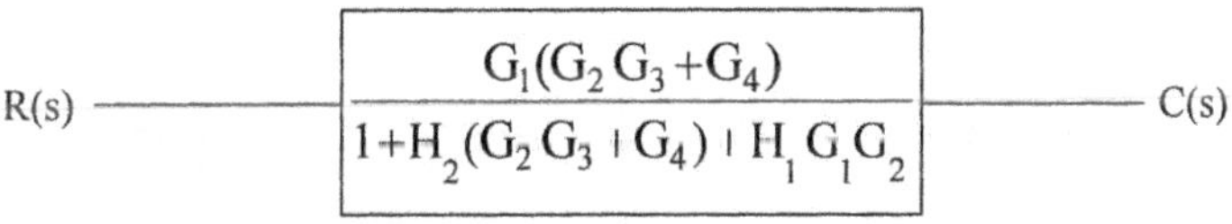

Fig. 2.35 (e) Simplified block of the sytem

$$\frac{C(s)}{R(s)} = \frac{G_1 (G_2 G_3 + G_4)}{1 + H_2 (G_2 G_3 + G_4) + H_1 G_1 G_2}$$

2.4 Signal Flow Graph Representation of Control Systems

Another useful way of representating a system is by a signal flow graph. Although block diagram representation of a system is a simple way of describing a system, it is rather cumbusome to use block diagram algebra and obtain its overall transfer function. A signal flow graph describes how a signal gets modified as it travels from input to output and the overall transfer function can be obtained very easily by using Mason's gain formula. Let us now see how a system can be represented by a signal flow graph. Before we describe a system using a signal flow graph, let us define certain terms.

1. ***Signal flow graph :*** It is a graphical representation of the relationships between the variables of a system.
2. ***Node :*** Every variable in a system is represented by a node. The value of the variable is equal to the sum of the signals coming towards the node. Its value is unaffected by the signals which are going away from the node.
3. ***Branch :*** A signal travels along a branch from one node to another node in the direction indicated on the branch. Every branch is associated with a gain constant or transmittance. The signal gets multiplied by this gain as it travels from one node to another.

   In the example shown in Fig. 2.36, there are four nodes representing variables x_1, x_2, x_3 and x_4. The nodes are hereafter referred by the respective variables for convenience. For example, nodes 1 and 2 are referred to as nodes x_1 and x_2 respectively. The transmittance or gain of the branches are a_{12}, a_{23} and a_{42}. In this example,

   $$x_2 = a_{12} x_1 + a_{42} x_4$$

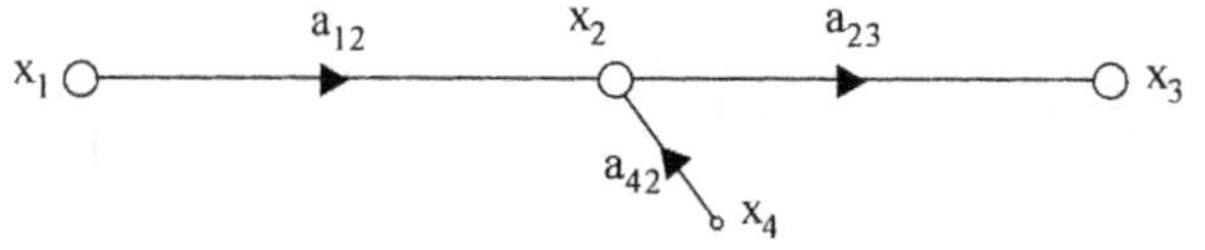

Fig. 2.36 Example shows nodes, branches and gains of the branches

The value x_2 is unaffected by signal going away from node x_1 to node x_3. Similarly

$$x_3 = a_{23} x_2$$

4. ***Input node :*** It is a node at which only outgoing branches are present. In Fig. 2.37 node x_1 is an input node. It is also called as source node.

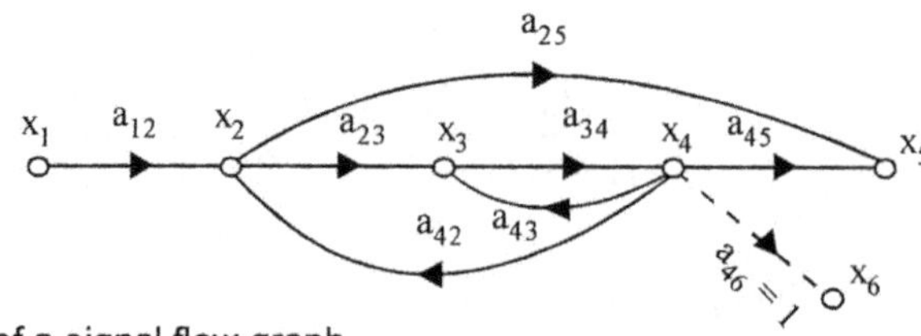

Fig. 2.37 An example of a signal flow graph

5. ***Output node :*** It is a node at which only incoming signals are present. Node x_5 is an output node. In some signal flow graphs, this condition may not be satisfied by any of the nodes. We can make any variable, represented by a node, as an output variable. To do this, we can introduce a branch with unit gain, going away from this node. The node at the end of this branch satisfies the requirement of an output node. In the example of Fig. 2.37, if the variable x_4 is to be made an output variable, a branch is drawn at node x_4 as shown by the dotted line in Fig. (2.37) to create a node x_6. Now node x_6 has only incoming branch and its gain is $a_{46} = 1$. Therefore x_6 is an output variable and since $x_4 = x_6$, x_4 is also an output variable.
6. ***Path :*** It is the traversal from one node to another node through the branches in the direction of the branches such that no node is traversed twice.
7. ***Forward path :*** It is a path from input node to output node.

   In the example of Fig. 2.37, $x_1 - x_2 - x_3 - x_4 - x_5$ is a forward path. Similarly $x_1 - x_2 - x_5$ is also a forward path
8. ***Loop :*** It is a path starting and ending on the same node. For example, in Fig. 2.37, $x_3 - x_4 - x_3$ is a loop. Similarly $x_2 - x_3 - x_4 - x_2$ is also a loop.
9. ***Non touching loops :*** Loops which have no common node, are said to be non touching loops.
10. ***Forward path gain :*** The gain product of the branches in the forward path is called *forward path gain.*
11. ***Loop gain :*** The product of gains of branches in the loop is called as *loop gain.*

2.4.1 Construction of a Signal Flow Graph for a System

A signal flow graph for a given system can be constructed by writing down the equations governing the variables. Consider the network shown in Fig. 2.38.

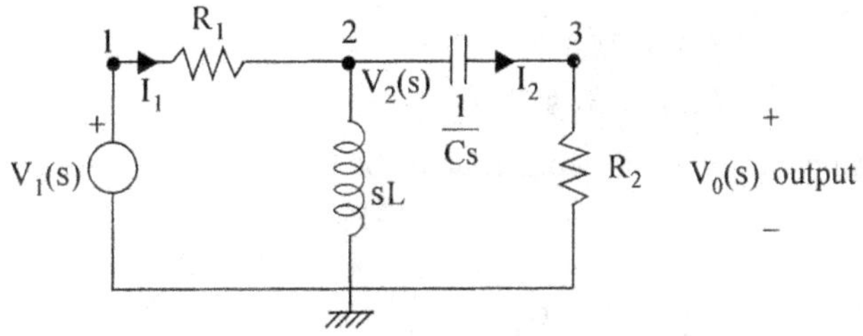

Fig. 2.38 A network for constructing a signal flow graph

Identifying the currents and voltages in the branches as shown in Fig. 2.38; we have

$$I_1(s) = \frac{V_1(s) - V_2(s)}{R_1}$$

$$V_2(s) = [I_1(s) - I_2(s)]\, sL$$

$$I_2(s) = [V_2(s) - V_0(s)]Cs$$

$$V_0(s) = I_2(s)\, R_2$$

The variables $I_1(s)$, $I_2(s)$, $V_1(s)$, $V_2(s)$ and $V_0(s)$ are represented by nodes and these nodes are interconnected with branches to satisfy the relationships between them. The resulting signal flow graph is shown in Fig. 2.39.

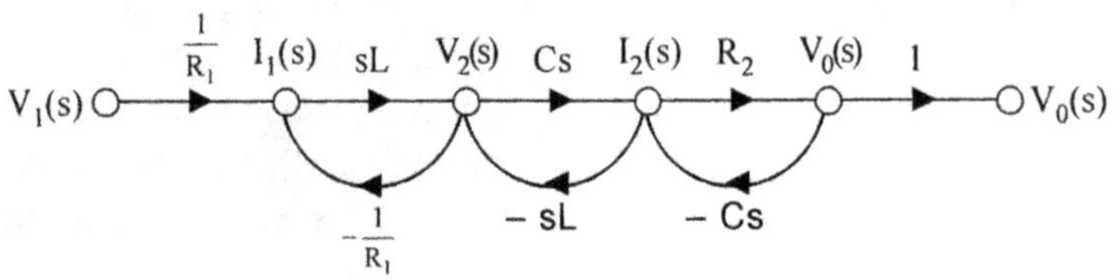

Fig. 2.39 Signal flow graph for the network in Fig. 2.38

Now, consider the block diagram given in Fig. 2.40.

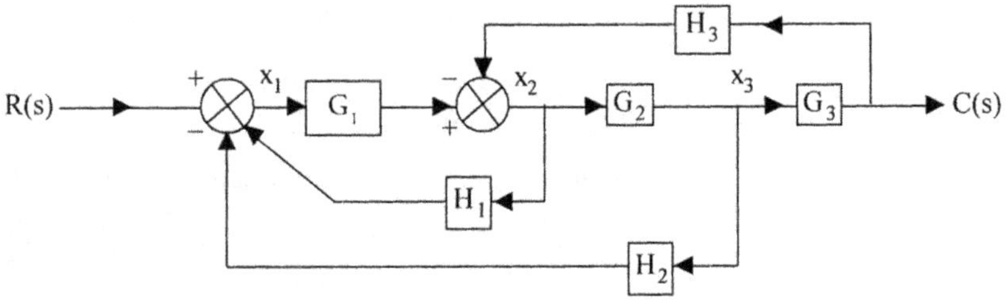

Fig. 2.40 Block diagram of a system

The relationship between the various variables indicated on the diagrams are :

$$X_1(s) = R(s) - H_1(s)\, X_2(s) - H_2(s)\, X_3(s)$$

$$X_2(s) = G_1(s)\, X_1(s) - H_3(S)\, C(s)$$

$$X_3(s) = G_2(s)\, X_2(s)$$

$$C(s) = G_3(s)\, X_2(s)$$

Signal flow graph can be easily constructed for this block diagram, as shown in Fig. 2.41.

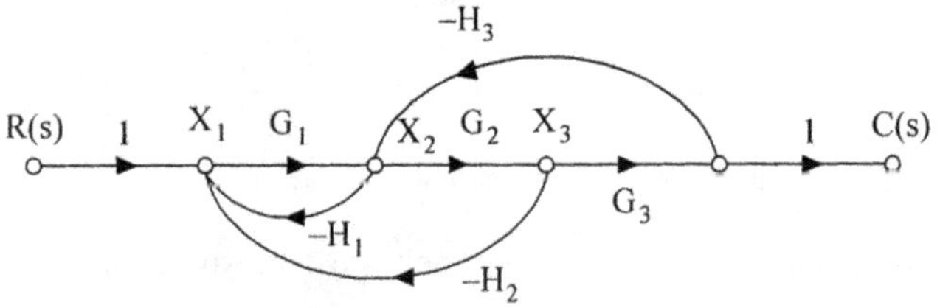

Fig. 2.41 Signal flow graph for the system in Fig. 2.40

2.4.2 Mason's Gain Formula

The transfer function (gain) of the given signal flow graph can be easily obtained by using Mason's gain formula. Signal flow graphs were originated by S.J. Mason and he has developed a formula to obtain the ratio of output to input, called as gain, for the given signal flow graph.

Mason's gain formula is given by,

$$T = \frac{\text{output variable}}{\text{input variable}} = \frac{\sum_k M_k \Delta_k}{\Delta} \qquad .....(2.65)$$

where M_k is the kth forward path gain, Δ is the determinant of the graph, given by,

$$\Delta = 1 - \Sigma P_{m1} + \Sigma P_{m2} \; ...... \; (-1)^r \; \Sigma P_{mr} \qquad .....(2.66)$$

where P_{mr} is the product of the gains of mth possible combination of r non touching loops.

or $\Delta = 1 -$ (sum of gains of individual loops) + sum of gain products of possible combinations of 2 non touching loops) - (sum of gain products of all possible combinations of 3 nontouching loops) + ...(2.67)

and Δ_k is the value of Δ for that part of the graph which is non touching with k^{th} forward path.

To explain the use of this formula, let us consider the signal flow graph shown in Fig. 2.42, and find the transfer function C(s)/R(s).

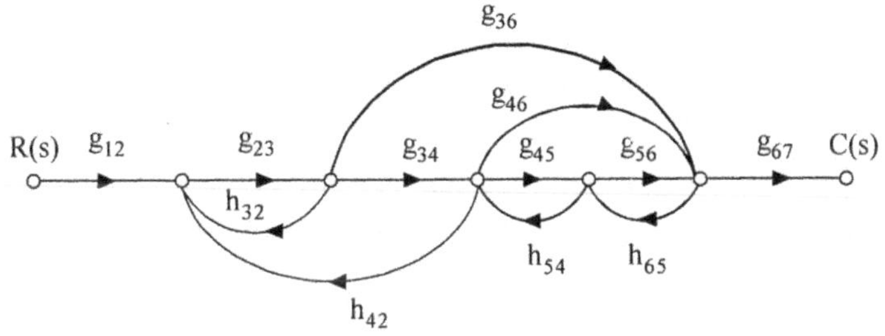

Fig. 2.42 Signal flow graph to illustrate Mason's gain formula

Let us enumerate the forward paths and loops of this flow graph.

1. *Forward paths*

$$M_1 = g_{12}\, g_{23}\, g_{34}\, g_{45}\, g_{56}\, g_{67}$$
$$M_2 = g_{12}\, g_{23}\, g_{36}\, g_{67}$$
$$M_3 = g_{12}\, g_{23}\, g_{34}\, g_{46}$$

2. *Loops*

$$L_1 = g_{23}\, h_{32}$$
$$L_2 = g_{23}\, g_{34}\, h_{42}$$
$$L_3 = g_{45}\, h_{54}$$
$$L_4 = g_{56}\, h_{65}$$
$$L_5 = g_{46}\, h_{65}\, h_{54}$$
$$L_6 = g_{23}\, g_{36}\, h_{65}\, h_{54}\, h_{42}$$

3. *Two non touching loops*

Out of the six loops there are 3 combinations of two non touching loops. Their gain products are:

$$P_{12} = L_1 L_3 = g_{23}\, g_{45}\, h_{32}\, h_{54}$$
$$P_{22} = L_1 L_4 = g_{23}\, g_{56}\, h_{32}\, h_{65}$$
$$P_{32} = L_1 L_5 = g_{23}\, g_{46}\, h_{32}\, h_{65}\, h_{54}$$

4. *Three non touching loops*

There is no combinations of 3 non touching loops or 4 nontouching loops etc.

$\therefore \qquad P_{m3} = P_{m4} = \ldots\ldots 0$

Let us now calculate the determinant of the flow graph.

$$\Delta = 1 - (L_1 + L_2 + L_3 + L_4 + L_5 + L_6) + P_{12} + P_{22} + P_{32}$$
$$= 1 - (g_{23}\, h_{32} + g_{23}\, g_{34}\, h_{42} + g_{45}\, h_{54} + g_{56}\, h_{65} + g_{46}\, h_{65}\, h_{54} + g_{23}\, g_{36}\, h_{65}\, h_{54}\, h_{42}) + (g_{23}\, g_{45}\, h_{32}\, h_{54} + g_{23}\, g_{56}\, h_{32}\, h_{65} + g_{23}\, g_{46}\, h_{32}\, h_{65}\, h_{54})$$

and Δ_1 = value of Δ which is non touching with M_1.

Eliminate all terms in Δ which have any node in common with forward path M_1. All loops L_1 to L_6 have at least one node in common with M_1 and hence,

$$\Delta_1 = 1$$

Similarly Δ_2 = value of Δ which is non touching with M_2.

The loop L_3 is not having any common node with M_2 and hence eliminating all loops except L_3 we have,

$$\Delta_2 = 1 - L_3 = 1 - g_{45}\, h_{54}$$

Similarly, $\qquad \Delta_3 = 1$

Applying Mason's gain formula (eqn. 2.65) we have

$$T(s) = \frac{C(s)}{R(s)} = \frac{M_1 \Delta_1 + M_2 \Delta_2 + M_3 \Delta_3}{\Delta}$$

Example 2.12

Find the transfer function $\dfrac{V_0(s)}{V_1(s)}$ for the network shown in Fig.2.38.

Solution : The network and its signal flow graph are reproduced in Fig. 2.43(a) and (b) for convenience.

Step 1 : *Forward paths :* There is only one forward path.

$$P_1 = \frac{1}{R_1} \,.\, sL.\, Cs.\, R_2$$
$$= \frac{LCs^2 R_2}{R_1}$$

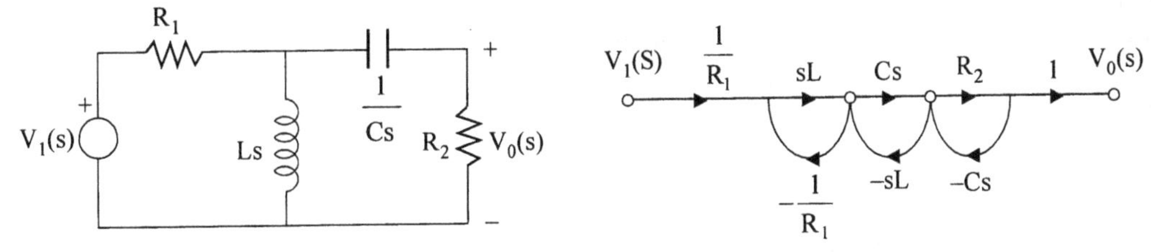

(a) The network

(b) Its signal flow graph

Fig. 2.43

Step 2 : (i) *Loop gains*

$$L_1 = -\frac{Ls}{R_1}$$

$$L_2 = Cs\,(-sL) = -s^2 LC$$

$$L_3 = -R_2 Cs$$

(ii) *Product of gains of two non touching loops*

$$P_{12} = L_1 . L_3$$

$$= -\frac{sL}{R_1}(-R_2 cs)$$

$$= \frac{R_2}{R_1} . LCs^2$$

Step 3 : *The determinant of the graph*

$$\Delta = 1 + \frac{Ls}{R_1} + s^2 LC + R_2 Cs + \frac{R_2}{R_1} LCs^2$$

and $\Delta_1 = 1$

Step 4 : *The transfer function is*

$$\frac{V_0(s)}{V_1(s)} = \frac{\dfrac{LCs^2 R_2}{R_1}}{1 + \dfrac{Ls}{R_1} + s^2 LC + R_2 Cs + \dfrac{R_2}{R_1} LCs^2}$$

$$= \frac{LCs^2 R_2}{s^2 LC(R_1 + R_2) + s(R_1 R_2 C + L) + R_1}$$

Example 2.13

Find the transfer function for the block diagram of Fig. 2.40.

Solution : The block diagram and its signal flow graph are reproduced in Fig. 2.44 (a) and (b).

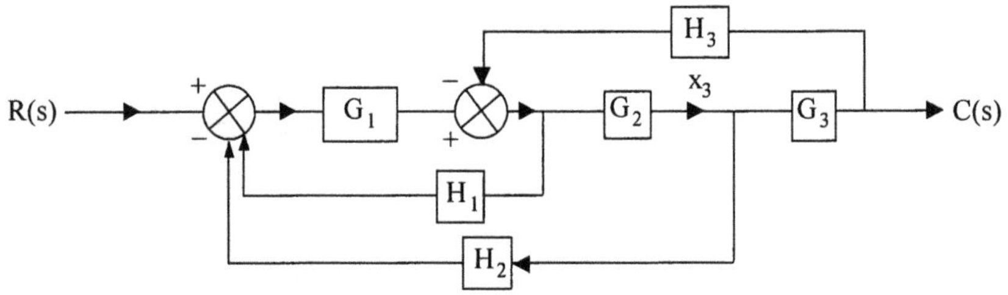

Fig. 2.44 (a) Block diagram of a system

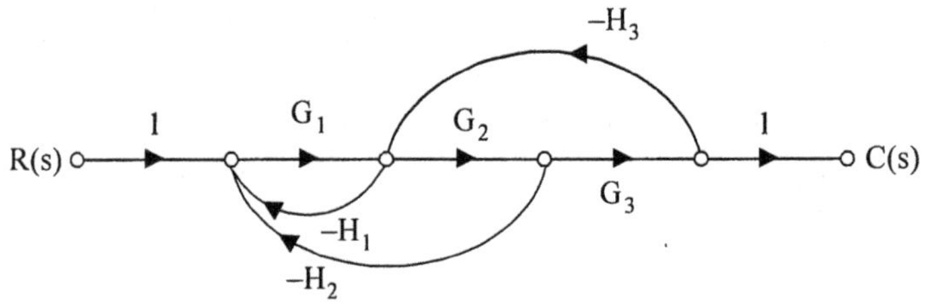

Fig. 2.44 (b) Signal flow graph of block diagram in Fig. 2.44 (a)

Step 1 : *Forward path gains*

$$P_1 = G_1\, G_2\, G_3$$

Step 2 : *Loop gain*

$$L_1 = -G_1\, H_1$$
$$L_2 = -\, G_1\, G_2\, H_2$$
$$L_3 = -\, G_2\, G_3\, H_3$$

Two or more non touching loops are not present hence

$$P_{mr} = 0 \quad \text{for} \quad r = 2, 3, \ldots..$$

Step 3 : *The determinant of the graph*

$$\Delta = 1 + G_1\, H_1 + G_1\, G_2\, H_2 + G_2\, G_3\, H_3$$
$$\Delta_1 = 1$$

Step 4 : *Transfer function*

$$\frac{C(s)}{R(s)} = \frac{G_1\, G_2\, G_3}{1 + G_1\, H_1 + G_1\, H_2 + G_2\, G_3\, H_3}$$

Example 2.14

Draw the signal flow graph for the block diagram given in Fig. 2.45 (a) and obtain the transfer function C(s)/R(s).

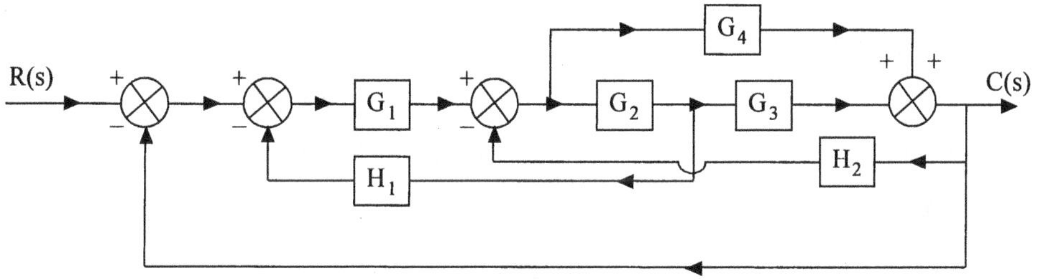

Fig. 2.45 (a) Block diagram of a system

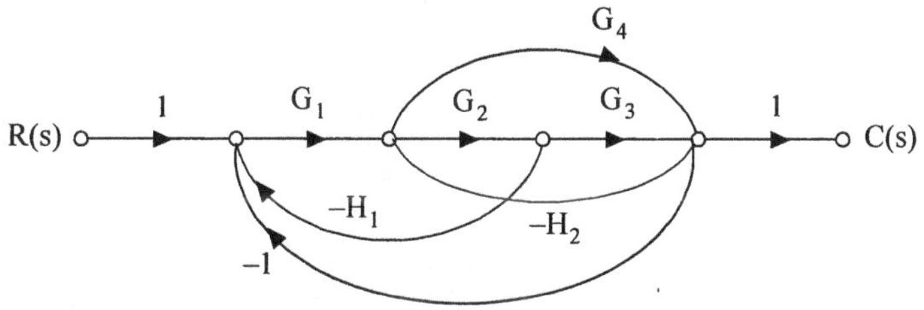

Fig. 2.45 (b) Signal flow graph of block diagram in Fig. 2.45 (a)

Solution : The signal flow graph can be easily written down as

Step 1 : *Forward path gains*

There are two forward paths

$$P_1 = G_1 G_2 G_3$$
$$P_2 = G_1 G_4$$

Step 2 : *Loop gains*

$$L_1 = - G_1 G_2 H_1$$
$$L_2 = - G_2 G_3 H_2$$
$$L_3 = - G_1 G_2 G_3$$
$$L_4 = - G_4 H_2$$
$$L_5 = - G_1 G_4$$

No two or more non touching loops. Hence $P_{mr} = 0$ for r = 2, 3, 4 .....

Step 3 : *The determinant of the graph*

$$\Delta = 1 + G_1 G_2 H_1 + G_2 G_3 H_2 + G_1 G_2 G_3 + G_4 H_2 + G_1 G_4$$
$$\Delta_1 = 1$$
$$\Delta_2 = 1$$

Step 4 : The transfer function is,

$$\frac{C(s)}{R(s)} = \frac{G_1(G_2G_3 + G_4)}{1 + G_1G_2H_1 + G_2G_3H_2 + G_1G_2G_3 + G_1G_4 + G_4H_2}$$

Example 2.15

Find the transfer function $\frac{C(s)}{R_1(s)}$ and $\frac{C(s)}{R_2(s)}$ in Fig. 2.46(a) using signal flow graph technique and assuming that only one input is present in each case.

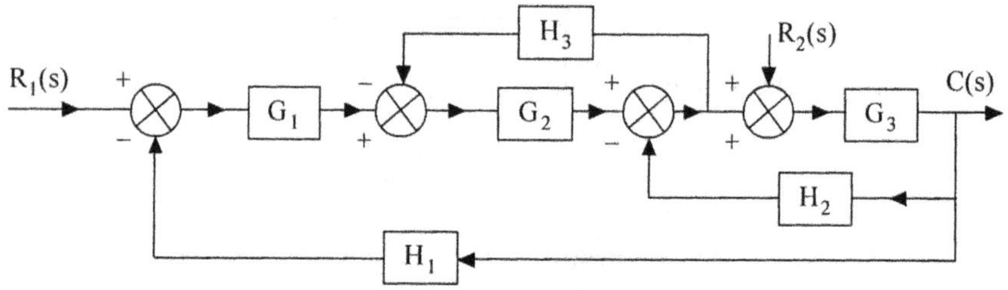

Fig. 2.46 (a) Block diagram of the system

Solution : The signal flow graph of the system can be easily written down as;

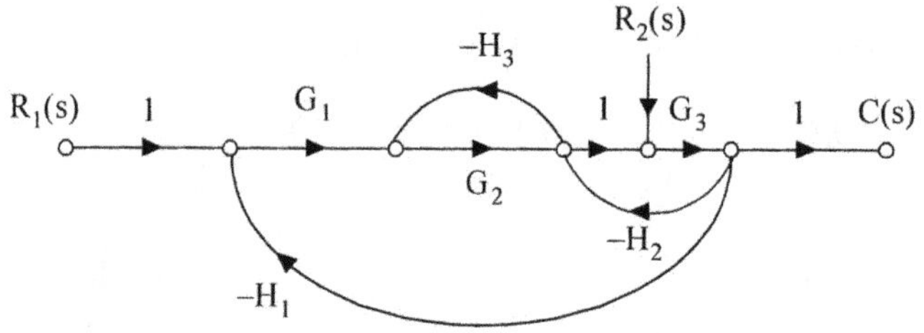

Fig. 2.46 (b) Signal flow graph of system in Fig. 2.46(a)

(a) Transfer function $\frac{C(s)}{R_1(s)}$; $R_2(s) = 0$

Step 1 : Forward path gain

$$P_1 = G_1 G_2 G_3$$

Step 2 : Loop gain

$$L_1 = -G_2 H_3$$
$$L_2 = -G_3 H_2$$
$$L_3 = -G_1 G_2 G_3 H_1$$

Two or more non touching loops are not present.

$$P_{mr} = 0 \qquad \text{for } r = 2, 3, 4 \ldots$$

Step 3 : Determinant of the graph

$$\Delta = 1 + G_2 H_3 + G_3 H_2 + G_1 G_2 G_3 H_1$$

$$\Delta_1 = 1$$

Step 4 : Transfer function $\frac{C(s)}{R_1(s)}$ is given by

$$\frac{C(s)}{R_1(s)} = \frac{G_1G_2G_3}{1+G_2H_3+G_3H_2+G_1G_2G_3H_1}$$

(b) Transfer function $\frac{C(s)}{R_2(s)}$; $R_1(s) = 0$

Step 1 : Forward path gains

$$P_1 = G_3$$

Step 2 : Loop gains

$$L_1 = -G_3 H_2$$

$$L_2 = -G_3 H_1 G_1 G_2$$

$$L_3 = -G_2 H_3$$

Two or more non touching loops are not present. Hence

$$P_{mr} = 0 \quad \text{for } r = 2, 3 \ .....$$

Step 3 : Determinant of the graph

$$\Delta = 1 + G_3 H_2 + G_3 H_1 G_1 G_2 + G_2 H_3$$

$$\Delta_1 = 1 + G_2 H_3$$

Step 4 : Transfer function $\frac{C(s)}{R_2(s)}$ is given by

$$\frac{C(s)}{R_2(s)} = \frac{G_3(1+G_2H_3)}{1+G_3H_2+G_3H_1G_1G_2+G_2H_3}$$

Example 2.15

Obtain $\frac{C(s)}{R(s)}$ for the signal flow graph of Fig. 2.47.

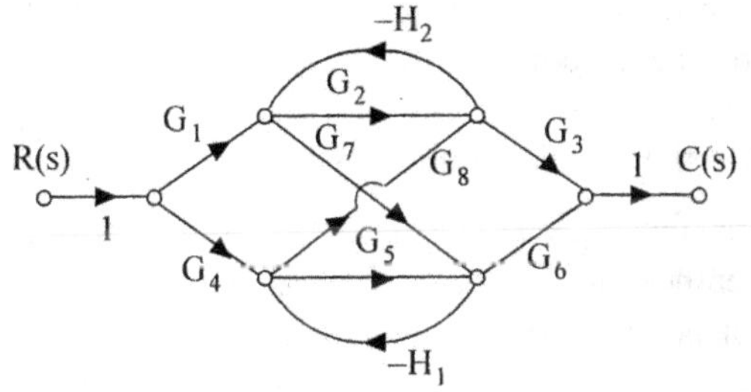

Fig. 2.47 Signal flow graph for ex. 2.15

Solution :

Step 1 : Forward path gains

There are six forward paths in this graph.

$$P_1 = G_1 G_2 G_3$$
$$P_2 = G_1 G_7 G_6$$
$$P_3 = -G_1 G_7 H_1 G_8 G_3$$
$$P_4 = -G_4 G_8 H_2 G_7 G_6$$
$$P_5 = G_4 G_5 G_6$$
$$P_6 = G_4 G_8 G_3$$

Step 2 : Loop gains

$$L_1 = -G_2 H_2$$
$$L_2 = -G_5 H_1$$
$$L_3 = G_7 H_1 G_8 H_2$$

There is one set of two non touching loops. They are L_1 and L_2

$$\therefore \quad P_{12} = G_2 G_5 H_1 H_2$$

Step 3 : The determinant of the graph is

$$\Delta = 1 + G_2 H_2 + G_5 H_1 - G_7 G_8 H_1 H_2 + G_2 G_5 H_1 H_2$$
$$\Delta_1 = (1 + H_1 G_5)$$
$$\Delta_2 = \Delta_3 = \Delta_4 = \Delta_6 = 1$$
$$\Delta_5 = 1 + H_2 G_2$$

Step 4 : The transfer function C(s)/R(s) is

$$\frac{C(s)}{R(s)} = \frac{G_1 G_2 G_3 (1 + H_1 G_5) + G_1 G_7 G_6 - G_1 G_7 H_1 G_8 G_3 - G_4 G_8 H_2 G_7 G_6 \\ G_4 G_5 G_6 (1 + H_2 G_2) + G_4 G_8 G_3}{1 + G_2 H_2 + G_5 H_1 - G_7 G_8 H_1 H_2 + G_2 G_5 H_1 H_2}$$

2.5 Effects of Feedback

The output of an open loop system depends on the input and the parameters of the system. For the same input, the output may be different if the parameters of the system change due to various reasons like, variation in environmental conditions, degradation of components etc. The output is also affected by any noise that may be present in the system. Also, if the system dynamics are to be altered to suit the requirements, the system parameters have to be altered.

To overcome these problems, feedback can be effectively used. The output is measured by suitable means, compared with the reference input and an actuating signal can be produced as an input to the system, so that desired output is obtained. It means that the input to the system is continuously changed in such a way that the system produces the required output. When the output is the desired output, the error signal and the actuating signal are zero. As long as this error signal is present, the output gets modified continuously. Even if the parameters of the system change due to any reason, the output will always tend towards the desired value. A feedback system is shown in Fig. 2.48.

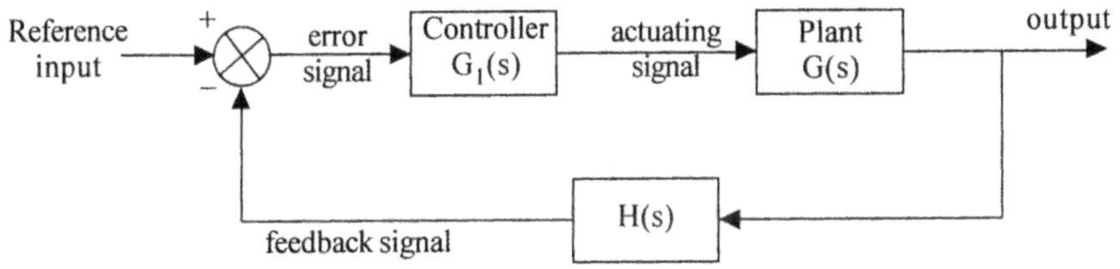

Fig. 2.48 A feedback system

Let us now consider the effects of feedback on the performance of the system.

2.5.1 Effect on system dynamics

The system dynamics can be changed without actually altering the system parameters. This can be illustrated with a simple example of a first order system. Consider the system shown in Fig. 2.49.

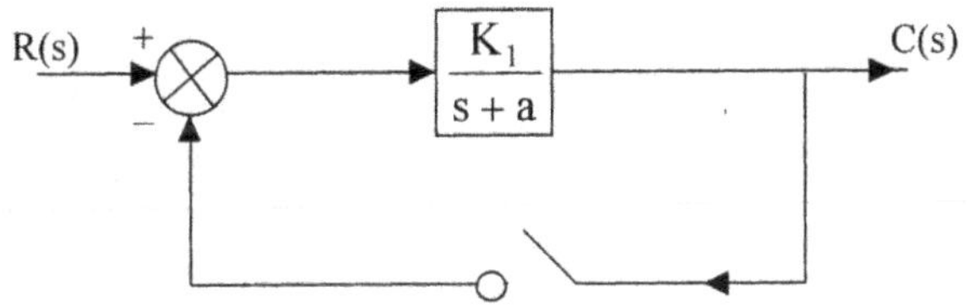

Fig. 2.49 A first order system

When the switch is open, the system is an open loop system and the open loop transfer function is given by

$$T(s) = \frac{C(s)}{R(s)} = \frac{K_1}{s+a} \qquad .....(2.67)$$

The impulse response of the system is given by $h(t) = L^{-1}\, T(s) = K_1 e^{-at}$ .....(2.68)

The time constant τ is given by

$$\tau = \frac{1}{a} \qquad .....(2.69)$$

and the d.c gain (value of T(s) when s = 0) is given by $K = \frac{K_1}{a}$ .....(2.70)

The transfer function has a pole at $s = -a$

If the gain K_1 is changed by using an amplifier in cascade with the system, it cannot change the time constant of the system. It can only alter the d.c. gain. So if the system speed is to be improved, only the parameter *a* has to be changed.

Let us now close the switch so that, the system becomes a feedback system. The closed loop transfer function is given by,

$$T(s) = \frac{C(s)}{R(s)} = \frac{\dfrac{K_1}{s+a}}{1+\dfrac{K_1}{s+a}} = \frac{K_1}{s+a+K_1} \quad .....(2.71)$$

The impulse response of this system is, $h(t) = K_1\, e^{-(a+K_1)t}$ .....(2.72)

The time constant and the dc gain are given by,

$$\tau = \frac{1}{a+K_1} \quad .....(2.73)$$

$$K = \frac{K_1}{a+K_1} \quad .....(2.74)$$

By a suitable choice of K_1, the time constant τ can be altered. The system response can be speeded up by decreasing the time constant by a suitable choice of the gain K_1 of the amplifier. The pole of the feedback system is located at $-(a + K_1)$ and this can be pushed to the left of the $j\omega$ axis to speed up the system as shown in Fig. 2.50.

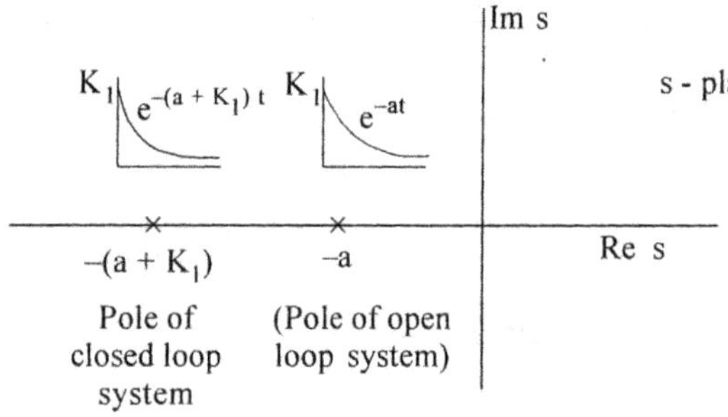

Fig. 2.50 Poles of open loop and closed loop systems and their responses

Thus without altering the system parameter 'a' the system time constant can be adjusted to a desired value by using an external amplifier with a suitable gain. However the d.c. gain is reduced as given by equation 2.74. This gain can be compensidated by having additional gain outside the loop.

2.5.2 Effect due to parameter variations

The output of an open loop system is affected by any variations in the parameter values. However, in a closed loop system, the variations in system parameters do not change the output significantly. This effect can be best studied by finding the sensitivity of the output to variations in system parameters. The term sensitivity of a variable with respect to a parameter is defined as

$$S_y^x = \text{Sensitivity} = \frac{\text{Percentage change in variable x}}{\text{Percentage change in parameter y}} \quad .....(2.75)$$

$$= \frac{\partial x/x}{\partial y/y} = \frac{\partial x}{\partial y}\cdot\frac{y}{x} \quad .....(2.76)$$

1. Sensitivity of overall transfer function for a small change in forward path transfer function G(s), S_G^T

   (a) For open loop system

$$C(s) = G(s)\, R(s)$$

$$\frac{C(s)}{R(s)} = T(s) = G(s)$$

$$S_G^T = \frac{\partial T}{\partial G}\cdot\frac{G}{T} = 1.\frac{G}{G} = 1 \qquad .....(2.77)$$

   (b) For closed loop system

$$\frac{C(s)}{R(s)} = T(s) = \frac{G(s)}{1+G(s)H(s)}$$

$$S_G^T = \frac{\partial T}{\partial G}\cdot\frac{G}{T} = \frac{(1+GH)-GH}{(1+GH)^2} \cdot \frac{G(1+GH)}{G}$$

$$= \frac{1}{1+GH} \qquad .....(2.78)$$

   From eqns. (2.77) and (2.78) it is clear that the sensitivity of over all transfer function for a small variation in forward path transfer function is reduced by a factor (1 + GH) in a closed loop system compared to the open loop system.

2. Sensitivity of closed loop transfer function T(s) with respect to variation in feedback transfer function H(s), S_G^T

$$S_G^T = \frac{\partial T}{\partial H}\cdot\frac{H}{T}$$

$$T(s) = \frac{G}{1+GH}$$

$$S_G^T = -\frac{G}{(1+GH)^2}\, G\, \frac{H(1+GH)}{G}$$

$$= -\frac{GH}{1+GH} \qquad .....(2.79)$$

From eqn. (2.79) it is clear that for large values of GH, the sensitivity approaches unity. Thus the over all transfer function is highly sensitive to variation in feedback elements and therefore the feedback elements must be properly chosen so that they do not change with environmental changes or do not degrade due to ageing. These elements are usually low power elements and hence high precision elements can be chosen at a relatively less cost. On the other hand, the forward path elements need not satisfy stringest requirements, as any variations in them do not affect the performance of the system to a significant extent. These are high power elements and hence less precise elements at lower cost can be chosen for the forward path.

Example 2.16

(a) For the open loop system in Fig. 2.51 (a) find the percentage change in the steadystate value of C(s) for a unit step input and for a 10% change in K.

(b) For the closed loop system in Fig. 2.51 (b), find the percentage change in the steadystate value of C(s) for a unit step input for the same increase in the value of K.

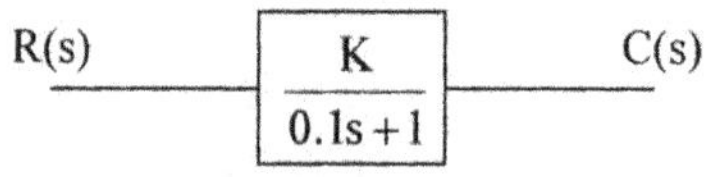

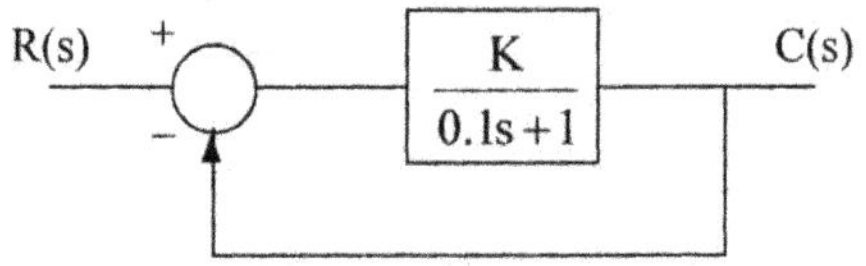

Fig. 2.51 (a) Open loop system **(b)** Closed loop system

Solution :

(a)

$$G(s) = \frac{K}{0.1s+1}$$

$$C(s) = G(s)\,R(s)$$

$$= \frac{K}{s\,(0.1s+1)}$$

$$c_{ss} = \underset{s \to 0}{\text{Lt}}\; sC(s) = \underset{s \to 0}{\text{Lt}}\; \frac{Ks}{s(0.1s+1)} = K$$

$$S_H^{c_{ss}} = \frac{\partial c_{ss}}{\partial K} \cdot \frac{K}{c_{ss}} = 1.\frac{K}{K} = 1$$

$\therefore$ % change in c_{ss} for 10% change in K = 10%

(b)

$$T(s) = \frac{K}{0.1s+1+K}$$

$$C(s) = \frac{K}{s(0.1s+1+K)}$$

$$c_{ss} = \frac{K}{1+K}$$

$$S_K^{c_{ss}} = \frac{\partial c_{ss}}{\partial_K} \cdot \frac{K}{c_{ss}} = \frac{(1+K)1 - K.1}{(1+K)^2}\;\frac{K(1+K)}{K}$$

$$= \frac{1}{1+K}$$

$\therefore$ % change in c_{ss} for 10% change in $K = \dfrac{10}{1+K}$

and for K = 10

$$\text{% change in } \quad c_{ss} = \frac{10}{11} = 0.91$$

2.5.3 Effect on Bandwidth

A control system is essentially a low pass filter. At some frequency ω_b the gain drops to $\frac{1}{\sqrt{2}}$ of its value at $\omega = 0$. This frequency ω_b is called as the bandwidth of the system. If the bandwidth is large, the system response is good for high frequencies also. To put in another way, if the bandwidth of the system is large, the speed of response is high.

Consider the open loop system with $G(s) = \frac{K}{\tau s + 1}$

At any given frequency $s = j\omega$

$$G(j\omega) = \frac{K}{j\omega\tau + 1}$$

If $\omega = \omega_b$, $$|G(j\omega_b)| = \frac{K}{|j\omega_b\tau + 1|}$$

But $$|G(j\omega_b)| = \frac{G(0)}{\sqrt{2}} \text{ and } G(0) = K$$

This is true if $\omega_b \tau = 1$

$$\therefore \quad \omega_b = \frac{1}{\tau} \qquad .....(2.80)$$

Now consider a unity feedback system with $G(s) = \frac{K}{\tau s + 1}$

The closed loop transfer function is given by

$$T(s) = \frac{K}{\tau s + 1 + K}$$

At a given frequency $s = j\omega$, we have

$$T(j\omega) = \frac{K}{\tau j\omega + 1 + K}$$

$$|T((j\omega_b)| = \frac{K}{|\tau j\omega_b + 1 + K|}$$

At $\omega = \omega_b$, $$T(j\omega_b) = \frac{T(0)}{\sqrt{2}} = \frac{1}{\sqrt{2}} \cdot \frac{K}{1+K}$$

This is true if, $\omega_b \tau = 1 + K$

or $$\omega_b = \frac{1+K}{\tau} \qquad .....(2.81)$$

From eqns. (2.80) and (2.81) it is clear that the bandwidth of closed loop system is (1 + K) times the bandwidth of open loop system. Hence the speed of response of a closed loop system is larger than that of an open loop system.

2.5.4 Effect on Noise Signals

(a) Noise in the forward paths

Consider the block diagram of the system shown in Fig. 2.52.

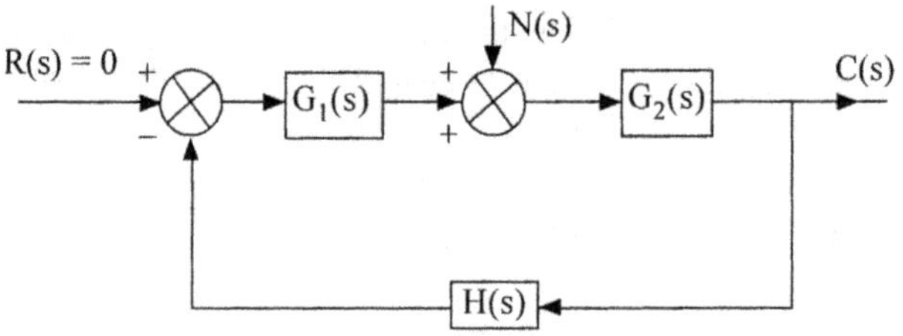

Fig. 2.52 Noise in the forward path

A noise signal is added between the forward path blocks $G_1(s)$ and $G_2(s)$. To evaluate the effect of N(s), we assume R(s) = 0. The transfer function between C(s) and N(s) can be easily derived as,

$$T_N(s) = \frac{C(s)}{N(s)} = \frac{G_2(s)}{1 + G_1(s)G_2(s)H(s)}$$

For the practical case of $|G_1(s)\ G_2(s)\ H(s)| >> 1$

$$T_N(s) = \frac{1}{G_1(s)H(s)}$$

or $$C(s) = \frac{N(s)}{G_1(s)H(s)} \qquad .....(2.81)$$

It can be seen from eqn. (2.81) that the effect of noise on output can be reduced by making $|(G_1(s)|$ sufficiently large.

(b) Noise in feedback path

Consider the block diagram in Fig. 2.53.

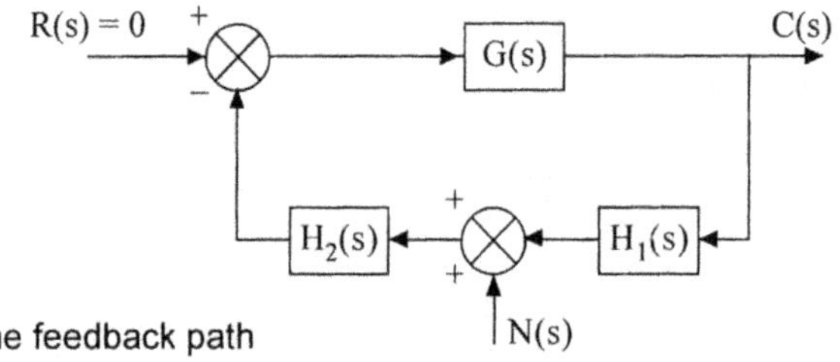

Fig. 2.53 Noise in the feedback path

The transfer function between output and the noise signal with R(s) = 0 is given by

$$T(s) = \frac{C(s)}{N(s)} = \frac{-G(s)H_2(s)}{1 + G(s)H_1(s)H_2(s)}$$

For $|G(s)\ H_1(s)\ H_2(s)| >> 1$, we have

$$T(s) = -\frac{1}{H_1(s)}$$

and $$C(s) = -\frac{1}{H_1(s)} N(s)$$

Thus by making $H_1(s)$ large the effects of noise can be effectively reduced.

2.6 Modelling of Elements of Control Systems

A feedback control system usually consists of several components in addition to the actual process. These are : error detectors, power amplifiers, actuators, sensors etc. Let us now discuss the physical characteristics of some of these and obtain their mathematical models.

2.6.1 DC Servo Motor

A DC servo motor is used as an actuator to drive a load. It is usually a DC motor of low power rating. DC servo motors have a high ratio of starting torque to inertia and therefore they have a faster dynamic response. DC motors are constructed using rare earth permanent magnets which have high residual flux density and high coercivity. As no field winding is used, the field copper losses are zero and hence, the overall efficiency of the motor is high. The speed torque characteristic of this motor is flat over a wide range, as the armature reaction is negligible. Moreover speed is directly proportional to the armature voltage for a given torque. Armature of a DC servo motor is specially designed to have low inertia.

In some application DC servo motors are used with magnetic flux produced by field windings. The speed of PMDC motors can be controlled by applying variable armature voltage. These are called armature voltage controlled DC servo motors. Wound field DC motors can be controlled by either controlling the armature voltage or controlling the field current. Let us now consider modelling of these two types of DC servo motors.

(a) Armature controlled DC servo motor

The physical model of an armature controlled DC servo motor is given in Fig. 2.54.

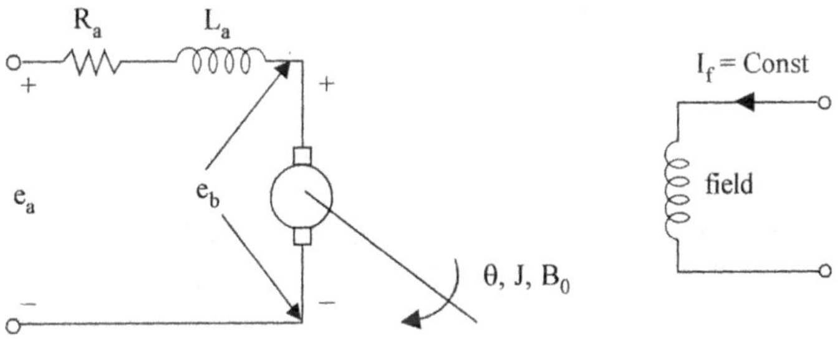

Fig. 2.54 Armature controlled DC servo motor.

The armature winding has a resistance R_a and inductance L_a. The field is produced either by a permanent magnet or the field winding is separately excited and supplied with constant voltage so that the field current I_f is a constant.

When the armature is supplied with a DC voltage of e_a volts, the armature rotates and produces a back e.m.f e_b. The armature current i_a depends on the difference of e_a and e_b. The armature has a moment of inertia J, frictional coefficient B_0. The angular displacement of the motor is θ.

The torque produced by the motor is given by,

$$T = K_T \, i_a \quad .....(2.84)$$

where K_T is the motor torque constant.

The back emf is proportional to the speed of the motor and hence

$$e_b = K_b \dot{\theta} \qquad .....(2.85)$$

The differential equation representing the electrical system is given by,

$$R_a i_a + L_a \frac{di_a}{dt} + e_b = e_a \qquad .....(2.86)$$

Taking Laplace transform of eqns. (2.84), (2.85) and (2.86) we have

$$T(s) = K_T I_a(s) \qquad .....(2.87)$$

$$E_b(s) = K_b s \theta(s) \qquad .....(2.88)$$

$$(R_a + s L_a) I_a(s) + E_b(s) = E_a(s) \qquad .....(2.89)$$

$$\therefore \quad I_a(s) = \frac{E_a(s) - K_b s \theta(s)}{R_a + sL_a} \qquad .....(2.90)$$

The mathematical model of the mechanical system is given by,

$$J \frac{d^2\theta}{dt^2} + B_0 \frac{d\theta}{dt} = T \qquad .....(2.91)$$

Taking Laplace transform of eqn. (2.91),

$$(Js^2 + B_0 s) \theta(s) = T(s) \qquad .....(2.92)$$

Using eqns. (2.87) and (2.90) in eqn. (2.92), we have

$$\theta(s) = K_T \frac{E_a(s) - K_b s \theta(s)}{(R_a + sL_a)(Js^2 + B_0 s)} \qquad .....(2.93)$$

Solving for θ(s), we get

$$\theta(s) = \frac{K_T E_a(s)}{s[(R_a + sL_a)(Js + B_0) + K_T K_b]} \qquad .....(2.94)$$

The block diagram representation of the armature controlled DC servo motor is developed in steps, as shown in Fig. 2.55. Representing eqns. (2.89), (2.87), (2.92) and (2.88) by block diagrams respectively, we have

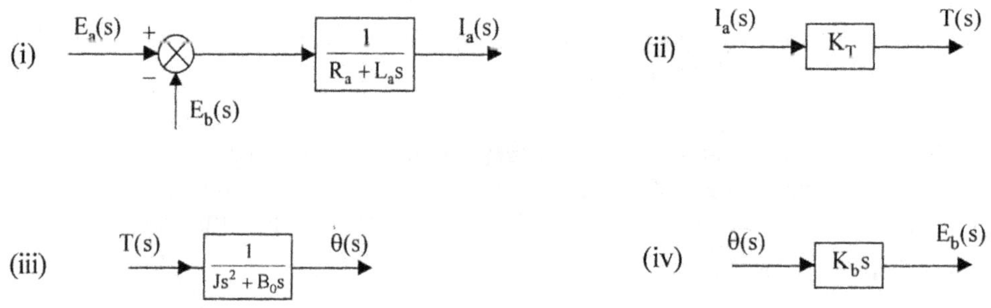

Fig. 2.55 Individual blocks of the armature controlled DC servo motor.

Combining these blocks suitably we have the complete block diagram as shown in Fig. 2.56.

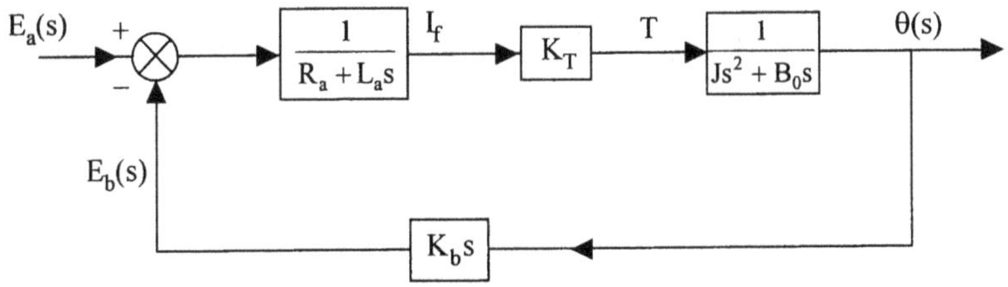

Fig. 2.56 Complete block diagram of an armature controlled DC servo motor

Usually the inductance of the armature winding is small and hence neglected. The overall transfer function, then, becomes,

$$T(s) = \frac{\theta(s)}{E_a(s)} = \frac{K_T / R_a}{s\left[Js + B_0 + \frac{K_b K_T}{R_a}\right]} \quad .....(2.95)$$

$$= \frac{K_T / R_a}{s(Js + B)} \quad .....(2.96)$$

where $B = B_0 + \frac{K_b K_T}{R_a}$ is the equivalent frictional coefficient.

It can be seen from eqn. (2.95) that the effect of back emf is to increase the effective frictional coeffcient thus providing increased damping. Eqn. (2.96) can be written in another useful form known as time constant form, given by,

$$T(s) = \frac{K_M}{s(\tau_m s + 1)} \quad .....(2.97)$$

where $K_M = \frac{K_T}{R_a J}$ is the motor gain constant

and $\tau_m = \frac{J}{B}$ is the motor time constant

(**Note :** K_b and K_T are related to each other and in MKS units $K_b = K_T$. K_b is measured in V/rad/sec and K_T is in Nm/A)

Armature controlled DC servo motors are used where power requirements are large and the additional damping provided inherently by the back emf is an added advantage.

(b) Field controlled DC servo motor

The field controlled DC servo motor is shown in Fig. 2.57.

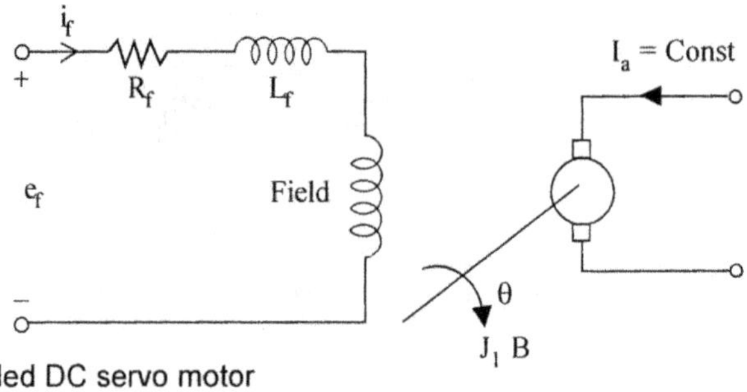

Fig. 2.57 Field controlled DC servo motor

The electrical circuit is modelled as,

$$I_f(s) = \frac{E_f(s)}{R_f + L_f s} \quad .....(2.98)$$

$$T(s) = K_T\, I_f(s) \quad .....(2.99)$$

and $$(Js^2 + B_0)\,\theta(s) = T(s) \quad .....(2.100)$$

Combining eqns. (2.98), (2.99) and (2.100) we have

$$\frac{\theta(s)}{E_f(s)} = \frac{K_T}{s(Js + B_0)(R_f + L_f s)}$$

$$= \frac{K_T / R_f B_0}{s\left(\frac{J}{B_0}s + 1\right)\left(\frac{L_f}{R_f}s + 1\right)}$$

$$= \frac{K_m}{s(\tau_m s + 1)(\tau_f s + 1)} \quad .....(2.101)$$

where $K_m = K_T/R_f B_0$ = motor gain constant

$\tau_m = J/B_0$ = motor time constant

$\tau_f = L_f/R_f$ = field time constant

The block diagram is as shown in Fig. 2.58.

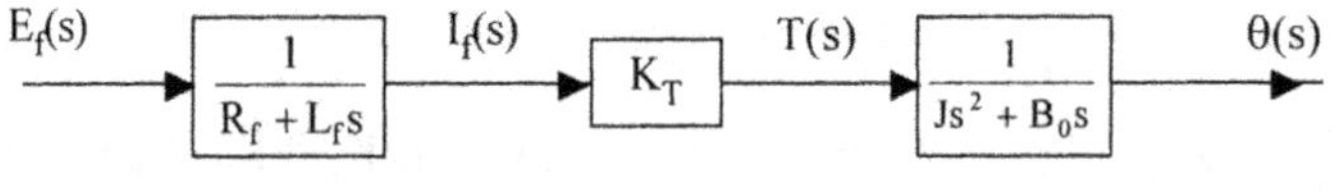

Fig. 2.58 Block diagram of a field controlled DC servo motor

Field controlled DC servo motors are economcial where small size motors are required. For the field circuit, low power servo amplifiers are sufficient and hence they are cheaper.

2.6.2 AC Servo Motors

An AC servo motor is essentially a two phase induction motor with modified constructional features to suit servo applications. The schematic of a two phase ac servo motor is shown in Fig. 2.59.

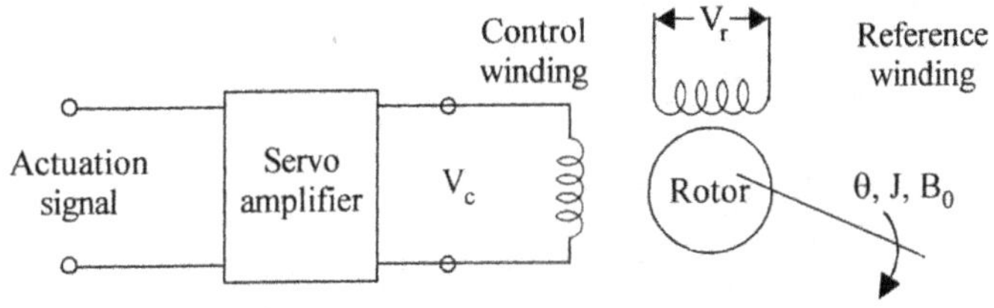

Fig. 2.59 Schematic of a 2 phase ac servo motor

It has two windings displaced by 90^0 on the stator. One winding, called as reference winding, is supplied with a constant sinusoidal voltage. The second winding, called control winding, is supplied with a variable control voltage which is diplaced by $\pm 90^0$ out of phase from the reference voltage.

The major differences between the normal induction motor and an AC servo motor are :

1. The rotor winding of an ac servo motor has high resistance (R) compared to its inductive reactance (X) so that its $\frac{X}{R}$ ratio is very low. For a normal induction motor, $\frac{X}{R}$ ratio is high so that the maximum torque is obtained in normal operating region which is around 5% of slip. The torque speed characteristics of a normal induction motor and an ac servo motor are shown in Fig. 2.60.

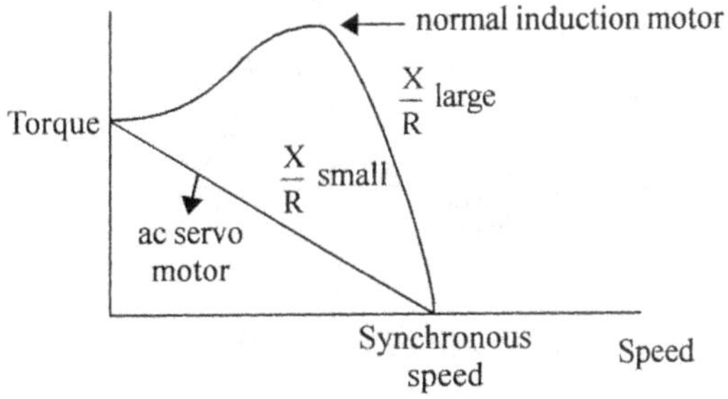

Fig. 2.60 Torque speed characteristics of normal induction motor and ac servo motor

The Torque speed characteristic of a normal induction motor is highly nonlinear and has a positive slope for some portion of the curve. This is not desirable for control applications, as the positive slope makes the systems unstable. The torque speed characteristic of an ac servo motor is fairly linear and has negative slope throughout.

2. The rotor construction is usually squirrel cage or drag cup type for an ac servo motor. The diameter is small compared to the length of the rotor which reduces inertia of the moving parts. Thus it has good accelerating characteristic and good dynamic response.

3. The supply to the two windings of ac servo motor are not balanced as in the case of a normal induction motor. The control voltage varies both in magnitude and phase with respect to the constant reference voltage applied to the reference winding. The direction of ratation of the motor depends on the phase ($\pm 90^0$) of the control voltage with respect to the reference voltage.

For different rms values of control voltage the torque speed characteristics are shown in Fig. 2.61. The torque varies approximately linearly with respect to speed and also control voltage. The torque speed characteristics can be linearised at the operating point and the transfer function of the motor can be obtained.

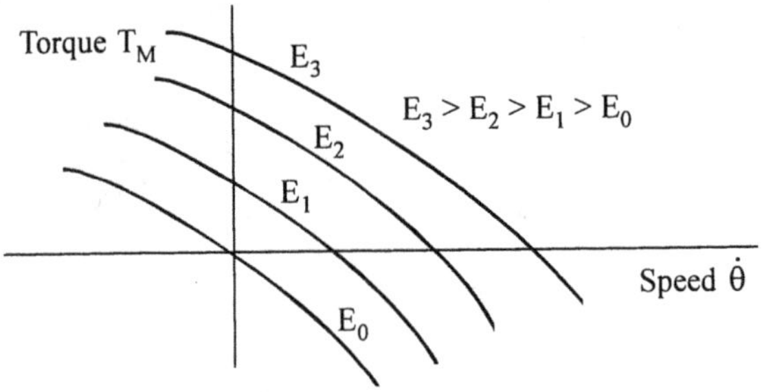

Fig. 2.61 AC servo motor speed torque characteristics

The torque is a function of speed $\dot{\theta}$ and the control voltage E. Thus

$$T_M = f(\dot{\theta}, E) \quad .....(2.102)$$

Expanding eqn. (2.102) in Taylor series around the operating point, $T_M = T_{MO}$, $E = E_0$ and $\dot{\theta} = \dot{\theta}_0$ and neglecting terms of order equal to and higher than two, we have

$$T_M = T_{MO} + \left.\frac{\partial T_M}{\partial E}\right|_{\substack{E=E_0 \\ \dot{\theta}=\dot{\theta}_0}} (E - E_o) + \left.\frac{\partial T_M}{\partial \dot{\theta}}\right|_{\substack{E=E_0 \\ \dot{\theta}=\dot{\theta}_0}} (\dot{\theta} - \dot{\theta}_o) \quad .....(2.103)$$

or

$$T_M - T_{MO} = K(E - E_o) - B(\dot{\theta} - \dot{\theta}_o) \quad .....(2.104)$$

where

$$K = \left.\frac{\partial T_M}{\partial E}\right|_{\substack{E=E_0 \\ \dot{\theta}=\dot{\theta}_0}} \text{ and } B = -\left.\frac{\partial T_M}{\partial \dot{\theta}}\right|_{\substack{E=E_0 \\ \dot{\theta}=\dot{\theta}_0}} \quad .....(2.105)$$

(**Note :** Since $\frac{\partial T_M}{\partial \dot{\theta}}$ is negative, B will be positive)

Eqn. (2.104) can be written as

$$\Delta T_M = K\,\Delta E - B\,\Delta\dot{\theta} \quad .....(2.106)$$

The mechanical equation is given by

$$J\,\Delta\ddot{\theta} + B_o\,\Delta\dot{\theta} = \Delta T_M = K\Delta E - B\Delta\dot{\theta} \quad .....(2.107)$$

Taking Laplace transform of eqn. (2.107), we get the transfer function of an ac servo motor as,

$$\frac{\Delta\theta(s)}{\Delta E(s)} = \frac{K}{Js^2 + (B_o + B)s} \quad .....(2.108)$$

In time constant form we can write eqn. (2.108) as

$$\frac{\Delta\theta(s)}{\Delta E(s)} = \frac{K_m}{s(\tau_m s + 1)} \quad .....(2.109)$$

where $$K_m = \frac{K}{B_o + B} = \text{Motor gain constant}$$

and $$\tau_m = \frac{J}{B_o + B} = \text{Motor time constant}$$

If the slope of the torque speed characteristic is positive, B, from eqn. (2.105) is negative and the effective friction coefficient B_o + B may become negative and the system may become unstable.

The transfer function of the motor around the operating point may be written as,

$$C(s) = \frac{\theta(s)}{E(s)} = \frac{K_m}{s(\tau_m s + 1)} \quad \text{......(2.110)}$$

The constants K and B can be obtained by conducting a no load test and blocked rotor test on the ac servo motor at the rated control voltage E_c.

On no load $T_M = 0$ and on blocked rotor, $\dot{\theta} = 0$.

These two points are indicated as P and Q respectively on the diagram of Fig. 2.62. The line joining P and Q represents the approximate speed torque characteristic at rated control voltage.

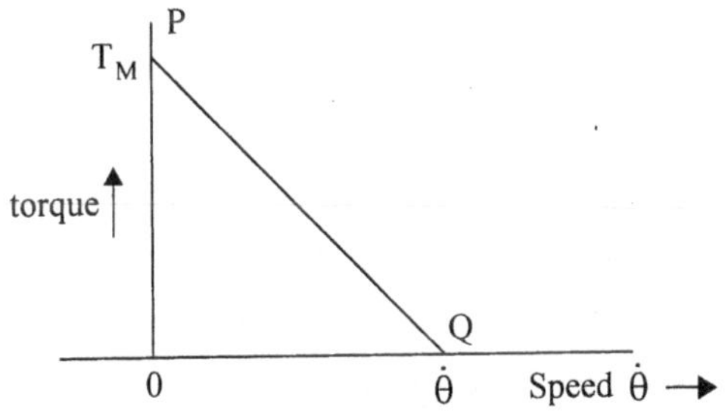

Fig. 2.62 Torque speed characteristic of servo motor at rated control voltage

$$K = \frac{T_M}{E_C} \quad \text{.....(2.111)}$$

and $$B = \frac{T_M}{\dot{\theta}} \quad \text{.....(2.112)}$$

To take into effect the nonlinearity of the torque speed curve, B is usually taken to be half of the value given by eqn. (2.112).

2.6.3 Synchros

A commonly used error detector of mechanical positions of rotating shafts in AC control systems is the Synchro. It consists of two electro mechanical devices. 1. Synchro transmitter 2. Synchro receiver or control transformer. The principle of operation of these two devices is same but they differ slightly in their construction.

The contruction of a Synchro transmitter is similar to a 3 phase alternator. The stator consists of a balanced three phase winding and is star connected. The rotor is of dumbbell type construction and is wound with a coil to produce a magnetic field. When an ac voltage is applied to the winding of the rotor, a magnetic field is produced. The coils in the stator link with this sinusoidally distributed magnetic flux and voltages are induced in the three coils due to transformer action. Thus the three voltages are in time phase with each other and the rotor voltage. The magnitudes of the voltages are proportional to the cosine of the angle between the rotor position and the respective coil axis. The position of the rotor and the coils are shown in Fig. 2.63.

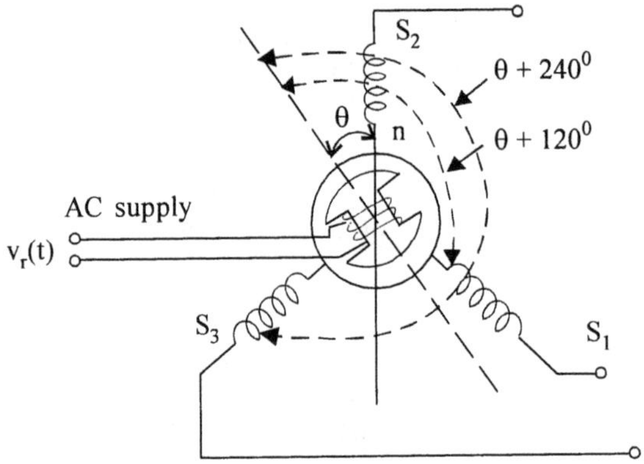

Fig. 2.63 Synchro transmitter

If the voltages induced in the three coils are designated as v_{s_1}, v_{s_2} and v_{s_3} and if the rotor axis makes angle θ with the axis of S_1 winding, we have,

$$v_R(t) = v_r \text{ Sin } \omega_r t$$

$$v_{s_{1n}} = KV_r \text{ Sin } \omega_r t \cos(\theta + 120) \quad .....(2.113)$$

$$v_{s_{2n}} = KV_r \text{ Sin } \omega_r t \cos\theta \quad .....(2.114)$$

$$v_{s_{3n}} = KV_r \text{ Sin } \omega_r t \cos(\theta + 240) \quad .....(2.115)$$

These are the phase voltages and hence the line voltages are given by,

$$v_{s_1 s_2} = v_{s_1 n} - v_{s_2 n} = \sqrt{3}\ KV_r \text{ Sin } (\theta + 240) \text{ Sin } \omega_r t \quad .....(2.116)$$

$$v_{s_2 s_3} = v_{s_2 n} - v_{s_3 n} = \sqrt{3}\ KV_r \text{ Sin } (\theta + 120) \text{ Sin } \omega_r t \quad .....(2.117)$$

$$v_{s_3 s_1} = v_{s_3 n} - v_{s_1 n} = \sqrt{3}\ KV_r \text{ Sin } \theta \text{ Sin } \omega_r t \quad .....(2.118)$$

When $\theta = 0$, the axis of the magnetic field coincides with the axis of coil S_2 and maximum voltage is induced in it as seen from eqn. (2.114). For this position of the rotor, the voltage $v_{s_3 s_1}$ is zero, as given by eqn. (2.118). This position of the rotor is known as the 'Electrical Zero' of the transmitter and is taken as reference for specifying the rotor position.

In summary, it can be seen that the input to the transmitter is the angular position of the rotor and the set of three single phase voltages is the output. The magnitudes of these voltages depend on the angular position of the rotor as given in eqn. (2.116) to (2.118).

Now consider these three voltages to be applied to the stator of a similar device called control transformer or synchro receiver. The construction of a control transformer is similar to that of the transmitter except that the rotor is made cylindrical in shape whereas the rotor of transmitter is dumbell in shape. Since the rotor is cylindrical, the air gap is uniform and the reluctance of the magnetic path is constant. This makes the output impedance of rotor to be a constant. Usually the rotor winding of control transformer is connected to an amplifier which requires signal with constant impedance for better performance. A synchro transmitter is usually required to supply several control transformers and hence the stator winding of control transformer is wound with higher impedance per phase.

Since the same currents flow through the stators of the synchro transmitter and receiver, the same pattern of flux distribution will be produced in the air gap of the control transformer. The control transformer flux axis is in the same position as that of the synchro transmitter. Thus the voltage induced in the rotor coil of control transformer is proportional to the cosine of the angle between the two rotors. Hence,

$$e_r(t) = K_1 V_r \text{Cos } \phi \sin \omega_r t \quad .....(2.119)$$

where ϕ is the angle between the two rotors. When $\phi = 90^o$, $e_r(t) = 0$ and the two rotors are at right angles. This position is known as the 'Electrical Zero' for the control transformer. In Fig. 2.64, a synchro transmitter receiver pair (usually called Synchro pair) connected as an error detector, is shown in the respective electrical zero positions.

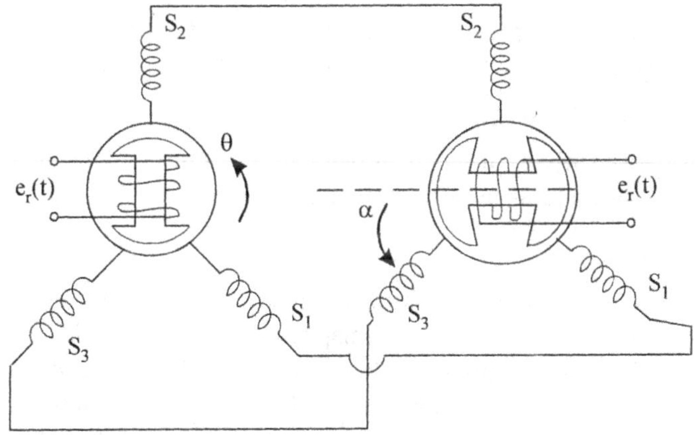

Fig. 2.64 Synchro pair connected as error detector

If the rotor of the transmitter rotates through an angle θ in the anticlockwise direction, and the rotor of control transformer rotates by an angle α in the anticlockwise direction, the net angular displacement between the two is (90 – θ + α). From eqn. (2.119),

$$e_r(t) = K_1 V_r \text{Sin } \omega_r t \text{ Cos } (90 - \theta + \alpha)$$

$$= K_1 V_r \text{Sin } (\theta - \alpha) \text{ Sin } \omega_r t \quad .....(2.120)$$

If $(\theta - \alpha)$ is small, which is usually the case,

$$e_r(t) = K_1 V_r (\theta - \alpha) \text{ Sin } \omega_r t \quad .....(2.121)$$

Thus the synchro pair acts as an error detector, by giving a voltage $e_r(t)$ proportional to the difference in the angles of the two rotors. If the angular position of synchro transmitter is used as the reference position, the transformer rotor can be coupled to the load to indicate the error between the reference and the actual positions.

The waveform of error voltage for a given variation of difference in the angular positions together with the reference voltage is shown in Fig. 2.65.

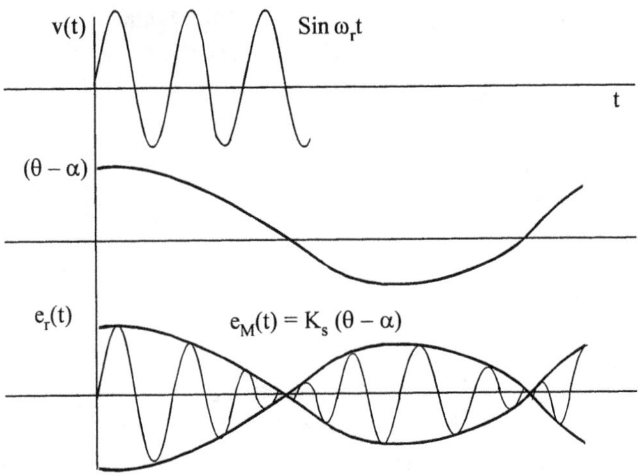

Fig. 2.65 Waveforms of voltages in synchro error detector

Thus, we see that the output of the error detector is a modulated signal with the ac input to the rotor of transmitter acting as carrier wave. The modulating signal $e_m(t)$ is

$$e_m(t) = K_s (\theta - \alpha) \qquad .....(2.122)$$

where K_s is known as the sensitivity of the error detector. Thus the synchro pair is modelled by the eqn. (2.122) when it is connected as an error detector.

2.6.4 AC Tacho Generator

An ac tacho generator consists of two coils on the stator displaced by 90^0. The rotor is a thin aluminium cup that rotates in the air gap in the magnetic flux produced by one of the stator coils called as reference coil. This coil is supplied with a sinusoidal voltage, $V_r \, Sin \, \omega_c t$, where ω_c is called the carrier frequency. The aluminium cup has low inertia and high conductivity and acts as a short circuited rotor. Because of the special construction of the rotor, this generator is also known as drag cup generator. The schematic of an ac tachometer is shown in Fig. 2.66.

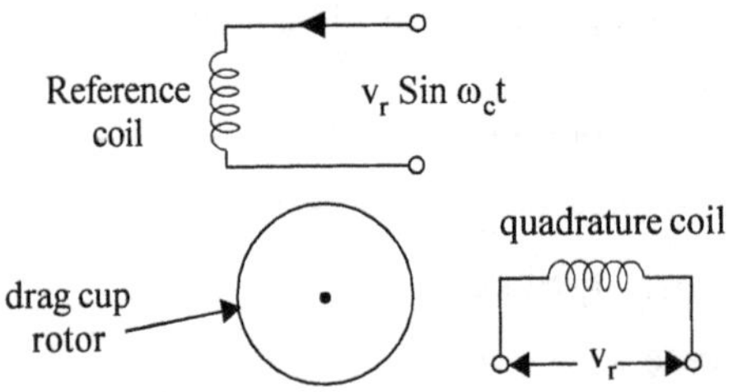

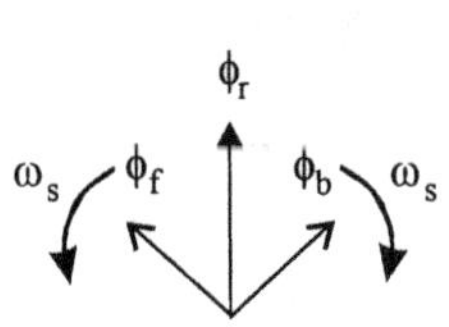

Fig. 2.66 (a) Schematic of AC tacho generator (b) Ferrari's principle

The voltage applied to the reference winding produces a main flux of $\phi_r \, Cos \, \omega_c t$. As per Ferrari's principle, this alternating flux can be considered as equivalent to two rotating fluxes ϕ_f and ϕ_b. These two fluxes are equal in magnitude but rotating in opposite directions with synchronous speed of ω_s.

If the rotor is stationary, these two fluxes induce equal and opposite voltages in the quadrature winding on the stator and therefore the voltage across this winding is zero. The two fluxes also produce induced currents in the drag cup which in turn produce reaction fluxes oppositing the corresponding main fluxes, resulting in the net fluxes ϕ_f and ϕ_b.

Now let the rotor rotate with a speed $\dot{\theta}$ in the direction of the forward flux ϕ_f. Since the relative speed of rotor conductors decreases, the induced currents decrease and therefore the main forward flux ϕ_f increases. The relative speed of rotor conductors increases with respect to backward flux ϕ_b and the induced current in the rotor due to this flux increases. This increases the reaction flux thereby decreasing backward flux ϕ_b. This imbalance in the fields, produces a voltage across the quadrature coil. This voltage will be in phase quadrature with the reference voltage and its amplitude is proportional to the speed of the rotor. If the rotor rotates in the opposite direction, reverse action takes place. Thus the voltage across the quadrature coil is given by,

$$v_f(t) = K_t\, \dot{\theta}\, \text{Cos}\, \omega_c t \qquad .....(2.123)$$

If the speed varies slowly with respect to the frequency of the reference voltage, eqn. (2.123) still applies. Thus,

$$v_f(t) = K_t\, \dot{\theta}\, (t)\, \text{Cos}\, \omega_c t \qquad .....(2.124)$$

The magnitude of this voltage is given by

$$v_t(t) = K_t\, \dot{\theta}\, (t) \qquad .....(2.125)$$

where K_t is called the tachometer constant. Thus the ac tachometer which is generally used as a feedback element can be modelled as a proportional element and can be represented by the block diagrams (Fig. 2.67).

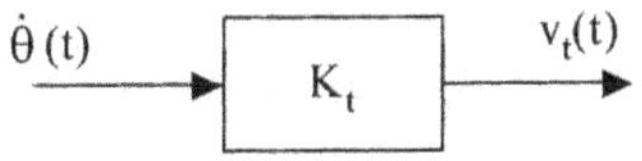

Fig. 2.67 Block diagram of an ac tacho meter

2.6.5 DC Tacho Generator

A DC tacho generator is used to feedback a voltage proportional to the speed in a DC control system. The construction of a DC tacho generator in simiar to a DC permanent magnet generator. The power rating of such a generator is very small and is designed to have low inertia.

2.6.6 Potentiometers

These devices are used commonly as feedback elements or error detectors. This is a highly reliable device used for measuring linear motion or angular motion. A potentiometer is a resistance with three terminals (Fig. 2.68 (a)). A reference voltage is applied to the two fixed terminals. The voltage between a movable terminals and one of the fixed terminals is an indication of the linear or angular displacement of the movable terminal. The total resistance of the potentiometer is uniformly distributed linearly for translatory motion or in helical form for rotatory motion. The output voltage is linearly related to the position of the movable arm as long as this voltage is connected to a device with high input resistance.

Otherwise loading effect will cause nonlinearity in the characteristic of the potentiometer as shown in Fig. 2.68 (b).

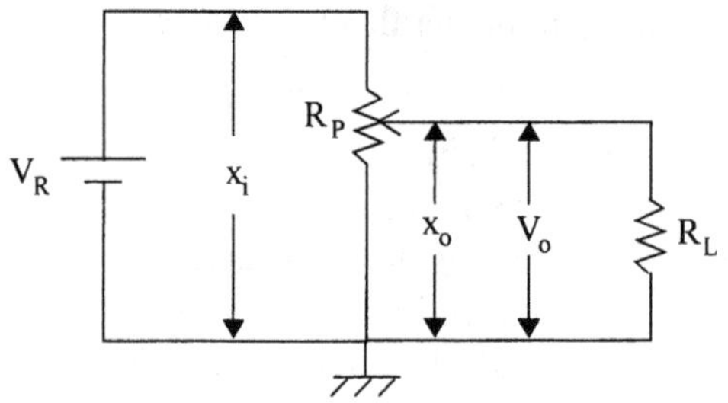

Fig. 2.68 (a) Potentiometer

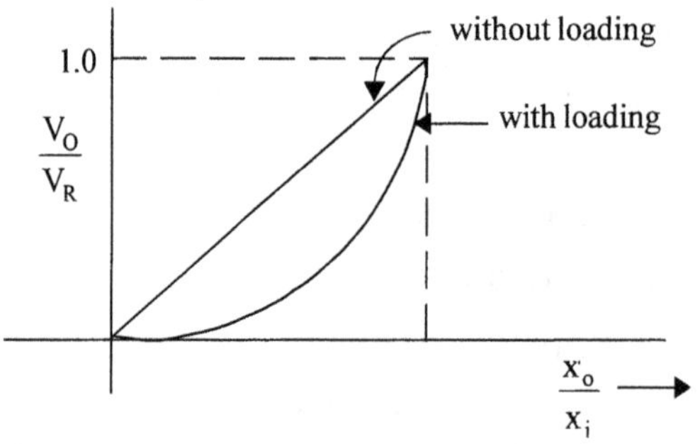

Fig. 2.68 (b) Effect of loading of a potentiometer

Potentiometer as an error detector.

Two resistors connected in parallel and supplied from a reference voltage with two movable terminals can be used as an error detector as shown in Fig. 2.69.

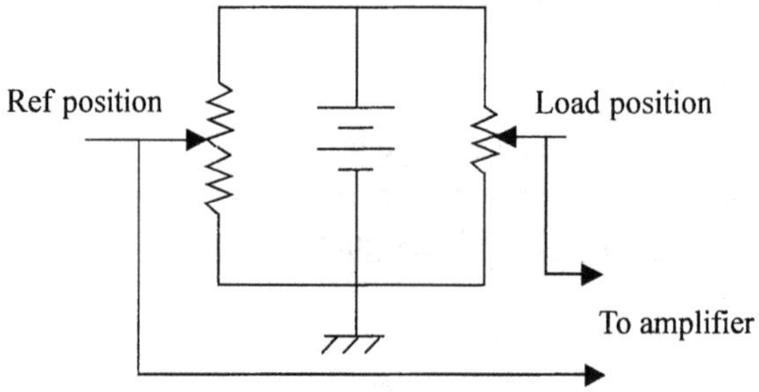

Fig. 2.69 Potentiometer as an error detector

One variable arm is used for setting the reference position or desired position. The second variable arm is connected to the load. When the position of the load is same as the reference position the voltage applied to the amplifier is zero. If the position of the load changes, the voltage between the two movable terminars will be non zero and this will be applied to the amplifier. Either positive or negative errors can be detected.

2.6.7 Stepper Motors

These motors are also called as stepping motors or simply step motors. This motor can be considered as an electromagnetic incremental actuator which converts electrical pulses into mechanical rotation. Stepper motors are used in digitally controlled position control systems. The input is a train of pulses. One advantage of stepper motor is that it does not require position or speed sensors, as this information can be directly obtained by counting the input pulses.

Depending on the principle of operation, there are two most common types of stepper motors.

1. Variable reluctance motor (VR motor) and
2. Permanent magnet motor (PM motor).

Variable reluctance motor

A VR motor has a wound stator and an unexcited rotor. A typical VR motor is shown in Fig. 2.70. This is a three phase stator with twelve stator teeth. For each phase there are four teeth distributed over the periphery of the stator and a winding is placed on these slots. The teeth are marked A B C and windings are not shown in the diagram. The rotor has eight teeth. Both the stator and rotor are constructed with high permeability material like silicon steel to allow large magnetic flux.

The principle of operation of VR motor is as follows. If any one phase is excited (say A phase), the rotor will align itself such that its teeth are directly under the teeth of excited winding so that the reluctance of the magnetic path is low. In this position the rotor is in stable equilibrium. Now if a pulse is given to the next phase (say B phase), the rotor teeth nearest to the B phase stator teeth will be pulled to align directly under B phase teeth. Thus the rotor moves one step in the anticlock wise direction. If now a pulse is given to the A phase, the rotor moves one more step in the counter clockwise direction. If the sequence of phases to which pulses are given is altered ie from A B C to A C B, the motor rotates in clockwise direction.

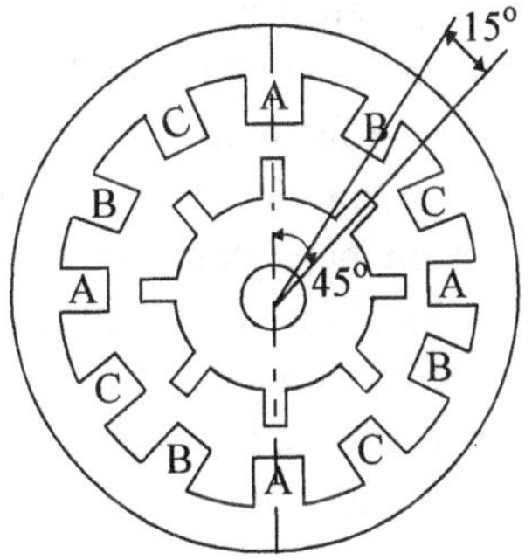

Fig. 2.70 Variable reluctance stepper motor

When a tooth of rotor is aligned with say, phase A, the next tooth on the rotor is $\frac{360}{8} = 45^0$ away from it. The tooth corresponding to the next phase ie phase B is at an angle of $\frac{360}{12} = 30^0$ from phase A.

Hence when phase B is excited, the tooth on the rotor nearest to the tooth corresponding to B phase is $45 - 30 = 15^0$ away from it and therefore the rotor rotates through 15^0 in one step. So for every pulse the rotor rotates by 15^0, either in clockwise direction or anticlockwise direction depending on the sequence of excitation of the three phases. The angle throgh which the rotor moves for one pulse is known as the step angle. The frequency of pulses applied to the stator windings will control the speed of rotation of the stopper motor.

Permanent Magnet motor (PM motor)

In permanent magnet motors, the rotor is made up of ferrite or rare earth material and is permanently magnetised. In the 4 pole, 2 phase PM motor shown in Fig. 2.71 the stator consists of two stacks, each with a phase winding, displaced electrically by 90^0. By exciting these two phases, phase a and phase b suitably, stepping action can be obtained. This is shown in Fig. 2.71.

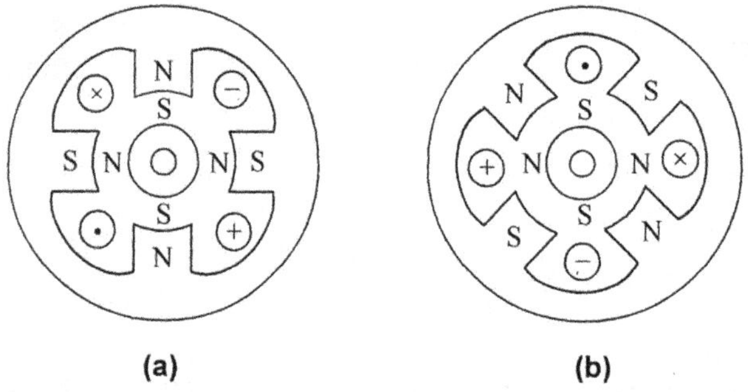

Fig. 2.71 (a) Phase a of a PM motor **(b)** Phase b of a PM motor

Let us suppose that the phase a is given a pulse so that the stator poles are as indicated in Fig. 2.71 (a). The rotor will align itself in the position shown in Fig. 2.71 (a). Now let us excite the phase b also by a pulse, the stator poles effectively shift counter clockwise by 22.5^0 as shown in Fig. 2.71 (b). Since the magnetic axis is shifted, the rotor also has to move by 22.5^0 in counter clockwise direction to align itself with the new magnetic axis. Now if phase a is unenergised, the stator magnetic axis moves by another 22.5^0 in the counter clockwise direction and hence the rotor also moves by the same angle in the same direction. To move another step in the same direction, phase a is given a pulse in the opposite direction. Rotation in the clockwise direction can be-obtained by suitably modifying the switching scheme. Generally, two coils are provided for each phase to facilitate easy switching by electronic circuitry.

Problems

2.1 Obtain the transfer function for the following networks.

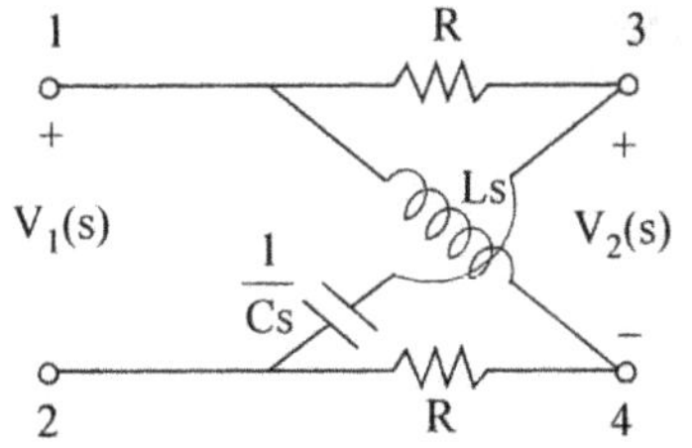

$$T(s) = \frac{V_2(s)}{V_1(s)}$$

Fig. P 2.1 (a)

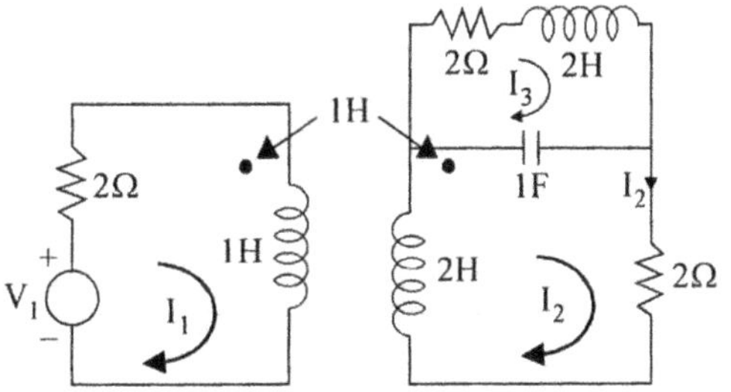

$$T(s) = \frac{I_2(s)}{V_1(s)}$$

Fig. P 2.1 (b)

2.2 Obtain the transfer functions for the following mechanical translational systems.

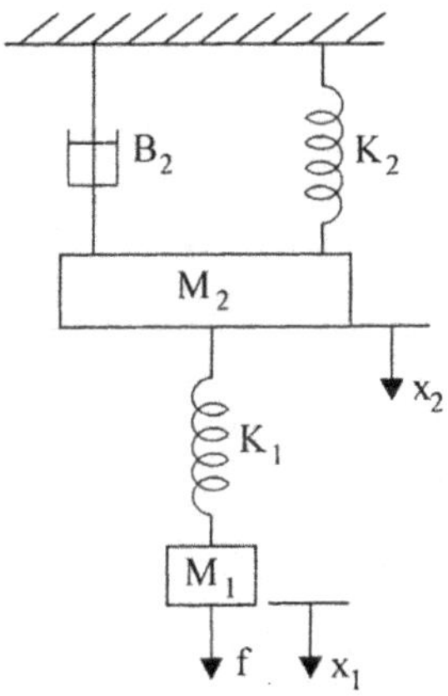

$$T(s) = \frac{X_2(s)}{F(s)}$$

Fig. P 2.2 (a)

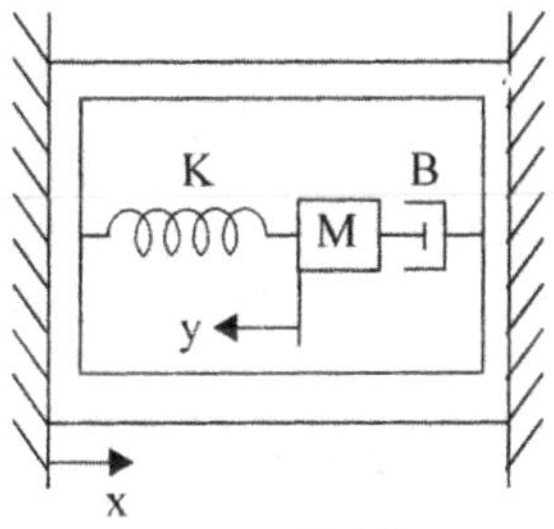

$$T(s) = \frac{Y(s)}{X(s)}$$

Fig. P 2.2 (b)

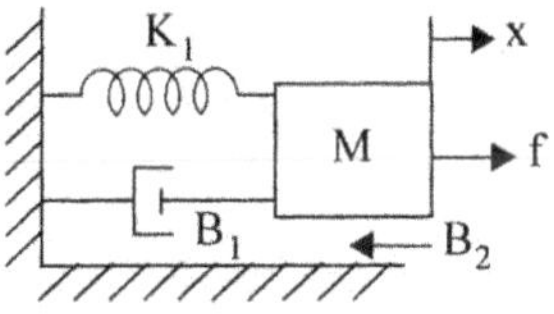

$$T(s) = \frac{X(s)}{F(s)}$$

Fig. P 2.2 (c)

2.3 Obtain the transfer function for the following mechanical rotational system.

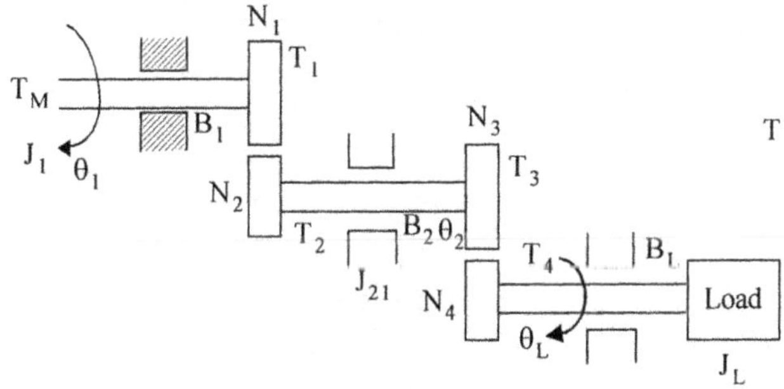

$$T = \frac{\theta_L(s)}{T_M(s)}$$

Fig. P 2.3

2.4 Draw F - v and F - i analogous circuits for the problem 2.2 (a), (b) and (c).

2.5 Obtain the transfer function $\frac{X(s)}{V(s)}$ for the electromechanical system shown in Fig P 2.5.

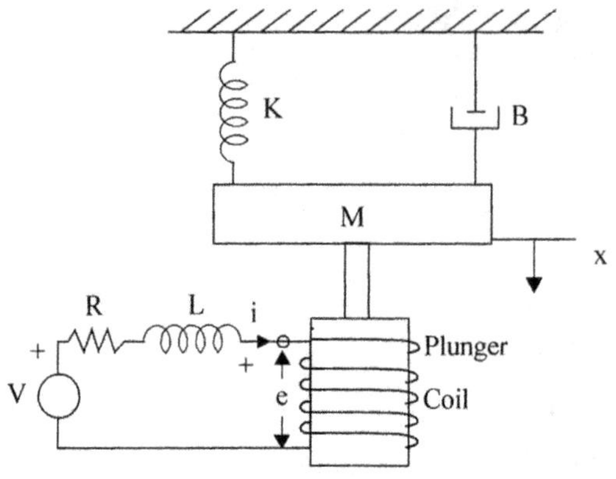

Fig. P 2.5

Assume $e = K_b \frac{dx}{dt}$ and the force produced on the mass $f = k_f i$.

2.6 In the thermal system shown, heater coil is used to heat the liquid entering the insulated tank at temperature θ_i to hot liquid at temperature θ. The liquid is thoroughly mixed to maintain uniform temperature of the liquid in the tank. If M is the mass of the liquid in the tank in Kg, C is the specific heat of liquid in J / Kg / o_K, W is the steady state liquid flow rate in kg/sec and h_i is the heat input rate in J/sec, obtain the transfer function of the system when,

(i) Heat input rate is changed, with inlet temperature of liquid kept constant and

(ii) inlet temperature is changed with heat input rate held constant. Also write the differential equation when heat input rate and inlet liquid temperature are charged.

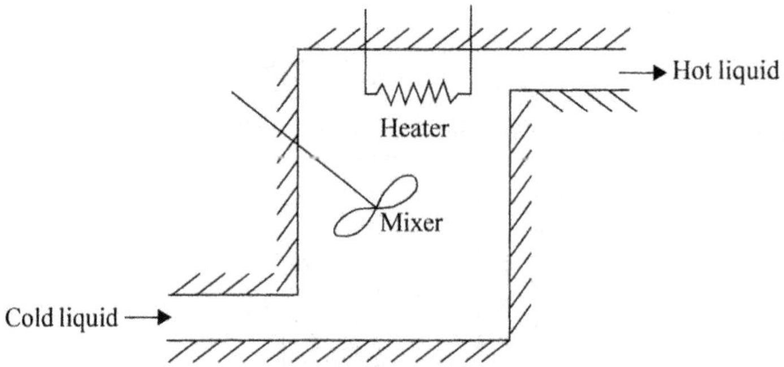

Fig. P 2.6

2.7 A thin glass bulb mercury thermometer is immersed suddenly in a hot both of temperature θ_b. Obtain expression for the temperature of the thermometer.

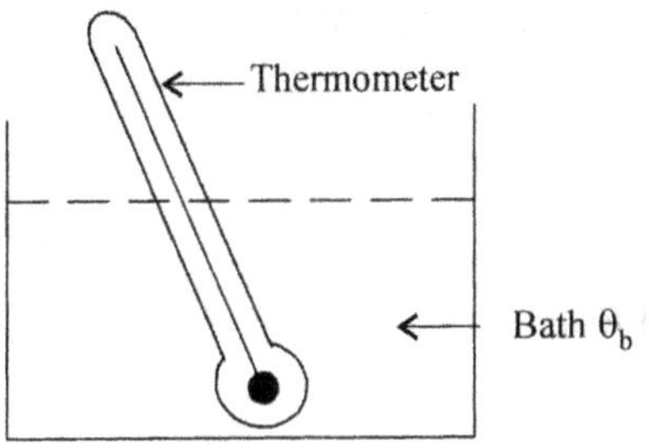

Fig. P 2.7

Thermal capacitance of thermometer is C

Thermal resistance of the thermometer is R

2.8 Obtain the transfer function C(s)/ M(s) for the liquid manometer system shown in Fig P 2.8

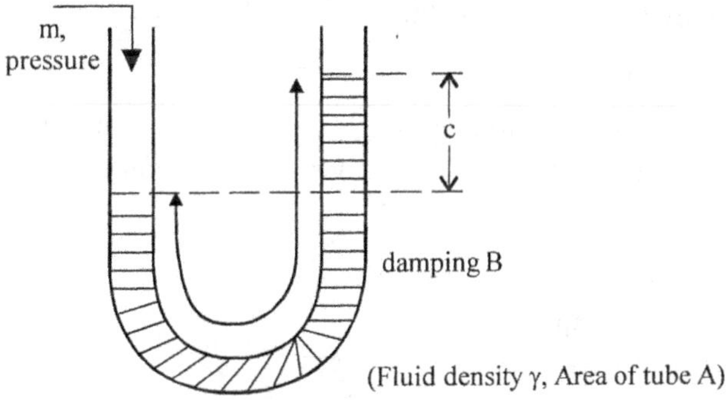

Fig. P 2.8

2.9 Obtain the transfer function of the liquid level system shown in Fig. P 2.9. The output is the liquid level *c* in the second vessel and the input is the inflow *m*.

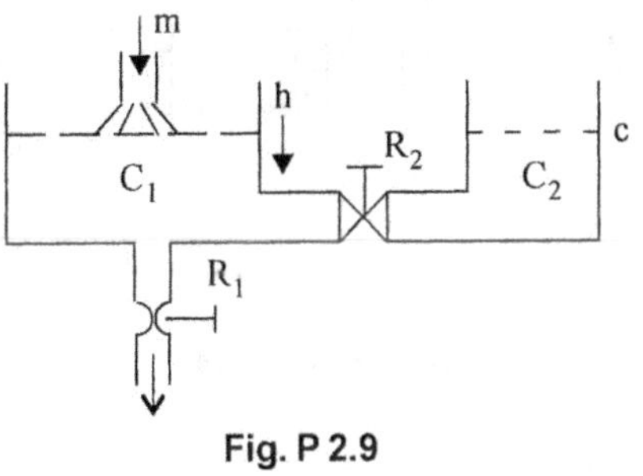

Fig. P 2.9

2.10 A d.c position control system consists of a permanent magnet armature controlled d.c srevomotor, a potentiometer error detector, a d.c Amplifier and a tachogenerator coupled to the motor. Assuming the motor and load frictions to be negligible and a fraction K of the tachogenerator output is fedback, draw the schematic of the control system and obtain its transfer function.

The parameters of the system are,

Moment of inertia of motor,	$J_M = 3 \times 10^{-3}$ kg - m^2
Moment of inertia of load,	$J_L = 4$ kg - m^2
Motor to load gear ratio	$\frac{\dot{\theta}_L}{\dot{\theta}_M} = \frac{1}{50}$
Load to potentiometer gear ratio	$\frac{\dot{\theta}_L}{\dot{\theta}_e} = 1$
Motor torque constant	$K_t = 2.5$ Nω m / Amp
Tachogenerator constant	$K_t = 0.1$ V/rad/sec
Sensitivity of error detector	$K_P = 0.5$ V/rad
Amplifier gain	K_A Amps/Volt

2.11 A position control system using a pair of synchros and an armature controlled dc servomotor is shown in Fig. P 2.11. The demodulator has a transfer function given by K_d. d.c volts/a.c volts.

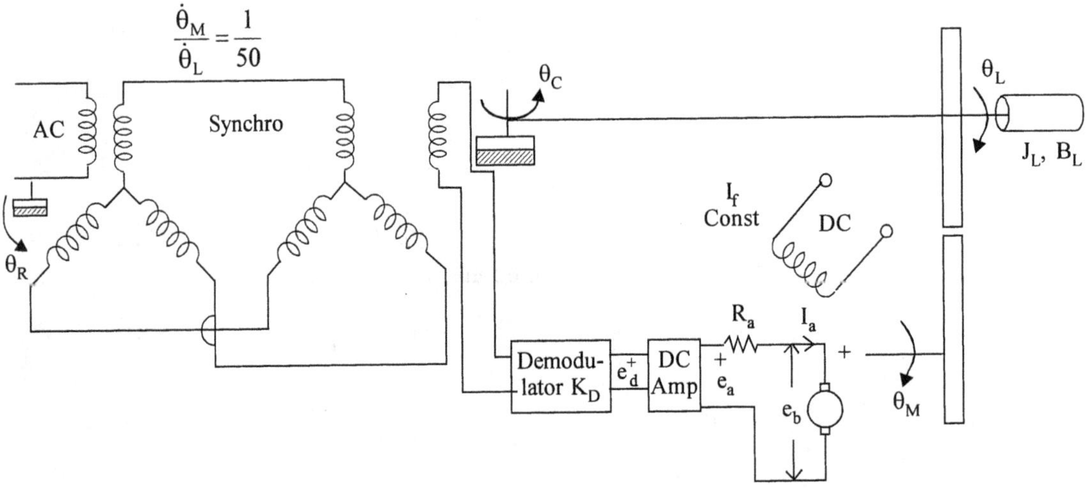

Fig. P 2.11

The sensitivity of error detector is K_s.v/rad

Draw (a) The block diagram

(b) The signal flow graph of of the system and obtain its transfer function. Use the following values for the parameters.

$K_d = 0.5$ V/V

$K_s = 50$V/rad

$K_A = 10$V/V

The back emf constant $K_b = 1.5$ V/rad/sec

$K_T = 1.5$ Nw m/A

$J_L = 0.5$ Kgm^2

$B_L = 1$ Nω/rad/sec.

$R_a = 1\Omega$

2.12 Reduce the block diagram shown in Fig. P 2.12 and obtain the over all transfer function.

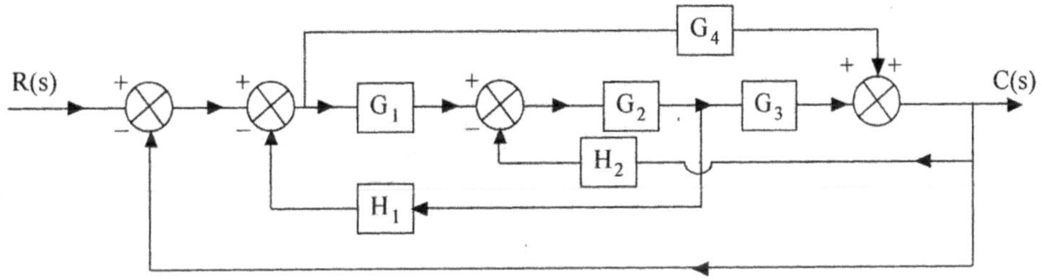

Fig. P 2.12

2.13 Obtain the overall transfer functions for the following signal flow graphs using mason's gain formula.

(a)

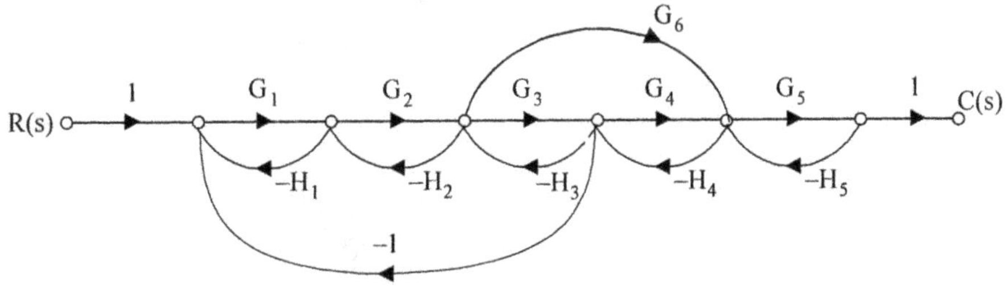

(b)

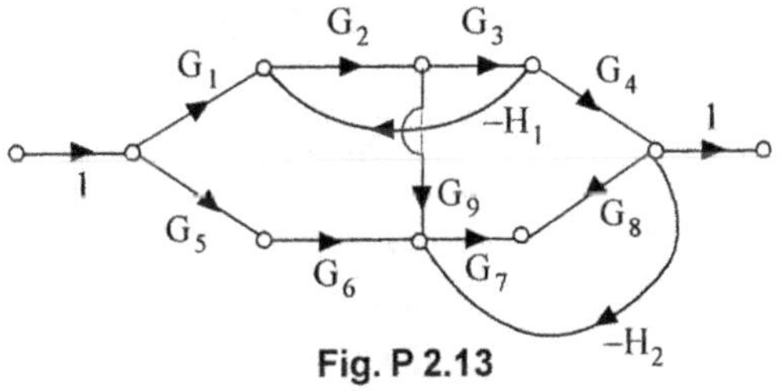

Fig. P 2.13

2.14 Reduce the block diagrams shown in Fig P 2.14 and obtain their transfer functions.

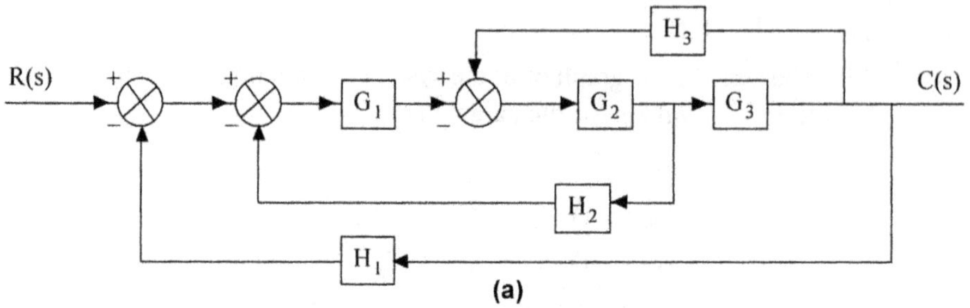

(a)

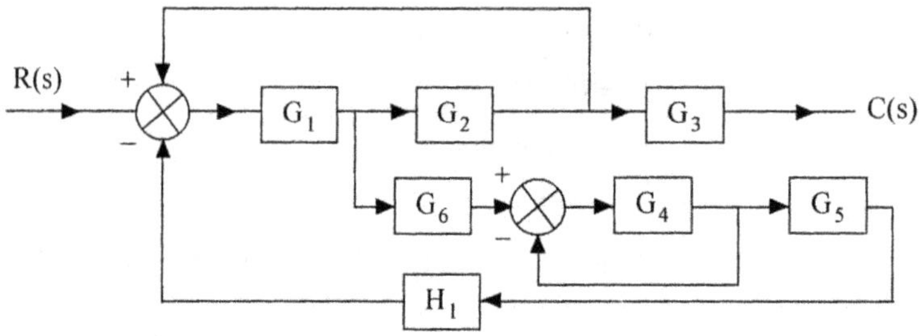

(b)

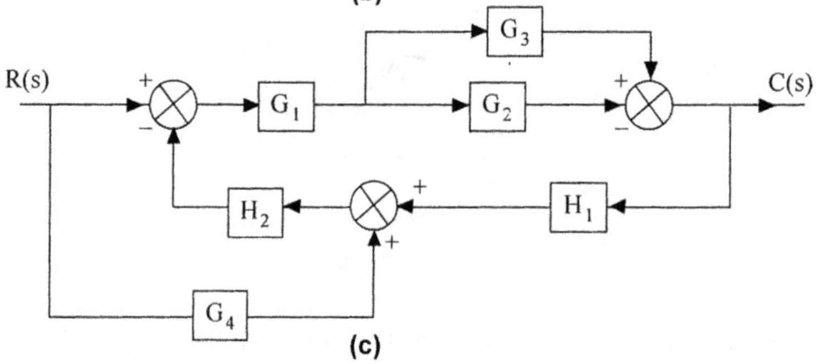

(c)

Fig. P. 2.14

2.15 Draw the signal flow graph for the network shown and obtain its transfer function,

$$T(s) = \frac{I_2(s)}{V_1(s)}$$

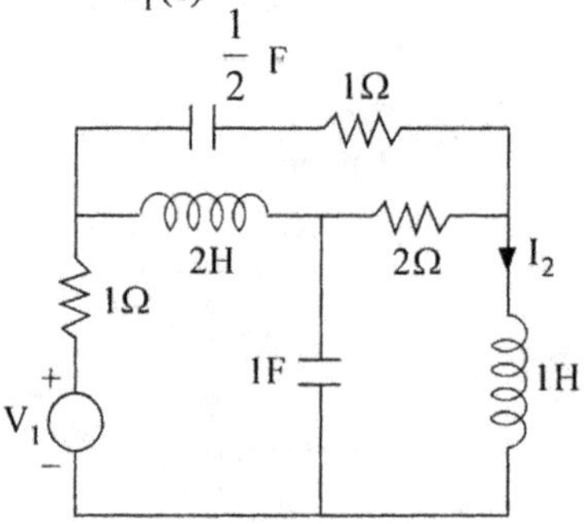

Fig. P 2.15

2.16 Draw the signal flow graphs of the systems given in problem 2.13 and obtain their transfer functions.

2.17 Consider the voltage regulator shown in Fig. P 2.17.

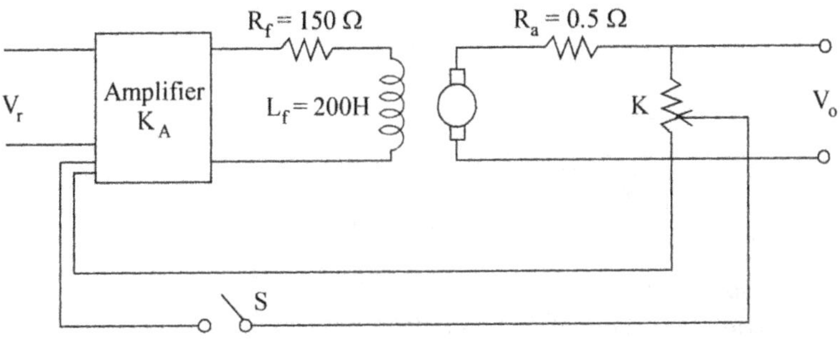

Fig. P 2.17

A voltage of KV_o is fedback to the amplifier. The amplifier gain K_A = 10V/volt and the generator constant K_g = 50V/Amp.

(a) With switch *S* open, what should be the reference voltage to get an output voltage of 200V on no load. If the same reference voltage is maintained, what will be the output voltage when the generator delivers a steady current of 20A.

(b) With reference voltage of 80V and switch *S* closed what part of the output voltage is to be fedback to get a steady terminal voltage of 10V on no load.

(c) With the system as in part *b*, if the generator supplies a load of 20A, what will be the terminal voltage.

(d) What should be the reference voltage to maintain an output voltage of 200V at 20A load under closed loop condition.

2.18 For the system shown in Fig P 2.18 determine the sensitivity of the closed loop transfer function with respect to G, H and K. Take ω = 2 rad/sec.

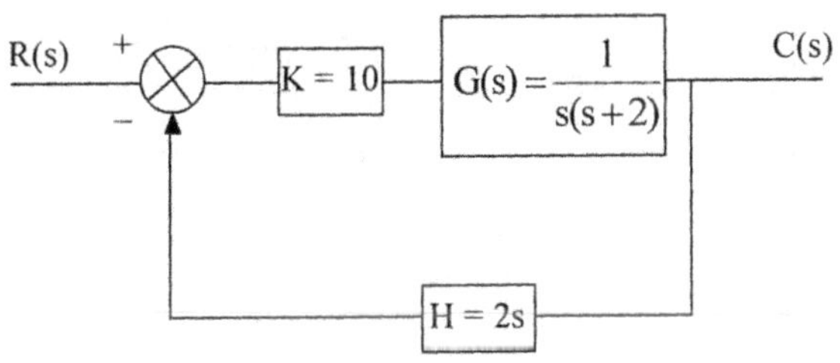

Fig. P. 2.18

2.19 In the position control system shown in Fig P. 2.19, derivative error compensation is used. Find the sensitivity of the closed loop transfer function at $\omega = 1\text{rad/sec}$ with respect to the derivative error coefficient K_d with nominal value of $K_d = 2$

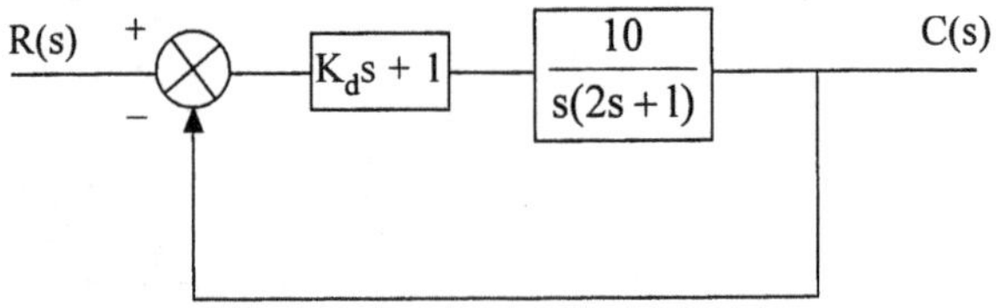

Fig. P. 2.19

2.20 The torque speed characteristic of a two phase ac servo motor is as shown in Fig P. 2.20 with rated control voltage of 115V at 50Hz applied to the control winding. The moment of Inertia of the motor is 7.5×10^{-6} kg-m^2. Neglecting the friction of the motor, obtain its transfer function $\frac{\theta(s)}{V_C(s)}$ where $\theta(s)$ is the shaft position and $V_C(s)$ is the control voltage.

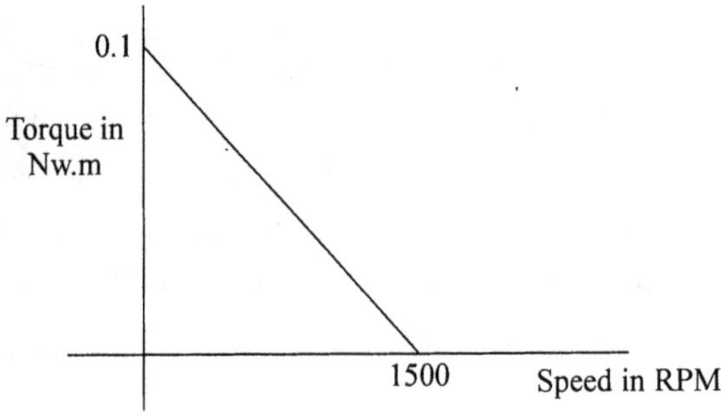

Fig. P. 2.20

3 Time Response Analysis of Control Systems

3.1 Introduction

The first step in the analysis of a control system is, describing the system in terms of a mathematical model. In chapter 2 we have seen how any given system is modelled by defining its transfer function. The next step would be, to obtain its response, both transient and steadystate, to a specific input. The input can be a time varying function which may be described by known *mathematical functions* or it may be a *random signal*. Moreover these input signals may not be known *apriori*. Thus it is customary to subject the control system to some standard input test signals which strain the system very severely. These standard input signals are : an impulse, a step, a ramp and a parabolic input. Analysis and design of control systems are carried out, defining certain performance measures for the system, using these standard test signals.

It is also pertinent to mention that any arbitary time function can be expressed in terms of linear combinations of these test signals and hence, if the system is linear, the output of the system can be obtained easily by using supersition principle. Further, convolution integral can also be used to determine the response of a linear system for any given input, if the response is known *for a step or an impulse input*.

3.2 Standard Test Signals

3.2.1 Impulse Signal

An impulse signal is shown in Fig. 3.1.

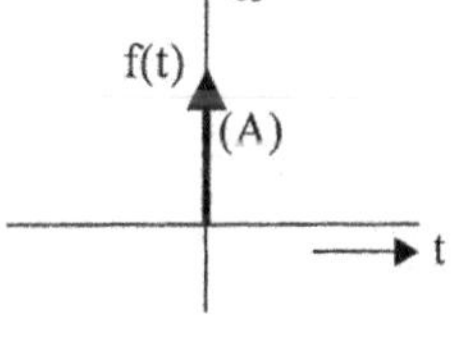

Fig. 3.1 An Impulse signal.

The impulse function is zero for all $t \neq 0$ and it is infinity at $t = 0$. It rises to infinity at $t = 0^-$ and comes back to zero at $t = 0^+$ enclosing a finite area. If this area is A it is called as an impulse function of strength A. If A = 1 it is called a unit impulse function. Thus an impulse signal is denoted by $f(t) = A\ \delta(t)$.

3.2.2 Step Signal

A step signal is shown in Fig. 3.2.

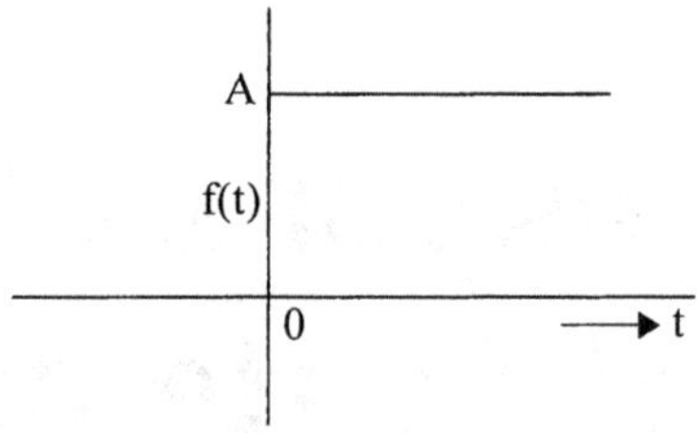

Fig. 3.2 A Step Signal.

It is zero for $t < 0$ and suddenly rises to a value A at $t = 0$ and remains at this value for $t > 0$: It is denoted by $f(t) = Au(t)$. If A = 1, it is called a *unit step function.*

3.2.3 Ramp signal

A ramp signal is shown in Fig. 3.3.

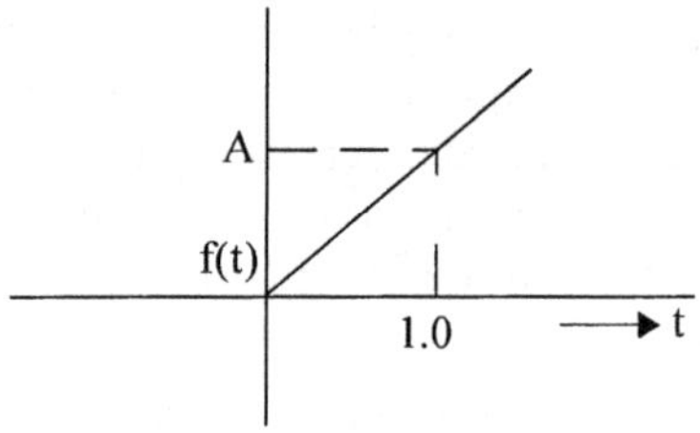

Fig. 3.3 A Ramp Signal.

It is zero for $t < 0$ and uniformly increases with a slope equal to A. It is denoted by $f(t) = At$. If the slope is unity, then it is called a *unit ramp signal.*

3.2.4 Parabolic signal

A parabolic signal is shown in Fig. 3.4.

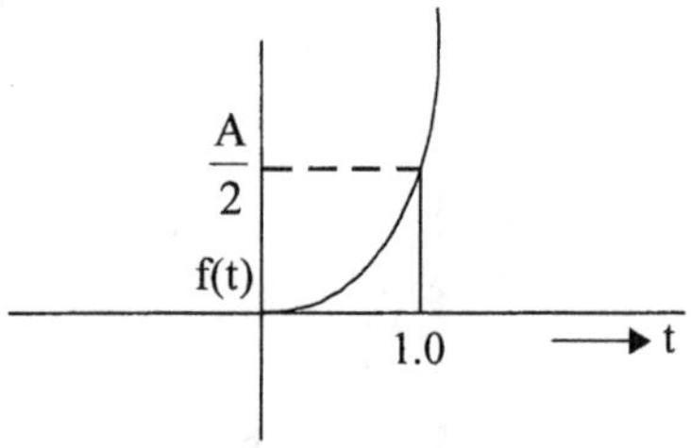

Fig. 3.4 A unit parabolic signal.

A parabolic signal is denoted by $f(t) = \frac{At^2}{2}$. If A is equal to unity then it is known as a *unit parabolic signal.*

It can be easily verified that the step function is obtained by integrating the impulse function from o to ∞; a ramp function is obtained by integrating the step function and finally the prabolic function is obtained by integrating the ramp function. Similarly ramp function, step function and impulse function can be to obtained by successive differentiations of the parabolic function.

Such a set of functions which are derived from one another are known as *singularity functions*. If the response of a linear system is known for any one of these input signals, the response to any other signal, out of these singularity functions, can be obtained by either differentiation or integration of the known *response*.

3.3 Representation of Systems

The input output description of the system is mathematically represented either as a differential equation or a transfer function.

The differential equation representation is known as a time domain representation and the transfer function is said to be a frequency domain representaiton. We will be considering the transfer function representation for all our analysis and design of control systems.

The open loop transfer function of a system is represented in the following two forms.

1. Pole-zero form

$$G(s) = K_1 \frac{(s+z_1)(s+z_2)...(s+z_m)}{(s+p_1)(s+p_2)...(s+p_n)} \quad .....(3.1)$$

Zeros occur at $s = -z_1, -z_2, ---, -z_m$

Poles occur at $s = -p_1, -p_2, ---, -p_m$

The poles and zeros may be simple or repeated. Poles and zeros may occur at the origin. In the case where some of the poles occur at the origin, the transfer function may be written as

$$G(s) = \frac{K_1(s+z_1)(s+z_2)...(s+z_m)}{s^r(s+p_{r+1})(s+p_{r+2})...(s+p_n)} \quad .....(3.2)$$

The poles at the origin are given by the term $\frac{1}{s^r}$. The term $\frac{1}{s}$ indicates an integration in the system and hence $\frac{1}{s^r}$ indicates the number of integrations present in the system. Poles at origin influence the steadystate performance of the system as will be explained later in this chapter. Hence the systems are classified according to the number of poles at the origion.

If $r = 0$, the system has no pole at the origin and hence is known as a type – 0 system. If $r = 1$, there is one pole at the origin and the system is known as a type - 1 system. Similarly if $r = 2$, the system is known as type - 2 system. Thus it is clear that the type of a system is given by the number of poles it has at the origin.

2. Time Constant Form

The open loop transfer function of a system may also be written as,

$$G(s) = \frac{K(\tau_{z_1}s+1)(\tau_{z_2}s+1)...(\tau_{z_m}s+1)}{(\tau_{p_1}s+1)(\tau_{p_2}s+1)...(\tau_{p_n}s+1)} \qquad .....(3.3)$$

The poles and zeros are related to the respective time constants by the relation

$$z_i = \frac{1}{\tau_{zi}} \qquad \text{for } i = 1, 2, ...m$$

$$p_j = \frac{1}{\tau_{pj}} \qquad \text{for } j = 1, 2, ...n$$

The gain constans K_1 and K are related by

$$K = K_1 \frac{\prod_{i=1}^{m} z_i}{\prod_{j=1}^{n} p_j}$$

The two forms described above are equivalent and are used whereever convenience demands the use of a particular form.

In either of the forms, the degree of the denomination polynomial of G(s) is known as the order of the system. The complexity of the system is indicated by the order of the system. In general, systems of order greater than 2, are difficult to analyse and hence, it is a practice to approximate higher order systems by second order systems, for the purpose of analysis.

Let us now find the response of first order and second order systems to the test signals discussed in the previous section.

The impulse test signal is difficult to produce in a laboratory. But the response of a system to an impulse has great significance in studying the behavior of the system. The response to a unit impulse is known as *impulse response of the system*. This is also known as the *natural response of the system.*

For a unit impulse function, R(s) = 1

and $C(s) = T(s).1$

and $c(t) = \mathcal{L}^{-1}[T(s)]$

The Laplace inverse of T(s) is the impulse response of the system and is usually denoted by *h(t)*.

$\therefore$ $\mathcal{L}^{-1}[T(s)] = h(t)$

If we know the impulse response of any system, we can easily calculate the response to any other arbitrary input *v(t)* by using convolution integral, namely

$$c(t) = \int_0^t h(\tau)\, v(t-\tau)\, d\tau$$

Since the impulse function is difficult to generate in a laboratory at is seldon used as a test signal. Therefore, we will concentrate on other three inputs, namely, unit step, unit velocity and unit acceleration inputs and find the response of first order and second order systems to these inputs.

3.4 First Order System

3.4.1 Response to a Unit Step Input

Consider a feedback system with $G(s) = \dfrac{1}{\tau s}$ as show in Fig. 3.5.

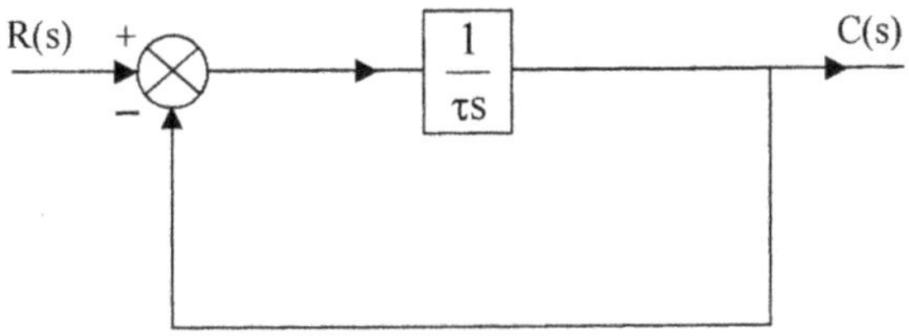

Fig. 3.5 A first order feedback system.

The closed loop transfer function of the system is given by

$$T(s) = \frac{C(s)}{R(s)} = \frac{1}{\tau s + 1} \qquad .....(3.4)$$

For a unit step input $R(s) = \dfrac{1}{s}$ and the output is given by

$$C(s) = \frac{1}{s(\tau s + 1)} \qquad .....(3.5)$$

Inverse Laplace transformation yields

$$c(t) = 1 - e^{-t/\tau} \qquad .....(3.6)$$

The plot of c(t) Vs t is shown in Fig. 3.6.

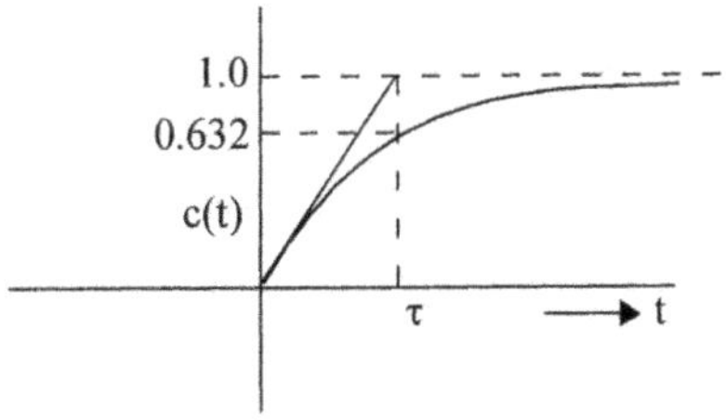

Fig. 3.6 Unit step response of a first order system.

The response is an exponentially increasing function and it approaches a value of unity as $t \to \infty$. At $t = \tau$ the response reaches a value,

$$c(\tau) = 1 - e^{-1} = 0.632$$

which is 63.2 percent of the steady value. This time, τ, is known as the *time constant of the system.* One of the characteristics which we would like to know about the system is its speed of response or how fast the response is approaching the final value. The time constant τ is indicative of this measure and the speed of response is inversely proportional to the time constant of the system.

Another important characteristic of the system is the error between the desired value and the actual value under steady state conditions. This quantity is known as the steady state error of the system and is denoted by e_{ss}.

The error E(s) for a unity feedback system is given by

$$E(s) = R(s) - C(s)$$

$$= R(s) - \frac{G(s)R(s)}{1+G(s)}$$

$$= \frac{R(s)}{1+G(s)} \qquad .....(3.7)$$

For the system under consideration $G(s) = \frac{1}{\tau s}$, $R(s) = \frac{1}{s}$ and therefore

$$E(s) = \frac{\tau}{\tau s+1}$$

$$\therefore \quad e(t) = e^{-t/\tau}$$

As $t \to \infty$ $e(t) \to 0$. Thus the output of the first order system approaches the reference input, which is the desired output, without any error. In other words, we say a first order system tracks the step input without any steadystate error.

3.4.2 Response to a Unit Ramp Input or Unit Velocity Input

The response of the system in Fig. 3.4 for a unit ramp input, for which,

$$R(s) = \frac{1}{s^2},$$

is given by,

$$C(s) = \frac{1}{s^2(\tau s+1)} \qquad .....(3.9)$$

The time response is obtained by taking inverse Laplace transform of eqn. (3.9).

$$c(t) = t - \tau(1 - e^{-t/\tau}) \qquad .....(3.10)$$

If eqn. (3.10) is differentiated we get

$$\frac{dc(t)}{dt} = 1 - e^{-t/\tau} \qquad .....(3.11)$$

Eqn. (3.11) is seen to be identical to eqn. (3.6) which is the response of the system to a step input. Thus no additional information about the speed of response is obtained by considering a ramp input. But let us see the effect on the steadystate error. As before,

$$E(s) = \frac{1}{s^2} \cdot \frac{\tau s}{\tau s+1} = \frac{\tau}{s(\tau s+1)}$$

$$\therefore \quad e(t) = \tau(1 - e^{-t/\tau})$$

and $$e_{ss} = \underset{t \to \infty}{\text{Lt}} \; e(t) = \tau \qquad \text{.....(3.12)}$$

Thus the steady state error is equal to the time constant of the system. The first order system, therefore, can not track the ramp input without a finite steady state error. If the time constant is reduced not only the speed of response increases but also the steady state error for ramp input decreases. Hence the ramp input is important to the extent that it produces a finite steady state error. Instead of finding the entire response, it is sufficient to estimate the steady state value by using the final value theorem. Thus

$$e_{ss} = \underset{s \to 0}{\text{Lt}} \; s\,E(s)$$

$$= \underset{s \to 0}{\text{Lt}} \; \frac{\tau s}{s(\tau s + 1)}$$

$$= \tau$$

which is same as given by eqn. (3.12)

The response of a first order system for unit ramp input is plotted in Fig. 3.7.

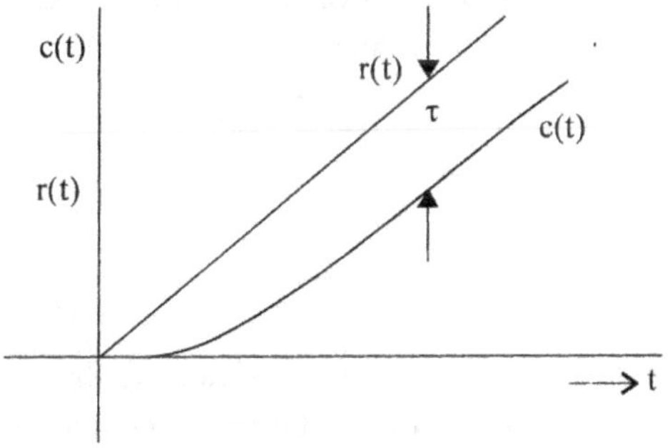

Fig. 3.7 Unit ramp response of a first order system.

3.4.3 Response to a Unit Parabolic or Acceleration Input

The response of a first order system to a unit parabolic input, for which

$$R(s) = \frac{1}{s^3} \text{ is given by,}$$

$$C(s) = \frac{1}{s^3(\tau s + 1)}$$

$$c(t) = \tau^2 - \tau t + \frac{t^2}{2} - \tau^2 e^{-\frac{1}{\tau}t} \qquad \text{.....(3.13)}$$

Differentiating eqn. (3.13), we get,

$$\frac{dc(t)}{dt} = -\tau + t + \tau\, e^{-\frac{1}{\tau}t}$$

$$= t - \tau\left(1 - e^{-\frac{1}{\tau}t}\right) \qquad .....(3.14)$$

Eqn. (3.14) is seen to be same as eqn. (3.10), which is the response of the first order system to unit velocity input. Thus subjecting the first order system to a unit parabolic input does not give any additional information regarding transient behaviour of the system. But, the steady state error, for a prabolic input is given by,

$$e\,(t) = r\,(t) - c\,(t)$$

$$= \frac{t^2}{2} - \tau^2 + \tau\, t - \frac{t^2}{2} + \tau^2\, e^{-\frac{1}{\tau}t}$$

$$e_{ss} = \underset{t \to \infty}{Lt}\; e\,(t) = \infty$$

Thus a first ordr system has infinite state error for a prabolic input. The steady state error can be easily obtained by using the final value theorem as :

$$e_{ss} = \underset{s \to 0}{Lt}\; s\,E(s) = \underset{s \to 0}{Lt}\; \frac{R(s)}{\tau + 1}$$

$$= \underset{s \to 0}{Lt}\; \frac{s.1}{s^3(\tau s + 1)} = \infty$$

Summarizing the analysis of first order system, we can say that the step input yields the desired information about the speed of transient response. It is observed that the speed of response is inversely proportional to the time constant τ of the system. The ramp and parabolic inputs do not give any additional information regarding the speed of response. However, the steady state errors are different for these three different inputs. For a step input, the steadystate error e_{ss} is zero, for a velocity input there is a finite error equal to the time constant τ of the system and for an acceleration input the steadystate error is infinity.

It is clear from the discussion above, that it is sufficient to study the behaviour of any system to a unit step input for understanding its transient response and use the velocity input and acceleration input for understanding the steadystate behaviour of the system.

3.5 Second Order System

3.5.1 Response to a Unit Step Input

Consider a Type 1, second order system as shown in Fig. 3.8. Since G(s) has one pole at the origin, it is a type one system.

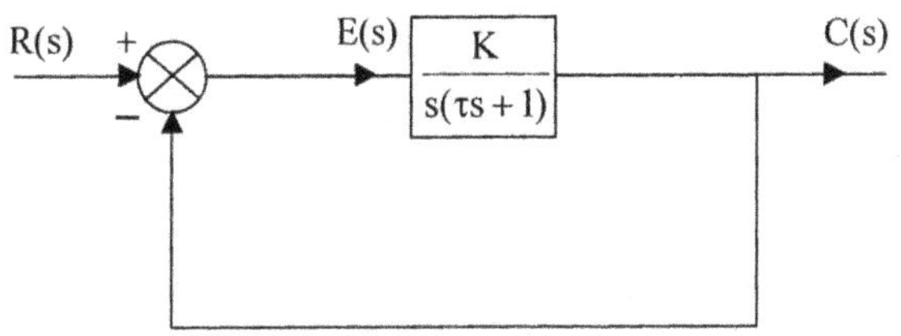

Fig. 3.8 Second Order System.

The closed loop transfer function is give by,

$$T(s) = \frac{C(s)}{R(s)} = \frac{K}{\tau s^2 + s + K} \qquad .....(3.15)$$

The transient response of any system depends on the poles of the transfer function T(s). The roots of the denominator polynomial in *s* of *T*(*s*) are the poles of the transfer function. Thus the denominator polynomial of T(s), given by

$$D(s) = \tau s^2 + s + K$$

is known as the *characteristic polynomial* of the system and D(s) = 0 is known as the *characteristic* equation of the system. Eqn. (3.15) is normally put in standard from, given by,

$$T(s) = \frac{K/\tau}{s^2 + \frac{1}{\tau}s + K/\tau}$$

$$= \frac{\omega_n^2}{s^2 + 2\delta\omega_n s + \omega_n^2} \qquad .....(3.16)$$

Where,

$$\omega_n = \sqrt{\frac{K}{\tau}} = \text{natural frequency}$$

$$\delta = \frac{1}{2\sqrt{K\tau}} = \text{damping factor}$$

The poles of T(s), or, the roots of the characteristic equation

$$s^2 + 2\,\delta\,\omega_n\,s + \omega_n^2 = 0$$

are given by,

$$s_{1,2} = \frac{-2\delta\omega_n \pm \sqrt{4\delta^2\omega_n^2 - 4\omega_n^2}}{2}$$

$$= -\delta\,\omega_n \pm j\,\omega_n\,\sqrt{1-\delta^2} \qquad \text{(assuming } \delta < 1\text{)}$$

$$= -\delta\,\omega_n \pm j\,\omega_d$$

Where $\omega_d = \omega_n\sqrt{1-\delta^2}$ is known as the *damped natural frequency* of the system. If $\delta > 1$, the two roots s_1, s_2 are real and we have an over damped system. If $\delta = 1$, the system is known as a *critically damped system*. The more common case of $\delta < 1$ is known as the *under damped system*.

If ω_n is held constant and δ is changed from 0 to ∞, the locus of the roots is shown in Fig. 3.9. The magnitude of s_1 or s_2 is ω_n and is independent of δ. Hence the locus is a semicircle with radius ω_n until $\delta = 1$. At $\delta = 0$, the roots are purely imaginary and are given by $s_{1,2} = \pm j\omega_n$. For $\delta = 1$, the roots are purely real, negative and equal to $-\omega_n$. As δ increases beyond unity, the roots are real and negative and one root approached the origin and the other approaches infinity as shown in Fig. 3.9.

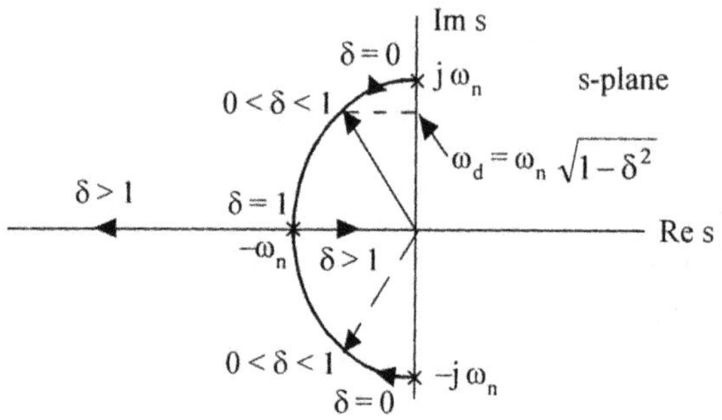

Fig. 3.9 Locus of the roots of the characteristic equation.

For a unit step input $R(s) = \frac{1}{s}$ and eqn. 3.16 can be written as

$$C(s) = T(s).\ R(s) = \frac{\omega_n^2}{s^2 + 2\delta\omega_n s + \omega_n^2} \cdot \frac{1}{s} \qquad .....(3.17)$$

Splitting eqn. (3.17) in to partial fractions, assuming δ to be less than 1, we have

$$C(s) = \frac{K_1}{s} + \frac{K_2 s + K_3}{s^2 + 2\delta\omega_n s + \omega_n^2}$$

Evaluating K_1, K_2 and K_3 by the usual procedure, we have,

$$C(s) = \frac{1}{s} - \frac{s + 2\delta\omega_n}{(s+\delta\omega_n)^2 + \omega_n^2(1-\delta^2)}$$

$$= \frac{1}{s} - \frac{s + \delta\omega_n}{(s+\delta\omega_n)^2 + \omega_n^2(1-\delta^2)} - \frac{\delta\omega_n\sqrt{1-\delta^2}}{\sqrt{1-\delta^2}\,(s+\delta\omega_n)^2 + \omega_n^2(1-\delta^2)} \qquad .....(3.18)$$

Taking inverse Laplae transform of eqn. (3.18), we have

$$c(t) = 1 - e^{-\delta\omega_n t}\left[\cos\omega_n\sqrt{1-\delta^2}\,t + \frac{\delta}{\sqrt{1-\delta^2}}\sin\omega_n\sqrt{1-\delta^2}\,t\right] \qquad .....(3.19)$$

Eqn. (3.19) can be put in a more convenient from as,

$$c\,(t) = 1 - \frac{e^{-\delta\omega_n t}}{\sqrt{1-\delta^2}}\ \text{Sin}\,(\omega_d t + \phi)$$

Where $\omega_d = \omega_n \sqrt{1-\delta^2}$

and $$\tan\phi = \frac{\sqrt{1-\delta^2}}{\delta} \qquad .....(3.20)$$

This response is plotted in Fig. 3.10. The response is oscillatory and as $t \to \infty$, it approaches unity.

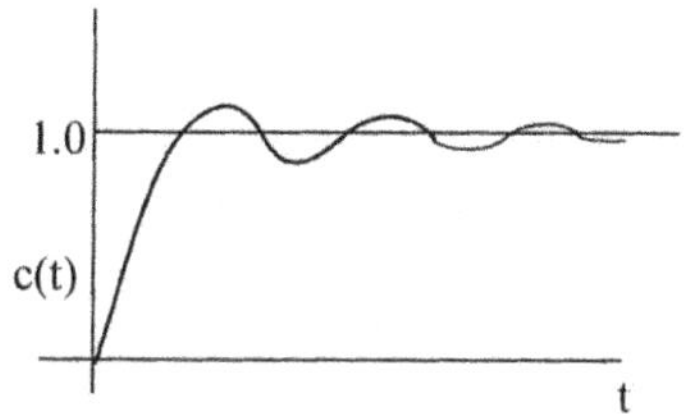

Fig. 3.10 Step response of an underdamped second order system.

If $\delta = 1$, the two roots of the characteristic equations are $s_1 = s_2 = -\omega_n$ and the response is given by

$$C\,(s) = \frac{\omega_n^{\,2}}{(s+\omega_n)^2} \cdot \frac{1}{s}$$

and $$c\,(t) = 1 - e^{-\omega_n t} - t\,\omega_n\, e^{-\omega_n t} \qquad .....(3.21)$$

This is plotted in Fig. 3.11.

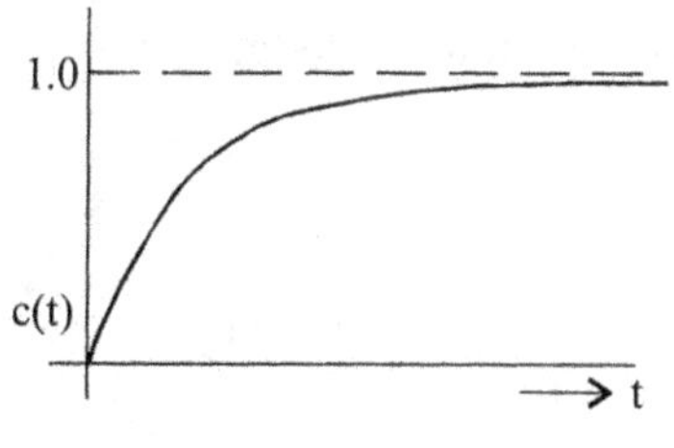

Fig. 3.11 Response of a critically damped second order system.

As the damping is increased from a value less than unity, the oscillations decrease and when the damping factor equals unity the oscillations just disappear. If δ is increaed beyond unity, the roots of the characteristic equation are real and negative and hence, the response approaches unity in an exponential way. This response is known as overdamped response and is shown in Fig. 3.12.

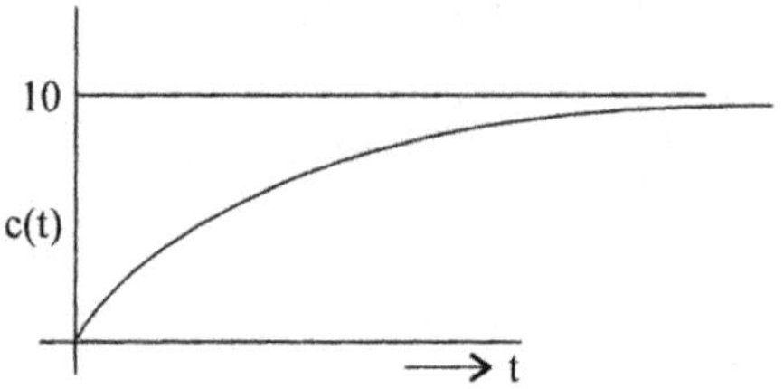

Fig. 3.12 Step response of an overdamped second order system.

$$c(t) = K_1 e^{-s_1 t} + K_2 e^{-s_2 t} \qquad .....(3.22)$$

Where s_1 and s_2 are given by,

$$s_{1,2} = -\delta\omega_n \pm \omega_n \sqrt{\delta^2 - 1}$$

and K_1 and K_2 are constants.

3.5.2 Response to a Unit Ramp Input

For a unit ramp input,

$$R(s) = \frac{1}{s^2}$$

and the output is given by,

$$C(s) = \frac{\omega_n^2}{s^2(s^2 + 2\delta\omega_n s + \omega_n^2)}$$

Taking inverse Laplace transform, we get the time response $c(t)$ as,

$$c(t) = t - \frac{2\delta}{\omega_n} + \frac{e^{-\delta\omega_n t}}{\omega_n\sqrt{1-\delta^2}} \text{Sin}\left(\omega_n\sqrt{1-\delta^2}\, t + \phi\right) \text{ for } \delta < 1 \qquad .....(3.23)$$

The time response for a unit ramp input is plotted in Fig. 3.13.

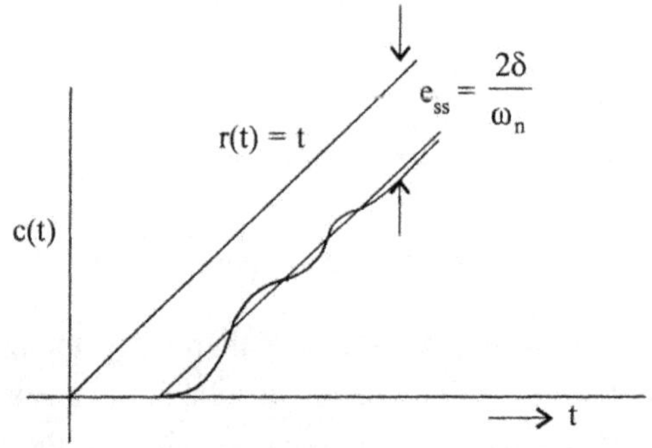

Fig. 3.13 Unit ramp response of a second order system.

The response reveals two aspects of the system.

1. The transient response is of the same form as that of a unit step response. No new information is obtained regarding speed of response or oscillations in the system.

2. It has a steadystate error $e_{ss} = \dfrac{2\delta}{\omega_n}$, unlike the step response, where the steady state error was zero. Thus, no new information is gained by obtaining the transient response of the system for a ramp input. The steadystate error could be easily calculated using final value theorem instead of laboriously solving for the entire reponse. For the given system, the error E (s) is given by

$$E(s) = R(s) - C(s)$$

$$= \frac{1}{s^2} - \frac{\omega_n^2}{s^2(s^2 + 2\delta\omega_n s + \omega_n^2)} = \frac{s^2 + 2\delta\omega_n s + \omega_n^2 - \omega_n^2}{s^2(s^2 + 2\delta\omega_n s + \omega_n^2)}$$

and from the final value theorem,

$$e_{ss} = \underset{s \to 0}{\text{Lt}}\; s\,E(s) = \underset{s \to 0}{\text{Lt}}\; \frac{s^2(s + 2\delta\omega_n)}{s^2(s^2 + 2\delta\omega_n s + \omega_n^2)}$$

$$= \frac{2\delta\omega_n}{\omega_n^2} = \frac{2\delta}{\omega_n} \qquad \text{.....(3.24)}$$

In a similar manner, the unit parabolic input does not yield any fresh information about the transient response. The steadystate error can be obtained using final value theorem in this case also. For the given system, for a unit acceleration input,

$$e_{ss} = \infty \qquad \text{.....(3.25)}$$

3.5.3 Time Domain Specifications of a Second Order System

The performance of a system is usually evaluated in terms of the following qualities.

1. How fast it is able to respond to the input,
2. How fast it is reaching the desired output,
3. What is the error between the desired output and the actual output, once the transients die down and steady state is achieved,
4. Does it oscillate around the desired value,

and 5. Is the output continuously increasing with time or is it bounded.

The last aspect is concerned with the stability of the system and we would require the system to be stable. This aspect will be considered later. The first four questions will be answered in terms of time domain specifications of the system based on its response to a unit step input. These are the specifications to be given for the design of a controller for a given system.

In section 3.5, we have obtained the response of a type 1 second order system to a unit step input. The step response of a typical underdamped second order system is plotted in Fig. 3.14.

It is observed that, for an underdamped system, there are two complex conjugate poles. Usually, even if a system is of higher order, the two complex conjugate poles nearest to the $j\omega$ – axis (called dominant poles) are considered and the system is approximated by a second order system. Thus, in designing any system, certain design specifications are given based on the typical underdamped step response shown as Fig. 3.14.

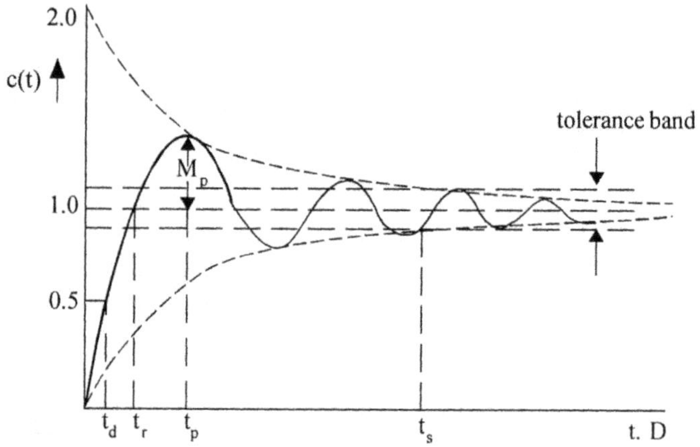

Fig. 3.14 Time domain specifications of a second order system.

The design specifications are :

1. *Delay time* t_d: It is the time required for the response to reach 50% of the steady state value for the first time
2. *Rise time* t_r : It is the time required for the response to reach 100% of the steady state value for under damped systems. However, for over damped systems, it is taken as the time required for the response to rise from 10% to 90% of the steadystate value.
3. *Peak time* t_p : It is the time required for the response to reach the maximum or Peak value of the response.
4. *Peak overshoot* M_p : It is defined as the difference between the peak value of the response and the steady state value. It is usually expressed in percent of the steady state value. If the time for the peak is t_p, percent peak overshoot is given by,

$$\text{Percent peak overshoot } M_p = \frac{c(t_p) - c(\infty)}{c(\infty)} \times 100. \qquad .....(3.26)$$

   For systems of type 1 and higher, the steady state value $c(\infty)$ is equal to unity, the same as the input.

5. *Settling time* t_s : It is the time required for the response to reach and remain within a specified tolerance limits (usually $\pm$ 2% or $\pm$ 5%) around the steady state value.
6. *Steady state error* e_{ss} : It is the error betwen the desired output and the actual output as $t \to \infty$ or under steadystate conditions. The desired output is given by the reference input $r(t)$ and

$$\text{therefore, } e_{ss} = \underset{t \to \infty}{\text{Lt}} [r(t) - c(t)]$$

From the above specifications it can be easily seen that the time response of a system for a unit step input is almost fixed once these specifications are given. But it is to be observed that all the above specifications are not independent of each other and hence they have to be specified in such a way that they are consistent with others.

Let us now obtain the expressions for some of the above design specifications in terms of the damping factor δ and natural frequency ω_n.

1. ***Rise time (t_r)***

If we consider an underdamped system, from the definition of the rise time, it is the time required for the response to reach 100% of its steadystate value for the first time. Hence from eqn. (3.20).

$$C(t_r) = 1 = 1 - \frac{e^{-\delta\omega_n t_r}}{\sqrt{1-\delta^2}} \operatorname{Sin}\left(\omega_n\sqrt{1-\delta^2}\, t_r + \phi\right)$$

Or $$\frac{e^{-\delta\omega_n t_r}}{\sqrt{1-\delta^2}} \operatorname{Sin}\left(\omega_n\sqrt{1-\delta^2}\, t_r + \phi\right) = 0$$

Since $\dfrac{e^{-\delta\omega_n t_r}}{\sqrt{1-\delta^2}}$ cannot be equal to zero,

$$\operatorname{Sin}(\omega_d t_r + \phi) = 0$$

$$\therefore \quad \omega_d t_r + \phi = \pi$$

and $$t_r = \frac{\pi - \phi}{\omega_n\sqrt{1-\delta^2}} = \frac{\pi - \tan^{-1}\dfrac{\sqrt{1-\delta^2}}{\delta}}{\omega_n\sqrt{1-\delta^2}} \qquad .....(3.27)$$

2. ***Peak time (t_P)***

At the peak time, t_P, the response attains its maximum value and this can be obtained by differentiating *c* (*t*) and equating it to zero. Thus,

$$\frac{dc(t)}{dt} = \frac{\delta\omega_n}{\sqrt{1-\delta^2}} e^{-\delta\omega_n t} \operatorname{Sin}(\omega_d t + \phi) - \frac{e^{-\delta\omega_n t}}{\sqrt{1-\delta^2}} \cos(\omega_d t + \phi).\ \omega_d = 0$$

Simplifying we have,

$$\delta \operatorname{Sin}(\omega_d t + \phi) - \sqrt{1-\delta^2} \cos(\omega_d t + \phi) = 0$$

This can be written as,

$$\cos\phi \sin(\omega_d t + \phi) - \sin\phi \cos(\omega_d t + \phi) = 0$$

where $$\tan\phi = \frac{\sqrt{1-\delta^2}}{\delta}$$

$\therefore$ $$\sin(\omega_d t + \phi - \phi) = \sin\omega_d t = 0$$

or $$\omega_d t = n\pi \qquad \text{for } n = 0, 1, 2, \ldots$$

Here

n = 0 Corresponds to its minimum value at t = 0

n = 1 Corresponds to its first peak value at $t = t_p$

n = 2 Corresponds to its first undershoot

n = 3 Corresponds to its second overshoot and so on

Hence for n = 1

$$t_p = \frac{\pi}{\omega_n\sqrt{1-\delta^2}} \qquad \text{.....(3.28)}$$

Thus, we see that the peak time depends on both ω_n and δ. If we consider the product of ω_n and t_p, which may be called as *normalised peak time*, we can plot the variation of this normalised peak time with the damping factor δ. This is shown in Fig. 3.15.

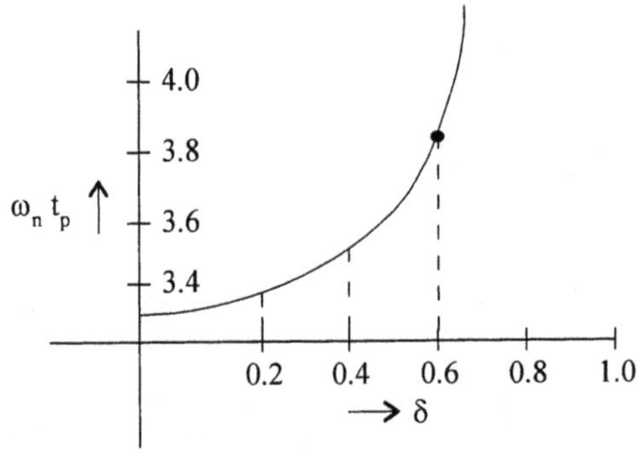

Fig. 3.15 Normalised peak time $\omega_n t_p$ Vs δ for a second order system.

3. ***Peak overshoot (M_p)***

The peak overshoot is defined as

$$M_p = c(t_p) - 1$$

$$= 1 - \frac{e^{-\delta\omega_n t_p}}{\sqrt{1-\delta^2}} \sin(\omega_d t_p + \phi) - 1$$

$$= -\frac{e^{-\delta\omega_n t_p}}{\sqrt{1-\delta^2}} \sin\left(\omega_d \cdot \frac{\pi}{\omega_d} + \phi\right)$$

$$M_p = \frac{e^{-\frac{\delta \omega_n \pi}{\omega_n \sqrt{1-\delta^2}}}}{\sqrt{1-\delta^2}} \text{Sin}\,\phi \quad (\because \text{Sin}\,(\pi+\phi) = -\text{Sin}\,\phi)$$

$$= e^{\frac{-\pi\delta}{\sqrt{1-\delta^2}}} \qquad \left(\because \text{Sin}\,\phi = \sqrt{1-\delta^2}\right)$$

Hence, peak overshoot, expressed as a percentage of steady state value, is given by,

$$M_p = 100\, e^{\frac{-\pi\delta}{\sqrt{1-\delta^2}}}\ \% \qquad .....(3.29)$$

It may be observed that peak overshoot M_p is a function of the damping factor δ only. Its variation with damping factor is shown in Fig. 3.16.

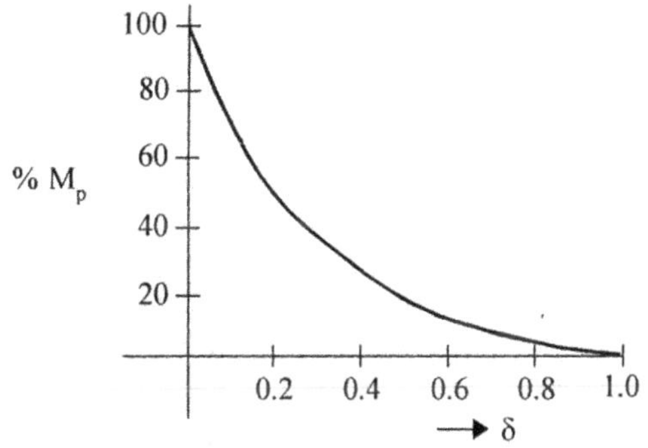

Fig. 3.16 Percent overshoot M_p Vs δ for a second order system.

4. ***Settling time*** (t_s)

The time varying term in the step response, *c* (*t*), consists of a product of two terms; namely, an exponentially delaying term, $\frac{e^{-\delta\omega_n t}}{\sqrt{1-\delta^2}}$ and a sinusoidal term, *Sin* ($\omega_d t + \phi$). It is clear that the response is a decaying sinusoid, the envelop of which is given by $\frac{e^{-\delta\omega_n t}}{\sqrt{1-\delta^2}}$. Thus, the response reaches and remains within a given band, around the steadystate value, when this envelop crosses the tolerance band. Once this envelop reaches this value, there is no possibility of subsequent oscillations to go beyond these tolerane limits. Thus for a 2% tolerance band,

$$\frac{e^{-\delta\omega_n t_s}}{\sqrt{1-\delta^2}} = 0.02$$

For low values of δ, $\delta^2 << 1$ and therefore $e^{-\delta\omega_n t_s} \simeq 0.02$

$$\therefore \qquad t_s \simeq \frac{4}{\delta\omega_n} = 4\,\tau \qquad .....(3.30)$$

where τ is the time constant of the exponential term.

Eqn. (3.30) shows that the settling time is a function of both δ and ω_n. Since damping factor is an important design specification, we would like to know the variation of the setting time with δ, with ω_n fixed. Or, in otherwords, we can define a normalised time $\omega_n t_s$, and find the variation of this quantity with respect to δ. The step response of a second order system is plotted in Fig. 3.17 for different values of δ, taking normalised time $\omega_n t$, on x-axis. The curves are magnified around the steady state value for clarity.

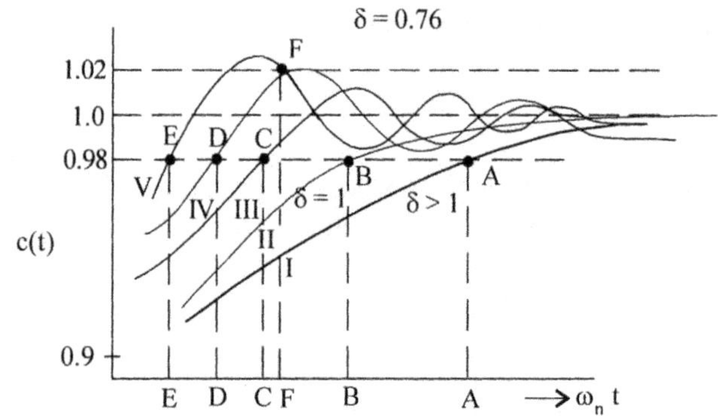

Fig. 3.17 C (t) plotted for different value of δ.

The settling time monotonically decreases as the damping is decreased from a value greater than one (over damped) to less than one (under damped). For 2% tolerance band, it decreases until the first peak of the response reaches the tolerance limit of 1.02 as shown by the curve IV in Fig. 3.17. Points A, B, C, and D marked on the graph give the values $\omega_n t_s$, for decreasing values of δ. The peak value of the response reaches 1.02 at a damping factor $\delta = 0.76$. The settling time for this value of δ is marked as point D on the curve. If δ is decreased further, since the response crosses the upper limit 1.02, the point E no longer represents the settling time. The settling time suddenly jumps to a value given by the point F on the curve. Thus there is a discontinuity at $\delta = 0.76$. If δ is decreased further the setting time increases until the first undershoot touches the lower limit of 0.98. Similarly, the third discontinuity occurs when the second peak touches the upper limit of 1.02 and so on. The variation of $\omega_n t_s$ with δ for a tolerance band of 2% is plotted in Fig. 3.18.

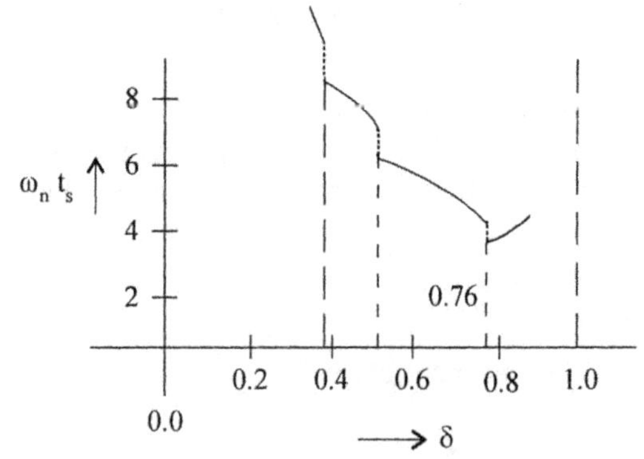

Fig. 3.18 Variation of normalised settling time $\omega_n t_s$ Vs δ.

From Fig. 3.18 it is observed that the least settling time is obtained for a damping factor of $\delta = 0.76$. Since settling time is a measure of how fast the system reaches a steady value, control systems are usually designed with a damping factor of around 0.7. Sometimes, the systems are designed to have even lesser damping factor because of the presence of certain nonlinearities which tend to produce an error under steadystate conditions. To reduce this steadystate error, normally the system gain K is increased, which in turn decreases the damping $\left(\because \delta = \frac{1}{2\sqrt{KT}}\right)$. However, for robotic control, the damping is made close to and slightly higher than unity. This is because the output of a robotic system should reach the desired value as fast as possible, but it should never overshoot it.

5. *Steady state error (e_{ss})*

For a type 1 system, considered for obtaining the design specifications of a second order control system, the steady state error for a step input is obviously zero. Thus

$$e_{ss} = \underset{t \to \infty}{\text{Lt}} \; 1 - c(t) = 0$$

The steady state error for a ramp input was obtained in eqn. (3.24) as $e_{ss} = \frac{2\delta}{\omega_n}$.

As the steadystate error, for various test signals, depends on the type of the system, it is dealt in the next section in detail.

3.6 Steady State Errors

One of the important design specifications for a control system is the steadystate error. The steady state output of any system should be as close to desired output as possible. If it deviates from this desired output, the performance of the system is not satisfactory under steadystate conditions. The steadystate error reflects the accuracy of the system. Among many reasons for these errors, the most important ones are the type of input, the type of the system and the nonlinearities present in the system. Since the actual input in a physical system is often a random signal, the steady state errors are obtained for the standard test signals, namely, step, ramp and parbaolic signals.

3.6.1 Error Constants

Let us consider a feedback control system shown in Fig. 3.19.

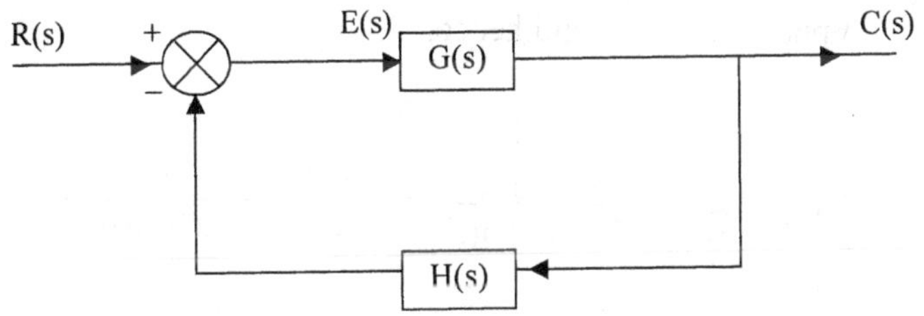

Fig. 3.19 Feedback Control System.

The error signal E (s) is given by

$$E(s) = R(s) - H(s)\,C(s) \qquad .....(3.31)$$

But

$$C(s) = G(s)\,E(s) \qquad .....(3.32)$$

From eqns. (3.31) and (3.32) we have

$$E(s) = \frac{R(s)}{1+G(s)H(s)}$$

Applying final value theorem, we can get the steady state error e_{ss} as,

$$e_{ss} = \underset{s \to 0}{\text{Lt}}\; s\,E(s) = \underset{s \to 0}{\text{Lt}}\; \frac{sR(s)}{1+G(s)H(s)} \qquad .....(3.33)$$

Eqn. (3.33) shows that the steady state error is a function of the input R(s) and the open loop transfer function G(s). Let us consider various standard test signals and obtain the steadystate error for these inputs.

1. Unit step or position input.

For a unit step input, $R(s) = \frac{1}{s}$. Hence from eqn. (3.33)

$$e_s = \underset{s \to 0}{\text{Lt}}\; \frac{s.\frac{1}{s}}{1+G(s)H(s)}$$

$$= \frac{1}{1+\underset{S \to 0}{\text{Lt}}\; G(s)\,H(s)} \qquad .....(3.34)$$

Let us define a useful term, *position error constant* K_p as,

$$K_p \triangleq \underset{s \to 0}{\text{Lt}}\; G(s)\,H(s) \qquad .....(3.35)$$

In terms of the position error constant, e_{ss} can be written as,

$$e_{ss} = \frac{1}{1+K_p} \qquad .....(3.36)$$

2. Unit ramp or velocity input.

For unit velocity input, $R(s) = \frac{1}{s^2}$ and hence,

$$e_{ss} = \underset{s \to 0}{\text{Lt}}\; \frac{s.\frac{1}{s}}{1+G(s)H(s)} = \underset{s \to 0}{\text{Lt}}\; \frac{1}{s+sG(s)H(s)}$$

$$= \frac{1}{\underset{s \to 0}{\text{Lt}}\; sG(s)H(s)} \qquad .....(3.37)$$

Again, defining the *velocity error constant* K_v as,

$$K_v = \underset{s \to 0}{Lt}\ s\,G(s)\,H(s) \quad .....(3.38)$$

$$\therefore \quad e_{ss} = \frac{1}{K_v} \quad .....(3.39)$$

3. Unit parabolic or acceleration input.

For unit acceleration input $R(s) = \frac{1}{s^3}$ and hence

$$e_{ss} = \underset{s \to 0}{Lt} \frac{s}{s^3[1+G(s)H(s)]} = \underset{s \to 0}{Lt} \frac{1}{s^2 + s^2G(s)H(s)}$$

$$= \frac{1}{\underset{s \to 0}{Lt}\ s^2G(s)H(s)} \quad .....(3.40)$$

Defining the acceleration error constant K_a as,

$$K_a = \underset{s \to 0}{Lt}\ s^2\,G(s)\,H(s) \quad .....(3.41)$$

$$\therefore \quad e_{ss} = \frac{1}{K_a} \quad .....(3.42)$$

For the special case of unity of feedback system, H (s) = 1 and eqns. (3.35) (3.38) or (3.41) are modified as,

$$K_p = \underset{s \to 0}{Lt}\ G(s) \quad .....(3.43)$$

$$K_v = \underset{s \to 0}{Lt}\ sG(s) \quad .....(3.44)$$

and

$$K_a = \underset{s \to 0}{Lt}\ s^2\,G(s) \quad .....(3.45)$$

In design specifications, instead of specifying the steady state error, it is a common practice to specify the error constants which have a direct bearing on the steadystate error. As will be seen later in this section, if the open loop transfer function is specified in time constant form, as in eqn. (3.3), the error constant is equal to the gain of the open loop system.

3.6.2 Dependence of Steadystate Error on Type of the System

Let the loop transfer function G (s) H (s) or the open loop transfer function G (s) for a unity feedback system, be given is time constant form.

$$G(s) = \frac{K(T_{z1}s+1)(T_{z2}s+1)----}{S^r(T_{p1}s+1)(T_{p2}s+1)----} \quad .....(3.46)$$

As $s \to 0$, the poles at the origin dominate the expression for $G(s)$. We had defined the type of a system, as the number of poles present at the origin. Hence the steady state error, which depends on $\underset{s\to 0}{\text{Lt}}\ G(s)$, $\underset{s\to 0}{\text{Lt}}\ s\,G(s)$ or $\underset{s\to 0}{\text{Lt}}\ s^2\,G(s)$, is dependent on the type of the system. Let us therefore obtain the steady state error for various standard test signals for type–0, type–1 and type–2 systems.

1. ***Type –0 system***

From eqn. (3.46) with $r = 0$, the error constants are given by

$$K_p = \underset{s\to 0}{\text{Lt}}\ G(s) = \underset{s\to 0}{\text{Lt}}\ \frac{K(\tau_{Z1}s+1)(\tau_{Z2}s+1)--}{(\tau_{p1}s+1)(\tau_{p2}s+1)--} = K$$

$$K_v = \underset{s\to 0}{\text{Lt}}\ s\,G(s) = \underset{s\to 0}{\text{Lt}}\ \frac{sK(\tau_{Z1}s+1)(\tau_{Z2}s+1)--}{(\tau_{p1}s+1)(\tau_{p2}s+1)--} = 0$$

Similarly
$$K_a = \underset{s\to 0}{\text{Lt}}\ s^2\,G(s) = 0 \qquad .....(3.47)$$

The steady state errors for unit step, velocity and acceleration inputs are respectively, from eqns. (3.34), (3.37) and (3.40),

$$e_{ss} = \frac{1}{1+K_P} = \frac{1}{1+K} \quad \text{(step input)}$$

$$e_{ss} = \frac{1}{K_v} = \infty \quad \text{(velocity input)}$$

$$e_{ss} = \frac{1}{K_a} = \infty \quad \text{(acceleration input)}$$

2. ***Type 1 system***

For type 1 system, $r = 1$ in eqn. (3.46) and

$$K_p = \underset{s\to 0}{\text{Lt}}\ G(s) = \text{Lt}\ \frac{K}{s} = \infty$$

$$K_v = \underset{s\to 0}{\text{Lt}}\ s\,G(s) = \underset{s\to 0}{\text{Lt}}\ s.\frac{K}{s} = K$$

and
$$K_a = \underset{s\to 0}{\text{Lt}}\ s\,G(s) = \underset{s\to 0}{\text{Lt}}\ s^2.\frac{K}{s} = 0$$

The steady state error for unit step, unit velocity and unit acceleration inputs are respectively,

$$e_{ss} = \frac{1}{1+K_p} = \frac{1}{\infty} = 0 \qquad \text{(position)}$$

$$e_{ss} = \frac{1}{K_v} = \frac{1}{K} \qquad \text{(velocity)}$$

and $$e_{ss} = \frac{1}{K_a} = \frac{1}{0} = \infty \qquad \text{(acceleration)}$$

3. ***Type 2-system***

For a type – 2 system $r = 2$ in eqn. (3.46) and

$$K_p = \underset{s \to 0}{\text{Lt}}\ G(s) \quad = \underset{s \to 0}{\text{Lt}}\ \frac{K}{s^2} = \infty$$

$$K_v = \underset{s \to 0}{\text{Lt}}\ s\,G(s) \quad = \underset{s \to 0}{\text{Lt}}\ \frac{sK}{s^2} = \infty$$

and $$K_a = \underset{s \to 0}{\text{Lt}}\ s^2\,G(s) \quad = \underset{s \to 0}{\text{Lt}}\ \frac{s^2 K}{s^2} = K$$

The steady state errors for the three test inputs are,

$$e_{ss} = \frac{1}{1+K_P} = \frac{1}{1+\infty} = 0 \quad \text{(position)}$$

$$e_{ss} = \frac{1}{K_v} = \frac{1}{\infty} = 0 \qquad \text{(velocity)}$$

and $$e_{ss} = \frac{1}{K_a} = \frac{1}{K} \qquad \text{(acceleration)}$$

Thus a type zero system has a finite steady state error for a unit step input and is equal to

$$e_{ss} = \frac{1}{1+K} = \frac{1}{1+K_p} \qquad \text{.....(3.47)}$$

Where K is the system gain in the time constant from. It is customary to specify the gain of a type zero system by K_p rather than K.

Similarly, a type –1 system has a finite steady state error for a velocity input only and is given by

$$e_{ss} = \frac{1}{K} = \frac{1}{K_v} \qquad \text{.....(3.48)}$$

Thus the gain of type –1 system in normally specified as K_v.

A type –2 system has a finite steady state error only for acceleration input and is given by

$$e_{ss} = \frac{1}{K} = \frac{1}{K_a} \qquad \text{.....(3.49)}$$

As before, the gain of type –2 system is specified as K_a rather than K.

The steady state errors, for various standard inputs for type – 0, type – 1 and type – 2 are summarized in Table. 3.1.

Table. 3.1 Steady state errors for various inputs and type of systems

Standard input	Steadystate error e_{ss}		
	Type - 0	**Type - 1**	**Type - 2**
	$K_p = \underset{s \to 0}{Lt} G(s)$	$K_v = \underset{s \to 0}{Lt} s\,G(s)$	$K_a = \underset{s \to 0}{Lt} s^2 G(s)$
Unit step	$\frac{1}{1+K_P}$	0	0
Unit velocity	∞	$\frac{1}{K_v}$	0
Unit acceleration	∞	∞	$\frac{1}{K_a}$

If can be seen from Table. 3.1, as the type of the system and hence the number of integrations increases, more and more steady state errors become zero. Hence it may appear that it is better to design a system with more and more poles at the origin. But if the type of the system is higher than 2, the systems tend to be more unstable and the dynamic errors tend to be larger. The stability aspects are considered in chapter 4.

3.6.3 Generalized Error Coefficients - Error Series

The main disadvantage of defining the steadystate error in terms of error constants is that, only one of the constants is finite and non zero for a particular system, where as the other constants are either zero or infinity. If any error constant is zero, the steady state error is infinity, but we do not have any clue as to how the error is approaching infinity.

If the inputs are other than step, velocity or acceleration inputs, we can extend the concept of error constants to include inputs which can be represented by a polynomial. Many functions which are analytic can be represented by a polynomial in *t*. Let the error be given by,

$$E(s) = \frac{R(s)}{1+G(s)} \qquad .....(3.50)$$

Eqn. (3.50) may be written as

$$E(s) = Y(s).\,R(s) \qquad .....(3.51)$$

Where

$$Y(s) = \frac{1}{1+G(s)} \qquad .....(3.52)$$

Using Convolution theorem eqn. (3.51) can be written as

$$e(t) = \int_0^t y(\tau)\, r(t-\tau)\, d\tau \qquad .....(3.53)$$

Assuming that *r* (*t*) has first n deriratives, *r* (*t* – *τ*) can be expanded into a Taylor series,

$$r(t-\tau) = r(t) - \tau r'(t) + \frac{\tau^2}{2!} r''(t) - \frac{\tau^3}{3!} r'''(t) ---- \qquad .....(3.54)$$

where the primes indicate time derivatives. Substituting eqn. (3.54) into eqn. (3.53), we have,

$$e(t) = \int_0^t y(\tau) \left[r(t) - \tau r'(t) + \frac{\tau^2}{2!} r''(t) - \frac{\tau^3}{3!} r'''(t) ---- \right] d\tau$$

$$= r(t) \int_0^t y(\tau)\, d\tau - r'(t) \int_0^t \tau y(\tau)\, d\tau + r''(t) \int_0^t \frac{\tau^2}{2!} y(\tau)\, d\tau + ... \qquad .....(3.55)$$

To obtain the steady state error, we take the limit $t \to \infty$ on both sides of eqn. (3.55)

$$e_{ss} = \underset{t\to\infty}{Lt}\; e(t) = \underset{t\to\infty}{Lt} \left[r(t) \int_0^t y(\tau) d\tau - r'(t) \int_0^t \tau y(\tau) d\tau + r''(t) \frac{\tau^2}{2!} y(\tau) d\tau..... \right] \qquad .....(3.56)$$

$$e_{ss} = r_{ss}(t) \int_0^\infty y(\tau)\, d\tau - r'_{ss}(t) \int_0^\infty \tau y(\tau)\, d\tau + r_{ss}''(t) \int_0^\infty \frac{\tau^2}{2!} y(\tau)\, d\tau + ... \qquad .....(3.57)$$

Where the suffix *ss* denotes steady state part of the function. It may be further observed that the integrals in eqn. (3.57) yield constant values. Hence eqn. (3.57) may be written as,

$$e_{ss} = C_0 r_{ss}(t) + C_1 r'_{ss}(t) + \frac{C_2}{2!} r''_{ss}(t) + ... + \frac{C_n}{n!} r^{(n)}_{ss}(t) + \qquad ......(3.58)$$

Where,

$$C_0 = \int_0^\infty y(\tau)\, d\tau \qquad .....(3.59)$$

$$C_1 = -\int_0^\infty \tau y(\tau)\, d\tau \qquad .....(3.60)$$

$$C_n = (-1)^n \int_0^\infty \tau^n y(\tau)\, d\tau \qquad .....(3.61)$$

The coefficients $C_0, C_1, C_2, \ldots C_n, \ldots$ are defined as generalized error coefficients. Eqn. (3.58) is known as generalised error series. It may be observed that the steady state error is obtained as a function of time, in terms of generalised error coefficients, the steady state part of the input and its derivatives. For a given transfer function G(s), the error coefficients can be easily evaluated as shown in the following.

Let $$y(t) = \mathcal{L}^{-1}\, Y(s)$$

$$\therefore \quad Y(s) = \int_0^\infty y(t)\, e^{-s\tau}\, d\tau \qquad .....(3.62)$$

$$\underset{s \to 0}{\text{Lt}}\ Y(s) = \underset{s \to 0}{\text{Lt}} \int_0^\infty y(\tau)\, e^{-s\tau}\, d\tau$$

$$= \int_0^\infty y(\tau)\ \underset{s \to 0}{\text{Lt}}\ e^{-s\tau}\, d\tau$$

$$= \int_0^\infty y(\tau)\, d\tau$$

$$= C_0 \qquad .....(3.63)$$

Taking the derivative of eqn. (3.62) with respect to *s*,

We have,

$$\frac{dY(s)}{ds} = \int_0^\infty y(\tau)\,(-\tau)\, e^{-s\tau}\, d\tau \qquad .....(3.64)$$

Now taking the limit of equation (3.64) as $s \to 0$, we have,

$$\underset{s \to 0}{\text{Lt}}\ \frac{dY(s)}{ds} = \int_0^\infty y(\tau)\,(-\tau)\ \underset{s \to 0}{\text{Lt}}\ e^{-s\tau}\, d\tau$$

$$= -\int_0^\infty \tau y(\tau)\, d\tau$$

$$= C_1 \qquad .....(3.65)$$

Similarly,

$$C_2 = \underset{s \to 0}{\text{Lt}}\ \frac{d^2Y(s)}{ds^2} \qquad .....(3.66)$$

$$C_3 = \underset{s \to 0}{\text{Lt}}\ \frac{d^3Y(s)}{ds^3} \qquad .....(3.67)$$

$$C_n = \underset{s \to 0}{\text{Lt}}\ \frac{d^nY(s)}{ds^n} \qquad .....(3.68)$$

Thus the constants can be evaluated using eqns. (3.63), (3.65) and (3.66) and so on and the time variation of the steadystate error can be obtained using eqn. (3.58).

The advantages of error series can be summarized as,

1. It provides a simple way of obtaining the nature of steadystate response to almost any arbitrary input.
2. We can obtain the complete steadystate response without actually solving the system differential equation.

Example 3.1

The angular position θ_C of a mass is controlled by a servo system through a reference signal θ_r. The moment of intertia of moving parts referred to the load shaft, J, is 150 kgm^2 and damping torque coefficient referred to the load shaft, B, is 4.5×10^3 Nwm / rad / sec. The torque developed by the motor at the load is 7.2×10^4 Nw–m per radian of error.

(a) Obtain the response of the system to a step input of 1 rad and determine the peak time, peak overshoot and frequency of transient oscillations. Also find the steadystate error for a constant angular velocity of 1 revolution / minute.

(b) If a steady torque of 1000 Nwm is applied at the load shaft, determine the steadystate error.

Solution :

The block diagram of the system may be written as shown in Fig. 3.20.

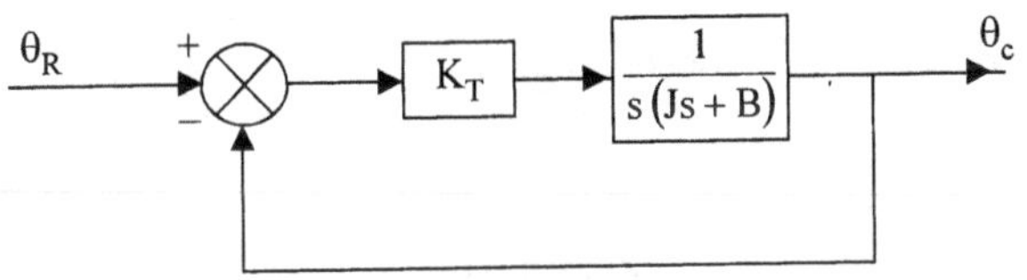

Fig. 3.20 Block diagram of the given system

From the block diagram, the forward path transfer function G (s) is given by,

$$G(s) = \frac{K_T}{s(Js+B)}$$

For the given values of K_T, J and B, we have

$$G(s) = \frac{7.2\times10^4}{s(150s+4.5\times10^3)}$$

$$= \frac{16}{s(0.333s+1)}$$

Thus $\quad K_V = 16$

$\tau = 0.333$ sec.

and

$$\delta = \frac{1}{2\sqrt{K_V\tau}} = \frac{1}{2\sqrt{16\times0.0333}}$$

$$= 0.6847$$

$$\omega_n = \sqrt{\frac{K_V}{\tau}} = \sqrt{\frac{16}{0.0333}}$$
$$= 21.91 \text{ rad/sec}$$

(a)
$$\theta(t) = 1 - \frac{e^{-\omega_n t}}{\sqrt{1-\delta^2}} \text{Sin}\left(\omega_n\sqrt{1-\delta^2}\, t + \tan^{-1}\sqrt{\frac{1-\delta^2}{\delta}}\right)$$
$$= 1 - 1.372\, e^{-15\,t} \text{ Sin } (15.97\, t + 46.8^0)$$

Peak time,
$$t_P = \frac{\pi}{\omega_n\sqrt{1-\delta^2}} = \frac{\pi}{\omega_d}$$
$$= \frac{\pi}{15.97} = 0.1967 \text{ sec}$$

Peak over shoot,
$$M_P = 100\, e^{-\frac{\pi\delta}{\sqrt{1-\delta^2}}}$$
$$= 5.23\%$$

Frequency of transient oscillations, $\omega_d = 15.97$ rad/sec

Steady state error $\dot{\theta}_R = \frac{2\pi}{60}$ rad/sec

$K_v = 16$

$$e_{ss} = \frac{2\pi}{60 \times 16} = 6.54 \times 10^{-3} \text{ rad}$$

(b) When a load torque of 1000 Nwm is applied at the load shaft, using super position theorem, the error is nothing but the response due to this load torque acting as a step input with

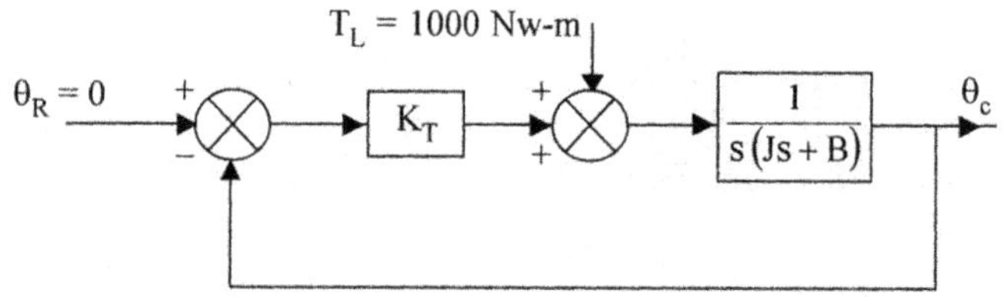

Fig. 3.21 Block diagram of the system with load torque applied

$\theta_R = 0$. The block diagram may be modified as shown in Fig. 3.21.

From Fig. 3.21, we have

$$\frac{\theta_C(s)}{T_L(s)} = \frac{\frac{1}{s(Js+B)}}{1+\frac{K_T}{s(Js+B)}} = \frac{1}{Js^2 + Bs + K_T}$$

$$\theta_C(s) = \frac{1000}{s(150s^2 + 4.5\times10^3 s + 7.2\times10^4)}$$

Using final value theorem,

$$\theta_{Css} = \underset{s \to 0}{Lt} \; s\,\theta_C(s) = \frac{1000}{7.2 \times 10^4}$$

$$= 0.01389 \text{ rad}$$

$$= 0.796^0$$

Example 3.2

The open loop transfer function of a unity feedback system is given by,

$$G(s) = \frac{K}{s(\tau s + 1)} \qquad K, \tau > 0$$

With a given value of K, the peak overshoot was found to be 80%. It is proposed to reduce the peak overshoot to 20% by decreasing the gain. Find the new value of K in terms of the old value.

Solution :

Let the gain be K_1 for a peak overshoot of 80%

$$\therefore \qquad e^{-\frac{\pi\delta_1}{\sqrt{1-\delta_1^{\,2}}}} = 0.8$$

$$\frac{\pi\delta_1}{\sqrt{1-\delta_1^2}} = ln\,\frac{1}{0.8} = 0.223$$

$$\pi^2\,\delta_1^{\,2} = 0.223^2\,(1 - \delta^2)$$

Solving for δ_1, we get

$$\delta_1 = 0.07$$

Let the new gain be K_2 for a peak overshoot of 20%

$$e^{-\frac{\pi\delta_2}{\sqrt{1-\delta_2^{\,2}}}} = 0.2$$

$$\frac{\pi\delta_2}{\sqrt{1-\delta_2^{\,2}}} = 1.61$$

Solving for δ_2,

$$\delta_2 = 0.456$$

But

$$\delta = \frac{1}{2\sqrt{K\tau}}$$

$$\frac{\delta_1}{\delta_2} = \frac{1}{2\sqrt{K_1\tau}}.2\sqrt{K_2\tau} = \sqrt{\frac{K_2}{K_1}}$$

$$\frac{\delta_1^2}{\delta_2^2} = \frac{K_2}{K_1}$$

$$\therefore \qquad K_2 = \frac{\delta_1^2}{\delta_2^2} . K_1 = 0.0236\ K_1$$

Example 3.3

Find the steadystate error for unit step, unit ramp and unit acceleration inputs for the following systems.

1. $\frac{10}{s(0.1s+1)(0.5s+1)}$ 2. $\frac{1000(s+1)}{(s+10)(s+50)}$ 3. $\frac{1000}{s^2(s+1)(s+20)}$

Solution :

1. $$G(s) = \frac{10}{s(0.1s+1)(0.5s+1)}$$

(a) Unit step input

$$K_p = \underset{s \to 0}{Lt}\ G(s) = \infty$$

$$e_{ss} = \frac{1}{1+K_p} = 0$$

(b) Unit ramp input

$$K_v = \underset{s \to 0}{Lt}\ s\ G(s)$$

$$= \underset{s \to 0}{Lt}\ \frac{10}{(0.1s+1)(0.5s+1)} = 10$$

$$e_{ss} = \frac{1}{K_V} = \frac{1}{10} = 0.1$$

(c) Unit acceleration input

$$K_a = \underset{s \to 0}{Lt}\ s^2\ G(s) = \underset{s \to 0}{Lt}\ \frac{10s}{(0.1s+1)(0.5s+1)} = 0$$

$$e_{ss} = \frac{1}{K_a} = \infty$$

2. $$G(s) = \frac{1000(s+1)}{(s+10)(s+50)}$$

The transfer function is given in pole zero form. Let us put this in time constant form.

$$G(s) = \frac{1000(s+1)}{500(0.1s+1)(0.02s+1)} = \frac{2(s+1)}{(0.1s+1)(0.02s+1)}$$

Since this is a type zero system we can directly obtain

$$K_p = 2, \qquad K_v = 0 \qquad K_a = 0$$

The steadystate errors are,

(a) Unit step input

$$e_{ss} = \frac{1}{1+K_p} = \frac{1}{1+2} = \frac{1}{3}$$

(b) Unit ramp input

$$e_{ss} = \frac{1}{K_v} = \frac{1}{0} = \infty$$

(c) Unit acceleration input

$$e_{ss} = \frac{1}{K_a} = \frac{1}{0} = \infty$$

3. $$G(s) = \frac{1000}{s^2(s+1)(s+20)}$$

Expressing G (s) in time constant form,

$$G(s) = \frac{1000}{20s^2(s+1)(0.05s+1)} = \frac{50}{s^2(s+1)(0.05s+1)}$$

The error constants for a type 2 system are

$$K_p = \infty \qquad K_v = \infty \qquad K_a = 50$$

The steadystate errors for,

(a) a unit step input

$$e_{ss} = \frac{1}{1+K_p} = \frac{1}{\infty} = 0$$

(b) a unit ramp input

$$e_{ss} = \frac{1}{K_v} = \frac{1}{\infty} = 0$$

(c) a unit acceleration input

$$e_{ss} = \frac{1}{K_a} = \frac{1}{50} = 0.02$$

Example 3.4

The open loop transfer function of a servo system is given by,

$$G(s) = \frac{10}{s(0.2s+1)}$$

Evaluate the error series for the input,

$$r(t) = 1 + 2t + \frac{3t^2}{2}$$

Solution :

$$G(s) = \frac{10}{s(0.2s+1)}$$

$$Y(s) = \frac{1}{1+G(s)} = \frac{1}{1+\dfrac{10}{s(0.2s+1)}} = \frac{s(0.2s+1)}{0.2s^2+s+10}$$

The generalised error coefficients are given by,

$$C_0 = \underset{s \to 0}{Lt}\; Y(s)$$

$$= \underset{s \to 0}{Lt}\; \frac{s(0.2s+1)}{0.2s^2+s+10} = 0$$

$$C_1 = \underset{s \to 0}{Lt}\; \frac{dY(s)}{ds}$$

$$C_1 = \underset{s \to 0}{Lt}\; \frac{(0.2s^2+s+10)(0.4s+1) - s(0.2s+1)(0.4s+1)}{(0.2s^2+s+10)^2}$$

$$= \underset{s \to 0}{Lt}\; \frac{10(0.4s+1)}{(0.2s^2+s+10)^2}$$

$$= \frac{10}{10^2} = 0.1$$

$$C_2 = \underset{s \to 0}{Lt}\; \frac{d^2Y(s)}{ds^2}$$

$$= \underset{s \to 0}{Lt}\; \frac{(0.2s^2+s+10)^2(4) - 10(0.4s+1)\left[2(0.2s^2+s+10)(0.4s+1)\right]}{(0.2s^2+s+10)^4}$$

$$= \frac{400-10(20)}{(10)^4} = 0.02$$

The input and its deriatives are,

$$r(t) = 1 + 2\,t + \frac{3t^2}{2}$$

$$r'(t) = 2 + \frac{6t}{2} = 2 + 3t$$

$$r''(t) = 3$$

$$r'''(t) = 0 = r^{iv}(t) = r^{v}(t)$$

∴ The error series is given by,

$$e_{ss}(t) = C_0\, r_{ss}(t) + C_1\, r_{ss}'(t) + \frac{C_2}{2!}\, r_{ss}''(t)$$

$$e_{ss}(t) = 0\left(1 + 2t + \frac{3t^2}{2}\right) + 0.1\,(2 + 3t) + \frac{0.02}{2}\,(3)$$

$$= 0.23 + 0.3t$$

3.7 Design Specifications of a Control System

A second order control system is required to satisfy three main specifications, namely, peak overshoot to a step input (M_P), settling time (t_s) and steadystate accuracy. Peak overshoot is indicative of damping (δ) in the system and for a given damping settling time indicates the undamped natural frequency of the system. The steadystate accuracy is specified by the steadystate error and error can be made to lie within given limits by choosing an appropriate error constant K_P, K_V or K_a depending on the type of the system. If any other specifications like rise time or delay time are also specified, they must be specified consistent with the other specifications. Most control systems are designed to be underdamped with a damping factor lying between 0.3 and 0.7. Let us examine the limitations in choosing the parameters of a type one, second order system to satisfy all the design specifications.

The expressions for δ, t_s and e_{ss} are given by,

$$\delta = \frac{1}{2\sqrt{K_v \tau}} \qquad \text{.....(3.69)}$$

$$t_s = \frac{4}{\delta\omega_n} \qquad \text{.....(3.70)}$$

$$e_{ss} = \frac{1}{K_v} \qquad \text{.....(3.71)}$$

In a second order system, the only variables are K_v and τ. Even if both of them are variable, we can satisfy only two out of the three specifications namely, δ, t_s and e_{ss}. Generally, we are given a system for which a suitable controller has to be designed. This means that the system time constant is fixed and the only variable available is the system gain K_v. By using a proportional controller, the gain can be adjusted to suit the requirement of the steadystate accuracy. If K_v is adjusted for an allowable limit on steadystate error, this value of K_v is usually large enough to make the system damping considerably less, as given by eqn. (3.69). Thus the transient behaviours of the system is not satisfactory. Hence suitable compensation schemes must be designed so that the dynamic response improves. Some control schemes used in industry are discussed in the next section.

3.7.1 Proportional Derivative Error Control (PD control)

A general block diagram of a system with a controller is given in Fig. 3.22.

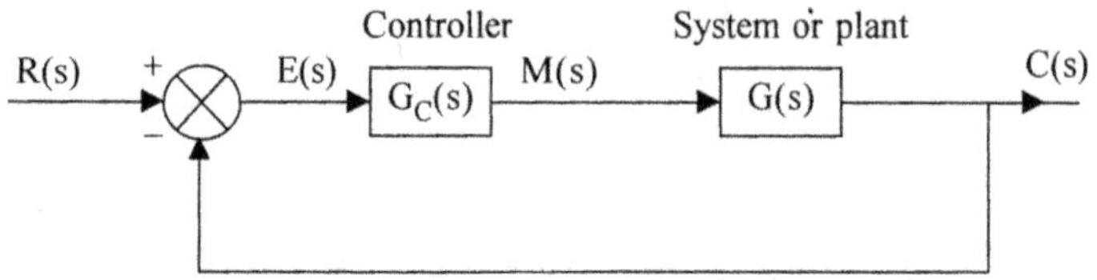

Fig. 3.22 General block diagram of a system with controller and unity feed back

For a second order, Type 1 system,

$$G(s) = \frac{K'_v}{s(\tau s+1)}$$

By choosing different configurations for the controller transfer function G_C (s) we get different control schemes. The input to the controller is termed as error signal or most appropriately, actuating signal. The output of the controller is called as the manipulating variable, m (t) and is the signal given as input to the system or plant. Thus, we have,

$$m(t) = K_P\left(e(t) + K_D\frac{de(t)}{dt}\right)$$

and $$M(s) = K_P\,(1 + K_D\,s)\,E\,(s)$$

The open loop transfer function with PD controller is given by,

$$G_0\,(s) = G_C\,(s)\,G\,(s)$$

$$= \frac{K_P(1+K_D s)K'_v}{s(\tau s+1)}$$

The closed loop transfer function of the system is given by,

$$T(s) = \frac{\frac{K_P.K'_v}{\tau}(K_D s+1)}{s^2 + s\left(\frac{1+K_P K_D K'_v}{\tau}\right) + \frac{K_P K'_v}{\tau}}$$

If we define $$K_v = K_P\,K'_v,$$

$$T(s) = \frac{\frac{K_v}{\tau}(K_D\,s+1)}{s^2 + s\left(\frac{1+K_v K_D}{\tau}\right) + \frac{K_v}{\tau}}$$

The damping and natural frequency of the system are given by,

$$\delta' = \frac{1+K_v K_D}{2\sqrt{K_v \tau}} = \delta + \frac{K_D}{2}\sqrt{\frac{K_v}{\tau}} \qquad .....(3.72)$$

$$\omega_n' = \sqrt{\frac{K_v}{\tau}} = \omega_n \qquad .....(3.73)$$

By a suitable choice of the proportional controller gain K_P (Amplifier gain), the steadystate error requirements can be met. As seen earlier, such a choice of K_P usually results in a low value of damping and hence this can be increased to a suitable value by of a proper choice of K_D, the gain of the derivative term in the controller, as given by the eqn. (3.72). It can be observed from eqn. (3.73) that the natural frequency is not altered for a given choice of K_P. Hence the settling time is automatically reduced since ω_n is fixed and δ for the compensated system has increased.

It may also be observed that, adding a derivative term in the controller introduces a zero in the forward path transfer function and we have seen that the effect of this is to increase the damping in the system.

3.7.2 Proportional Integral Controller (PI Control)

If the amplifier in the forward path is redesigned to include an integrator so that the output of the controller is given by,

$$m(t) = K_P\left(e(t) + K_1\int_0^t e(t)dt\right) \qquad .....(3.74)$$

or

$$M(s) = K_P\left(1 + \frac{K_1}{s}\right) E(s)$$

We have,

$$G_0(s) = G_C(s)\,G(s)$$

$$= \frac{K_P(s + K_1)K_v^1}{s^2(\tau s + 1)} = \frac{K_v(T_1 s + 1)}{s^2(\tau s + 1)} \qquad .....(3.75)$$

Where $T_1 = \dfrac{1}{K_1}$

and $K_v = \dfrac{K_P K'_v}{K_1}$

and

$$T(s) = \frac{K_v(T_1 s + 1)}{\tau s^3 + s^2 + K_v T_1 s + K_v} \qquad .....(3.76)$$

From eqn. (3.75) we observe that the type of the system is changed from Type 1 to type 2 and hence the steadystate error for a unit velocity input is reduced to zero. Hence an integral controller is usually preferred wherever the steadystate accuracy is important. But the dynamics of the system can not be easily obtained, as the system order is increased from two to three, because of the introduction of integral control.

Moreover, the stability of the sytem (as will be discussed in chapter 4) may be affected adversely if the system order is increased. Since the system is of third order, it is usually designed to have two complex poles nearer to the imaginary axis and one real pole, as for away from origin as desirable. The response due to the complex poles dominate the overall response and hence the damping factor of these poles will have to be properly chosen to get a satisfactory transient response.

3.7.3 Proportional, Integral and Derivative Controller (PID Control)

An integral control eliminates steadystate error due to a velocity input, but its effect on dynamic response is difficult to predict as the system order increases to three. We have seen in sector 3.7.1 that a derivative term in the forward path improves the damping in the system. Hence a suitable combination of integral and derivative controls results in a proportional, integral and derivate control, usually called PID control. The transfer function of the PID controller is given by,

$$G_C(s) = K_P\left(1 + K_D s + \frac{K_I}{s}\right)$$

The overall forward path transfer function is given by,

$$G_o(s) = \frac{K_P K_v'\left(1 + K_D s + \frac{K_I}{s}\right)}{s(\tau s + 1)}$$

and the overall transfer function is given by,

$$T(s) = \frac{K_P K_v' \; (K_D s^2 + s + K_I)}{\tau s^3 + s^2(1 + K_p K_D) + sK_P K_v' \; s + K_P K_I}$$

Proper choice of K_P, K_D and K_I results in satisfactory transient and steadystate responses. The process of choosing proper K_P, K_D, at K_I for a given system is known as *tuning of a PID controller*.

3.7.4 Derivative Output Control

So far, we have discussed controllers in the forward path for which the input is the error. Some times control is provided by taking a signal proportional to the rate at which the output is changing and feeding back to the amplifier in the forward path. A typical block diagram fo such a system, employing rate feedback, as it is often known, is given in Fig. 3.23.

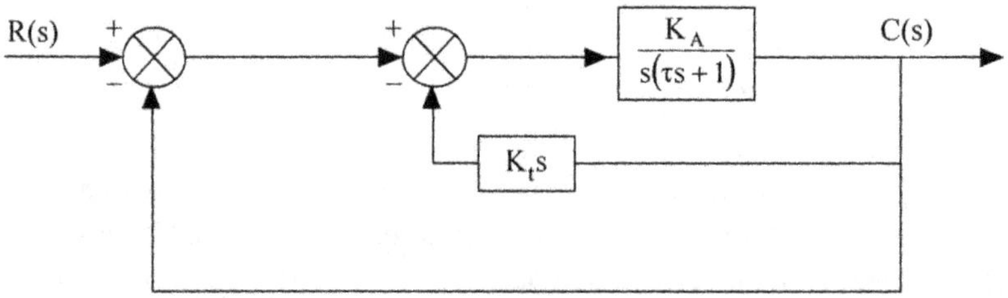

Fig. 3.23 (a) Block diagram of a system employing rate feedback

The inner loop provides the desired rate feedback as its output is proportional to $\frac{dc(t)}{dt}$. Simplifying the inner loop, we have,

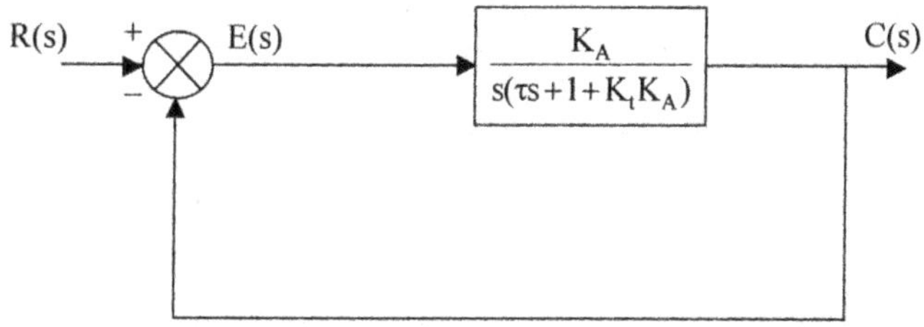

Fig. 3.23 (b) Simplified block diagram of Fig. 3.23 (a)

The forward path transfer function is given by,

$$G(s) = \frac{\frac{K_A}{1+K_t K_A}}{s\left(\frac{\tau}{1+K_t K_A}s+1\right)} = \frac{K'_v}{s(\tau' s+1)} \qquad .....(3.77)$$

Where, $$K'_v = \frac{K_A}{1+K_t K_A} \qquad .....(3.78)$$

$$\tau' = \frac{\tau}{1+K_t K_A} \qquad .....(3.79)$$

Thus the new damping factor is given by,

$$\delta' = \frac{1}{2\sqrt{K_v'\tau'}} = \frac{1}{2\sqrt{\frac{K_A}{1+K_t K_A}\cdot\frac{\tau}{1+K_t K_A}}} \qquad .....(3.80)$$

$$= \frac{1+K_t K_A}{2\sqrt{K_A \tau}}$$

$$= (1 + K_t\, K_A)\,\delta \qquad .....(3.81)$$

$$\omega'_n = \sqrt{\frac{K'_V}{\tau'}} = \sqrt{\frac{K_A}{\tau}} = \omega_n \qquad .....(3.82)$$

The product, $K'_v\,\delta'$ is given by,

$$K'_v\,\delta' = \frac{K_A}{2\sqrt{K_A \tau}} = \frac{1}{2}\sqrt{\frac{K_A}{\tau}}$$

or $$K_A = 4\,(K'_v\,\delta')^2\,\tau \qquad .....(3.83)$$

If the values of K'_v and δ' are specified, the amplifier gain (K_A) can be adjusted to get a suitable value using eqn. (3.82). For this vlaue of K_A, the rate feedback constant K_t is given by, using eqn. 3.78,

$$K_t = \left(\frac{1}{K'_v} - \frac{1}{K_A}\right) \quad .....(3.84)$$

If rate feedback is not present,

$$K_v = K_A \quad .....(3.85)$$

and
$$\delta = \frac{1}{2\sqrt{K_A \tau}}$$

With rate feedback, if same velocity error constant is specified, comparing eqn. (3.78) with eqn. (3.85), we see that the amplifier gain, K_A has to be more. Thus ω_n' given by eqn. (3.82) will be more. Hence the derivative output compensation increases both damping factor and natural frequency, thereby reducing the settilng time.

3.5 Example

Consider the position control system shown in Fig. 3.24 (a). Draw the block diagram of the system. The particulars of the system are the following.

Total Moment of Inertia referred to motor shaft, $J = 4 \times 10^{-3}$ kgm^2.

Total friction coefficient referred to motor shaft, $f = 2 \times 10^{-3}$ Nwm|rad|sec

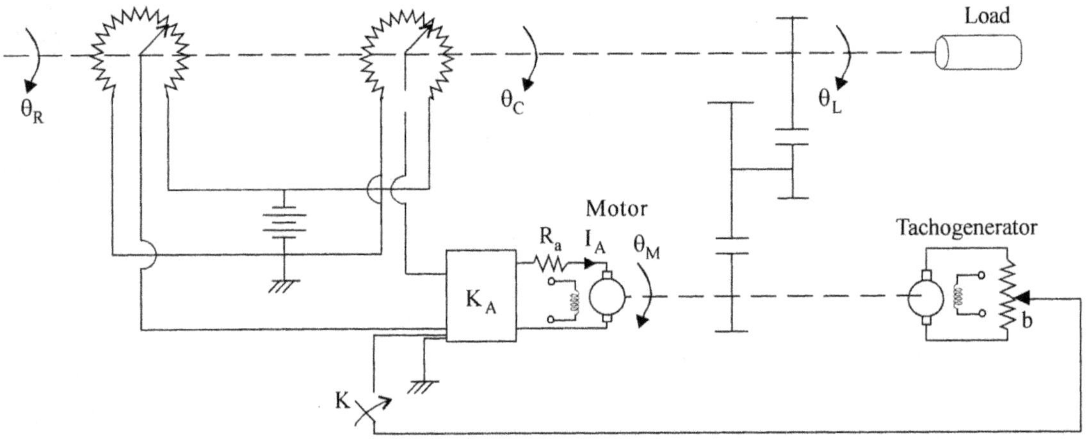

Fig. 3.24 (a) Schematic of a position control system.

Motor to load Gear ratio, $n = \frac{\theta_L}{\theta_M} = \frac{1}{50}$

Load to potentiometer gear ratio, $\frac{\dot{\theta}_L}{\dot{\theta}_C} = 1$

Motor torque constant,	$K_T = 2$ Nw-m | amp
Tachogenerator constant,	$K_t = 0.2$V | rad | sec
Sensitivity of error detector,	$K_P = 0.5$V | rad.
Amplifier gain,	K_A Amps/V. (variable)

(a) With switch K open, obtain the vlaue of K_A for a steadystate error of 0.02 for unit ramp input. Calculate the values of damping factor, natural frequency, peak overshoot and settling time.

(b) With switch K open, the amplifier is modified to include a derivative term, so that the armature current i_a (t) is given by

$$i_a(t) = K_A\left(e(t) + K_D \frac{de(t)}{dt}\right)$$

Find the vlaues of K_A and K_D to give a steadystate error within 0.02 for a unit ramp input and damping factor of 0.6. Find the natural frequency and settling time in this case.

(c) With switch K closed and with proportional control only, find the portion of tachogenerator voltage to be fedback, b, to get a peak overshort of 20%. Steadystate error should be less than 0.02 for a unit ramp input. Find the settling time and natural frequency.

Solution : The block diagram of the system is given in Fig. 3.24 (b).

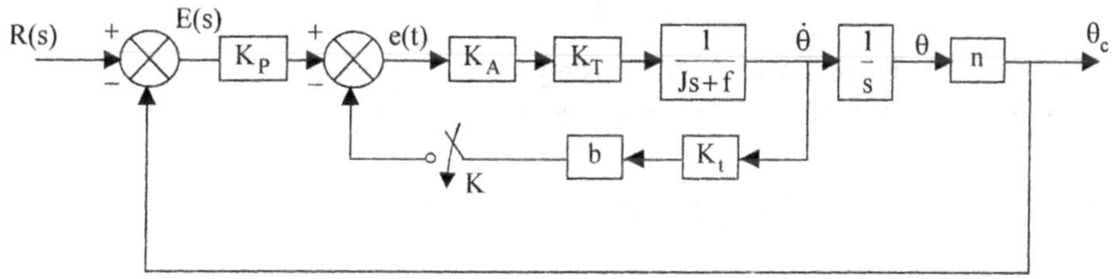

Fig. 3.24 (b) Block diagram of the position control system of Fig. 3.24 (a).

With switch K open,

$$G(s) = K_P K_A K_T \frac{n}{s(Js+f)}$$

$$= \frac{0.5 \times 2K_A}{50} \cdot \frac{1}{s(4\times10^{-3}s + 2\times10^{-3})}$$

$$= \frac{K_A}{50\times2\times10^{-3}s(2s+1)} = \frac{10K_A}{s(2s+1)}$$

$$= \frac{K_v}{s(2s+1)}$$

Now $e_{ss} = 0.02$

$$e_{ss} = \frac{1}{K_v} = \frac{1}{10K_A}$$

$$K_A = \frac{1}{10 \times 0.02} = 5$$

$$K_v = 10\,K_A = 50$$

damping factor, $\delta = \dfrac{1}{2\sqrt{K_v\tau}} = \dfrac{1}{2\sqrt{50 \times 2}} = \dfrac{1}{20}\ 0.05$

Natural frequency, $\omega_n = \sqrt{\dfrac{K_v}{\tau}} = \sqrt{\dfrac{50}{2}}$

$= 5.0$ rad/sec.

Peak overshoot, $M_P = e^{-\frac{\pi\delta}{\sqrt{1-\delta^2}}} = 85.45\%$

Settling time, $t_s = \dfrac{4}{\delta\omega_n} = \dfrac{4}{0.05 \times 5} \times 5 = 16$ sec

Thus, it is seen that, using proportional control only (Adjusting the amplifier gain K_A) the steadystate error is satisfied, but the damping is poor, resulting in highly oscillatory system. The settling time is also very high.

(b) With the amplifier modified to include a derivative term,

$$i_a(t) = K_A\left[e(t) + K_D\frac{de(t)}{dt}\right]$$

The forward path transfer function becomes,

$$G(s) = \frac{K_P K_A(1 + K_D s)K_T n}{s(Js + f)} = \frac{0.5 \times K_A(1 + K_D s)2}{2 \times 10^{-3} \times 50s(2s + 1)} = \frac{10K_A(1 + K_D s)}{s(2s + 1)}$$

To satisfy steadystate error requirements, K_A is again chosen as 5. The damping factor is given by,

$$\delta = \frac{1 + K_v K_D}{2\sqrt{K_v\tau}}$$

$$0.6 = \frac{1 + 50K_D}{2\sqrt{50 \times 2}}$$

From which we get,

$$K_D = 0.22$$

$$\omega_n = \sqrt{\frac{K_v}{\tau}} = \sqrt{\frac{50}{2}} = 5 \text{ rad/sec}$$

$$t_s = \frac{4}{0.6 \times 5} = 1.333 \text{ sec}$$

The derivative control increases the damping and also reduces the setlling time. The natural frequency is unaltered.

(c) With switch K closed and with only proportional control, we have,

$$G(s) = \frac{K_P.n}{s}\left[\frac{K_A K_T}{Js + f + K_A K_T K_t b}\right]$$

$$= \frac{0.5}{50s}\left[\frac{2K_A}{4 \times 10^{-3} s + 2 \times 10^{-3} + 2 \times 0.2 \times K_A b}\right]$$

The closed loop transfer function is given by,

$$T(s) = \frac{K_A}{0.2s^2 + (0.1 + 20K_A b)s + K_A}$$

The steadystate error, $e_{ss} = 0.02$

$$\therefore \qquad K_v = \frac{1}{0.02} = 50$$

The peak over shoot,

$$M_P = e^{-\frac{\pi\delta}{\sqrt{1-\delta^2}}} = 0.2$$

$$\therefore \qquad \delta = 0.456$$

From the expression for G (s), we have,

$$K_v = \underset{s \to 0}{Lt}\ s\,G(s)$$

$$K_v = \frac{K_A}{50(2 \times 10^{-3} + 0.4K_A b)} = \frac{K_A}{0.1 + 20K_A b}$$

From the expression for T (s), we have

$$\omega_n = \sqrt{\frac{K_A}{0.2}}$$

$$\delta = \frac{0.1 + 20K_A b}{0.2} \cdot \frac{1}{2\sqrt{\frac{K_A}{0.2}}}$$

Taking the product of K_v and δ, we have

$$K_v\delta = \frac{K_A}{0.1+20K_A b}\times\frac{0.1+20K_A b}{2\sqrt{0.2K_A}}$$

$$= \frac{K_A}{2\sqrt{0.2K_A}}$$

But $K_v = 50$ and $\delta = 0.456$

$$(50 \times 0.456)^2 = \frac{K_A^2}{4\times 0.2K_A}$$

$\therefore$ $K_A = 4 \times 0.2\ (50 \times 0.456)^2 = 415.872$

We notice that the value of K_A is much larger, compared to K_A in part (a). Substituting the value of K_A in the expression for K_v, we have,

$$50 = \frac{415.872}{0.1+20\times 415.872\times b}$$

b can be calculated as,

$$b = 0.001$$

The natural frequency $\omega_n = \sqrt{\frac{K_A}{0.2}} = \sqrt{\frac{415.872}{0.2}} = 45.6$ rad/sec

Comparing ω_n in part (a), we see that the natural frequency has increased. Thus the setlling time is redued to a value give by,

$$t_s = \frac{4}{\delta\omega_n} = \frac{4}{0.456\times 45.6} = 0.1924 \text{ sec.}$$

This problem clearly illustrates the effects of P, P D and derivative output controls.

Example 3.6

Consider the control system shown in Fig. 3.25.

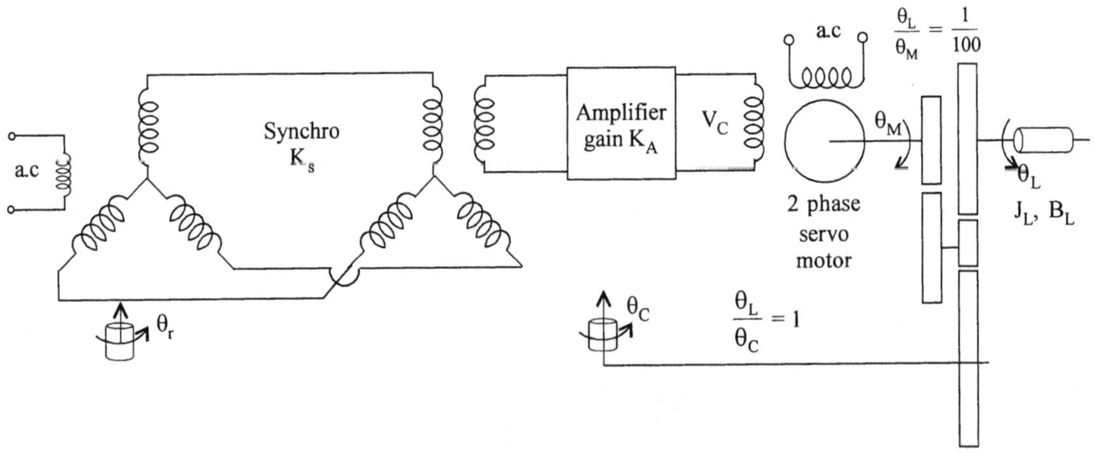

Fig. 3.25 Schematic of a control system for Ex. 3.6

The sensitivity of synchro error detector $K_S = 1$ V | deg. The transfer function of the two phase servo motor is given by,

$$\frac{\theta_M(s)}{V_C(s)} = \frac{10}{s(1+0.1s)}$$

(a) It is required that the load be driven at a constant speed of 25 rpm at steady state. What should be the gain of the amplifier, K_A, so that the error between output and input position does not exceed 2 deg under steadystate. For this gain what are the values of damping factor, natural frequency and settling time.

(b) To improve the transient behavior of the system, the amplifier is modified to include a derivative term, so that the output of the amplifier is given by,

$$v_C(t) = K_A\, e(t) + K_A\, T_d\, \dot{e}(t)$$

Determine the value of T_D so that the damping ratio is improved to 0.5. What is the settling time in this case.

Solution : The block diagram of the system is given in Fig. 3.26

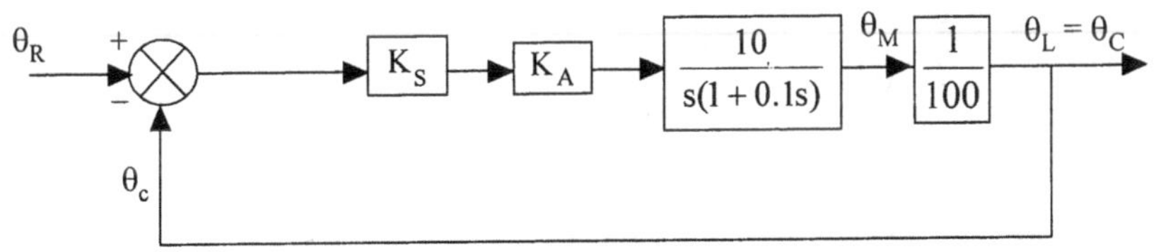

Fig. 3.26 The block diagram of system shown in Fig. 3.25

The forward path transfer function of the system is,

$$G(s) = \frac{K_S K_A}{10s\,(1+0.1s)}$$

$$K_S = 1\ \text{V/deg} = \frac{180}{\pi}\ \text{V/rad}$$

$$\therefore \quad G(s) = \frac{180 \times K_A}{10\pi s\,(1+0.1s)}$$

Steady state speed $= 25\text{ rpm} = \dfrac{25 \times 2\pi}{60}\ \text{rad | sec.} = \dfrac{5\pi}{6}\ \text{rad | sec}$

Steady state error, $e_{SS} = 2\text{ deg} = \dfrac{2}{180} \times \pi = \dfrac{\pi}{90}\ \text{rad}$

For the given system,

$$K_v = \underset{s \to 0}{Lt} \; s\,G(s)$$

$$= \underset{s \to 0}{Lt} \; \frac{s \times 180 \times K_A}{\pi s(1+0.1s)10}$$

$$= \frac{18K_A}{\pi}$$

$$e_{ss} = \frac{5\pi}{6K_v} = \frac{5\pi \times \pi}{6 \times 18K_A}$$

But $$e_{ss} = \frac{\pi}{90}$$

$$\therefore \quad \frac{5\pi^2}{108K_A} = \frac{\pi}{90}$$

From which, we get, $K_A = 13.1$

$$G_c\,(s) = \frac{75}{s(1+0.1s)}$$

$$T(s) = \frac{\theta_C(s)}{\theta_R(s)} = \frac{75}{0.1s^2 + s + 75}$$

$$= \frac{750}{s^2 + 10s + 750}$$

$$\omega_n = \sqrt{750} = 27.39$$

$$\delta = \frac{1}{2\sqrt{K_v \tau}} = \frac{1}{2\sqrt{75 \times 0.1}} = 0.1826$$

$$t_s = \frac{4}{\delta\omega_n} = \frac{4}{0.1826 \times 27.39} = 0.8 \text{ sec}$$

(b) When the amplifier is modified as,

$$v_C\,(t) = K_A e\,(t) + K_A\,T_D\,\dot{e}\,(t)$$

The open loop transfer function becomes,

$$G(s) = \frac{K_S K_A (1 + T_D s)}{10s(0.1s+1)}$$

$$T(s) = \frac{\theta_C(s)}{\theta_R(s)} = \frac{K_S K_A (1+T_D s)}{s^2 + (10 + K_S K_A T_D)s + K_S K_A}$$

$$= \frac{180 \times 13.1(1+T_D s)}{s^2 + \left(10 + \frac{180}{\pi} \times 13.1 T_D\right)s + \frac{180}{\pi} \times 13.1} \times \frac{1}{\pi}$$

$$= \frac{750(1+T_D s)}{s^2 + (10 + 750 T_D)s + 750}$$

$$\omega_n = \sqrt{750} = 27.39$$

$$2\,\delta\,\omega_n = 10 + 750\,T_D$$

$$T_D = \frac{2\delta\omega_n - 10}{750} = \frac{2 \times 0.5 \times 27.39 - 10}{750} = 0.0232$$

$$t_s = \frac{4}{\delta\omega_n} = \frac{4}{0.5 \times 27.39} = 0.292 \text{ see}$$

Thus, we see that, by including a derivative term in the amplifier, the transient perofrmance is improved. It may also be noted that K_v is not changed and hence the steady state error remains the same.

Problems

3.1 Draw the schematic of the position control system described below.

Two potentiometers are used as error detector with θ_R driving the reference shaft and the load shaft driving the secind potentiometer shaft. The error signal is amplified and drives a d c. Servomotor armature. Field current of the motor is kept constant. The motor drives the load through a gear.

Draw the block diagram of the system and obtain the closed loop transfer function. Find the natural frequency, damping factor, peak time, peak overshoot and settling time for a unit step input, when the amplifier gain K_A = 1500. The parameters of the system are as follows:

Potentiometer sensitivity	K_P = 1V/rad
Resistance of the armature	$R_a = 2\Omega$
Equivalent Moment of Inertia at motor shaft	$J = 5 \times 10^{-3}$ kg–m^2
Equivalent friction at the motor shaft	$B = 1 \times 10^{-3}$ NW/rad/sec
Motor torque constant	K_T = 1.5 Nω m/A
Gear ratio	$n = \frac{1}{10}$
Motor back e.m.f constant	K_b = 1.5 V/rad/sec

3.2 A position control system is shown in Fig. P 3.2

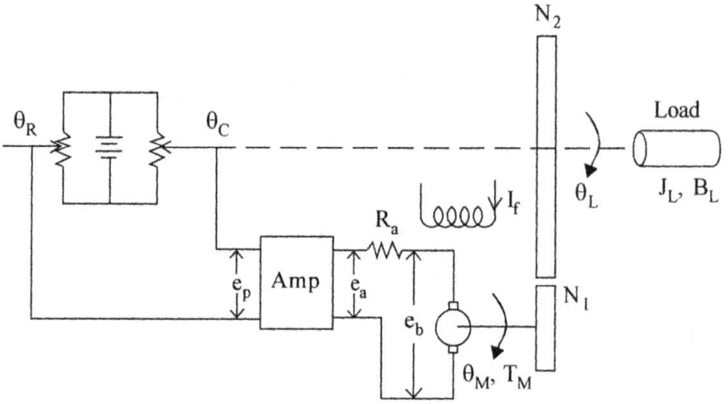

Fig. P 3.2

Sensitivity of the potentiometer error detection $K_P = \frac{24}{\pi}$ V/rad

Amplifier gain	K_A V/V
Armature resistance	$R_a = 0.2\ \Omega$
Motor back e.m.f constant	$K_b = 5.5 \times 10^{-2}$ V/rad/sec
Motor torque constant	$K_T = 6 \times 10^{-5}$ N-m/A
Moment of Inertia of motor referred to motor shaft	$J_m = 10^{-5}$ Kg m^2
Moment of Inertia of load referred to the output shaft	$J_L = 4.4 \times 10^{-3}$ Kgm^2
Friction coefficient of the load referred to the output shaft	$B_v = 4 \times 10^{-2}$ Nm/rad/sec

Gear ratio $\frac{N_1}{N_2} = \frac{1}{10}$

(i) If the amplifier gain is 10V/V obtain the transfer function of the system $\frac{C(s)}{R(s)} = \frac{\theta_L(s)}{\theta_R(s)}$.

Find the peak overshoot, peak time, and settling time of the system for a unit step input.

(ii) What values of K_A will improve the damping factor to 0.707

(iii) What value of K_A will give the frequency of oscillations of 9.23 rad/sec to a step input.

3.3 The open loop transfer function of a unity feed back control system is given by,

$$G(s) = \frac{K}{s(Ts+1)}$$

If the maximum response is obtained at t = 4 sec and the maximum value is 1.26, find the values if K and T.

3.4 A unity feedback system has the plant transfer function

$$G_1(s) = \frac{C(s)}{M(s)} = \frac{10}{s(s+2)}$$

A Proportional derivative control is employed to control the dynamics of the system. The controller characteristics are given by,

$$m(t) = e(t) + K_D \frac{de(t)}{dt}$$

where e(t) is the error.

Determine

(i) The damping factor and undamped natural frequency when $K_D = 0$

(ii) The value of K_D so that the damping factor is increased to 0.6.

3.5 Consider the system shown in Fig. P 3.5.

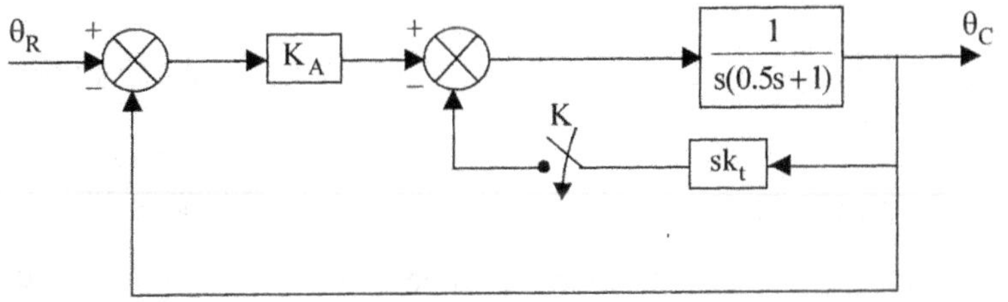

Fig. P 3.5

(i) With switch K open, determine the damping factor and the natural frequency of the system. If a unit ramp input is applied to the system, find the steady state output. Take $K_A = 5$

(ii) The damping factor is to be increases to 0.7 by including a derivative output compensation. Find the value of k_t to achieve this. Find the value of undamped natural frequency and the steady state error due to a unit ramp input.

(iii) It is possible to maintain the same steady state error for a unit ramp input as in part (i) by choosing proper values of K_A and k_t. Find these values.

3.6 In the system shown in Fig. P 3.6 find the values of *K* and *a* so that the peak overshoot for a step input is 25% and peak time is 2 sec.

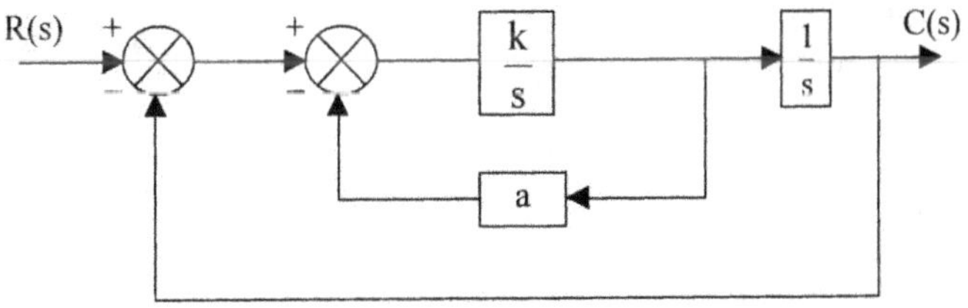

Fig. P 3.6

3.7 Determine the values of K and a such that the damping factor is 0.6 and a settling time of 1.67sec.

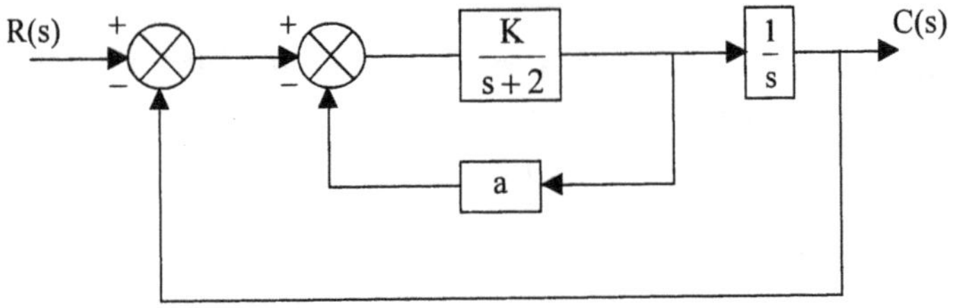

Fig. P 3.7

Also find the step response of the system.

3.8 Find the steadystate errors for unit step, unit velocity and unit acceleration inputs for the following systems.

(i) $\dfrac{15}{s(s+1)(s+5)}$

(ii) $\dfrac{10(0.1s+1)}{(0.02s+1)(0.2s+1)}$

(iii) $\dfrac{100}{s^2(0.1s+1)(.01s+1)}$

(iv) $\dfrac{(s+2)(s+5)}{s(0.2s+1)(0.6s+1)}$

3.9 In the system shown in Fig P 3.7 find the value of K and a such that the damping factor of the system is 0.6 and the steady state error due to a unit ramp input is 0.25.

3.10 For the unity feedback system with,

$$G(s) = \frac{10}{s+10}$$

Find the error series for the input,

$$v(t) = 1 + 2t + \frac{3t^2}{2}$$

3.11 Find the steadystate error as a function of time for the unity feedback system,

$$G(s) = \frac{100}{s(1+0.1s)}$$

for the following inputs.

(a) $r(t) = \dfrac{t^2}{2}\, u(t)$

(b) $r(t) = 1 + 2t + \dfrac{t^2}{2}$.

4 Stability of Systems

4.1. Introduction

It is usually not desirable that a small change in the input, initial condition, or parameters of the system produces a very large change in the response of the system. If the response increases indefinitely with time the system is said to be unstable. Stability is an important property that a system is required to possess. It is not only essential to design a system to obtain the desired response, it is also necessary to design a system which is stable. Suppose we have a control system which is designed to drive a load at a desired speed. For some reason if there is a small change in the load on the motor, the motor speed should not increase indefinitely.

In general there are two concepts of stability. The first one is known as a Bounded Input Bounded Output stability (BIBO). According to this concept, if a bounded input is given to the system, the output should be bounded. The second one is defined with no input to its system. If an initial condition is applied to the system, the system should return to its equilibrium condition, which is usually the origin. For linear time invariant systems the two concepts are equivalent. For non linear systems, the determination of stability is more complicated. Even if a system is found to be stable for a certain bounded input it may not be stable for another bounded input. If a nonlinear system is found to be stable for certain initial condition, it may not be stable when a bounded input is given. Usually nonlinear system stability is studied for autonomous systems i.e., systems without input. In contrast to this, for a linear, time invariant system, there are simple criteria for determining the stability. In this chapter, we will deal with an algebraic criterion for determining the stability of linear, time invariant systems.

4.2 Stability Preliminaries

The two concepts of stability are equivalent for linear time invariant systems. To see this, let us consider the output of a linear, time invariant system, given by,

$$c(t) = \int_0^t h(\tau)\ r(t-\tau)d\tau \qquad .....(4.1)$$

where h(t) is the impulse response of the system,

and r(t) is the input to the system.

We know that $\qquad h(t) = \mathcal{L}^{-1}[T(s)] = \mathcal{L}^{-1}\left[\dfrac{C(s)}{R(s)}\right]$

and
$$T(s) = \frac{b_0 s^m + b_1 s^{m-1} + ... + b_m}{a_0 s^n + a_1 s^{n-1} + ... a_n} \qquad .....(4.2)$$

If the input is bounded, i.e., $|r(t)| \leq R_1$, we have from eqn. (4.1).

$$|c(t)| = \left|\int_0^t h(\tau)\ r(t-\tau)\ d\tau\right|$$

$$\leq \int_0^t |h(\tau)|\ |r(t-\tau)|\ d\tau$$

$$\leq R_1\ 1\int_0^t |h(\tau)|d\tau \qquad .....(4.3)$$

For a stable system, bounded input should produce bounded output. Hence from eqn. (4.3).

$$|c(t)| = R_1 \int_0^t |h(\tau)|d\tau \leq R_2 \qquad .....(4.4)$$

Thus the system is BIBO stable, if $\int_0^t |h(\tau)|d\tau$ is finite or h(t) is absolutely integrable. If $|h(t)|$ is plotted with respect to time, this condition means that the area bounded by this curve and time axis must be finite between the limits $t = 0$ and $t = \infty$. Thus the stability of the system can be ascertained from the impulse response or the natural response of the system which is independent of the input.

If $\int_0^t |h(\tau)|d\tau$ is bounded the response for any initial condition will also be bounded and the system will return to its equilibrium condition.

Since the nature of impulse response h(t) depends on the location of poles of T(s), the transfer function T(s) is given by eqn. (4.2) and can be written as,

$$T(s) = \frac{N(s)}{D(s)} \qquad .....(4.5)$$

D(s) = 0 is known as the characteristic equation of the system and the poles of T(s) are the roots of D(s) = 0. The roots of D(s) may be classified as :

(i) Real roots at $s = \sigma$

The transfer function T(s) contains terms like $\frac{K}{s-\sigma}$ which give rise to terms like $Ke^{\sigma t}$ in the impulse response. If σ is positive, the exponential term increases indefinitely with time and h(t) is unbounded. If σ is negative, the response h(t) decreases with time and $h(t) \rightarrow 0$ as $t \rightarrow \infty$. The area under the curve of h(t) is bounded.

If real roots have multiplicity m, T(s) contains terms of the form,

$$\frac{K_{1i}}{(s-\sigma)^i} \quad i = 1, 2, ...m$$

which on inversion, gives rise to terms of the form $A_i\, t^{i-1}\, e^{\sigma t}$, i = 1, 2, ...m

If σ is positive, these terms make h(t) unbounded. On the other hand if σ is negative, the response dies down as $t \rightarrow \infty$ and $\int_0^t |h(t)|\, dt$ is bounded.

(ii) Complex conjugate roots at $s = \sigma + j\omega$. In this case T(s) contains terms like $\frac{As+B}{(s-\sigma)^2+\omega^2}$ and therefore, h(t) contains terms like $Ke^{\sigma t}\,\text{Sin}\,(\omega t + \phi)$

If σ is positive i.e., if the real part of the root is positive, the ainplitude of sinusoidal oscillations increases indefinitely and h(t) will be unbounded. If σ is negative h(t) will have oscillations whose amplitudes tend to zero as $t \rightarrow \infty$ and thus, h(t) will be bounded.

If the complex conjugate roots are repeated with multiplicity m at $s = \sigma + j\omega$,

T(s) will contain terms like $\frac{A_i s + B_i}{\left[(s-\sigma)^2+\omega^2\right]^i}$ $\quad$ i = 1, 2, ...m.

which yields, on inversion,

$$C_1\, t^{i-1} K e^{\sigma t}\,\text{Sin}\,(\omega t + \phi_i), \qquad i = 1, 2, \; .....m$$

In this case also, if the real part of the complex root σ is positive, the response is unbounded and if it is negative, the response is bounded.

(iii) Roots at origin.

T(s) will have a term $\frac{K}{s}$, which gives rise to a constant term Ku(t), as the response. In this case, although h(t) = Ku(t) is bounded, $\int_0^\infty |h(t)|\, dt$ is not bounded. If repeated roots at origin are present in T(s), it will have terms like,

$$\frac{K_{1i}}{s^i} \qquad i = 1, 2, \; ..... \; m$$

which give rise to terms in the impulse response h(t), of the form,

$$C_i\, t^{i-1} \qquad \text{for } i = 1, 2, \; .... \; m$$

which clearly results in unbounded h(t).

(iv) Purely imaginary roots at $s = j\omega$

Roots on imaginary axis will contribute terms like $\frac{As + B}{s^2 + \omega^2}$ in T(s) and their inversion yields terms of the form,

$$C \sin(\omega t + \phi)$$

which have constant amplitude. In this case also even though h(t) is bounded $\int_0^\infty |h(t)|\, dt$ is not bounded.

If the imaginary roots are repeated, they yield terms like $\frac{A_i s + B_i}{(s_i^2 + \omega^2)^i}$ for i = 1, 2, ...m in h (t).

These terms clearly make $h(t)$ unbounded as $t \to \infty$.

To summarise, if the roots are negative, or have negative real parts if they are complex, the impulse response $h(t)$ is bounded and $\int_0^\infty |h(t)|\, dt$ is also bounded. The system will be BIBO stable for these roots occuring in the left half of s-plane even if they are repeated. If the roots are positive, or they have positive real parts if complex, the response is unbounded and hence the system is unstable. The response is bounded if the roots are purely imaginary but simple. The system is classified as marginally or limitedly stable, since the response will be oscillatory with constant amplitude. The location of the roots and corresponding responses are given in Fig. 4.1. The response is indicated near the corresponding root.

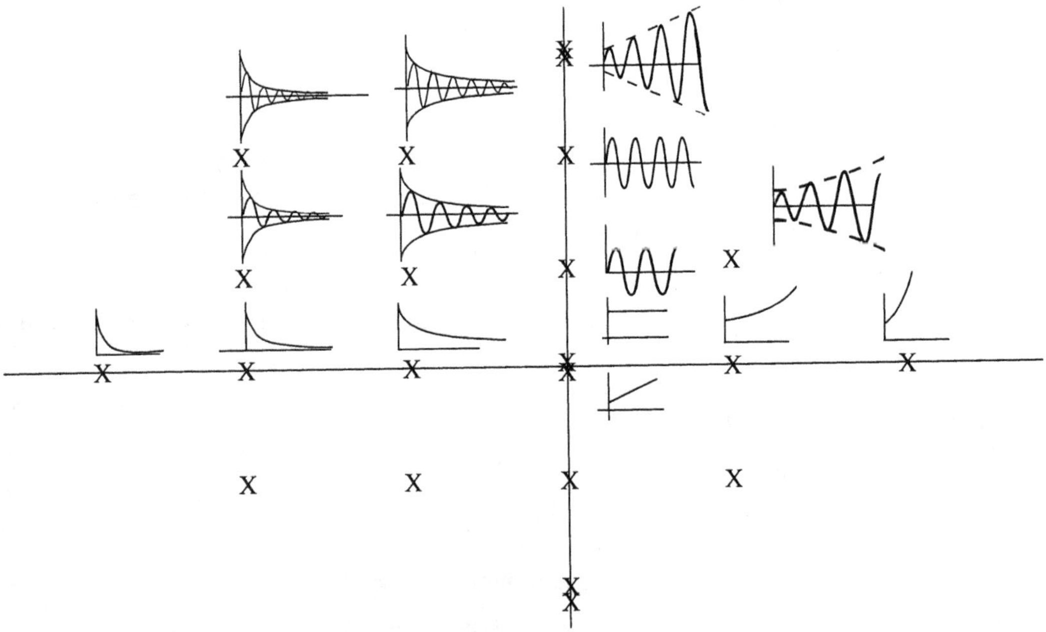

Fig. 4.1 Location of roots of characteristic equation and corresponding responses.

Based on the above observations regarding location of roots of the characteristic equation and stability of the system, the *s*-plane can be divided into two regions. The left side of imaginary axis is the stable region and if roots of characteristic equation occur in this half of *s*-plane, the system is stable. The roots may be simple or may occur with any multiplicity. On the other hand, even if simple roots occur in the right half of *s*-plane, the system is unstable. If simple roots occur on the $j\omega$-axis, which is the boundary for these two regions, the system is said to be marginally or limitedly stable. If roots on $j\omega$-axis are repeated, the system is unstable. The stable and unstable regions of *s*-plane are shown in Fig. 4.2.

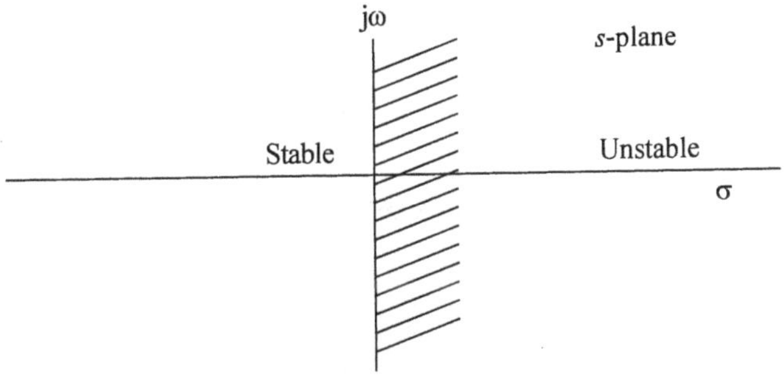

Fig. 4.2 Demarcation of stable and unstable regions of s-plane.

Hence, determination of stability of a linear time invariant system boils down to determining whether any roots of the characteristic equation or poles of transfer function lie in the right half of *s*-plane. One obvious method of determining stability of a system is to find all the roots of $D(s) = a_0 s^n + a_1 s^{n-1} + ... + a_n$.

For a polynomial of degree $n > 2$, it is difficult to find the roots analytically. Numerical methods for root determination for higher order polynomials is quite cumbersome. Hence, algebraic criteria are developed to find out if any roots lie in the right half of *s*-plane. The actual location of the roots is unimportant.

4.3 Necessary Conditions for Stability

In section 4.2 we have seen that the system will be stable if the roots of the characteristic equation lie in the left half of *s*-plane. The factors of the characteristic polynomial D(s) can have terms like

$$(s + \sigma_i),\ (s + \sigma_k)^2 + \omega_k^2$$

where σ_i are positive and real. Thus

$$D(s) = a_0 s^n + a_1 s^{n-1} + ... + a_n \quad .....(4.4)$$

$$= a_0 \Pi (s + \sigma_i) \Pi \{(s + \sigma_k)^2 + \omega_k^2\} \quad .....(4.5)$$

Since σ_i and σ_k are all positive and real the product in eqn. (4.5) results in all positive and real coefficients in the polynomial of *s*. Thus if the system is stable all the coefficients, a_i, must be positive and real.

Further, since there are no negative terms involved in the product of eqn. (4.5), no cancellations can occur and hence no coefficient can be zero. Thus none of the powers of *s* in between the highest and lowest powers of *s* must be missing. But, if a root is present at the origin a_n is zero.

There is one exception to this condition, namely, all the odd powers of *s*, or all the even powers of *s*, may be missing. This special case occurs if the characteristic equation contains roots only on the $j\omega$-axis. If there is no root at the origin, the characteristic polynomial is given by

$$D(s) = \Pi\,(s^2 + \omega_i^2)$$

which will yield a polynomial with even powers of *s* only. On the other hands, if it has a root at the origin, in addition to roots on $j\omega$-axis only, D(s) is given by,

$$D(s) = s\Pi\,(s^2 + \omega_i^2)$$

and D(s) will have only odd powers of *s*. Since simple roots on $j\omega$-axis are permitted, we may conclude that if any power of *s* is missing in D(s), all even powers or all odd powers of *s* may be missing for stable systems. In conclusion, it can be stated that the necessary conditions for stability of a system are :

1. The characteristic polynomial D(s) = 0 must have all coefficients real and positive.
2. None of the coefficients of the polynomial should be zero except the constant a_n.
3. None of the powers of *s* between the highest and lowest powers of *s* should be missing. However all odd powers of *s*, or all even powers of *s* may be missing.

It is to be emphasised that these are only necessary conditions but not sufficient. All stable systems must satisfy these conditions but systems satisfying all these conditions need not be stable.

For example,

$$D(s) = s^3 + s^2 + 3s + 24 = 0$$

$$= (s - 1 + j2\sqrt{2})\,(s - 1 - j\,2\sqrt{2})\,(s + 3) = 0$$

Eventhough, D(s) has all its coefficients positive and no power of *s* is missing, the complex roots have real parts which are positive. Hence the system with the above characteristic equation is not stable.

The necessary conditions help us to eliminate polynomials with negative coefficients or missing powers of *s* by visual inspection only. If the characteristic equation satisfies all the necessary conditions, it is a possible candidate for examining further, for stability. A. Hurwitz and E.J. Routh have independently established the conditions for stability of a system without actually finding out the roots. The criteria is known as Routh-Hurwitz criterion for stability.

4.4 Routh - Hurwitz Stability Criterion

This criteria determines how many roots of the characteristic equation lie in the right half of s-plane. This test also determines all the roots on the $j\omega$-axis, so that their multiplicity can be found out. The polynomial coefficients are arranged in an array called Routh array. Let the characteristic equation be given by,

$$D(s) = a_0\, s^n + a_1\, s^{n-1} + a_2\, s^{n-2} + \ldots\ldots a_1 s + a_0 = 0$$

The Routh array is constructed as follows. Each row is identified in the descending order of powers of s. The first row contains coefficients of alternate powers of s starting with the highest power s^n. The second row contains coefficients of alternate powers of s starting with the second highest power s^{n-1}. The other rows are constructed in a systematic way as indicated in the procedure given below.

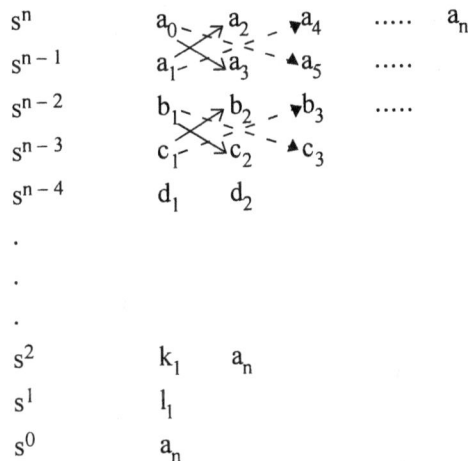

$$
\begin{array}{llllll}
s^n & a_0 & a_2 & a_4 & \ldots\ldots & a_n \\
s^{n-1} & a_1 & a_3 & a_5 & \ldots\ldots & \\
s^{n-2} & b_1 & b_2 & b_3 & \ldots\ldots & \\
s^{n-3} & c_1 & c_2 & c_3 & & \\
s^{n-4} & d_1 & d_2 & & & \\
\cdot & & & & & \\
\cdot & & & & & \\
\cdot & & & & & \\
s^2 & k_1 & a_n & & & \\
s^1 & l_1 & & & & \\
s^0 & a_n & & & &
\end{array}
$$

The coefficients in s^{n-2} row are obtained as follows :

$$b_1 = \frac{a_1 a_2 - a_0 a_3}{a_1}$$

$$b_2 = \frac{a_1 a_4 - a_0 a_5}{a_1}$$

$$b_3 = \frac{a_1 a_6 - a_0 a_7}{a_1} \quad \text{and so on.}$$

The coefficients of s^{n-3} row are obtained in a similar way, considering the coefficients of the previous two rows as follows :

$$c_1 = \frac{b_1 a_3 - a_1 b_2}{b_1}$$

$$c_2 = \frac{b_1 a_5 - a_1 b_3}{b_1} \quad \text{and so on.}$$

Similarly, the coefficients of any particular row can be obtained by considering the previous two rows and forming the products as before. If any element in a row is missing, it is regarded as zero. There will be only 2 entries in the s^2 row and one element each in s^1 and s^0 rows. The Routh array is constructed until the s^0 row is computed. The Routh Hurwitz criterion is stated as follows :

For a system to be stable, it is necessary and sufficient that each entry in the first column of Routh array, constructed from the characteristic equation, be positive. If any entry in the first column is negative, the system is unstable and the number of roots of the characteristic equation lying in the right half of s-plane is given by, the number of changes in the sign of entries in the first column of Routh array.

Observe that the Routh Hurwitz criterion tells us whether the system is stable or not. It does not give any indication of the exact location of the roots.

Example 4.1

Consider the characteristic equation,

$$D(s) = s^4 + 2s^3 + 8s^2 + 4s + 3 = 0$$

Comment on its stability.

Solution :

Let us construct the Routh array.

s^4	1	8	3	
s^3	2	4	0	($\because$ Since there is no entry in the 3rd column it is taken as zero)
s^2	$\frac{2\times8-1\times4}{2} = 6$	$\frac{2\times3-1\times0}{2} = 3$		
s^1	$\frac{6\times4-2\times3}{6} = 3$			
s^0	3			(The entry in this row will always be a_n)

Consider the entries in the first column, 1, 2, 6, 3, 3. All are positive and therefore, the system is stable.

Example 4.2

Examine the characteristic equation

$$D(s) = s^4 + 2s^3 + s^2 + 4s + 2 = 0$$

for stability.

Solution :

Constructing the Routh array, we have

s^4	1	1	2
s^3	2	4	
s^2	$\frac{2-4}{2} = -1$	2	
s^1	8		
s^0	2		

The first column entries are 1, 2, –1, 8 and 2. One of the coefficients is negative and hence the system is unstable. Also, there are two sign changes, 2 to – 1 and –1 to + 8. Hence there are two roots of the characteristic equation in the right half of *s*-plane.

4.4 Special Cases

There are two special cases which occur in the construction of Routh array. Whenever they occur it is not possible to complete the table, in a routine way. Let us see how these special cases can be tackled in completing the table.

(i) First case :

Sometimes, the first entry in a particular row may turn out to be a zero. Since the calculation of next row entries involves division by this zero, the process of constructing the Routh array stops there. To overcome this difficulty, the following procedures may be adapted.

(a) ***First method :***

Replace the zero in that row by $\in$. Proceed with the construction of the table. Consider the entries of the first column of the array by letting $\in \to 0$ from the positive side.

Example 4.3

Consider the characteristic equation

$$D(s) = s^5 + s^4 + 3s^3 + 3s^2 + 6s + 4$$

Comment on the stability.

Solution :

Construct the Routh array.

s^5	1	3	6
s^4	1	3	4
s^3	$\frac{1\times3-1\times3}{1}=0$	$\frac{6-4}{1}=2$	

Now there is a zero in the s^3 row in the first column. Replace this by $\in$ and proceed.

s^5	1	3	6
s^4	1	3	4
s^3	$\in$	2	
s^2	$\frac{3\in-2}{\in}$	4	
s^1	$\frac{\frac{6\in-4}{\in}-4\in}{\frac{3\in-2}{\in}}=\frac{6\in-4-4\in^2}{3\in-2}$		
s^0	4		

Letting $\varepsilon \to 0$ for the entries in the first column we have,

$$s^3 \qquad \lim_{\epsilon \to 0} \epsilon = 0$$

$$s^2 \qquad \lim_{\epsilon \to 0} \frac{3\epsilon - 2}{\epsilon} = -\infty$$

$$s^1 \qquad \lim_{\epsilon \to 0} \frac{6\epsilon - 4 - 4\epsilon^2}{3\epsilon - 2} = 2$$

$$s^0 \qquad 4$$

Hence the elements of 1st column of Routh array are 1, 1, 0, $-\infty$, 2 and 4. There are two sign changes and hence there are two roots in the right half of s-plane. The system therefore is unstable.

(b) ***Second method :***

(i) Replace s by $\frac{1}{z}$ in D(s) and apply the Routh criterion for the resulting equation in z.

Considering Ex. 4.3 again and replacing s by $\frac{1}{z}$, we have,

$$\frac{1}{z^5} + \frac{1}{z^4} + \frac{3}{z^3} + \frac{3}{z^2} + \frac{6}{z} + 4 = 0$$

Rearranging the terms,

$$4z^5 + 6z^4 + 3z^3 + 3z^2 + z + 1 = 0$$

Routh array.

z^5	4	3	1
z^4	6	3	1
z^3	$\frac{18-12}{6} = 1$	$\frac{1}{3}$	
z^2	$\frac{3-2}{1} = 1$	1	
z^1	$\frac{\frac{1}{3} - 1}{1} = -2$		
z^0	1		

There are two sign changes in the first column of the Routh array for this modified characteristic equation. Hence there are two roots in the right half of z-plane. This implies that there are also two roots in the right half of s-plane for the original system.

(ii) Second case :

For some systems, a particular row may contain all zero entries. This happens when the characteristic equation contains roots which are symmetrically located about real and imaginary axes, namely :

(i) one or more pairs of roots on the $j\omega$-axis

(ii) one or more pairs of real roots with opposite signs, and

(iii) one or more pairs of complex roots with their mirror images about the $j\omega$-axis, together forming quadrates in the s-plane.

The polynomial whose coefficients are the entries in the row above the row of zeros is called an auxiliary polynomial and the roots of this polynomial give these symmetrically located roots.

Since a row contains all zero entries, the Routh table cannot be constructed further. To overcome this, the row of zeros is replaced with the coefficients of the differential of the auxiliary equation. This auxiliary equation is always a polynomial with even powers of s, since the roots of this polynomial occur always in pairs. The procedure is illustrated with an example.

Example 4.4

Consider

$$D(s) = s^6 + s^5 + 6s^4 + 5s^3 + 10s^2 + 5s + 5$$

Obtain the number of roots in the RHS of s-plane.

Solution :

Routh Table

s^6	1	6	10	5
s^5	1	5	5	
s^4	$\frac{6-5}{1} = 1$	5	5	
x^3	0	0		

The Routh table construction procedure breaks down here, since the s^3 row has all zeros. The auxiliary polynomial coefficients are given by the s^4 row. Therefore the auxiliary polynomial is,

$$A(s) = s^4 + 5s^2 + 5$$

$$\frac{dA(s)}{ds} = 4s^3 + 10s$$

Replacing the s^3 row in the Routh table with the coefficients of $\frac{dA(s)}{ds}$, we have

s^6	1	6	10	5
s^5	1	5	5	
s^4	1	5	5	
s^3	4	10		
s^2	$\frac{20-10}{4} = 2.5$	5		
s^1	$\frac{25-20}{2.5} = 2$			
s^0	5			

Examining the first column of this table we see that there are no sign changes. But since there are symmetrically located roots we have to find these roots for concluding about the stability. Factoring the auxiliary polynomial, we have the root as,

$$\pm j\,1.1756 \text{ and } \pm j\,1.902$$

These are roots on the $j\omega$-axis and are simple. Therefore the system has no roots in the right half plane and the roots on $j\omega$-axis are simple. Hence the system is limitedly stable.

Example 4.5

Comment on the stability of the system with the following characteristic equation.

$$D(s) = s^6 + s^5 + 7s^4 + 6s^3 + 31s^2 + 25s + 25$$

Solution :

Since D(s) satisfies all necessary conditions let us construct the Routh table.

s^6	1	7	31	25
s^5	1	6	25	
s^4	1	6	25	
s^3	0	0		

Since the Routh table terminates prematurely and there is a row of zeros, let us construct the auxiliary polynomial.

$$A(s) = s^4 + 6s^2 + 25$$

$$\frac{dA(s)}{ds} = 4s^3 + 12s$$

Continuing the Routh table

s^6	1	7	31	25
s^5	1	6	25	
s^4	1	6	25	
s^3	4	12		
s^2	3	25		
s^1	$\frac{-64}{3}$			
s^0	25			

There are two sign changes in the first column of the Routh table and hence the system is unstable. Let us find the symmetrical roots present, by factoring the auxiliary polynomial $A(s)$.

$$\begin{aligned} A(s) &= s^4 + 6s^2 + 25 \\ &= (s^2 + 5)^2 + 6s^2 - 10s^2 \\ &= (s^2 + 5)^2 - 4s^2 \\ &= (s^2 + 2s + 5)(s^2 - 2s + 5) \\ &= (s + 1 + j2)(s + 1 - j2)(s - 1 + j2)(s - 1 - j2) \end{aligned}$$

Hence we have symmetrically placed roots out of which two are in the right half of s-plane. The location of the roots are shown in Fig. 4.3.

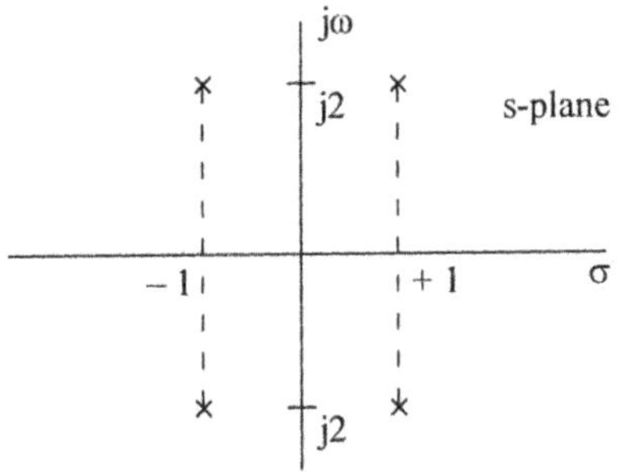

Fig. 4.3 Roots forming a quadrate

If after completing the Routh table, there are no sign changes, the auxiliary polynomial will have roots only on the $j\omega$-axis. They can be found out by factoring the auxiliary polynomial. The procedure to determine stability of a system and find the number of roots in the right half of s-plane, using Routh Hurwitz criterion, is summarised as follows.

Step 1 : The characteristic equation is examined for necessary conditions viz.

(i) All the coefficients must be positive

(ii) No coefficient of D(s) is zero between the highest and lowest powers of s. If these conditions are satisfied, go to step 2.

Step 2 : Construct the Routh table. Examine the first column of the table. If there are 'm' sign changes in the column, there are m roots in the right half of s-plane and hence the system is unstable.

If in any row, the first entry is a zero and at least one other element is not a zero, it is not possible to proceed further in completing the table. Once a zero or a negative entry is present in the first column, it can be concluded that the system is unstable. If it is required to find the number of roots in the RHS of s-plane go to step 3. If all the entries in a row are zero, go to step 4.

Step 3 : Replace the zero by a small positive number ϵ. Complete the table. Let $\epsilon \to 0$ in all the elements of first column entries involving ϵ. Find the signs of these elements. The number of sign changes gives the number of roots in the RHS of s-plane.

Step 4 : If all the elements of a row are zeros, it indicates that there are roots which are symmetrically situated in the s-plane. The location of these roots can be determined by considering the row above the row of zeros. An auxiliary polynomial $A(s)$ is constructed with the coefficient as the entries in this row. It is invariably an even polynomial in s. The roots of this polynomial gives the symmetrically situated roots. Replace the row of zeros with the coefficients of the differential of the polynomial $A(s)$. Complete the table now.

Step 5 : If there are m sign changes in the first column of the table, there are m roots in the RHS of s-plane and the system is unstable. There is no need to find the symmetrically situated roots. If there are no sign changes go to step 6.

Step 6 : Factorise the polynomial $A(s)$. Since there are no roots in the RHS of s-plane. $A(s)$ will contain roots on the $j\omega$-axis only. Find these roots. If these roots are simple, the system is limitedly stable. If any of these roots is repeated, the system is unstable. This concludes the procedure.

In many situations, the open loop system is known and even if the open loop system is stable, once the feedback loop is closed there is a chance for losing the stability. The forward loop invariably includes an amplifier whose gain can be controlled. It is therefore desirable to know the range of the values of this gain K for maintaining the system in stable condition. An example is considered to explain the procedure.

Example 4.6

Find the range of values of K for the closed loop system in Fig. 4.4 to remain stable. Find the frequency of sustained oscillations under limiting conditions.

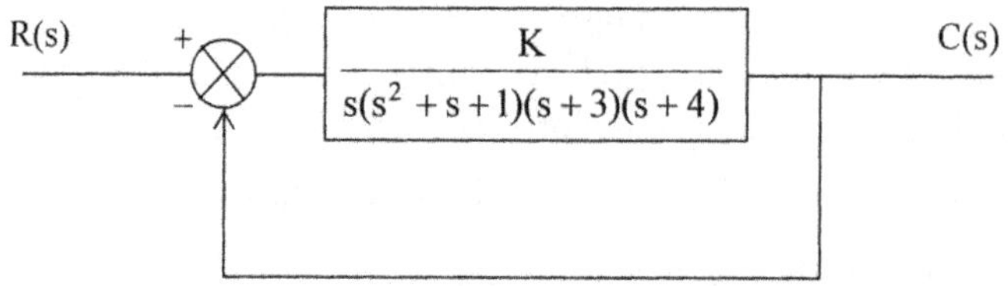

Fig. 4.4 A closed loop system for Ex. 4.6.

Solution :

The closed loop transfer function is given by

$$T(s) = \frac{C(s)}{R(s)} = \frac{G(s)}{1+G(s)H(s)}$$

Here $$G(s) = \frac{K}{s(s^2+s+1)(s+3)(s+4)}; \qquad H(s) = 1$$

$$T(s) = \frac{N(s)}{D(s)} = \frac{K}{s(s^2+s+1)(s+3)(s+4)+K}$$

Therefore, the characteristic equation is given by

$$D(s) = s\,(s^2+s+1)\,(s+3)\,(s+4)+K$$
$$= s^5+8s^4+20s^3+19s^2+12s+K$$

Routh table :

s^5	1	20	12
s^4	8	19	K
s^3	17.625	$\frac{96-K}{8}$	
s^2	$\frac{238.875+K}{17.625}$	K	
s^1	$\dfrac{\frac{238.875+K}{17.625}\times\frac{96-K}{8}-17.625K}{\frac{238.875+K}{17.625}}$		
s^0	K		

Examining the first column the system will be stable if,

(i) $$\frac{238.875+K}{17.625} > 0$$

(ii) $$\frac{(238.875+K)(96-K)}{141} - 17.625\,K > 0$$

(iii) $K > 0$

If condition (iii) is satisfied condition (i) is automatically satisfied. Let us find out for what values of *K*, condition (ii) will be satisfied.

$$-K^2 - 142.875\,K + 22932 - 2485.125\,K > 0$$

or
$$K^2 + 2628\,K - 22932 < 0$$

$$(K - 8.697)\,(K + 2636.7) < 0$$

Since K > 0, the above condition is satisfied for K < 8.697. Thus the range of values of *K* for stability is 0 < K < 8.697.

If K > 8.697 the entry in s^1 row will be negative and hence these will be two roots in the RHS of *s*-plane. The system will be unstable. If K < 8.697 the s^1 row entry will be positive and hence the system will be stable. If K = 8.697, this entry will be zero and since the only entry in this row is a zero, it indicates roots on the imaginary axis. s^2 row will give these roots.

$$A(s) = \left(\frac{238.875 + 8.697}{17.625}\right) s^2 + 8.697$$

$$= 14.047s^2 + 8.697$$

The roots of this polynomial are $\pm$ j 0.7869. If K = 8.697, the closed loop system will have a pair of roots at $\pm$ j 0.7869 and the response will exhibit subtained oscillations with a frequency of 0.7869 rad/sec.

Let us consider another example.

Example 4.7

Examine the stability of the characteristic polynomial for K ranging from 0 to ∞.

$$D(s) = s^4 + 20\,Ks^3 + 5s^2 + 10s + 15$$

Solution :

Routh Table

s^4	1	5	15
s^3	20 K	10	
s^2	$\frac{100K - 10}{20K}$	15	
s^1	$\frac{\frac{100K - 10}{2K} - 300K}{\frac{100K - 10}{20K}}$		
s^0	15		

The system will be stable if,

(i) $K > 0$

(ii) $\dfrac{100K - 10}{20K} > 0$ or $K > 0.1$

(iii) $\left(\dfrac{100K - 10 - 600K^2}{10K - 1}\right) > 0$

or $$600\ K^2 - 100\ K + 10 < 0$$

$$60\ K^2 - 10\ K + 1 < 0$$

This is not satisfied for any real value of *K* as can be seen by factoring the expression.

Hence for no value of *K*, the system is stable.

Let us consider an open loop system which is unstable and find the values of the amplifier gain to obtain closed loop stability.

Example 4.8

Comment on the stability of the closed loop system as the gain *K* is changed in Fig. 4.5.

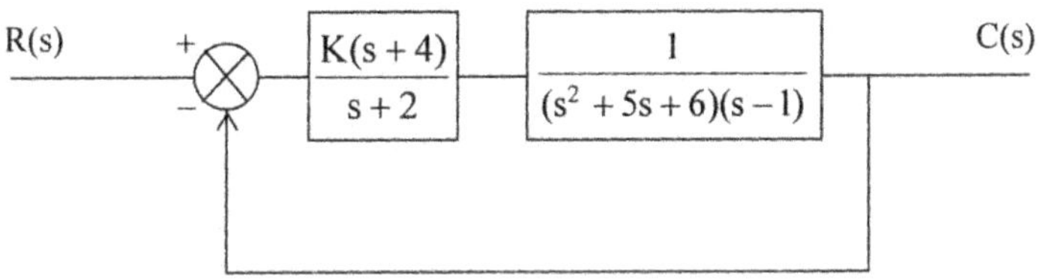

Fig. 4.5 Closed loop system for Ex. 4.8

Solution :

The open loop system is clearly unstable as it has a pole $s = 1$ in the RHS of *s*-plane. Let us examine whether it can be stabilised under closed loop operation by a suitable choice of the amplifier gain *K*.

The characteristic equation is given by,

$$D(s) = 1 + G(s)\ H(s) = 0$$

$$1 + \frac{K(s+4)}{(s+2)(s^2+5s+8)(s-1)} = 0$$

$$s^4 + 76s^3 + 11s^2 + s(K - 2) + 4\ (K - 4) = 0$$

Routh table :

s^4	1	11	4(K – 4)
s^3	6	(K – 2)	
s^2	$\frac{68-K}{6}$	4(K – 4)	
s^1	$\dfrac{\left(\frac{68K-10}{6}\right)(K-2)-24(K-4)}{\frac{68-K}{6}}$		
s^0	4(K – 4)		

For stability

(i) $68 - K > 0$ i.e., $K < 68$

(ii) $(68 - K)(K - 2) - 144(K - 4) > 0$

(iii) $4(K - 4) > 0$ or $K > 4$

From condition (ii)

$$-K^2 + 70K - 136 - 144K + 576 > 0$$

or

$$K^2 + 74K - 440 < 0$$

Finding the roots of

$$K^2 + 74K - 440 = 0$$

$$K = \frac{-74 \pm \sqrt{74^2 + 1760}}{2}$$

$$= 5.532, -79.532$$

$$\therefore (K - 5.532)(K + 79.532) < 0$$

This inequality will be satisfied for values of *K*.

$$-79.532 < K < 5.532$$

Combining the conditions (i), (ii) and (iii) we have, for stability

$$4 < K < 5.532$$

It can be observed that for $K = 4$, the constant term in the characteristic equation is zero and therefore there will be a pole at the origin. For $K = 5.532$, the elements in the s^1 row will all be zeros and hence there will be an imaginary pair of roots. To find these roots factorise the auxiliary polynomial.

$$A(s) = \left(\frac{68-K}{6}\right) s^2 + 4(K - 4)$$

for

$$K = 5.532$$

$$A(s) = 10.411 s^2 + 6.128$$

This has a pair of roots at $s = \pm j\ 0.7672$. Therefore the frequency of oscillation of response for K = 5.532 is 0.7672 rad/sec.

Systems which are stable, for a range of values of a parameter of the system, are said to be conditionally stable systems.

4.6 Relative Stability

By applying Routh-Hurwitz criteria, we can determine whether a system is stable or not. This is known as absolute stability. If there are no roots in the RHS of *s*-plane, we conclude that the system is stable. But imagine a system to have some roots with a very small negative real part i.e., the roots are very near to the $j\omega$-axis. $j\omega$-axis is the border for stable and unstable regions. Due to some environment conditions, let us assume that the parameter values have changed, which cause the roots nearest to the $j\omega$-axis to cross the threshold and enter unstable region. The system obviously becomes unstable. Hence it is necessary that the system's dominant poles (poles nearest to the $j\omega$-axis) are reasonably away from the $j\omega$-axis. Systems with more negative real parts of the dominant poles are relatively more stable than systems with less negative real parts of the dominants poles. Hence the distance from the $j\omega$-axis of the dominant poles is a measure of the relative stability of a system. We are more concerned with the relative stability of a system rather than its absolute stability. Routh-Hurwitz criterion gives us absolute stability of the system only. However we can modify the procedure to determine how far the dominant pole is from the $j\omega$-axis by shifting the $j\omega$-axis to the left by a small amount. Apply Routh Hurwitz criterion to find out if any roots lie on the right side of this shifted axis. By trial and error the negative real part of the dominant pole can be located.

Let the dominant poles of the system be given by $-\delta\omega_n \pm j\omega_n\sqrt{1-\delta^2}$. The response due to these poles is of the form $Ae^{-\delta\omega_n t}\,\text{Sin}\,(\omega_n\sqrt{1-\delta^2}\ t + \phi)$ the time constant of the exponentially decaying term is given by,

$$T = \frac{1}{\delta\omega_n}$$

Thus we see that the time constant is inversely proportional to the real part of the pole and hence the setting time t_s, which is approximately equal to 4T is also inversely proportional to the real part of the dominant pole. For reasonably small value of the settling time, the real part of the dominant pole must be at a suitable distance away from the $j\omega$-axis. For the largest time constant of the system (real part of the dominant pole) to be greater than τ seconds, the real part of the dominant root must be $\sigma = \frac{1}{\tau}$ units to the left of $j\omega$-axis. Thus shifting the $j\omega$-axis by σ to the left and applying Routh-Hurwitz criterion, we can ascertain whether the largest time constant is indeed greater than τ. To do this let $s = z - \sigma$ in the characteristic equation and we get an equation in *z*. Applying Routh criterion to *z*-plane polynomial, we can find if any roots are lying in the right half of *z*-plane. If there are no right half of *z*-plane roots, it means that the system has no roots to the right of $s = -\sigma$ line and therefore the time constant of the dominant pole is greater than τ sec.

Example 4.9

Determine whether the largest time constant of the system with characteristic equation given below is greater than 1 sec.

$$D(s) = s^4 + 6s^3 + 14s^2 + 16s + 8$$

Solution :

For the largest time constant, the dominant root must be away from $j\omega$-axis by an amount equal to

$$\sigma = \frac{1}{1.0} = 1$$

Shifting the $j\omega$-axis by 1.0 to the left by taking,

$$s = z - 1$$

and substituting in D(s), we have

$$D_1(z) = (z-1)^4 + 6(z-1)^3 + 14(z-1) + 8$$
$$= z^4 + 2z^3 + 2z^2 + 2z + 1$$

Routh table :

z^4	1	2	1
z^3	2	2	
z^2	1	1	
z^1	0	0	

Since we have a row of zeros, the polynomial in z has roots on the imaginary axis. It means that the dominant roots in the s-plane is having a real part equal to – 1. Therefore the largest time constant of the given system is equal to 1 sec.

Example 4.10

For what value of K in the following characteristic equation the dominant root will have a real part equal to – 1.

$$D(s) = s^3 + Ks^2 + 2Ks + 48$$

Solution :

Shift the imaginary axis by $s = -1$ ie., put $s = z - 1$

The characteristic equation becomes,

$$D_1(z) = (z-1)^3 + K(z-1)^2 + 2K(z-1) + 48$$
$$= z^3 + (K-3)z^2 + 3z + (47-K)$$

Routh table

z^3	1	3
z^2	$K-3$	$47-K$
z^1	$\frac{3(K-3)-47+K}{K-3}$	
z^0	$47-K$	

For the negative part of the dominant root of D(s) to be –1, the dominant roots of $D_1(z)$ must lie on the imaginary axis of *z*-plane. It means that in the Routh table z^1 entry must be zero.

$\therefore \quad 4K - 56 = 0$

(i) $K = \dfrac{56}{4} = 14$

For no other root of $D_1(z)$ to lie on RH side of imaginary axis, the following conditions must also be satisfied.

(ii) $K > 3$

(iii) $K < 47$

From (i), (ii) and (iii) we see that

$K = 14$ satisfies the conditions (ii) and (iii) and therefore if $K = 14$, the dominant root of D(s) will have a real part equal to – 1.

From the examples worked out, it is clear that Routh Hurwitz criterion is suitable to determine absolute stability only. It is rather cumbersome to assess relative stability using Routh Hurwitz criterion. Frequency response methods are used to determine relative stability of systems. These methods will be discussed in chapter 7.

Problems

4.1 Find the conditions on the coefficients of the following polynomials so that all the roots are in the left half of the *s*-plane.

(a) $s^2 + a_1 s + a_2$

(b) $s^3 + a_1 s^2 + a_2 s + a_3$

(c) $s^4 + a_1 s^3 + a_2 s^2 + a_3 s + a_4$

4.2 For the open loop systems given below, find the poles of the unity feedback closed loop system.

(a) $G(s) = \dfrac{10}{(s+1)(s+4)}$

(b) $G(s) = \dfrac{1}{(s+1)(s+3)}$

(c) $G(s) = \dfrac{9(3s+2)}{s^2(s+10)}$

4.3 For the open loop systems given below find the roots of the characteristic equation of *a* unity feedback system.

(a) $G(s) = \frac{4}{s^4}$

(b) $G(s) = \frac{10}{s^2(s+4)}$

(c) $G(s) = \frac{10s}{(s-1)(s+2)}$

4.4 Find the number of roots in the right half of *s*-plane. Comment on the stability.

(a) $s^4 + 8s^3 + 18s^2 + 16s + 4$

(b) $3s^4 + 10s^3 + 5s^2 + 5s + 1$

(c) $s^5 + s^4 + 2s^3 + 2s^2 + 4s + 6$

(d) $s^5 + 2s^4 + 6s^3 + 12s^2 + 8s + 16 = 0$

(e) $2s^6 + 2s^5 + 3s^4 + 3s^3 + 2s^2 + s + 1 = 0$

4.5 The characteristic equations of certain control systems are give below. Determine the range of values of k for the system to be stable.

(a) $s^3 + 4ks^2 + (k+3)s + 4 = 0$

(b) $s^4 + 20ks^3 + 5s^2 + (10+k)\, s + 15 = 0$

(c) $s^4 + ks^3 + (k+4)\, s^2 + (k+3)s + 4 = 0$

4.6 The open loop transfer function of a unity feed back control system is given by,

$$G(s) = \frac{K(s+1)}{s^2(s+4)(s+5)}$$

Discuss the stability of the closed loop system as a function of K.

4.7 Determine whether the largest time constant of the system, with the following characteristic equation, is greater than, less than or equal to 2 seconds.

(a) $s^4 + 5s^3 + 8s^2 + 7s + 3$

(b) $s^4 + 6s^3 + 16s^2 + 26s + 15$

(c) $s^4 + 7.2s^3 + 16.4s^2 + 28s + 15$

4.8 Determine all the roots of the characteristic equation given by,

$$D(s) = s^4 + 6s^3 + 18s^2 + 30s + 25$$

It is given that the nearest root to the imaginary axis has a real part equal to –1.

5 Root Locus Analysis

5.1 Introduction

A control system is designed in terms of the performance measures discussed in chapter 3. Therefore, transient response of a system plays an important role in the design of a control system. The nature of the transient response is determined by the location of poles of the closed loop system. Usually the loop gain of the system is adjustable and the value of this gain determines the location of poles of the closed loop system. It will be very informative if we can determine how these poles change their location as the gain is increased. The locus of these roots as one parameter of the system, usually the gain, is varied over a wide range, is known as the root locus plot of the system. Quite often, the adjustment of the system gain enables the designer to place the poles at the desired locations. If this is not possible, a compensator or a controller has to be designed to place the closed loop poles at desired locations.

The closed loop poles are the roots of the characteristic equation and hence finding closed loop poles amounts to finding the roots of the characteristic equation. For degree higher than 3, finding out the roots of the characteristic equation is tedius, more so, if one parameter is changing.

A systematic and simple method was developed by W. R. Evans, which is extensively used by the control engineers, for finding the locus of the roots of the characteristic equation when one of the parameters changes. This method is known as Root Locus Technique. In this method, the locus of the roots of the characteristic equation is plotted for all values of the parameter. Usually this parameter is the gain of the system, but it could be any other parameter also. Once the complete locus is obtained, all the roots for a given value of the parameter can be determined. We will develop the method assuming that the gain is the variable parameter and it can be varied from 0 to ∞.

5.2 Basic Idea

Consider the characteristic equation of a second order system, given by

$$s^2 + as + K = 0$$

Let us assume that a is a constant and K is the variable. We would like to obtain the locus of the roots, as K is changed from 0 to ∞.

The roots of the characteristic equation are given by,

$$s_{1,2} = \frac{-a}{2} \pm \sqrt{\frac{a^2 - 4K}{2}} \qquad .....(5.1)$$

Since a is a constant, when K = O, the roots are given by

$$s_{1,2} = -\frac{a}{2} \pm \frac{a}{2} = 0, -a$$

These are plotted on the s - plane in Fig. 5.1.

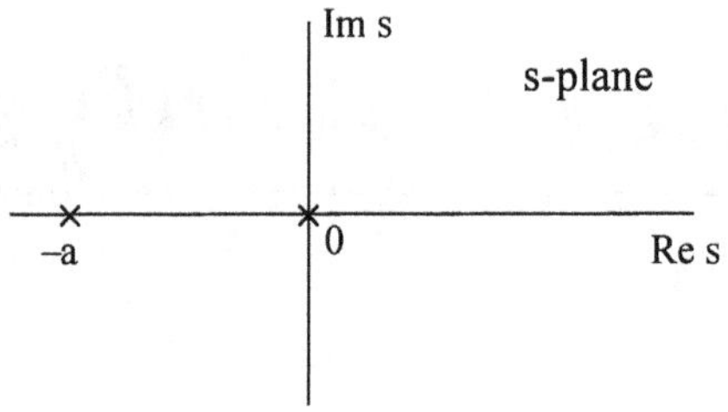

Fig. 5.1 Location of the roots when K = 0

The two roots are real and if a is positive, one root is zero and the second root is on the negative real axis. As K is increased upto a value,

$$K = \frac{a^2}{4} \qquad .....(5.2)$$

the two roots are real and negative. They lie on the negative real axis and always lie between 0 and – a i.e., the roots at K = 0, move along the negative real axis as shown in Fig. 5.2, until they meet at a point, $-\frac{a}{2}$ for $K = \frac{a^2}{4}$. At this value of K, the characteristic equation will have two equal, real and negative roots.

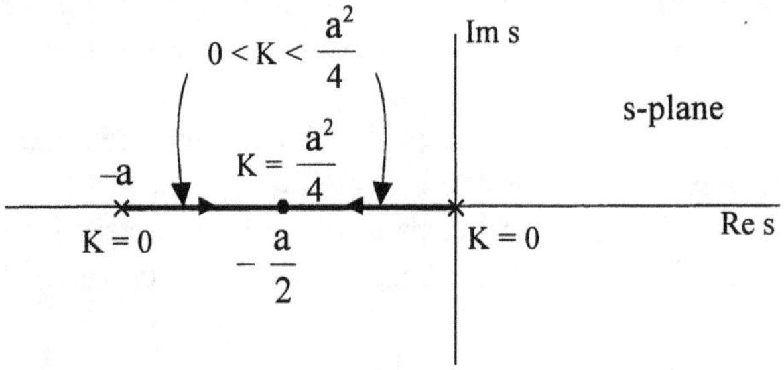

Fig. 5.2 Locus of roots for $0 \le K \le \frac{a^2}{4}$.

If K is increased further, the quantity under the radical sign becomes negative, real part remains the same and hence the roots become complex conjugate. As K value is increased further and further, the roots move on a line perpendicular to the real axis at $s = -\frac{a}{2}$, as shown in Fig. 5.3.

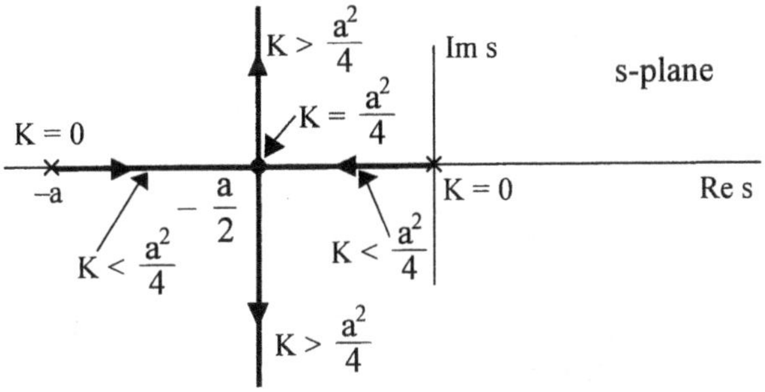

Fig. 5.3 Locus of the roots for K > 0.

The roots are given by,

$$s_{1,2} = -\frac{a}{2} \pm j\sqrt{\frac{4K - a^2}{2}} \quad .....(5.3)$$

For K = ∞, the two roots are complex conjugate with real part equal to $-\frac{a}{2}$ and imaginary part equal to $\pm j\,\infty$. The plot of the locus of the roots as K is changed from 0 to ∞ is known as the root locus plot for the characteristic equation.

For a given value of K, two points can be located on the root locus, which give the location on the root locus, which give the values of roots of the characteristic equation. For a simple second order system, the root locus could be plotted easily but for a characteristic equation of higher order it is not so straight forward.

5.3 Development of Root Locus Technique

Consider the closed loop system shown in Fig. 5.4.

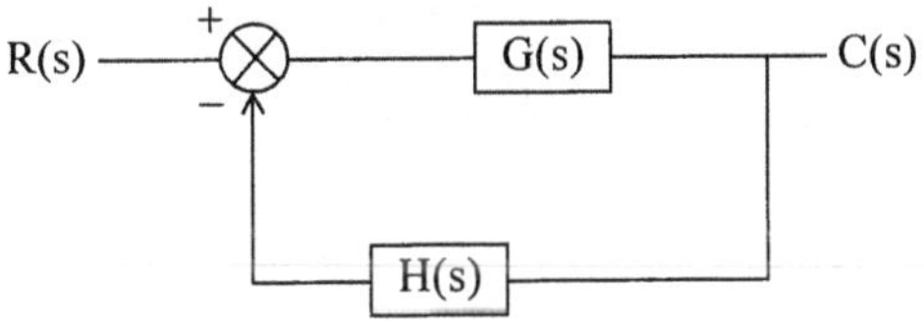

Fig. 5.4 A closed loop system.

The closed loop transfer function is given by

$$T(s) = \frac{C(s)}{R(s)} = \frac{G(s)}{1 + G(s)H(s)} \quad .....(5.4)$$

The characteristic equation is obtained by setting the denominator of right hand side of eqn. (5.4) to zero. Thus,

$$D(s) = 1 + G(s)\,H(s) = 0 \qquad .....(5.5)$$

Any value of *s*, which satisfies eqn. (5.5) is a root of this equation. Eqn. (5.5) can also be written as,

$$G(s)\,H(s) = -1 \qquad .....(5.6)$$

Since *s* is a complex variable, G(s) H(s) is a complex quantity and eqn. (5.6) amounts to,

$$|G(s)\,H(s)| = 1 \qquad .....(5.7)$$

and

$$\angle G(s)\,H(s) = \pm 180\,(2k+1) \qquad k = 0, 1, 2 \ldots \qquad .....(5.8)$$

The condition given by eqn. (5.7) is known as magnitude criterion and that given by eqn. (5.8) is known as angle criterion. Values of *s* - which satisfy both magnitude and angle criterion are the roots of the characteristic equation and hence the poles of the closed loop system. All points on the root locus must satisfy angle criterion and a particular point is located by applying magnitude criterion. In other words, the root locus is the plot of points satisfying angle criterion alone. To obtain the roots corresponding to particular value of the gain, we use magnitude criterion.

The loop transfer function G(s) H(s) can usually be written involving a gain parameter K, as

$$G(s)\,H(s) = \frac{K(s+z_1)(s+z_2).....(s+z_m)}{(s+p_1)(s+p_2).....(s+p_n)} \qquad .....(5.9)$$

and the characteristic equation becomes,

$$1 + G(s)\,H(s) = 1 + \frac{K(s+z_1)(s+z_2).....(s+z_m)}{(s+p_1)(s+p_2).....(s+p_n)} = 0 \qquad .....(5.10)$$

$s = -z_i$ for $i = 1, 2 \ldots m$ are the open loop zeros

and $s = -p_i$ for $i = 1, 2 \ldots m$ are the open loop poles

of the system. To check whether a point $s = s_1$ satisfies angle criterion or not, we have to measure the angles made by open loop poles and zeros at the test point s_1 as shown in Fig. 5.5. Four poles and two zeros are taken for illustration.

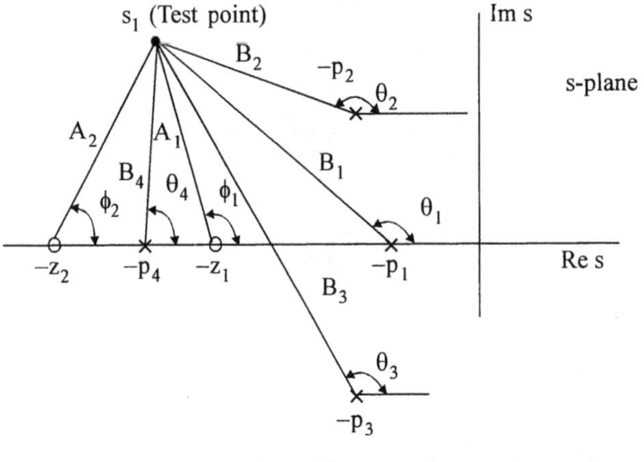

Fig. 5.5 Measurement of angles and magnitudes of the open loop poles and zeros at the test point

From Fig. 5.5 we can easily obtain the angle and magnitude of G(s) H(s) for $s = s_1$ as

$$\angle G(s_1)\,H(s_1) = \phi_1 + \phi_2 - \theta_1 - \theta_2 - \theta_3 - \theta_4 \qquad .....(5.11)$$

and

$$|G(s)\,H(s)| = \frac{A_1A_2}{B_1B_2B_3B_4} \qquad .....(5.12)$$

Note that the angles are measured in anticlockwise direction with respect to the positive direction of the real axis.

If the angle given by eqn. (5.11) is equal to $\pm(2k+1)\,180^{\circ}$, the point s_1 is on the root locus. Value of K for which the point s_1 is a closed loop pole is determined by

$$K\,|G(s_1)\,H(s)_1| = \frac{KA_1A_2}{B_1B_2B_3B_4} = 1 \qquad .....(5.13)$$

or

$$K = \frac{B_1B_2B_3B_4}{A_1A_2} \qquad .....(5.14)$$

5.4 Properties of Root Locus

If a root locus plot is to be drawn, a test point must be selected and checked for angle criterion. It becomes difficult to check all the infinite points in the *s*-plane for locating the points on the root locus.

If an approximate root locus can be sketched, it will save lot of effort to locate the points. Further, a rough sketch also helps the designer to visualize the effects of changing the gain parameter K or effect of introducing a zero or pole on the closed loop pole locations. Now software tools like MATLAB are available for plotting the root locus exactly.

To draw a rough sketch of the root locus, we study certain properties of the root locus plot.

Property 1

The root locus is symmetrical about the real axis.

The roots of the closed loop system are either real or complex. If they are complex, they must occur in conjugate pairs, i.e., if $\sigma_1 + j\omega_1$ is a root $\sigma_1 - j\omega_1$ must also be a root. Therefore the roots are either on the real axis or they lie symmetrically about the real axis.

Property 2

The closed loop poles are same as open loop poles at $K = 0$. Similarly closed loop poles will be same as open loop zeros or they occur at infinity when $K = \infty$.

Proof

The characteristic equation is given by,

$$1 + \frac{K(s+z_1)(s+z_2).....(s+z_m)}{(s+p_1)(s+p_2).....(s+p_n)} = 0$$

or

$$\prod_{i=1}^{n} (s+p_i) + K \prod_{j=1}^{m} (s+z_j) = 0 \qquad .....(5.15)$$

If K = 0 in eqn. (5.15), we have

$$\prod_{i=1}^{n} (s + p_i) = 0 \quad .....(5.16)$$

This is the characteristic equation if K = 0. Thus at K = 0, the open loop poles at $s = -p_i$ are the same as closed loop poles.

Dividing eqn. (5.15) by K, we have,

$$\frac{1}{K} \prod_{i=1}^{n} (s + p_i) + \prod_{j=1}^{m} (s + z_j) = 0 \quad .....(5.17)$$

If K = ∞ in eqn. (5.17), this equation yields,

$$\prod_{j=1}^{m} (s + z_j) = 0 \quad .....(5.18)$$

i.e., the closed loop poles are same as the open loop zeros and occur at $s = -z_j$ when $K \to \infty$.

Thus, we see that, as K changes from zero to infinity, n - branches of the root locus start at the n open loop poles. m of these branches terminate on m - open loop zeros for K = ∞. If $m < n$, the remaining $n - m$ branches go to infinity as $K \to \infty$ as shown below.

The characteristic eqn. (5.15) can be written as,

$$\frac{\prod_{j=1}^{m}(s+z_j)}{\prod_{i=1}^{n}(s+p_i)} = -\frac{1}{K} \quad .....(5.19)$$

$$\frac{s^m \prod_{j=1}^{m}\left(1+\frac{z_j}{s}\right)}{s^n \prod_{i=1}^{n}\left(1+\frac{p_i}{s}\right)} = -\frac{1}{K}$$

$$\frac{1}{s^{n-m}} \frac{\prod_{j=1}^{m}\left(1+\frac{z_j}{s}\right)}{\prod_{i=1}^{n}\left(1+\frac{p_i}{s}\right)} = -\frac{1}{K} \quad .....(5.20)$$

Let $s \to \infty$ and $K \to \infty$, then eqn. (5.20) yields

$$\underset{s\to\infty}{\text{lt}} \frac{1}{s^{n-m}} = \underset{K\to\infty}{\text{lt}} -\frac{1}{K} = 0 \quad .....(5.21)$$

Thus we see that $(n - m)$ points at infinity satisfy the condition given by eqn. (5.21) and therefore they are the roots of the characteristic equation at $K = \infty$.

Thus we see that n branches of the root locus start at open loop poles for $K = 0$ and m of these locii reach m open loop zeros as $K \to \infty$ and the remaining $n - m$ branches go to infinity. The next property helps us to locate these roots at infinity.

Property 3

The $(n - m)$ roots of the characteristic equation go to infinity as $K \to \infty$ along asymptotes making angles,

$$\phi = \frac{(2k+1)180^0}{n-m}; \qquad k = 0, 1, 2, \ldots (n - m - 1) \qquad \ldots\ldots(5.22)$$

Proof

Consider a point on the root locus at infinity. Since the poles and zeros of the open loop system are at finite points, for a point at infinity, all these poles and zeros appear to be at the same point. The angle made by the point at infinity at all these poles and zeros will essentially be the same and let it be equal to ϕ. Since there are n poles and m zeros in G(s) H(s), the net angle of G(s) H(s) at the point is,

$$\angle G(s)\,H(s) = -(n-m)\,\phi$$

Since the point is on the root locus, we have

$$-(n-m)\,\phi = \pm\,(2k+1)\,180^0$$

or

$$\phi = \pm\,\frac{(2k+1)180^0}{n-m}$$

As there are $(n - m)$ branches going to infinity, k can take on values 0, 1, 2 ... upto $(n - m - 1)$. Thus,

$$\phi = \pm\,\frac{(2k+1)180^0}{n-m} \qquad \text{for } k = 0, 1, 2, \ldots (n - m - 1)$$

Thus the closed loop poles at infinity lie on asymptotic lines making angles given by eqn. (5.22) at a point on the real axis. (For a point at infinity, the open loop poles and zeros appear to be at a single point on the real axis, since the open loop poles and zeros are finite and are symmetrically located with respect to the real axis)

Property 4

The $(n - m)$ asymptoes, along which the root locus branches go to infinity, appear to emanate from a point on the real axis, called centroid, given by

$$\sigma_a = \frac{\Sigma \text{ real parts of poles} - \Sigma \text{ real parts of zeros}}{n-m}$$

Proof

Consider the loop transfer function

$$G(s)\,H(s) = \frac{K \overset{m}{\underset{j=1}{\pi}}(s+z_j)}{\overset{n}{\underset{i=1}{\pi}}(s+p_i)} = \frac{N_1(s)}{D_1(s)} \qquad m \leq n \qquad .....(5.23)$$

The $N_1(s)$ is of power m and $D_1(s)$ is of power n. From the properties of the roots of polynomials, we have,

$$\text{Coefficient of } s^{m-1} \text{ in } N_1(s) = \sum_{j=1}^{m} z_j$$

$$\text{and coefficient of } s^{n-1} \text{ in } D_1(s) = \sum_{i=1}^{n} p_i$$

Thus eqn. (5.23) can be written as

$$G(s)\,H(s) = K.\frac{s^m + \sum\limits_{j=1}^{m} z_j s^{m-1} + ...}{s^n + \sum\limits_{i=1}^{m} p_i s^{n-1} + ...} \qquad .....(5.24)$$

Dividing $D_1(s)$ by $N_1(s)$ in eqn. (5.24), we have

$$G(s)\,H(s) = \frac{K}{s^{n-m} + \left(\sum\limits_{i=1}^{n} p_i - \sum\limits_{j=1}^{m} z_j\right)s^{n-m-1} + ...} \qquad .....(5.25)$$

As $s \to \infty$, the significant terms in G(s) H(s) are given by

$$\underset{s\to\infty}{\text{lt}}\ G(s)\,H(s) \simeq \frac{K}{s^{n-m} + \left(\sum\limits_{i=1}^{n} p_i - \sum\limits_{j=1}^{m} z_j\right)s^{n-m-1}} \qquad .....(5.26)$$

Let us consider an open loop transfer function

$$G_1(s)\,H_1(s) = \frac{K}{(s+\sigma_a)^{n-m}}$$

Since this has $(n - m)$ repeated open loop poles at $s = -\sigma_a$ and no zeros, $(n - m)$ branches of the root locus must start from open loop poles at $s = -\sigma_a$ and go to infinity. Let s_0 be a point on one of the branches of the root locus of $G_1(s)$ H(s). Let it make an angle ϕ at $s = -\sigma_a$ as shown in Fig. 5.6.

From angle criterion.

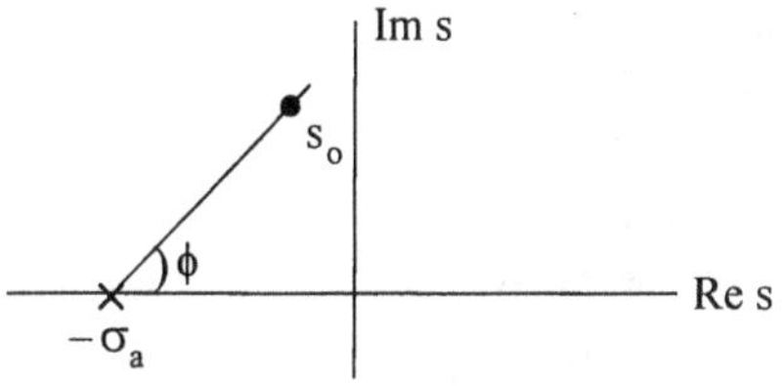

Fig. 5.6 A point on the root locus of $\dfrac{K}{(s+\sigma_a)^{n-m}}$

$$G_1(s_o)\,H_1(s_o) = -(n-m)\,\phi = \pm(2k+1)\,180^0$$

or $$\phi = \pm\frac{2k+1}{n-m}\,180^0; \qquad k = 0, 1, 2, \ldots n-m-1 \qquad .....(5.27)$$

Thus all points lying on straight lines making angles given by eqn. (5.27) are on the root locus. Thus the root locus of $1 + G_1(s)\,H_1(s) = 0$ is a set of straight lines starting at $s = -\sigma_a$ and going to infinity along lines making angles given by eqn. (5.27). For $n - m = 4$, the root locus of $1 + G_1(s)\,H_1(s)$ is shown in Fig. 5.7.

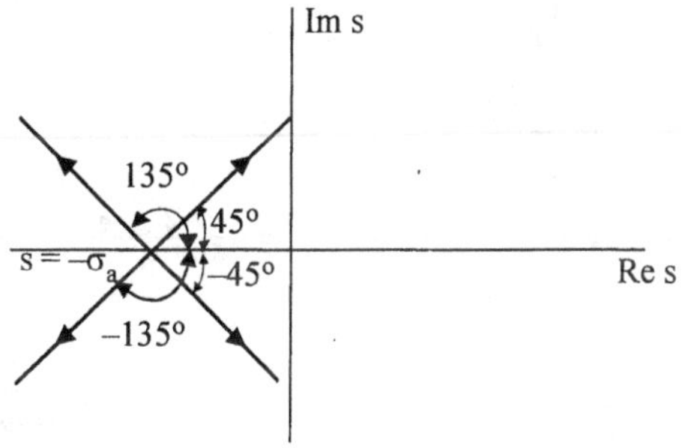

Fig. 5.7 Root locus of $\dfrac{K}{(s+\sigma_a)^4}$

But $G_1(s)\,H_1(s)$ can also be written as,

$$G_1(s)\,H_1(s) = \frac{K}{s^{n-m} + (n-m)\,\sigma_a s^{n-m-1} + \ldots..} \qquad .....(5.28)$$

and as $s \to \infty$ $$\underset{s\to\infty}{\text{lt}}\; G_1(s)\,H_1(s) \simeq \frac{K}{s^{n-m} + (n-m)\,\sigma_a s^{n-m-1}}$$

If $\left(\sum_{i=1}^{n} p_i - \sum_{j=1}^{m} z_j\right)$ in eqn. (5.26) and $(n-m)\,\sigma_a$ in eqn. (5.28) are equal, both G(s) H(s) and $G_1(s)$ $H_1(s)$ behave in the same way as $s \to \infty$. The root locus of $1 + G_1(s)\,H_1(s)$, which is a set of straight lines making angles give by eqn. (5.27) at $s = -\sigma_a$, will form asymptoes to the root locus branches of the characteristic equation 1 + G(s) H(s). Thus the asymptotes intersect at a point on the root locus given by,

$$(n-m)\,\sigma_a = \sum_{i=1}^{n} p_i - \sum_{j=1}^{m} z_j$$

or $$\sigma_a = \frac{\sum_{i=1}^{n} p_i - \sum_{j=1}^{m} z_j}{n-m} \qquad .....(5.29)$$

Since the poles and zeros occur as conjugate pair, the imaginary parts cancel each other when summed, and hence eqn. (5.29) can also be written as

$$\sigma_a = \frac{\Sigma \text{ real parts of poles} - \Sigma \text{ real parts of zeros}}{n-m}$$

Property 5

A point on the real axis lies on the root locus if the number of real poles and zeros to the right of this point is odd. This property demarcates the real axis into segments which form a part of root locus and which do not form a part of root locus.

Proof

Consider the poles and zeros of an open loop transfer function G(s) H(s) as shown in Fig. 5.8.

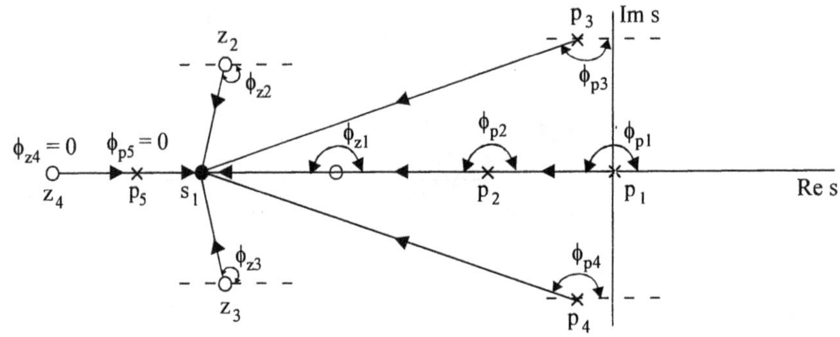

Fig. 5.8 Angles subtended by poles and zeros at a test point s_1

Consider any point $s = s_1$ on the real axis as shown in Fig. 5.8. To see whether this point is on root locus or not, we have to find out the total angle at this point due to all the poles and zeros at this point. Draw vectors from all the poles and zeros to this point. It is easy to verify that the complex poles or zeros subtend equal and opposite angles at any point on the real axis. So $\phi_{p3} = -\phi_{p4}$; $\phi_{z2} = -\phi_{z3}$. Hence the net angle contributed by the complex poles and zeros at any point on the real axis is zero. Coming to the real poles and zeros, each pole and zero to the right of the point $s = s_1$ substends an angle equal to 180^0. In the figure ϕ_{p1}, ϕ_{p2} and ϕ_{z1} are each equal to 180^0. Each pole and zero to the left of the point s_1 subtends an angle of zero at $s = s_1$. Hence these poles and zeros to the left of the point can be disregarded. The point $s = s_1$ on the real axis will be a point on the root locus if the total angle ϕ of G(s) H(s) at $s = s_1$,

$$\phi = \pm (2k + 1)\, 180^0$$

i.e., if it is an odd multiple of 180^0. Since each pole and zero to the right of $s = s_1$ contributes -180^0 and $+180^0$ respectively, the point $s = s_1$ satisfies angle criterion, if the total number of poles and zeros to the right of the point is odd.

For the given example, the root locus on real axis is as shown in Fig. 5.9.

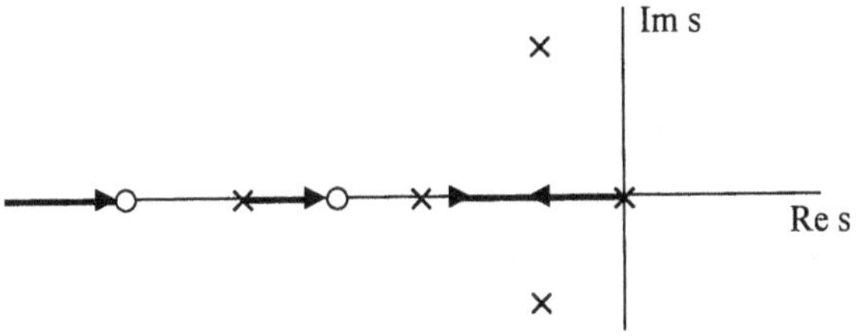

Fig. 5.9 Root locus branches on real axis.

Property 6

Two root locus branches starting from two open loop poles on real axis for K = O, meet at a point on the real axis for some particular value of $K = K_1$. At this value of K, the two closed loop poles will be equal. Such points on root locus where multiple closed loop poles occur, are known as breakaway points. The root locus branches meet at this point and for values of $K > K_1$, the closed loop poles become complex. Complex repeated roots may also occur in a closed loop system and hence the breakaway points can be real or complex. A real breakaway point and a complex break away point are shown in Fig. 5.10.

In Fig. 5.10, the position of real axis between the two poles, is a part of the root locus since the number of poles and zeros to the left of any point on this portion, is odd. Both the root locus branches start from the pole and go along the arrows indicated. They meet at a point s_1 on the real axis at $K = K_1$. This point represents a double pole of the closed loop system. If K is increased further, the root locus branches break away from the real axis at $s = s_1$ as shown in Fig. 5.10.

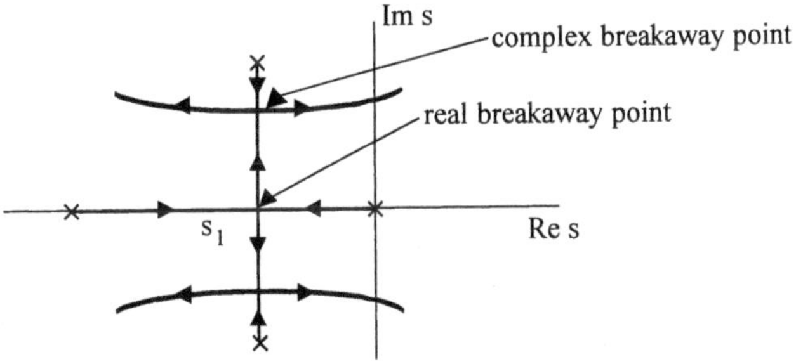

Fig. 5.10 Real and s complex breakaway point.

Similarly, a set of complex root locus branches may join at a point on the real axis for some $K = K_2$ and if $K > K_2$, they approach open loop zeros lying on the real axis at $K = \infty$, as shown in Fig. 5.11. The point s_2 is a double pole of the closed loop system for $K = K_2$.

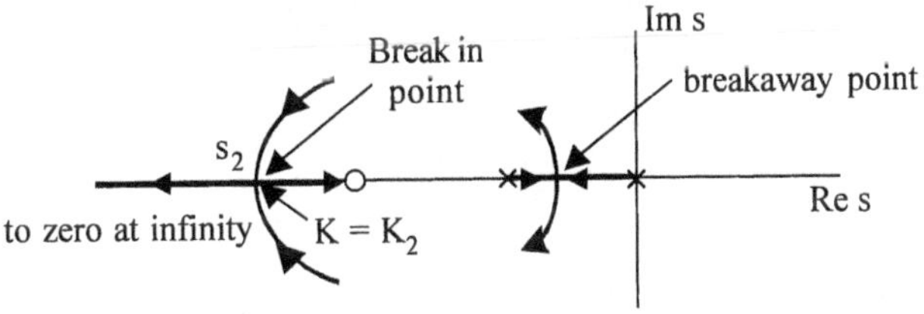

Fig. 5.11 Breakin point on the root locus.

For $K > K_2$ one root locus branch approaches the zero on the right and the other branch goes to zero at infinity. The point where multiple roots occur is such cases, is known as a breakin point. The complex root locus branches break into real axis at $s = s_2$ and there after remain on real axis.

To determine these breakaway or breakin points let us consider the characteristic equation.

$$D(s) = 1 + G(s)\,H(s) = O$$

Let this characteristics equation have a repeated root of order r at $s = s_1$. Hence D(s) can be written as

$$D(s) = (s - s_1)^r\, D_1(s) \qquad .....(5.30)$$

Differentiating eqn. (5.30), we have

$$\frac{d\,D(s)}{ds} = r\,(s - s_1)^{r-1}\, D_1(s) + (s - s_1)^r\, \frac{d\,D_1(s)}{ds} \qquad .....(5.31)$$

Since eqn. (5.31) is zero for $s = s_1$

$$\frac{d\,D(s)}{ds} = 0 \qquad .....(5.32)$$

Hence, the breakaway points are these points which satisfy $\frac{d\,D(s)}{ds} = 0$. This is only a necessary condition but not sufficient. Hence a root of $\frac{d\,D(s)}{ds}$ is a breakaway point but not all roots of $\frac{d\,D(s)}{ds}$ are breakaway points. Out of the roots of $\frac{d\,D(s)}{ds}$ those which also satisfy the angle criterion, are the breakaway or breakin points.

Let us reconsider eqn. (5.32). Since

$$D(s) = 1 + G(s)\,H(s)$$

$$\frac{d\,D(s)}{ds} = \frac{d}{ds}\,[G(s)\,H(s)] = 0 \qquad .....(5.33)$$

for a breakaway point. Further, if

$$G(s)\,H(s) = K.\,\frac{N_1(s)}{D_1(s)}$$

$$\frac{d[G(s)H(s)]}{ds} = K.\,\frac{D_1(s)\frac{dN_1(s)}{ds} - N_1(s)\frac{dD_1(s)}{ds}}{[D_1(s)]^2} = 0$$

In other words,

$$D_1(s)\,\frac{dN_1(s)}{ds} - N_1(s).\,\frac{dD_1(s)}{ds} = 0 \qquad .....(5.34)$$

Since $\quad 1 + G(s)\,H(s) = 0$

We have $$1 + K\frac{N_1(s)}{D_1(s)} = 0$$

or $$K = -\frac{D_1(s)}{N_1(s)} \qquad .....(5.35)$$

Considering K as a function of s in eqn. (5.35), if we differentiate K (s)

$$\frac{dK}{ds} = \frac{-\left[N_1(s)\frac{dD_1(s)}{ds} - D_1\frac{dN_1(s)}{ds}\right]}{[N_1(s)]^2} \qquad .....(5.36)$$

Comparing eqn. (5.34) and (5.36) we have

$$\frac{dK}{ds} = 0$$

Hence breakaway points are the roots of $\frac{dK}{ds}$ which satisfy the angle criterion.

In order to find out breakaway or breakin points, the procedure is as follows :

(i) Express K as

$$K = -\frac{D_1(s)}{N_1(s)} \text{ where } G(s)\,H(s) = \frac{N_1(s)}{D_1(s)}$$

(ii) Find $\frac{dK}{ds} = 0$

(iii) Find the points which satisfy the equation in step (ii).

(iv) Out of the points so found, points which satisfy the angle criterion are the required breakaway or breakin points.

Real axis breakaway or breakin points can be found out easily since we know approximately where they lie. By trial and error we can find these points. The root locus branches must approach or leave the breakin or breakaway point on real axis at an angle of $\pm\frac{180}{r}$, where r is the number of root locus branches approaching or leaving the point. This can be easily shown to be true by taking a point close to the breakaway point and at an angle of $\frac{180}{r}$. This point can be shown to satisfy angle criterion.

An example will illustrate the above properties of the root locus.

Example 5.1

Sketch the root locus of a unity feedback system with

$$G(s) = \frac{K(s+2)}{s(s+1)(s+4)}$$

Solution :

Step 1

Mark the open loop poles and zeros in *s* - plane as shown in Fig. 5.12 (a).

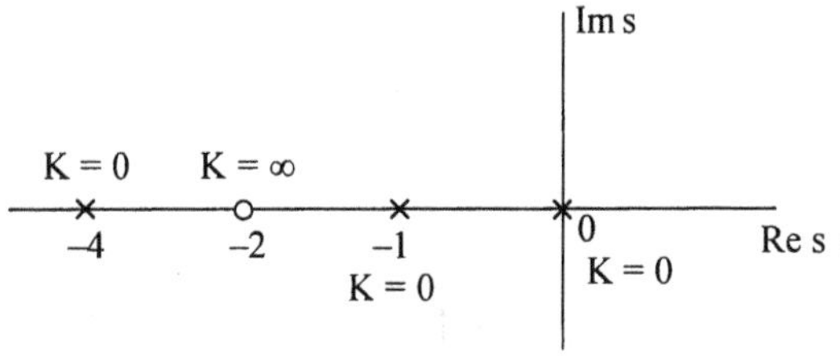

Fig. 5.12 (a) Poles and zeros of G(s).

Step 2

The number of root locus branches is equal to the number of open loop poles. Thus there are 3 root locus branches starting from $s = 0$, $s = -1$ and $s = -4$ for K = 0. Since there is only one open loop zero, one of the root locus branches approaches this zero for K = ∞. The other two branches go to zero at infinity along asymptotic lines.

Step 3

Angles made by asymptotes.

$$\phi_a = \frac{2k+1}{n-m} 180 \qquad k = 0, 1, 2, \ldots (n - m - 1)$$

Since $n = 3$ and $m = 1$, there are $n - m = 2$ asymptotes along which the root locus branches go to infinity

$$\phi_1 = \frac{180}{2} = 90^0 \qquad k = 0$$

$$\phi_2 = \frac{3 \times 180}{2} = 270 \qquad k = 1$$

Step 4

Centroid

$$\sigma_a = \frac{\Sigma \text{R.P of poles} - \Sigma \text{R.P of zeros}}{n - m}$$

$$= \frac{0 - 1 - 4 - (-2)}{2} = -\frac{3}{2}$$

Draw the asymptotes making angles 90^0 and 270^0 at $\sigma_a = -\frac{3}{2}$ on the real axis as shown in Fig. 5.12 (b).

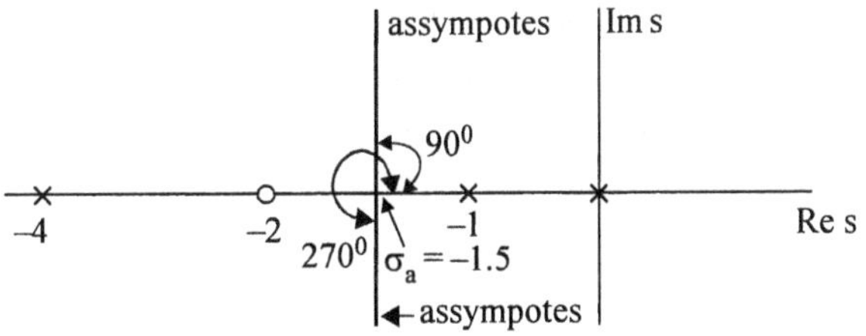

Fig. 5.12 (b) Asymptotes at $\sigma_a = -1.5$ making angles 90^0 and 270^0

Step 5

Root locus on real axis :

Using the property 5 of root locus, the root locus segments are marked on real axis as shown in Fig 5.12 (c).

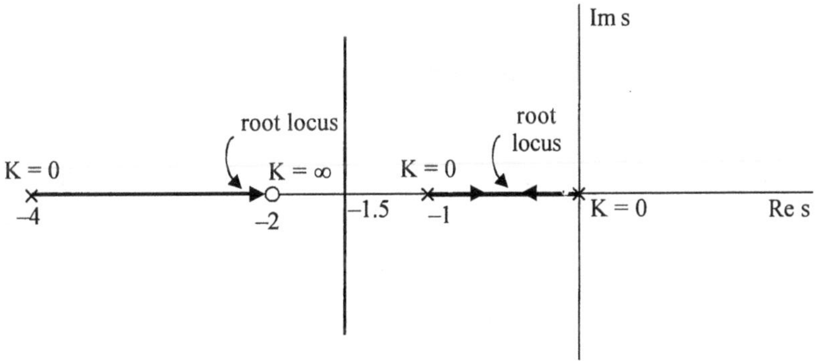

Fig. 5.12 (c) Real axis root locus branches of the system in Ex. 5.1

Step 6

There is one breakaway point between $s = 0$ and $s = -1$, since there are two branches of root locus approaching each other as K is increased. These two branches meet at one point $s = s_1$ for $K = K_1$ and for $K > K_1$ they breakaway from the real axis and approach infinity along the asymptotes.

To find the breakaway point,

$$G(s)\,H(s) = \frac{K(s+2)}{s(s+1)(s+4)}$$

$$\therefore \qquad K -- \frac{s(s+1)(s+4)}{s+2}$$

$$\frac{dK}{ds} = -\frac{(s+2)[3s^2+10s+4]-s(s+1)(s+4).1}{(s+2)^2} = 0$$

or $\qquad 2s^3 + 11s^2 + 20s + 8 = 0$

Out of the three roots of this equation, one which lies between 0 and –1 is the breakaway point. By trial and error, we can find the breakaway point to be $s = -0.55$.

Since the root locus branches should go to infinity along the asymptotes, a rough sketch of root locus is as shown in Fig. 5.12 (d).

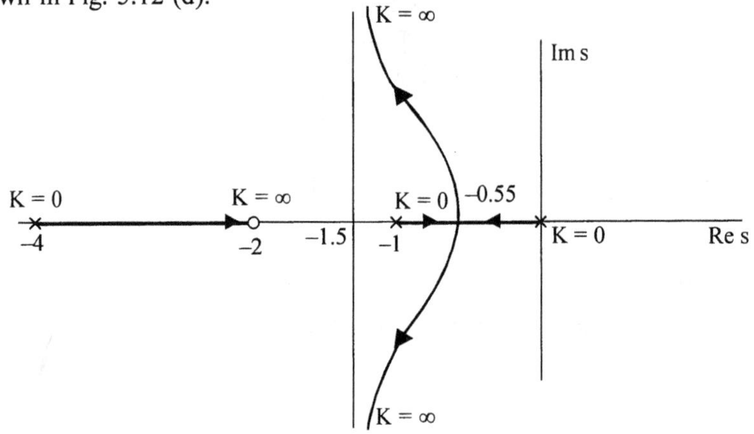

Fig. 5.12 (d) Complete root locus sketch for Ex. 5.1.

Property 7

The root locus branches emanating at complex open loop poles do so at an angle, called angle of departure from a complex pole. To determine this angle let us consider a system with pole zero plot given in Fig. 5.13.

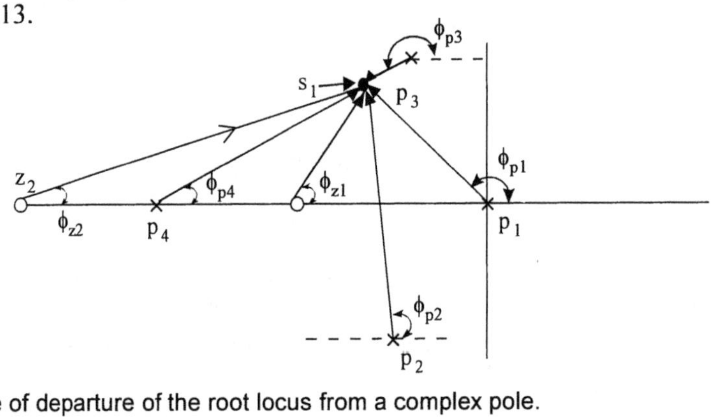

Fig. 5.13 Angle of departure of the root locus from a complex pole.

Consider a point, s_1 on the root locus very close to the complex pole. At this point angle criterion must be satisfied.

$$\angle G(s_1)\,H(s_1) = (\phi_{z1} + \phi_{z2}) - (\phi_{p1} + \phi_{p2} + \phi_{p3} + \phi_{p4})$$

$$= (2k + 1)\,180^0 \qquad .....(5.38)$$

If the point s_1 is very close to the complex pole, angle of departure of the root locus branch can be taken to be equal to ϕ_{p3}.

Let the angle of departure be ϕ_p.

$$\therefore \qquad \phi_p = \phi_{p3}$$

$\therefore$ From eqn. (5.38)

$$(\phi_{z1} + \phi_{z2} - \phi_{p1} - \phi_{p2} - \phi_{p4}) - \phi_p = \pm (2k + 1)\, 180^0 \quad .....(5.39)$$

The first term in parenthesis on the left hand side of eqn. (5.39) is the net angle contributed by all other poles and zeros at s_1 or the complex pole p_3 (since s_1 is very close to p_3). Denoting this net angle by ϕ, we have

$$\phi - \phi_p = \pm (2k + 1)\, 180^0 \qquad k = 0, 1, 2 ...$$

or

$$\phi_p = \phi \pm (2k + 1)\, 180^0 \qquad k = 0, 1, 2 ... \quad .....(5.40)$$

At a complex zero, root locus branches approach at an angle called angle of arrival. Angle of arrival at a complex zero can be shown to be equal to

$$\phi_z = \pm (2k + 1)\, 180 - \phi \qquad k = 0, 1, 2 ... \quad .....(5.41)$$

Where ϕ is the net angle contributed by all other poles and zeros at that complex zero.

Property 8

If a root locus branch crosses the imaginary axis, the cross over point can be obtained by using Routh - Hurwitz criterion.

As discussed in chapter 4, Routh Hurwitz criterion can be applied to the characteristic equation and the value of K for which a row of zeros is obtained in the Routh array can be determined. Whenever a row of zeros is obtained in Routh's array, it indicates roots on imaginary axis. These roots can be determined by solving the auxiliary equation.

Using these eight properties, a rough sketch of the root locus of a system can be plotted. These properties or rules are summerised in Table. 5.1.

Table 5.1 Properties of root locus

1. The root locus is symmetrical about the real axis
2. There are *n* root locus branches each starting from an open loop pole for K = 0. *m* of these branches terminate on *m* open loop zeros. The remaining *n* – *m* branches go to zero at infinity.
3. The *n* – *m* branches going to zeros at infinity, do so along asymptotes making angles

$$\phi = \frac{(2k \pm 1)180}{n - m}; \qquad k = 0, 1, 2 .... (n - m - 1)$$

with the real axis.

Note : For different values of (*n* – *m*) the angles of asymptotes are fixed. For example if

(i) $n - m = 1$ $\qquad \phi = 180^0$

(ii) $n - m = 2$ $\qquad \phi = 90, -90$

(iii) $n - m - 3$ $\qquad \phi = 60, 180, -60$

(iv) $n - m = 4$ $\qquad \phi = 45, 135, -135, -45$ and so on.

4. The asymptotes meet the real axis at

$$\sigma_a = \frac{\Sigma \text{ real parts of poles} - \Sigma \text{ real parts of zeros}}{n - m}$$

Table. 5.1 *Contd.....*

5. Segments of real axis are parts of root locus if the total number of real poles and zeros together to their right is odd.
6. Breakaway or Breakin points.

   These are points in s-plane where multiple closed loop poles occur. These are the roots of the equation,

$$\frac{dK}{ds} = 0$$

   Only those roots which satisfy the angle criterion also, are the breakaway or breakin points. If r root locus branches break away at a point on real axis, the breakaway directions are given by $\pm \frac{180^0}{r}$.
7. The angle of departure of the root locus at a complex pole is given by,

$$\phi_p = \pm (2k + 1)\ 180 + \phi$$

   where ϕ is the net angle contributed by all other open loop zeros and poles at this pole.

   Similarly the angle of arrival at a complex zero is given by

$$\phi_z = \pm (2k + 1)\ 180 - \phi$$

   where ϕ is the net angle contributed by all other open loop poles and zeros at this zero.
8. The cross over point of the root locus on the imaginary axis is obtained by using Routh Hurwitz criterion.

After drawing the root locus for a given system, if the value of K is desired at any given point $s = s_1$ on the root locus, magnitude criterion can be used. We know that,

$$|G(s)\,H(s)| = 1 \text{ at } s = s_1$$

$$\therefore \qquad K \left| \frac{\prod_{j=1}^{m} (s + z_j)}{\prod_{i=1}^{n} (s + p_i)} \right| = 1 \text{ at } s = s_1$$

or

$$K = \frac{\prod_{i=1}^{n} |s_1 + p_i|}{\prod_{j=1}^{m} |s_1 + z_j|} \qquad .....(5.42)$$

This can be evaluated graphically.

$|s_1 + p_i|$ is the length of the vector drawn from p_i to s_1 and $|s_1 + z_j|$ is the length of the vector drawn from z_j to s_1.

$$\therefore \qquad K = \frac{\text{Product of lengths of vectors drawn from open loop poles to } s_1}{\text{Product of length of vectors drawn from open loop zeros to } s_1}$$

Let us now consider some examples to illustrates the method of sketching a root locus.

Example 5.2

Sketch the root locus of the system with loop transfer function

$$G(s)\,H(s) = \frac{K}{s(s+2)(s^2+s+1)}$$

Step 1

Plot the poles and zeros of G(s) H(s)

Zeros : nil

$$\text{Poles} : 0, -2, -0.5 \pm j\,\frac{\sqrt{3}}{2}$$

Step 2

There are 4 root locus branches. Since there are no zeros all these branches go to infinity along asymptotes.

Step 3

Angles of asymptotes

$$\phi = \frac{(2k+1)180}{n-m} \qquad k = 0, 1, 2, 3$$

$$\phi_1 = \frac{180}{4-0} = 45^0$$

$$\phi_2 = 3.\ \frac{180}{4-0} = 135^0$$

$$\phi_3 = 5.\ \frac{180}{4-0} = 225^0$$

$$\phi_4 = 7.\ \frac{180}{4-0} = 315^0$$

Step 4

Centroid

$$\sigma_a = \frac{\Sigma\text{ real parts of poles} - \Sigma\text{ real parts of zeros}}{n-m}$$

$$= \frac{0-2-0.5-0.5}{4} = -\frac{3}{4}$$

Step 5

Root locus on real axis.

Root locus lies between 0 and – 2.

A sketch of the root locus upto this point is given in Fig. 5.14 (a)

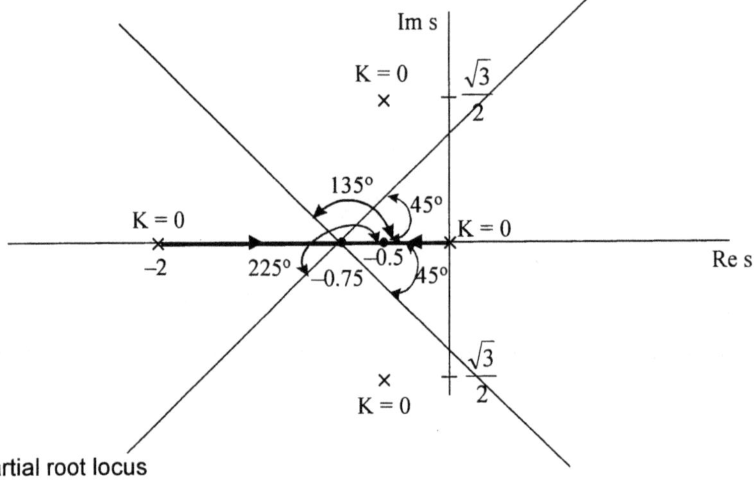

Fig. 5.14 (a) Partial root locus

Step 6

Breakaway points

There is one breakaway point lying between 0 and 2. Its location can be obtained by using,

$$\frac{dK}{ds} = 0$$

$$K = -s(s+2)(s^2+s+1)$$

$$= -s^4 - 3s^3 - 3s^2 - 2s$$

$$\frac{dK}{ds} = -4s^3 - 9s^2 - 6s - 2 = 0$$

$$4s^3 + 9s^2 + 6s + 2 = 0$$

Solving this for a root in the range 0 to 2 by trial and error we get the breakaway point as,

$$s = -1.455$$

Step 7

Angle of departure from complex poles

$$s = -0.5 \pm j\frac{\sqrt{3}}{2}$$

Draw vectors from all other poles and zeros to complex pole $s = -0.5 \pm j\frac{\sqrt{3}}{2}$ as shown in Fig. 5.14 (b).

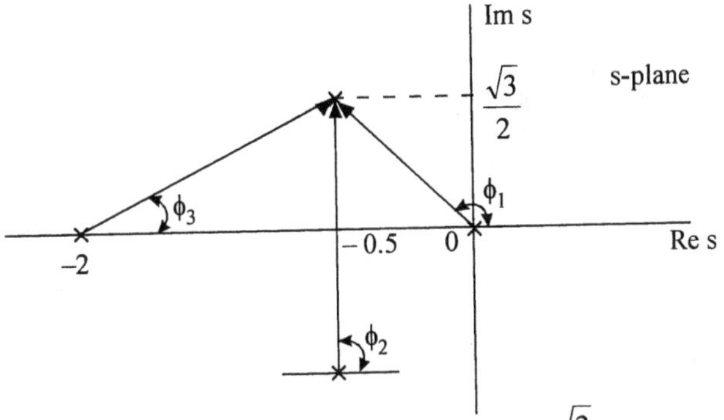

Fig. 5.14 (b) Calculation of angle of departure from $s = -0.5 \pm j\,\frac{\sqrt{3}}{2}$

The net angle at the complex pole due to all other poles and zeros is

$$\phi = -\phi_1 - \phi_2 - \phi_3$$

$$= -\left(180 - \tan^{-1}\frac{\sqrt{3}}{2\times\frac{1}{2}}\right) - 90 - \tan^{-1}\frac{\sqrt{3}}{2\times 1.5}$$

$$= -120 - 90 - 30 = -240$$

Angle of departure

$$\phi_p = \pm(2k+1)\,180 + \phi \qquad k = 0, 1, 2$$

$$= 180 - 240$$

$$= -60$$

Step 8

Crossing of jω-axis

The characteristic equation is,

$$1 + G(s)\,H(s) = 0$$

$$1 + \frac{K}{s(s+2)(s^2+s+1)} = 0$$

$$s^4 + 3s^3 + 3s^2 + 2s + K = 0$$

Constructing the Routh Table :

s^4	1	3	K
s^3	3	2	
s^2	$\frac{7}{3}$	K	
s^1	$\frac{14/3-3K}{7/3}$		
s^0	K		

s^1 row becomes zero for

$$3K = \frac{14}{3} \text{ or } K = \frac{14}{9}$$

Auxiliary equation for this value of K is,

$$\frac{7}{3}s^2 + \frac{14}{9} = 0$$

$$s^2 = -\frac{2}{3}$$

$$s = \pm j\sqrt{\frac{2}{3}} = \pm j\,0.8165$$

The root locus crosses jω-axis at $s = \pm j\,0.8165$ for $K = \frac{14}{9}$

From these steps the complete root locus is sketched as shown in Fig. 5.14 (c)

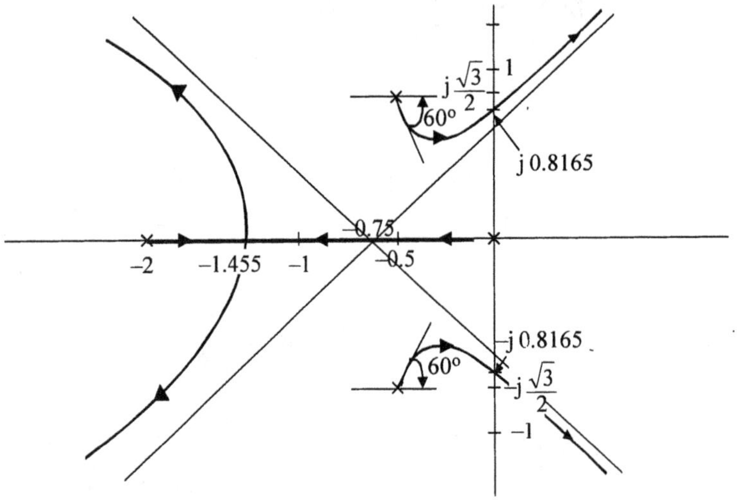

Fig. 5.14 (c) Complete root locus sketch for $G(s)\,H(s) = \frac{K}{s(s+2)(s^2+s+1)}$

Example 5.3

Sketch the root locus of the following unity feedback system with

$$G(s) = \frac{K}{s(s+2)(s^2+2s+4)}$$

(a) Find the value of K at breakaway points

(b) Find the value of K and the closed loop poles at which the damping factor is 0.6.

Solution :

Step 1

Plot the poles and zeros

Zeros : nil

Poles : $0, -2, -1 \pm j\sqrt{3}$

Step 2

There are 4 root locus branches starting from the open loop poles. All these branches go to zeros at infinity.

Step 3

Angles of asymptotes.

Since $n - m = 4,$

$\phi = 45, 135, 225, \text{ and } 315^0$

Step 4

Centroid

$$\sigma_a = \frac{0-2-1-1}{4} = -1$$

Step 5

The root locus branch on real axis lies between 0 and – 2 only.

Step 6

Breakaway points

$$\frac{dK}{ds} = 0$$

$$K = -s(s+2)(s^2+2s+4)$$

$$= (s^4 + 4s^3 + 8s^2 + 8s)$$

$$\frac{dK}{ds} = 4s^3 + 12s^2 + 16s + 8 = 0$$

It is easy to see that $(s + 1)$ is a root of this equation as the sum of the coefficients of odd powers of s is equal to the sum of the even powers of s. The other two roots can be obtained easily as.

$$s = -1 \pm j1$$

So the roots of $\frac{dK}{ds}$ are $s = -1, -1 \pm j1$.

$s = -1$ is a point on the root locus lying on the real axis and hence it is a breakaway point. We have to test whether the points $-1 \pm j1$ lie on the root locus or not.

$$\underline{/G(s)\,H(s)}|_{s=-1+j1} = \text{angle of } \frac{K}{(-1+j1)(-1+j1+2)[(-1+j1)^2+2(-1+j1)+4]}$$

$$\underline{/G(s)\,H(s)}|_{s=-1+j1} = (-135 - 45 - 0)$$

$$= -180^0.$$

Thus angle criterion is satisfied. Therefore $s = -1 \pm j1$ will be points on the root locus. All the roots of $\frac{dK}{ds} = 0$ are thus the breakaway points.

This is an example where all the roots of $\frac{dK}{ds} = 0$ are breakaway points and some breakaway points may be complex.

Step 7

Angles of departure

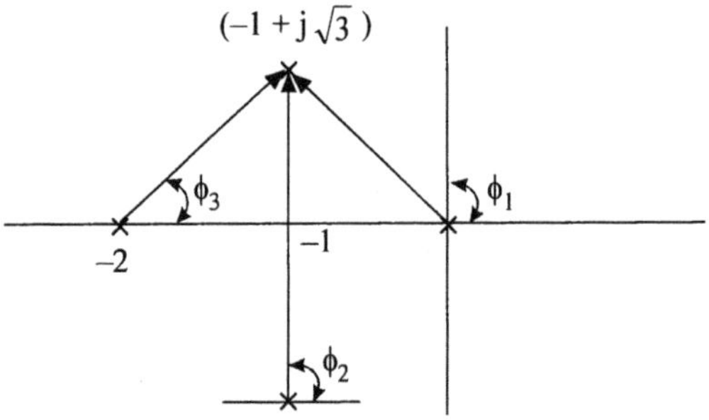

Fig. 5.15 (a) Angle of departure from complex poles.

The net angle at the complex pole $s = -1 + j\sqrt{3}$ due to all other poles is

$$\phi = -\phi_1 - \phi_2 - \phi_3$$
$$= -120 - 90 - 60$$
$$= 270^0$$
$$\therefore \phi_p = 180 - 270^0 = -90^0$$

Since there are two branches on the real axis and they breakaway from real axis at $s = -1$ at an angle $\frac{180}{2} = 90^0$, and the angles of departure from complex poles are also -90 and $+90$, the two root locus branches meet at $-1 + j1$ and break away from these as shown in Fig. 5.15 (b).

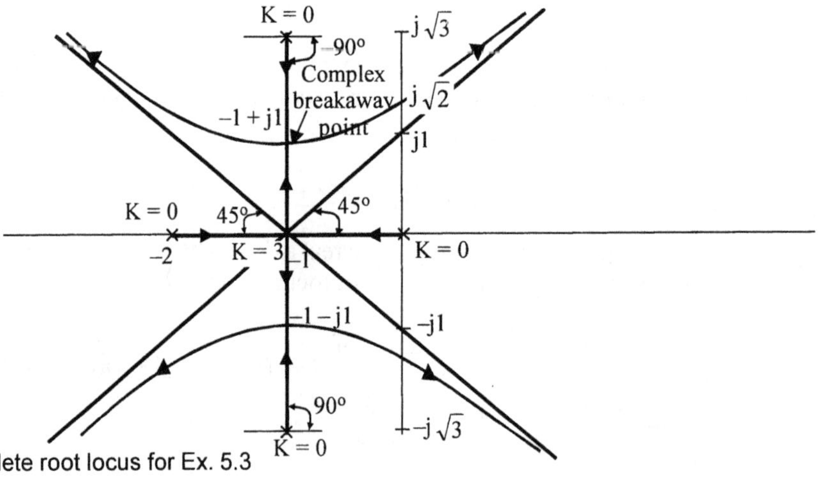

Fig. 5.15 (b) Complete root locus for Ex. 5.3

Step 8

Crossing of jω-axis.

The characteristic equation is

$$s^4 + 4s^3 + 8s^2 + 8s + K = 0$$

Applying Routh's criterion,

s^4	1	8	K
s^3	4	8	
s^2	6	K	
s^1	$\frac{48-4K}{6}$		
s^0	K		

Making s^1 row equal to zero, we have

$$48 - 4K = 0$$

$$K = 12$$

With K = 12 in s^2 row, the auxiliary equation is

$$6s^2 + 12 = 0$$

$$s = \pm j\sqrt{2}$$

The root locus branches cross the jω-axis at $\pm j\sqrt{2}$

The complete root locus is sketched in Fig. 5.15 (b)

(a) Let us find the value of K at which complex conjugate poles are repeated i.e., at $s = -1 + j1$.

$$K = -s(s+2)(s^2+2s+4)_{s=-1+j1}$$

$$= -(-1+j1)(-1+j1+2)\{(-1+j1)^2 + 2(-1+j1) + 4\}$$

$$= 4$$

The characteristic equation for this value of K is

$$(s+1+j1)^2(s+1-j1)^2 = 0$$

$$(s^2+2s+2)^2 = 0$$

At the real breakaway point $s = -1$

$$K = -s(s+2)(s^2+2s+4)_{s=-1}$$

$$= 1(1)(1-2+4)$$

$$= 3$$

(b) Recalling the fact that a complex pole can be written as

$$-\delta\omega_n + j\omega_n\sqrt{1-\delta^2}$$

the angle made by the vector drawn for origin to this pole is $\text{Cos}^{-1}\,\delta$, with the negative real axis as shown in Fig. 5.16.

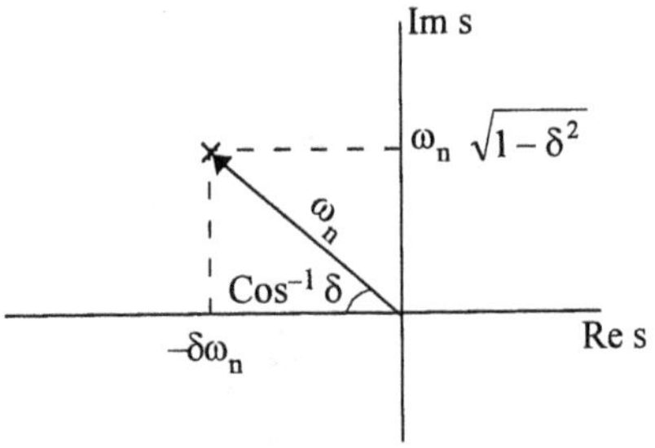

Fig. 5.16 Angle made by the vector from origin to a complex pole.

Let us find the value of K for which the damping factor of the closed loop system is 0.6 in example 5.3. Since $\text{Cos}^{-1}\,0.6 = 53.13^0$.

Let us draw a line making 53.13^0 with negative real axis as shown in Fig. 5.17.

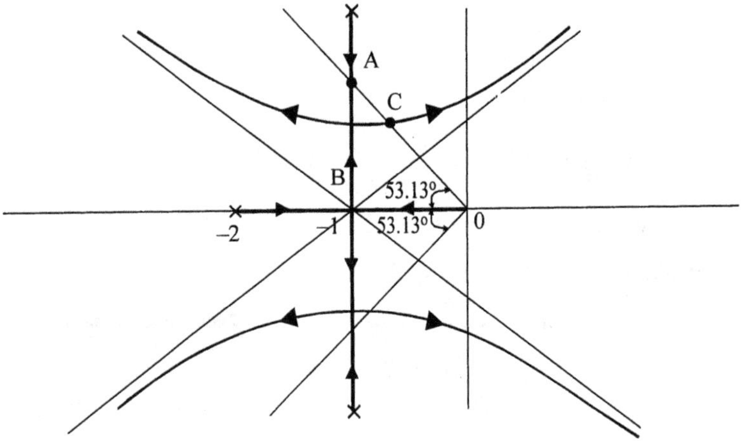

Fig. 5.17 Root locus of Ex. 5.3.

This line cuts the root locus at two points A and C. Let us find the value of K and the closed loop poles at the point A.

The point A is obtained as,

$$AB = OB \tan 53.13^0$$

$$= 1.333$$

The two complex roots of the closed loop system are,

$$s = -1 \pm j\,1.333$$

At this point, the value of K can be obtained as discussed earlier using eqn. (5.42).

$$K = \overset{4}{\underset{i=1}{\pi}} |(s_1 + p_i)|$$

$$= |(-1 + j1.333)(-1 + j1.333 + 2)(-1 + j1.333 + 1 + j\sqrt{3})(-1 + j1.333 + 1 - j\sqrt{3})|$$

$$= 1.666 \times 1.666 \times 3.06 \times 0.399$$

$$= 3.39$$

At this value of K, the other two closed loop poles can be found from the characteristic equation.

The characteristic equation is

$$s^4 + 4s^3 + 8s^2 + 8s + 3.39 = 0$$

The two complex poles are $s = -1 \pm j\ 1.333$

∴ The factor containing these poles is

$$[(s + 1)^2 + 1.777]$$

$$s^2 + 2s + 2.777$$

Dividing the characteristic equation by this factor, we get the other factor due to the other two poles. The factor is

$$s^2 + 2s + 1.223$$

The roots of this factor are

$$s = -1 \pm j\ 0.472$$

The closed loop poles with the required damping factor of $\delta = 0.6$, are obtained with $K = 3.39$. At this value of K, the closed loop poles are,

$$s = -1 \pm j\ 1.333,\ -1 \pm j\ 0.472$$

Note : The Examples 5.2 and 5.3 have the same real poles at $s = 0$ and $s = -2$. The complex poles are different. If the real part of complex poles is midway between the real poles, the root locus will have one breakaway point on real axis and two complex breakaway points. If real part is not midway between the real roots there is only one breakaway point. In addition, if the real part of the complex roots is equal to the imaginary part, the root locus will be as shown in Fig. 5.18.

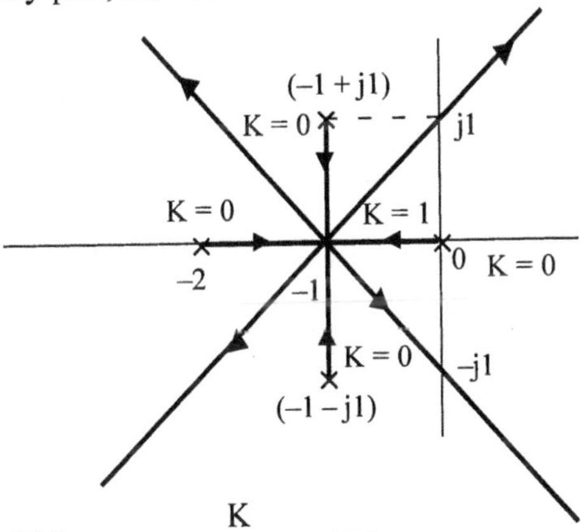

Fig. 5.18 Root locus of $G(s)\,H(s) = \dfrac{K}{s(s + 2)(s^2 + 2s + 2)}$

The breakaway point is $s = -1$ and it is a multiple breakaway point. The angles of departure from the complex poles is -90^0 and $+90^0$. Centroid is $\sigma_a = -1$. Hence four branches of root locus meet at $s = -1$ and break away at angles given by $\frac{180}{4} = 45^0$ along asymptotes. The asymptotes themselves are root locus branches after the breakaway point. For K > 1, all the roots are complex. In this case, the exact root locus is obtained and it is easy to locate roots for given K or K for a given dampling factor etc.

Example 5.4

Obtain the root locus of a unity feed back system with

$$G(s) = \frac{K(s+4)}{s^2 + 2s + 2}$$

Solution :

Step 1

The poles and zeros are plotted on s-plane

$$\text{Zeros : } s = -4$$

$$\text{Poles : } s = -1 \pm j1$$

Step 2

There are two root locus branches starting at $-1 + j1$ and one branch terminating on the finite zero $s = -4$ and the other on zero at infinity.

Step 3

Angle of asymptotes

Since $n - m = 1$

$$\phi = 180^0$$

Step 4

Centroid

$$\sigma_a = \frac{-1-1+4}{1} = 2$$

Step 5

Root locus branches on real axis.

Since there is only one zero on the real axis, the entire real axis to the left of this zero is a part of the root locus.

Step 6

Breakin point

$$K = -\frac{s^2 + 2s + 2}{s+4}$$

$$\frac{dK}{ds} = \frac{(s+4)(2s+2)-(s^2+2s+2)}{(s+4)^2} = 0$$

The roots of $s^2 + 8s + 6 = 0$ are,

$$s_{1,2} = \frac{-8 \pm \sqrt{64-24}}{2}$$

$$= -0.837, -7.162$$

Since $s_1 = -0.837$ is not a point on the root locus, $s = -7.162$ is the breakin point.

Step 7

Angle of departure

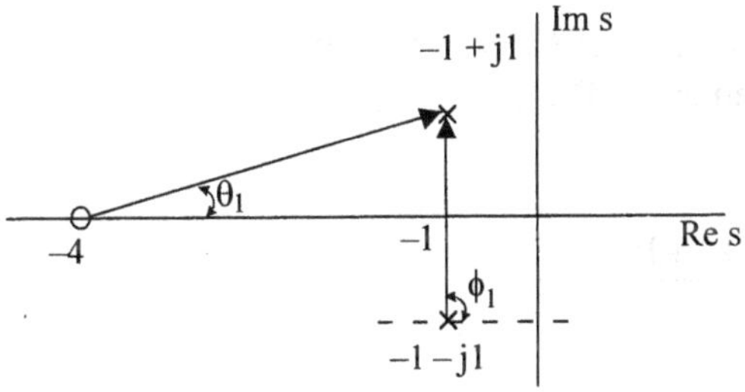

Fig. 5.19 (a) Calculation of angle of departure

Angle contribution at $-1 + j1$ by other poles and zeros

$$\phi = -90 + \tan^{-1}\frac{1}{3}$$

$$= -71.56^{\circ}$$

The angle of departure from $(-1 + j1)$

$$\phi_p = 180 - 71.56$$

$$= 108.43^{\circ}$$

From Fig. 5.19 (b) it can be seen that the root locus never crosses jω-axis. The complete root locus is given in Fig. 5.19 (b).

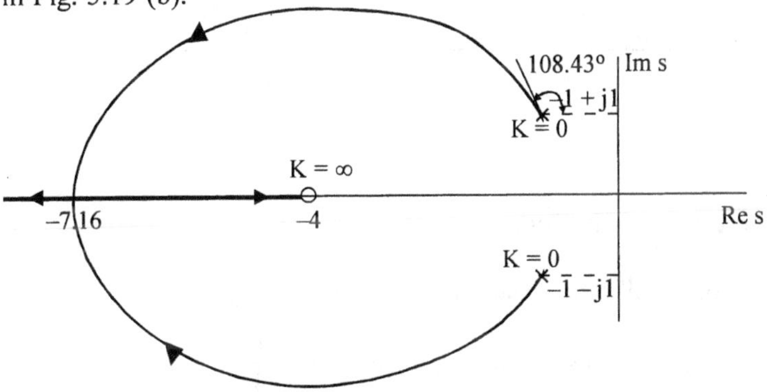

Fig. 5.19 (b) Root locus sketch for $G(s)\,H(s) = \dfrac{K(s+4)}{s^2+2s+2}$

Example 5.5

Sketch the root locus for

$$G(s)\ H(s) = \frac{K(s^2 + 2s + 2)}{s^2(s+4)}$$

Solution :

Step 1

Open loop zeros : $-1 \pm j1$

Poles : $s = 0, 0, -4$

Step 2

There are three root locus branches. Two of them approach the zeros at $1 + j1$. The third goes to infinity along the assymptote with an angle 180^0.

Step 3

$$\sigma_a = \frac{-4+1+1}{1} = -2$$

Step 4

Root locus on real axis lies between $-\infty$ and -4.

Step 5

Breakaway point.

Since $s = 0$ is a multiple root, the root locii break away at $s = 0$.

Step 6

Angle of arrival at complex zero $1 \pm j1$.

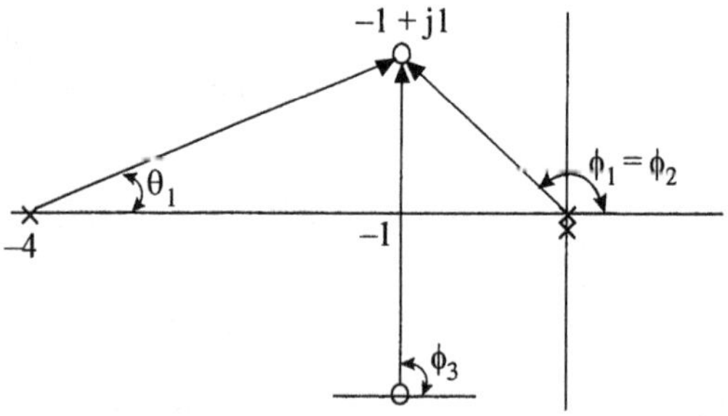

Fig. 5.20 (a) Angle of arrival at complex zero

Total angle contribution by all other poles and zeros at $s = 1 \pm j1$.

$$\phi = -2\,(180 - \tan^{-1} 1) + 90 - \tan^{-1}\frac{1}{3}$$

$$= -198.43^{\circ}$$

Angle of arrival at zero at $-1 \pm j1$ is

$$\phi_z = 180 - (-198.43)$$
$$= 18.43^0$$

Step 7

$j\omega$-axis crossing

The characteristic equation is

$$s^3 + s^2(K+4) + 2Ks + 2K = 0$$

Routh Table

s^3	1	2K
s^2	K + 4	2K
s^1	$\dfrac{2K^2 + 8K - 2K}{K+4}$	0
s^0	2K	

s^1 row will be zero if,

$$2K^2 + 6K = 0$$

or $$K = 0,\ K = -3$$

Since K = – 3 is not valid, the crossing of $j\omega$-axis is for K = 0 i.e., at s = 0.

The complete root locus plot is given in Fig. 5.20 (b).

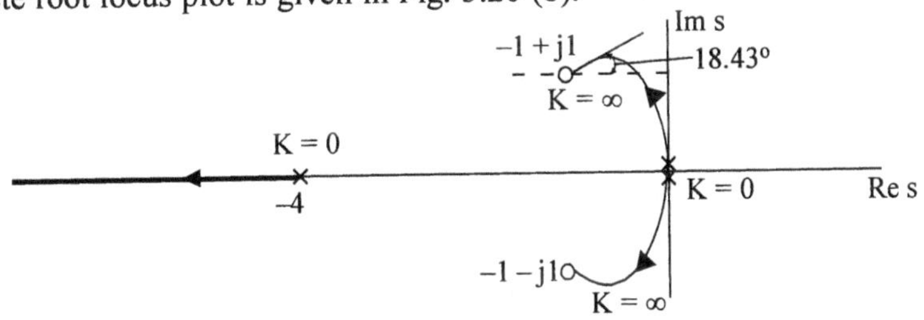

Fig. 5.20 (b) Complete root locus of $\dfrac{K(s^2+2s+2)}{s^2(s+4)}$

Example 5.6

Sketch the root locus of the system whose characteristic equation is given by

$$s^4 + 6s^3 + 8s^2 + Ks + K = 0$$

Solution :

Expressing the given characteristic equation in the form

$$1 + G(s)H(s) = 0$$

$$1 + \frac{K(s+1)}{s^4 + 6s^3 + 8s^2} = 0$$

$$\therefore\ G(s)H(s) = \frac{K(s+1)}{s^2(s+2)(s+4)}$$

Step 1

Open loop zeros : – 1

Poles : 0, 0, – 2, – 4

Step 2

There are 4 root locus branches starting from the open loop poles and one of them terminates on open loop zero at $s = -1$. The other three branches go to zeros at infinity.

Step 3

Angles of asymptotes

Since $n - m = 4 - 1 = 3$

$\phi = 60^0, 180^0, -60^0$

Step 4

Centroid

$$\sigma_a = \frac{-2-4+1}{3} = -\frac{5}{3}$$

Step 5

Root locus on real axis lies between – 1 and – 2, and – 4 to – ∞

Step 6

Break away point

The break away point is at s = 0 only.

Step 7

As there are no complex poles or zeros angle of arrival or departure need not be calculated.

Step 8

jω-axis crossing.

From the characteristic equation, Routh table is constructed.

Routh Table

s^4	1	8	K
s^3	6	K	
s^2	$\frac{48-K}{6}$	K	
s^1	$\dfrac{\frac{48K-K^2}{6}-6K}{\frac{48-K}{6}}$	0	
s^0	K		

A row of zeros is obtained when,

$$48\,K - K^2 - 36K^2 = 0$$

$$K = 0 \text{ or } K = \frac{48}{37}$$

K = 0 gives s = 0 as the cross over point.

For $K = \frac{48}{37}$ forming auxiliary equation using s^2 row,

$$\frac{48 - \frac{48}{37}}{6}\, s^2 + \frac{48}{37} = 0$$

$$7.784\, s^2 + 1.297 = 0$$

$$s = \pm j\, 0.408$$

The complete root locus is sketched in Fig. 5.21.

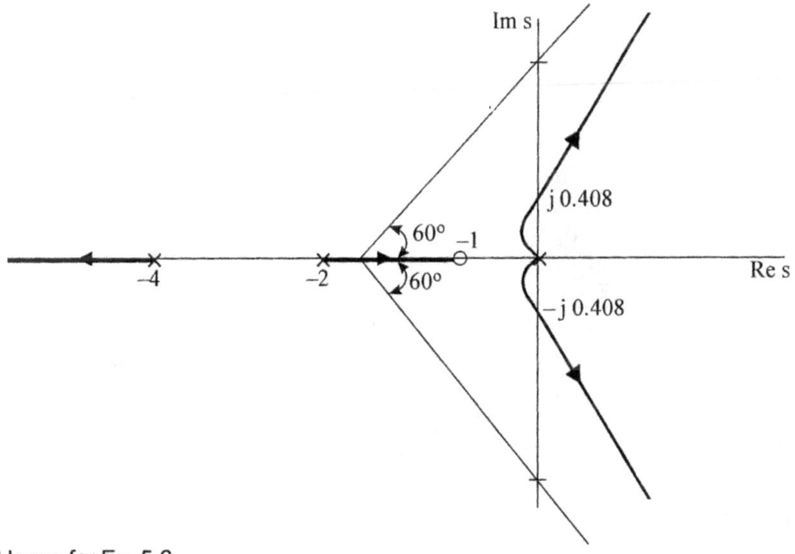

Fig. 5.21 Root locus for Ex. 5.6.

Example 5.7

If in the Example 5.7, an open loop zero is introduced at s = – 3, sketch the resulting root locus. What is the effect of this zero.

Solution :

Since a zero is introduced at s = – 3 on the real axis, the following changes occur.

1. The number of branches going to infinity is n – m = (4 – 2) = 2.
2. The angles of asymptotes are 90^0 and -90^0.

3. The centroid is

$$\sigma_a = \frac{-2-4+1+3}{2} = -1$$

4. Root locus on real axis will now be between (– 2 and – 1) and (– 4 and – 3)
5. Imaginary axis crossing is only at s = 0 for K = 0, as can be seen from Routh table.

The complete root locus is sketched in Fig. 5.22.

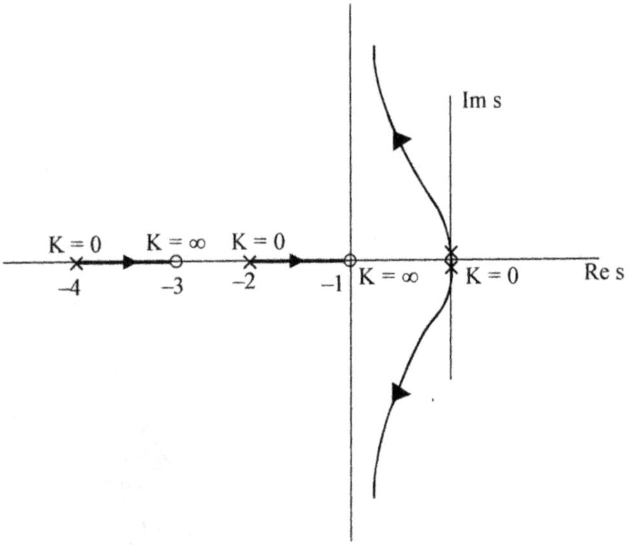

Fig. 5.22 Root locus for Ex. 5.7.

The effect of adding a zero at s = – 3 is to bend the root locus towards the left and make the system stable for all positive values of K.

If now a pole is added at s = – 2.5, draw the root locus and find the effect of adding a pole on the real axis.

Example 5.8

Sketch the root locus of the system shown in Fig. 5.23

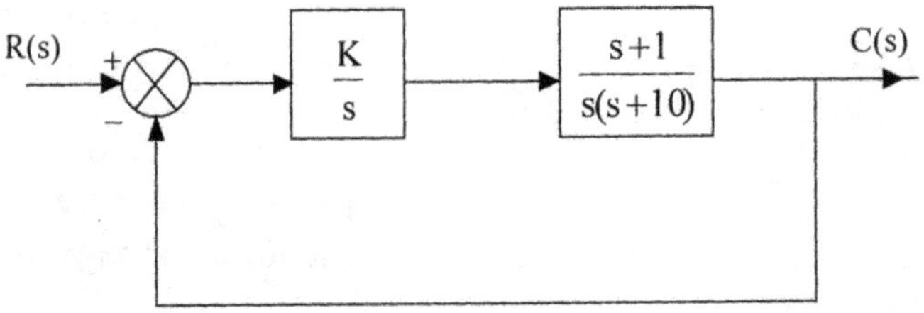

Fig. 5.23 Ststem for Ex. 5.8.

Solution :

$$G(s)\ H(s) = \frac{K(s+1)}{s^2(s+10)}$$

Step 1

Zeros : – 1

Poles : 0, 0, – 10

Step 2

3 root locus branches start from open loop poles and one branch goes to the open loop zero at s = – 1. The other two branches go to infinity.

Step 3

Since n – m = 2, the angles of asymptotes are

$$\phi = 90,\ -90$$

Step 4

Centroid

$$\sigma_a = \frac{-10+1}{2} = -4.5$$

Step 5

The root locus on real axis lies between, – 10 and – 1

Step 6

The breakaway points

$$K = -\frac{s^2(s+10)}{s+1}$$

$$\frac{dK}{ds} = -\frac{(s+1)(3s^2+20s)-s^2(s+10)}{(s+2)^2} = 0$$

$$2s^3 + 13\ s^2 + 20s = 0$$

The roots are

$$s = 0,\ -2.5 \text{ and } -4$$

Since all the roots are on the root locus segments on real axis, all of them are breakaway points. Let us calculate the values of K at the break away points.

At $s = 0\ \ K = 0$

At $s = -2.5$

$$K = -\left.\frac{s^2(s+10)}{s+1}\right|_{s=-2.5}$$

$$= -\frac{6.25 \times 7.5}{-1.5}$$

$$= 31.25$$

At $s = -4$ $$K = -\frac{16 \times 6}{-3}$$

$$= 32$$

Hence the complete sketch of the root locus is as shown in Fig. 5.24.

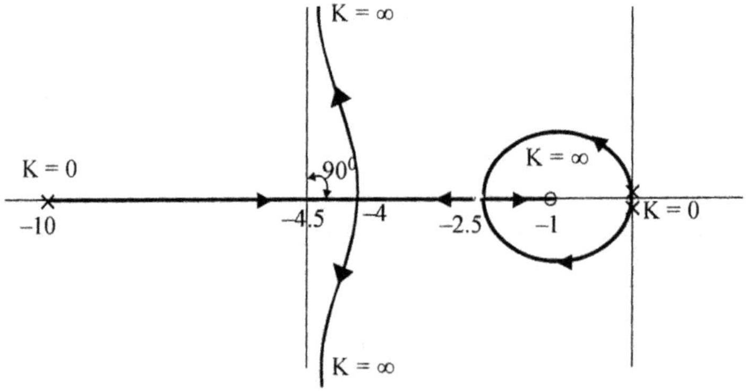

Fig. 5.24 Root locus of Ex. 5.8.

Plot the root locus if the pole at $s = -10$ in this example is changed to $s = -9$. Also plot the root locus if the pole at $s = -1$ is changed to $s = -2$.

Problems

5.1 Draw the root locus plot of the system with the following open loop transfer functions, with unity feedback. Determine the

(a) Centroid,

(b) angles of a asymptotes,

(c) break away / break in points, if any,

(d) Angles of departure / arrival, if any,

(e) Value of K, if any, for jω-axis crossing and frequency of sustained oscillations for this value of K.

(i) $\dfrac{K}{s(s+4)(s+11)}$

(ii) $\dfrac{K(s+1)}{s(s+4)(s+11)}$

(iii) $\dfrac{K}{(s+2)(s^2+s+2)}$

(iv) $\dfrac{K(s+1)}{s(s-1)}$

5.2 Draw the root locus of the system with,

$$G(s) = \frac{K}{(s+4)(s^2+2s+4)}$$

and $$H(s) = 1$$

Find the value of K for which the dominant poles of the closed loop system have real parts equal to – 0.5. For this value of K find all the roots of the closed loop system, damping factor of the dominant roots, frequency of oscillations, settling time for a unit step input.

5.3 Draw the root locus of the system, whose characteristic equation is given by,

$$s^3 + (4 + K)\, s^2 + (5 + 3K)s + 2K = 0$$

For what value of K, double roots occur in the closed loop system. Find all the closed loop poles for this value of K.

5.4 Determine the break away points, angles of departure and centroid of the root locus for the system,

$$G(s)\, H(s) = \frac{K(s+3)}{s(s+5)(s+6)(s^2+2s+2)}$$

Also sketch the root locus.

5.5 Consider the root locus of the system,

$$G(s)\ H(s) = \frac{K}{s(s+2)}$$

1. If a pole is added to the open loop system at s = –3 how does the root locus change. Is it stable for all values of K

2. If a set of complex conjugate poles is added to the open loop system, at s = –3 ± j2, how does the root locus change. Is this system stable for all values of K.

5.6 Sketch the root locus of the system,

$$G(s)\ H(s) = \frac{K}{s(s+1)(s+6)}$$

1. If a zero at s = –2 is added to the system sketch the root locus and comment on the effect of adding a zero. What happens to the break away point. Compare the values of K for which the original system and the modified system are stable.

2. Repeat part (i) With zero added at s = –0.5.

5.7 Obtain the root locus of the system shown in the Fig P 5.7. What value of K results in a damping factor of 0.707 for the dominant complex poles of the closed loop system. Find all the closed loop poles for this value of K.

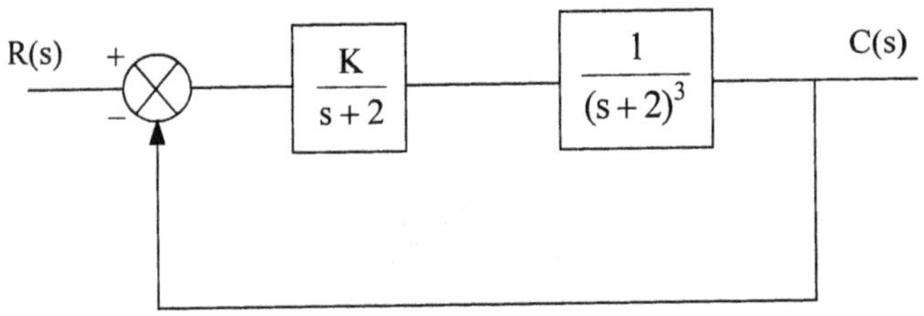

What value of K makes the system oscillatory. What is the frequency of oscillation

5.8 Sketch the root locus of the system,

$$G(s)\ H(s) = \frac{K(s+5)(s+40)}{s^3(s+200)(s+1000)}$$

Find

(a) Centroid and angles of asymptotes

(b) Angles at which the root town leaves the real axis at s = 0

(c) Find the values of K for which the root locus crosses the imaginary axis.

At what points the root locus crosses the jω-axis

5.9 Obtain the root locus of the following system as a varies from 0 to ∞.

$$G(s)\,H(s) = \frac{2(s+2)}{s(s+a)}$$

5.10 Sketch the root locus of the system,

$$G(s) = \frac{K}{s^2(s+2)(s+5)} \qquad H(s) = 1$$

Investigate the effect of changing the feedback element to

$$H(s) = 1 + 2s$$

on the root locus. Comment on the stability.

6 Frequency Response of Control Systems

6.1 Introduction

There are basically two methods of analysis of systems to determine certain properties, so that design procedures can be developed, using these properties as performance measures. If a time signal like step and ramp are used to excite the system and its time response is obtained, we call it a time response analysis or time domain analysis. On the other hand, if a sinusoidal signal of variable frequency is used to excite the system and the magnitude and phase of the steady state output from the system is measured, we call it frequency response analysis or frequency domain analysis. Both the methods have their own advantages and disadvantages.

6.2 Time Domain Analysis Vs Frequency Domain Analysis

In the following a comparison of time domain and frequency domain analysis is given.

(i) Variable frequency, sinusoidal signal generators are readily available and precision measuring instruments are available for measurement of magnitude and phase angle. The time response for a step input is more difficult to measure with accuracy.

(ii) It is easier to obtain the transfer function of a system by a simple frequency domain test. Obtaining transfer function from the step response is more tedious.

(iii) If the system has large time constants, it makes more time to reach steadystate at each frequency of the sinusoidal input. Hence time domain method is preferred over frequency domain method in such systems.

(iv) In order to do a frequency response test on a system, the system has to be isolated and the sinusoidal signal has to be applied to the system. This may not be possible in systems which can not be interrupted. In such cases, a step signal or an impulse signal may be given to the system to find its transfer function. Hence for systems which cannot be interrupted, time domain method is more suitable.

(v) The design of a controller is easily done in the frequency domain method than in time domain method. For a given set of performance measures in frequency domain, the parameters of the open loop transfer function can be adjusted easily using techniques to be discussed later in this chapter.

(vi) The effect of noise signals can be assessed easily in frequency domain rather than time domain.

(vii) The most important advantage of frequency domain analysis is the ability to obtain the relative stability of feedback control systems. The Routh Hurwitz criterion is essentially a time domain method which determines the absolute stability of a system. As discussed in Chapter 4, the determination of relative stability by Routh Hurwitz criterion is cumbersome. Nyquist criterion, which will be described in chapter 7, will not only give stability but also relative stability of the system without actually finding the roots of the characteristic equation.

Since the time response and frequency response of a system are related through Fourier transform, the time response can be easily obtained from the frequency response. The correlation between time and frequency response can be easily established so that the time domain performance measures can be obtained from the frequency domain specifications and vice versa.

6.3 Frequency Response of a Control System

In Chapter 3 we have discussed the time domain response of second order system to a unit step unit. In this section the frequency response of a second order system is obtained and a correlation between time domain at frequency domain response will be estbalisehd in section 6.4.

Consider a second order system with the transfer function,

$$T(s) = \frac{C(s)}{R(s)} = \frac{\omega_n^2}{s^2 + 2\delta\omega_n s + \omega_n^2} \qquad .....(6.1)$$

The steady state sinusoidal response is obtained by substituting $s = j\omega$ in eqn. (6.1).

$$T(s) = \frac{C(j\omega)}{R(j\omega)} = \frac{\omega_n^2}{-\omega^2 + 2j\delta\omega_n\omega + \omega_n^2} \qquad .....(6.2)$$

Normalising the frequency ω, with respect to the natural frequency ω_n by defining a variable

$$u = \frac{\omega}{\omega_n}$$

We have,

$$T(j\omega) = \frac{1}{1 - u^2 + 2j\delta u} \qquad .....(6.3)$$

From eqn. (6.3) the magnitude and angle of the frequency response is obtained as,

$$|T(j\omega)| = M = \frac{1}{\sqrt{(1-u^2)^2 + 4\delta^2u^2}} \qquad .....(6.4)$$

and

$$\angle T(j\omega) = \phi = -\tan^{-1}\frac{2\delta u}{1-u^2} \qquad .....(6.5)$$

The time response for a unit sinusoidal input with frequency ω is given by,

$$c(t) = \frac{1}{\sqrt{(1-u^2)^2 + 4\delta^2 u^2}} \text{ Sin } \left(\omega t - \tan^{-1} \frac{2\delta u}{1-u^2}\right) \qquad .....(6.6)$$

The magnitude and phase of steadystate sinusoidal response for variable frequency can be plotted from eqn. (6.4) and (6.5) and are given in Fig. 6.1 (a) and (b). It is to be noted that when,

$u = 0$ $\quad M = 1,$ $\quad \phi = 0$

$u = 1$ $\quad M = \frac{1}{2\delta},$ $\quad \phi = -90^0$

$u \to \infty,$ $\quad M \to 0$ and $\quad \phi \to -180^0$

Let us examine whether the magnitude response given by eqn. (6.4), monotonically decreases from 1 to 0 or it attains a maximum value and then decreases to zero. If it attains a maximum value at some frequency, its derivative should be zero at that frequency. Hence from eqn. (6.4).

$$\frac{dM}{du} = -\frac{1}{2} \frac{2(1-u^2)(-2u) + 8\delta^2 u}{\left[(1-u^2) + 4\delta^2 u^2\right]^{3/2}} = 0$$

$\therefore$
$$u^3 - u + 2\,\delta^2 u = 0$$

or
$$u = u_r = \sqrt{1-2\delta^2}$$

$$\frac{\omega_r}{\omega_n} = \sqrt{1-2\delta^2}$$

or
$$\omega_r = \omega_n \sqrt{1-2\delta^2} \qquad .....(6.7)$$

This frequency where the magnitude becomes maximum is known as the resonance frequency. Substituting $u = u_r$ in eqn. (6.4), we get the maximum value of the magnitude response. This value of $M = M_r$ is known as the resonance peak. Hence,

$$M_r = \frac{1}{\sqrt{(1-1+2\delta^2)^2 + 4\delta^2(1-2\delta^2)}}$$

$$= \frac{1}{\sqrt{4\delta^4 + 4\delta^2 - 8\delta^4}}$$

$$= \frac{1}{2\delta\sqrt{1-\delta^2}} \qquad .....(6.8)$$

Similarly, the phase angle at resonance frequency is given by,

$$\phi_r = \tan^{-1} \frac{2\delta\sqrt{1-2\delta^2}}{1-1+2\delta^2}$$

$$= -\tan^{-1} \frac{\sqrt{1-2\delta^2}}{\delta} \qquad .....(6.9)$$

From eqn. (6.7) it can be observed that ω_r becomes imaginary for values of $\delta > \frac{1}{\sqrt{2}}$ and hence if $\delta > 0.707$, the magnitude response does not have a peak and the response monotonically decreases from a value *1* at $u = 0$ to zero at $u = \infty$. If $\delta = 0$, the magnitude response goes to infinity and this occurs at $\omega_r = \omega_n$, the natural frequency of the system. This is depicted in Fig. 6.1.

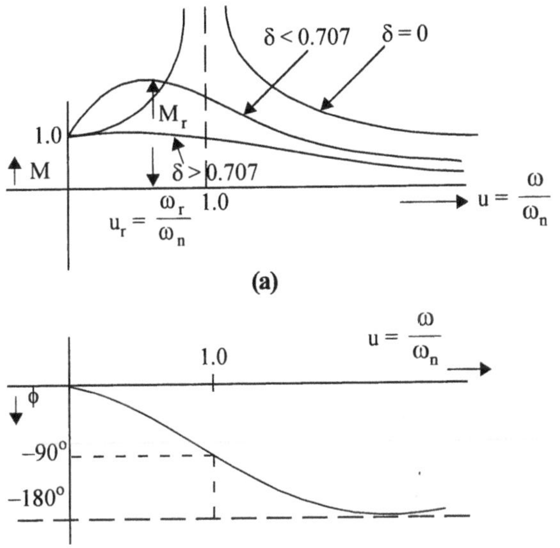

Fig. 6.1 (a) Frequency response : Magnitude Vs Normalised frequency
(b) Frequency response : Phase angle Vs Normalised frequency

One important observation can be made about the resonant peak. From eqn. (6.8) it is clear that the resonant peak depends only on the damping factor δ. For a given δ, the resonance frequency is indicative of the natural frequency of the system. In otherwords, resonant frequency is a measure of the speed of response of the system since settling time $t_s = \frac{4}{\delta\omega_n}$.

Just as peak overshoot and settling time are used as performance measures of a control system in time domain, the resonant peak and resonance frequency can be used as performance measures for a control system in frequency domain.

Another important performance measure, in frequency domain, is the bandwidth of the system. From Fig. 6.1, we observe that for $\delta < 0.707$ and $u > u_r$, the magnitude decreases monotonically. The frequency u_b where the magnitude becomes $\frac{1}{\sqrt{2}}$ is known as the cut off frequency. At this frequency, the magnitude will be $20 \log \frac{1}{\sqrt{2}} = -3\text{db}$.

As $M = 1$ at $u = 0$, control systems are considered as lowpass filters and the frequency at which the magnitude falls to – 3db is known as the bandwidth of the system. This is shown in Fig. 6.2. Frequencies beyond $u = u_b$ are greatly attenuated.

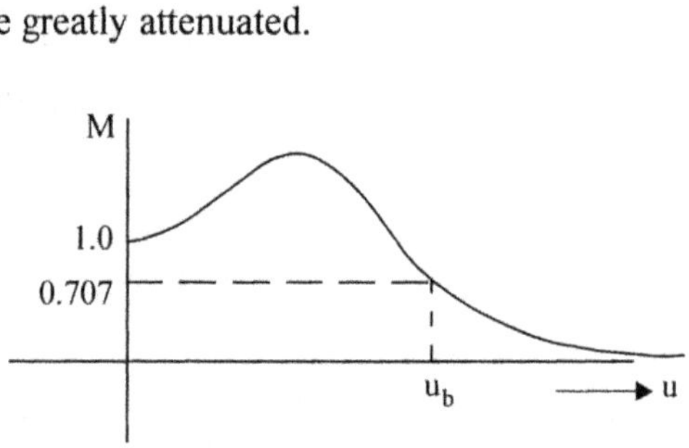

Fig. 6.2 Magnitude response of a second order system showing bandwidth.

An expression for bandwidth can be obtained by equating M = 0.707 at $u = u_b$ in eqn. (6.4).

$$\frac{1}{\sqrt{2}} = \frac{1}{\sqrt{\left(1-u_b^2\right)^2 + 4\delta^2 u_b^2}} \qquad .....(6.10)$$

Squaring on both sides of eqn. (6.10) and simplifying, we have

$$u_b^4 - 2u_b^2 + 1 + 4\,\delta^2 u_b^2 = 2$$

or

$$u_b^4 - 2\,(1 - 2\,\delta^2)\,u_b^2 - 1 = 0$$

Solving for u_b, we get,

$$u_b = \sqrt{\frac{2(1-2\delta)^2 \pm \sqrt{4(1-2\delta^2)^2 + 4}}{2}}$$

$$u_b = \sqrt{(1-2\delta^2) + \sqrt{2 - 4\delta^2 + 4\delta^4}} \qquad .....(6.11)$$

Only positive sign is considered in eqn. (6.11) because u_b must be positive and real.

$$\therefore\ \omega_b = \omega_n \sqrt{(1-2\delta^2) + \sqrt{2 - 4\delta^2 + 4\delta^4}} \qquad .....(6.12)$$

The bandwidth is indicative of the noise characteristics of the system. If the bandwidth is more, the system is more succeptible to noise signals. Also for a given δ, the bandwidth ω_b is a measure of ω_n and hence the speed of response. If the bandwidth is more, the speed of response is high.

6.4 Correlation Between Time Response and Frequency Response

The time doman specifications are obtained by subjecting the second order system to a unit step input.

The important time domain specification are,

Peak overshoot $\qquad M_p = e^{-\pi\delta/\sqrt{1-\delta^2}} \qquad$ for $0 \le \delta \le 1$ $\qquad .....(6.13)$

Damped frequency of oscillation, $\omega_d = \omega_n \sqrt{1-\delta^2}$ .....(6.14)

Settling time $t_s = \dfrac{4}{\delta\omega_n}$ .....(6.15)

In frequency domain, the second order system is subjected to a constant amplitude, variable frequency, sinusoidal input and the magnitude and phase response are obtained. The important frequency domain specifications are,

Resonant peak $M_r = \dfrac{1}{2\delta\sqrt{1-\delta^2}}$ for $\delta < 0.707$ .....(6.16)

Resonance frequency $\omega_r = \omega_n \sqrt{1-2\delta^2}$ .....(6.17)

Bandwidth $\omega_b = \omega_n \left[1-2\delta^2+\sqrt{2-4\delta^2+4\delta^4}\right]^{\frac{1}{2}}$ .....(6.18)

Correlation between M_p and M_r

One important observation can be made with regard to eqns. (6.13) and (6.16). Both M_p and M_r are dependent on the damping factor δ only and hence they are both indicative of damping in the system.

Given M_p, the resonant peak M_r can be evaluated provided δ is less than 0.707. This condition is usually satisfied by many practical control systems as δ is seldom greater than 0.707. Thus the resonant peak M_r and peak overshoot are well correlated. A plot of M_r and M_p with respect to damping factor is given in Fig. 6.3.

Comparison of eqn. (6.14) and (6.17) reveals the correlation between the time domain specification

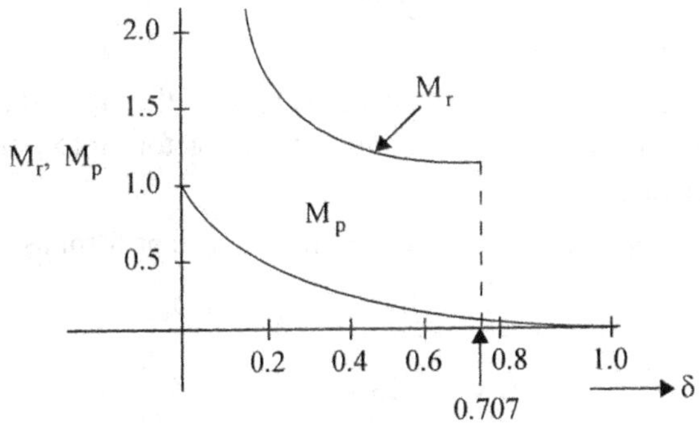

Fig. 6.3 Correlation of M_r and M_p

of damped natural frequency and the frequency domain specification of resonant frequency. For a given damping factor δ, the ratio $\dfrac{\omega_r}{\omega_d}$ is fixed and given the frequency domain specification, the corresponding time domain specification and vice versa, can be easily obtained.

Settling Time and Bandwidth

The speed of response is indicated by settling time in time domain as given in eqn. (6.15). The bandwidth, a frequency domain concept, given by eqn. (6.18), is also indicative of speed of response. Thus we can see that there is a perfect correlation between time domain and frequency domain performance measures : given one, the other can be obtained easily. Of course one should keep in mind that the correlation is valid only for $\delta < 0.707$, which is usually satisfied in many practical control systems.

6.5 Graphical Representation of Transfer Functions

In the study of control systems, the performance of a closed loop control system is often predicted from the open loop transfer function $G(s)$ or loop transfer function $G(s)\ H(s)$. Once the transfer function of an open loop system is known, its sinusoidal steady state response can be easily obtained by replacing s by $j\omega$ in the transfer function $G(s)$. The function $G(j\omega)$ is known as the sinusoidal transfer function. $G(j\omega)$ has magnitude and phase angle for a given value of ω. In this section we will study different types of graphical representations of $G(j\omega)$ which are useful in the design of control systems and also in ascertaining the stability of the systems.

6.5.1 Bode plots

One of the important representations of the sinusoidal transfer function is a Bode plot. In this type of representation the magnitude of $G(j\omega)$ in db, i.e., $20 \log |G(j\omega)|$ is plotted against '$\log \omega$'. Similarly phase angle of $G(j\omega)$ is plotted against $\log \omega$. Hence the abscissa is logarithm of the frequency and hence the plots are known as logarithmic plots. The plots are named after the pioneer in this field, H. W. Bode.

The transfer function $G(j\omega)$ can be written as

$$G(j\omega) = |G(j\omega)| \angle\phi(\omega) \qquad .....(6.19)$$

where $\phi(\omega)$ is the angle of $G(j\omega)$.

Since $G(j\omega)$ consists of many multiplicative factors in the numerator and denominator it is convenient to take logarithm of $|G(j\omega)|$ to convert these factors into additions and substractions, which can be carried out easily.

Let the open loop transfer function be given in time constant form as,

$$G(s) = \frac{K(1+sT_a)(1+sT_b)\ldots}{s^r(1+sT_1)(1+sT_2)\ldots\left(1+2\delta\frac{s}{\omega_n}+\frac{s^2}{\omega_n^2}\right)}\ldots.. \qquad .....(6.20)$$

The transfer function may have real zeros, complex zeros, real poles and complex poles. The sinusoidal transfer function is obtained by replacing s by $j\omega$ in eqn. (6.20). Thus,

$$G(j\omega) = \frac{K(1+j\omega T_a)(1+j\omega T_b)\ldots..}{(j\omega)^r(1+j\omega T_1)(1+j\omega T_2)\ldots..\left[1+2\delta\frac{j\omega}{\omega_n}+\left(\frac{j\omega}{\omega_n}\right)^2\right]}$$

$$|G(j\omega)| = \frac{K|1+j\omega T_a|\,|1+j\omega T_b| \;......}{|j\omega|^r\,|1+j\omega T_1|\;|1+j\omega T_2|...\left\|\left[1+2\delta\frac{j\omega}{\omega_n}+\left(\frac{j\omega}{\omega_n}\right)^2\right]\right\| \;......}$$

Taking 20 log of $|G(j\omega)|$

$$20 \log |G(j\omega)| = 20 \log K + 20 \log |1 + j\omega\ T_a| + 20 \log |1 + j\omega\ T_b| + ... - 20 \log \omega^r$$

$$- 20 \log |1 + j\omega\ T_1| - 20 \log |1 + j\omega\ T_2| \;.....- 20 \log \left|1+2\delta\frac{j\omega}{\omega_n}+\left(\frac{j\omega}{\omega_n}\right)^2\right| \quad .....(6.21)$$

Phase angle of $G(j\omega)$ is given by,

$$\phi(\omega) = \tan^{-1} \omega T_a + \text{Tan}^{-1} \omega T_b + ...$$

$$- r\,(90) - \tan^{-1} \omega T_1 - \tan^{-1} \omega T_2 - \tan^{-1} \frac{2\delta\omega_n}{\omega_n^2 - \omega^2} \;.... \quad .....(6.22)$$

The individual terms in eqns. (6.21) and (6.22) can be plotted w.r.t ω and their algebraic sum can be obtained to get the magnitude and phase plots. Let us now see how these individual terms can be plotted and from the individual plots how the overall plot can be obtained.

The transfer function mainly contains the following types of terms.

(i) Poles or zeros at the origin.

Factors like $\dfrac{K}{(j\omega)^r}$

where r could be positive or negative depending on whether poles or zeros are present at the origin respectively.

(ii) Real zeros

Factors of the form $(1 + j\omega\ T_a)$

(iii) Real poles

Factors of the form $\dfrac{1}{(1+j\omega T_1)}$

(iv) Complex conjugate poles

Factors of the form $\dfrac{\omega_n^2}{(j\omega)^2 + 2j\delta\omega\ \omega_n + \omega_n^2}$

Dividing by ω_n^2, we have,

$$\frac{1}{1+2j\delta\dfrac{\omega}{\omega_n}+\left(j\dfrac{\omega}{\omega_n}\right)^2}$$

(v) Complex conjugate zeros

Factors of the form $\left[1+2j\delta\frac{\omega}{\omega_n}+\left(\frac{j\omega}{\omega_n}\right)^2\right]$

Let us draw Bode plots for each of these terms.

(i) Factor $\frac{K}{(j\omega)^r}$

If $$G(j\omega) = \frac{K}{(j\omega)^r}$$

$$20 \log |G(j\omega)| = 20 \log \frac{K}{\omega^r}$$

$$db = 20 \log K - 20\, r\, (\log\omega) \qquad .....(6.23)$$

If log ω is taken an x - axis and db on y axis, eqn. (6.23) represents a st line $y = mx + c$. The slope m of this line is *20 r* and the intercept on y - axis is $c = 20 \log K$. Bode plots are drawn on semi log graph sheets in which the x - axis is in logarithmic scale and y - axis is in linear scale. A sample of semi log graph sheet is shown in Fig. 6.4.

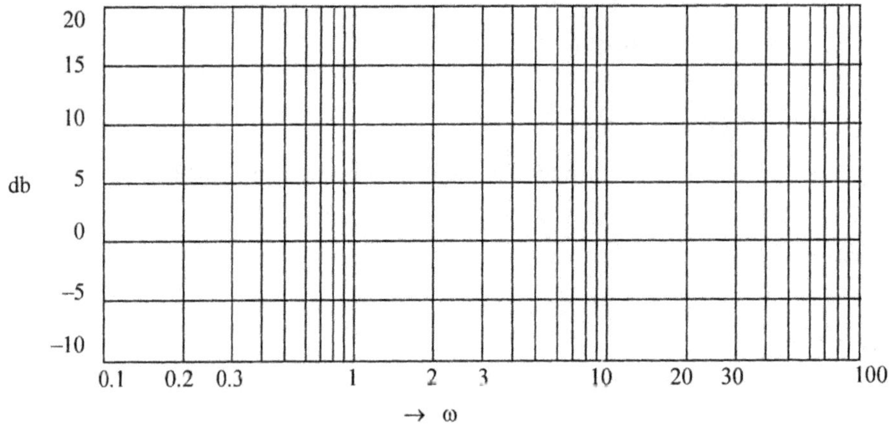

Fig. 6.4 A sample of semi log graph sheet

The slope is usually indicated as *db/decade* or *db/octave*. Let us calculate *db* at two different frequencies, ω_1 and ω_2. Since K is a constant.

$$db_1 = 20 \log K - 20r \log \omega_1$$

$$db_2 = 20 \log K - 20r \log \omega_2$$

$$db_2 - db_1 = 20r \log \omega_1 - 20r \log \omega_2$$

$$= 20\, r \log \frac{\omega_1}{\omega_2}$$

if $\omega_2 = 10\omega_1$, i.e., if there is a decade (10 times) change in frequency, the change in *db* is given by

$$db_2 - db_1 = 20r \log \frac{\omega_1}{10\omega_1} = -20r \log 10$$

Therefore, the slope is given as change in *db* for a given change in ω and is equal to *20 r db/decade*

On the other hand, if $\omega_2 = 2\omega_1$, the change in frequency is said to be one octave. In this case,

$$db_2 - db_1 = 20r \log \frac{\omega_1}{2\omega_1} = -20r \log 2$$

$$= -6r \text{ db/octave}$$

Depending on the multiplicity of poles we have,

(i) No pole at origin r = 0, slope is 0.

(ii) Single pole at origin r = 1.

Slope is – 20 db/dec or – 6 db/oct

(iii) Double pole at origin r = 2

Slope is – 40 db/dec or – 12 db/oct

(iv) Triple pole at origin r = 3

Slope is – 60 db/dec or – 18 db/oct.

If *r* is negative we have zeros at the origin.

(i) Simple zero at origin r = – 1

Slope is + 20 db/dec or + 6db/octave

(ii) Double zero at origin r = – 2

Slope is 40 db/dec or 12 db/octave and so on.

In order to draw a straight line we need a point and the slope.

From eqn. (6.23), at $\omega = 1$ we have

$$db = 20 \log K$$

Since K is known, this point can be marked at $\omega = 1$ and a line can be drawn with the required slope. If there is no pole or zero at the origin the Bode plot is

$$db = 20 \log K = \text{constant}$$

i.e., a line parallel to *x* - axis. It is clear that the phase angle is given by $\phi(\omega) = -r\,(90^0)$. Bode plots of $\frac{K}{(j\omega)^2}$ for various values of *r*, are plotted in Fig. 6.5 (a) and (b).

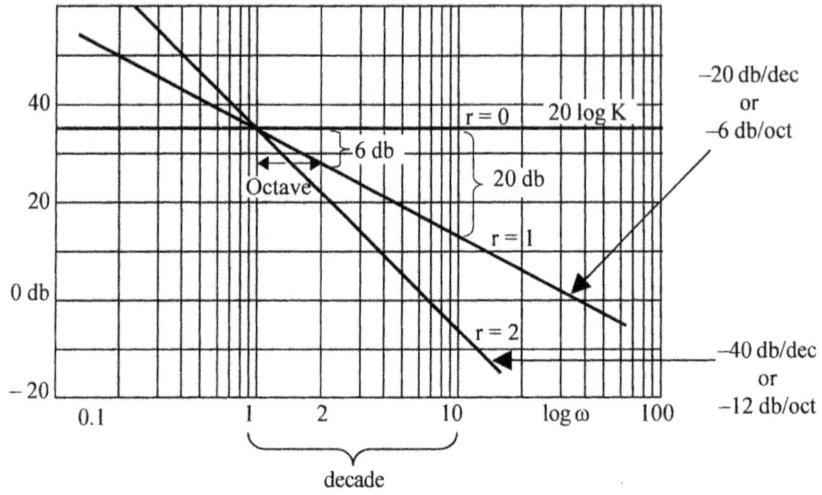

Fig. 6.5 (a) Magnitude plot of $\frac{K}{(j\omega)^r}$ for *r* positive.

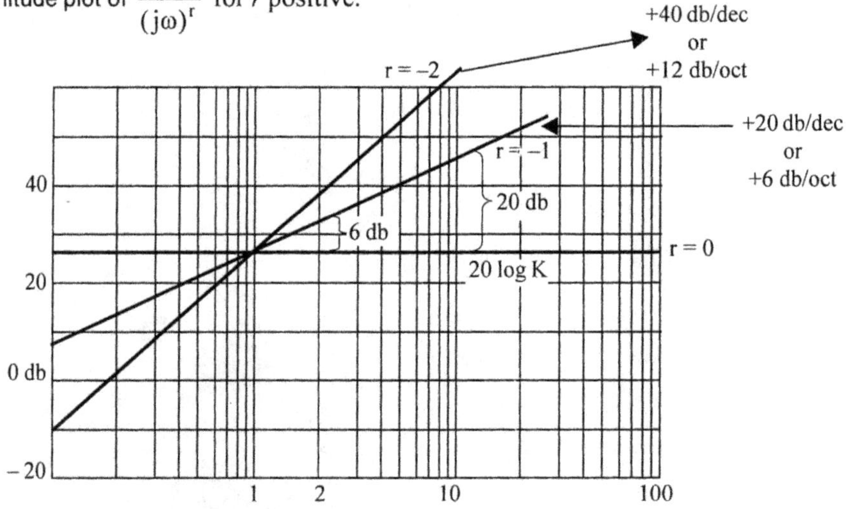

Fig. 6.5 (b) Magnitude plot of $\frac{K}{(j\omega)^r}$ for *r* negative.

(ii) Real zero

$$G(j\omega) = (1 + j\omega T_a)$$

$$20 \log |G(j\omega)| = 20 \log (1+\omega^2 T_a^2)^{\frac{1}{2}}$$
$$= 10 \log (1 + \omega^2 T_a^2) \qquad .....(6.25)$$

Let us first draw an approximate plot.

For very low frequencies $\omega T_a < 1$ and eqn. (6.25) can be approximated as,

$$10 \log 1 = 0$$

i.e., for small frequencies the log magnitude curve approaches asymptotically the *0* db line.

For very high frequencies $\omega^2 T_a^2 >> 1$ or $\omega >> \frac{1}{T_a}$

$$db = 10 \log \omega^2 T_a^2$$

$$= 20 \log \omega + 20 \log T_a \qquad .....(6.26)$$

This is a straight line with a slope of *20 db/decade*. This means the log magnitude curve asymptotically approaches the line with a slope of *20 db/dec*. The zero db line is called as low frequency asymptote and the line with *20 db/dec* slope is known as high frequency asymptote. If $\omega = \frac{1}{T_a}$ in eqn. (6.26), we get zero *db*, which means the high frequency asymptote intersects the low frequency asymptote at $\omega = \frac{1}{T_a}$.

The asymptotic approximation of Bode magnitude plot is shown in Fig. 6.6 (plot a).

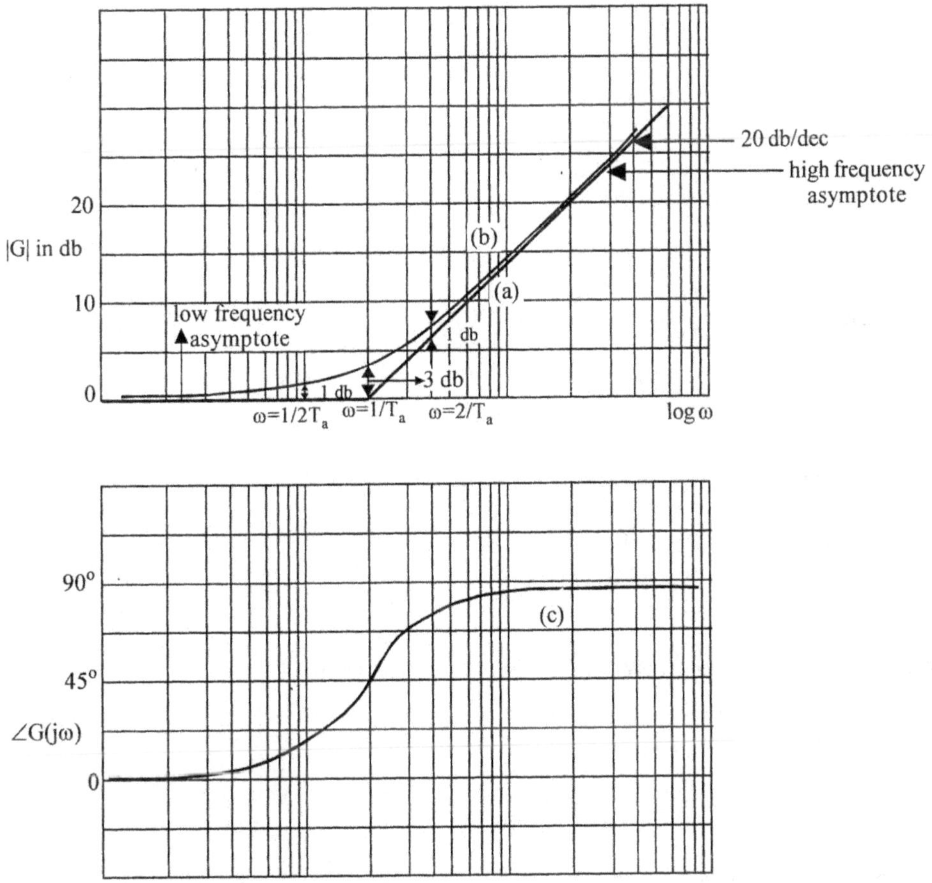

Fig. 6.6 Bode plot of $G(j\omega) = (1 + j\omega T_a)$

Plot (a) Asymptote plot of magnitude **Plot (b)** Exact plot of magnitude **Plot (c)** Phase angle plot

Thus the Bode magnitude plot of a real zero term can be approximated by the asymptotic plot consisting of two straight lines one with a zero slope and the other with a slope of *20 db/dec*. These two lines meet at $\omega = \frac{1}{T_a}$. This frequency is known as a corner frequency. If we want an exact plot we have to calculate the errors at various frequencies and apply the necessary corrections.

For $\omega < \frac{1}{T}$, the error is given by the difference between the actual value, which is $10 \log (1 + \omega^2 T^2)$ and the approximated value which is $10 \log 1$.

Thus error $= 10 \log (1 + \omega^2 T^2) - 10 \log 1$.

At the corner frequency, $\omega = \frac{1}{T}$, the error is 10 log 2 = 3 db.

At one octave below the corner frequency i.e., at $\omega = \frac{1}{2T}$,

$$\text{error} = 10 \log \left(1 + \frac{1}{4}\right) \cong 1 \text{ db}$$

Similarly, for $\omega > \frac{1}{T}$ the error is given by

$$10 \log (1 + \omega^2 T^2) - 10 \log \omega^2 T^2$$

At $\omega = \frac{1}{T}$, we have

$$\text{error} = 10 \log 2 - 10 \log 1$$
$$= 3 \text{ db}$$

and at $\omega = \frac{2}{T}$, i.e., one octave above the corner frequency,

$$\text{error} = 10 \log 5 - 10 \log 4$$
$$\cong 1 \text{ db}$$

The exact plot can be obtained by applying the necessary corrections at $\omega = \frac{1}{T}$, $\omega = \frac{1}{2T}$ and $\omega = \frac{2}{T}$.

At $\omega = \frac{1}{T}$, a point is marked 3 *db* above the asymptotic approximation and $\omega = \frac{1}{2T}$ and $\omega = \frac{2}{T}$, one *db* is marked above the asymptotic approximation as shown in Fig. 6.6 (plot b). These three points are joined by a smooth curve and extended on both sides asymptotically to the high frequency and low frequency asymptotes.

The phase angle is given by

$$\phi(\omega) = \tan^{-1} \omega T_a \qquad .....(6.27)$$

For $\omega << \frac{1}{T_a}$, $\phi(\omega) = 0$

$\omega = \frac{1}{T_a}$ $\quad \phi = 45^o$

$\omega >> \frac{1}{T_a}$ $\quad \phi = 90^o$

The Bode phase angle plot is shown in Fig. 6.6 (plot c).

(iii) Real poles

$$G(j\omega) = \frac{1}{1 + j\omega T_1}$$

$$20 \log |G(j\omega)| = -20 \ (1 + \omega^2 T_1^2)^{\frac{1}{2}}$$

$$db = -10 \log (1 + \omega^2 T_1)$$

For $\omega << \frac{1}{T_1}$

$$db = 0$$

Low frequency asymptote is the zero db line

For $\omega >> \frac{1}{T_1}$

$$db = -20 \log \omega - 20 \log T_1 \qquad .....(6.28)$$

High frequency asymptote is a straight line with *– 20 db/dec* and it intersects the zero *db* low frequency asymptote at $\omega = \frac{1}{T_1}$.

The phase angle is

$$\phi = -\tan^{-1} \omega T_1$$

$$\phi = 0 \qquad \text{for } \omega << \frac{1}{T_1}$$

$$\phi = -45^\circ \qquad \text{for } \omega = \frac{1}{T_1}$$

$$\phi = -90^\circ \qquad \text{for } \omega >> \frac{1}{T_1}$$

The Bode plots of magnitue and phase angle for real pole are given in Fig. 6.7 (plots a and c).

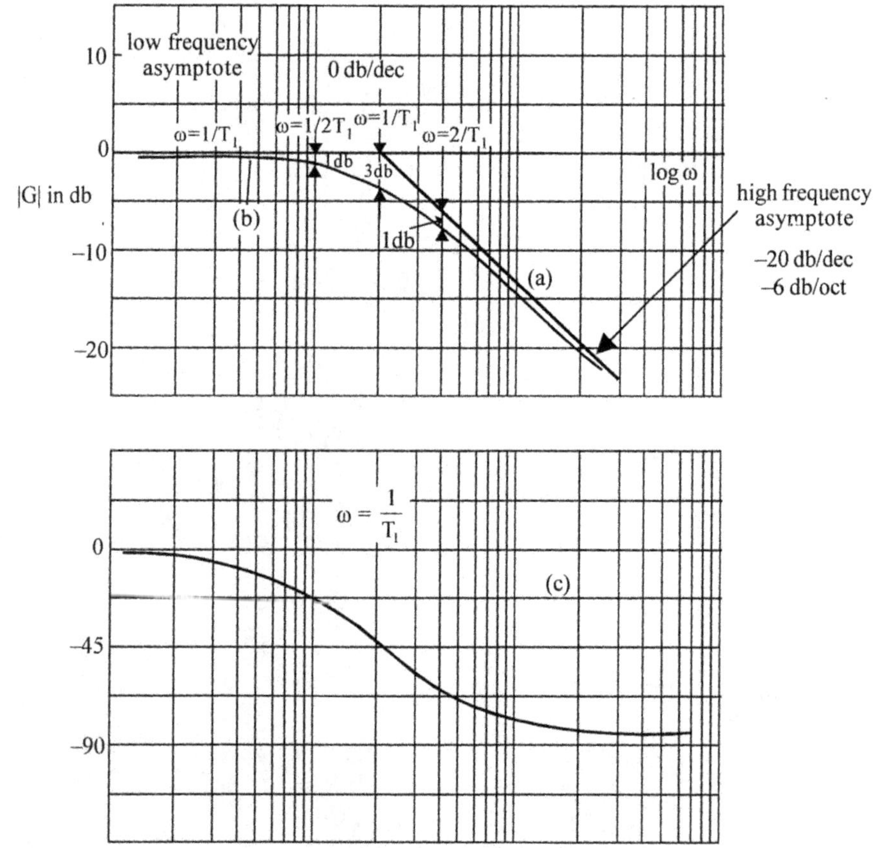

Fig. 6.7 Bode plot of $G(j\omega) = \frac{1}{1 + j\omega T_1}$

Plot (a) Asymptote magnitude plot **Plot (b)** Exact magnitude plot **Plot (c)** Phase angle plot

The exact plot for the real pole factor can be obtained in a similar fashion to that of a zero factor.

For $\omega < \frac{1}{T}$ the error is given by,

$$-10 \log (1 + \omega^2 T^2) + 10 \log 1$$

At $\omega = \frac{1}{T}$, the error is $-10 \log 2 = -3$ db.

At $\omega = \frac{1}{T}$, the error is $-10 \log \left(\frac{5}{4}\right) = -1$ db.

For $\omega > \frac{1}{T}$, the error is given by,

$$-10 \log (1 + \omega^2 T^2) + 10 \log (\omega^2 T^2)$$

At $\omega = \frac{1}{T}$, the error is,

$$-10 \log 2 = -3 \text{ db.}$$

and at $\omega = \frac{2}{T}$, the error is,

$$-10 \log 5 + 10 \log 4 = -1 \text{ db}$$

Marking the points, 3 db at $\omega = \frac{1}{T}$, 1 *db* at $\omega = \frac{1}{2T}$ and $\omega = \frac{2}{T}$, below the asymptotic plot of the real pole factor and joining them by a smooth curve gives the exact plot as shown in Fig. 6.7 (plot b).

(iv) Complex conjugate poles.

$$G(j\omega) = \frac{1}{1 + 2j\delta\frac{\omega}{\omega_n} + \left(\frac{j\omega}{\omega_n}\right)^2}$$

$$20 \log |G(j\omega)| = -20 \log \left[\left(1 - \frac{\omega^2}{\omega_n^2}\right) + 4\delta^2 \frac{\omega^2}{\omega_n^2}\right]^{\frac{1}{2}}$$

$$= -10 \log \left[\left(1 - \frac{\omega^2}{\omega_n^2}\right) + 4\delta^2 \frac{\omega^2}{\omega_n^2}\right] \quad .....(6.29)$$

For $\omega << \omega_n$

$$20 \log |G(j\omega)| = -10 \log 1 = 0$$

Thus the low frequency asymptote is the 0 *db* line.

For $\omega >> \omega_n$, $\left(\frac{\omega}{\omega_n}\right)^2 >> 1$ and also $\left(\frac{\omega}{\omega_n}\right)^4 >> 4\delta^2\left(\frac{\omega}{\omega_n}\right)^2$

$$20 \log |G(j\omega)| = -10 \log \left(\frac{\omega}{\omega_n}\right)^4$$

$$= -40 \log \omega + 40 \log \omega_n \quad .....(6.30)$$

Thus the high frequency asymptote is a line with slope – *40 db/dec* as given by eqn. (6.30). This high frequency asymptote intersects the low frequency asymptote at $\omega = \omega_n$ because for $\omega = \omega_n$ in eqn. (6.30) $20 \log |G(j\omega)| = 0$.

But for values of ω around the natural frequency ω_n, the gain depends on the damping factor as can be seen from eqn. (6.29). For various values of δ, the log magnitude curves are shown in Fig. 6.8.

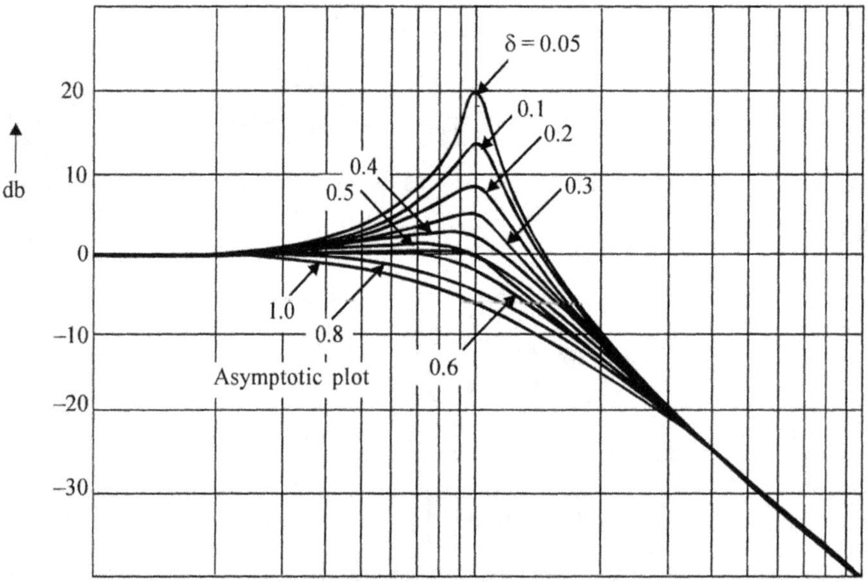

Fig. 6.8 Bode plot for complex conjugate pole for various values of δ

The phase angle is given by,

$$\phi = \angle G\,(j\omega) = -\tan^{-1}\frac{2\delta\dfrac{\omega}{\omega_n}}{1-\dfrac{\omega^2}{\omega_n^{\,2}}} \qquad .....(6.31)$$

For $\omega << \omega_n$, $\qquad \phi \to 0$

For $\omega >> \omega_n$ $\qquad \phi = -\tan^{-1}\left(-2\delta\dfrac{\omega_n}{\omega}\right) \to -180^0$

and for $\omega = \omega_n$ $\qquad \phi = -90^o$

When ω is around ω_n, the phase angle depends on the damping factor δ. The phase angle curves are given in Fig. 6.8.

(v) Complex conjugate zeros

These plots for complex conjugate zeros are same as for complex conjugate poles except that the slopes of asymptotes are positive.

Now having considered the Bode plots of individual terms, the total plot can be obtained by adding these plots at various frequencies as given by eqns. (6.21) and (6.22). Every simple pole or zero contributes 0 *db* to the plot for frequencies below and $\pm$ *20 db/dec* above the corresponding corner frequencies. We can follow the procedure given below to obtain the Bode plot of the given transfer function.

Procedure for Plotting Bode Plot

Step 1

Put the transfer function in the time constant form.

Step 2

Obtain the corner frequencies of zeros and poles.

Step 3

Low frequency plot can be obtained by considering the term $\dfrac{K}{(j\omega)^r}$. Mark the point *20 log K* at

$\omega = 1$. Draw a line with slope $-20\,r$ *db/dec* until the first corner frequency (due to a pole or zero) is encountered. If the first corner frequency is due to a zero of order *m*, change the slope of the plot by + *20 m db/dec* at this corner frequency. If the corner frequency is due to a pole, the slope changes by – *20 m db/dec*.

Step 4

Draw the line with new slope until the next corner frequency is encountered.

Step 5

Repeat steps 3 and 4 until all corner frequencies are considered. If a complex conjugate pole is encountered the slope changes by *– 40 db/dec* at $\omega = \omega_n$.

Step 6

To obtain the exact plot, corrections have to be applied at all the corner frequencies and one octave above and one octave below the corner frequencies. To do this, tabulate the errors at various corner frequencies and one octave above and one octave below the corner frequencies. Mark all these points and draw a smooth curve.

Step 7

The phase angle contributed at various frequencies by individual poles and zeros are tabulated and the resultant angle is found. The angle Vs, frequency is plotted to get phase angle plot.

These steps are illustrated by some examples.

Example 6.1

Draw the Bode magnitude and phase angle plots for the transfer function.

$$G(s) = \frac{2000(s+1)}{s(s+10)(s+40)}$$

Solution :

Step 1

Time constant form

$$G(s) = \frac{5(1+s)}{s(1+0.1s)(1+0.025s)}$$

Step 2

Corner frequencies are

Zero : $\omega = 1$ rad/sec

Poles : $\omega = 0$, $\omega = 10$ rad/sec, $\omega = 40$ rad/sec.

Step 3

Consider the term $\frac{5}{j\omega}$

Mark the point 20 log 5 = 14 db at $\omega = 1$. The low frequency plot is a straight line with slope *– 20 db/dec* passing through the point *14 db* at $\omega = 1$. The first corner frequency is $\omega = 1$ *rad/sec* and it is due to a zero. Continue the line with slope *– 20 db/dec* until $\omega = 1$ *rad/sec* and after this the slope will be 0 *db/dec* as the zero contributes + 20 *db/dec* for $\omega > 1$ *rad/sec*.

Step 4

Draw the line with slope *0 db/dec* until the next corner frequency of *10 rad/sec* is encountered. This corner frequency is due to a pole and hence it contributes – *20 db/dec* for $\omega > 10$ *rad/sec.*

Step 5

Change the slope of the plot at $\omega = 10$ *sec/sec* to – *20 db/dec.* Continue this line until the next corner frequency, i.e., $\omega = 40$ *rad/sec.*

At *40 rad/sec*, the slope of the plot changes by another – *20 db/dec* due to the pole. Hence draw a line with a slope of – *40 db/dec* at this corner frequency. Since there are no other poles or zeros, this is the asymptotic magnitude plot for the given transfer function.

Step 6

Make a tabular form for the corrections at various frequencies.

Consider all corner frequencies and one octave above and one octave below the corner frequencies as indicated Table. 6.1.

Table. 6.1 Error table

Frequency	**Error due to pole or zero factors in db**			**Total error in db**
ω	$1 + j\omega$	$\dfrac{1}{1+0.1j\omega}$	$\dfrac{1}{1+0.025j\omega}$	
0.5	1			1
1	3			3
2	1			1
5		– 1		– 1
10		– 3		– 3
20		– 1	– 1	– 2
40			– 3	– 3
80			– 1	– 1

Whenever error is positive at a frequency mark a point above the curve and whenever the error is negative, mark a point below the curve at that frequency. If there is a overlap due to various corner frequencies, the corresponding errors are algebraically added to get the total correction. For example the pole with corner frequency *40 rad/sec* and pole with corner frequency *10 rad/sec* contribute an error of -1 *db* each at $\omega = 20$ *rad/sec.* The total error at $\omega = 20$ *rad/sec* is therefore -2 *db.*

These points are marked on the asymptotic plot and a smooth curve is drawn to obtain the magnitude plot.

***Step** 7*

To obtain the phase plot, tabulate the angles at various frequencies due to different factors in the transfer function as shown in Table. 6.2. The frequencies are generally taken to include all corner frequencies and frequencies which are one octave above and one octave below the corner frequencies.

Table. 6.2 Calculation of angles for $G(s) = \dfrac{2000(s+1)}{s(s+10)(s+40)}$

Frequency	angles due to				
	$\frac{1}{j\omega}$	$1 + j\omega$	$\frac{1}{1+0.1j\omega}$	$\frac{1}{1+0.025j\omega}$	Total angle
	$\phi_1 = -90$	$\phi_2 = \tan^{-1}\omega$	$\phi_3 = -\tan^{-1} 0.1\omega$	$\phi_4 = -\tan^{-1} 0.025\omega$	$\phi_T = \phi_1 + \phi_2 + \phi_3 + \phi_4$
0.5	– 90	26.56	– 2.86	– 0.71	– 67.1
1	– 90	45.0	– 5.71	– 1.43	– 52.14
2	– 90	63.43	– 11.3	– 2.86	– 40.73
5	– 90	78.69	– 26.56	– 7.12	– 45.0
10	– 90	84.3	– 45.0	– 14.03	– 64.73
20	– 90	87.1	– 63.43	– 26.56	– 92.89
40	– 90	88.56	– 75.96	– 45.0	– 122.4
80	– 90	89.28	– 82.87	– 63.43	– 147.02
200	– 90	90.0	– 87.13	– 78.69	– 165.82
500	– 90	90.0	– 88.85	– 85.42	– 174.27
1000	– 90	90.0	– 90.0	– 87.7	– 177.7

These angles are plotted against $log\ \omega$ and a smooth curve joining these points gives the phase angle plot. The Bode plots are shown in Fig. 6.9.

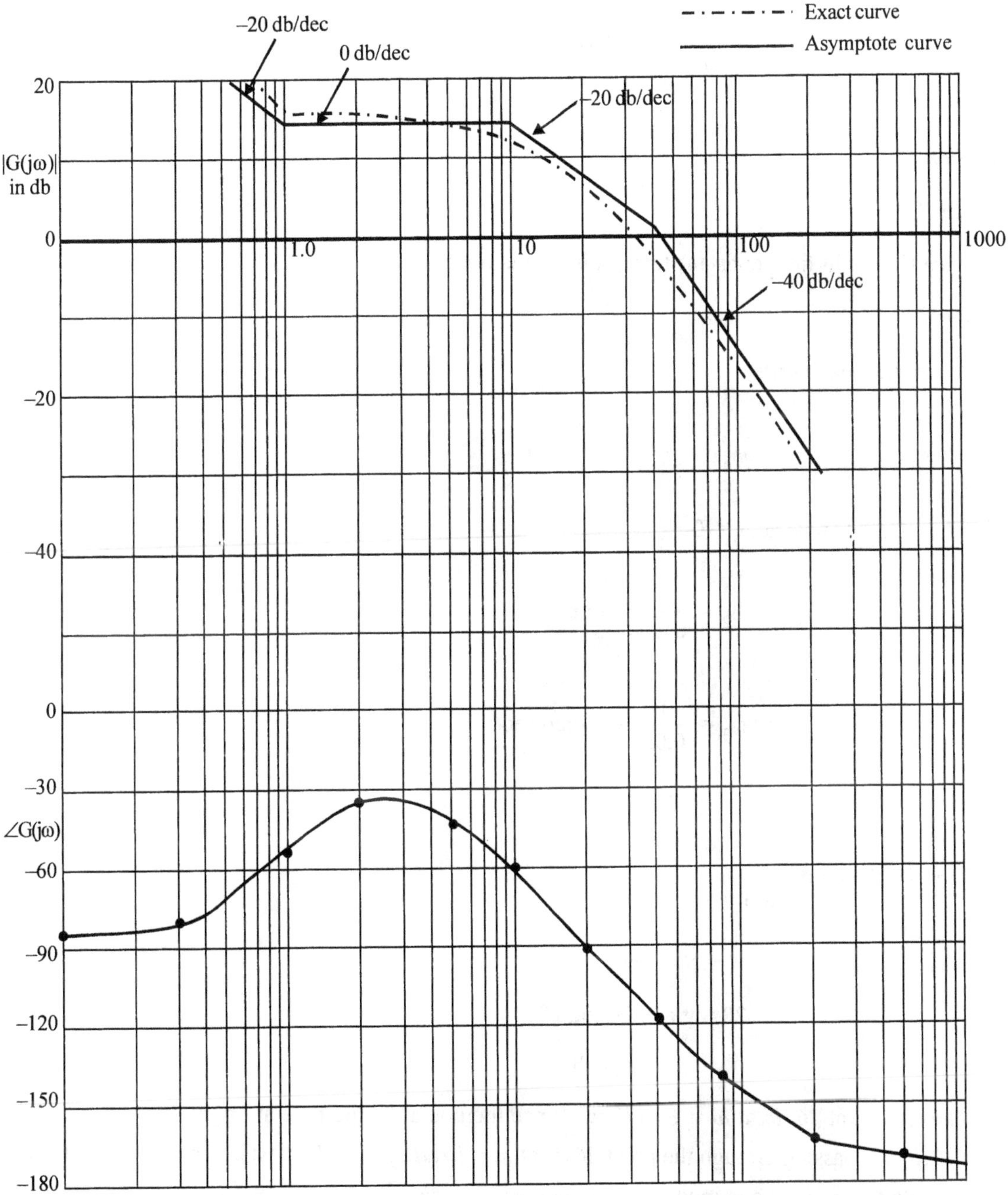

Fig. 6.9 Bode plot for Ex. 7.1

Example 6.2

Obtain magnitude and phase angle Bode plots for the system

$$G(s) = \frac{20(0.1s+1)}{s^2(0.2s+1)(0.02s+1)}$$

Solution :

Step 1

Since G(s) is in time constant form, this step is not necessary.

Step 2

Corner frequencies are

Zeros : $\omega_a = \frac{1}{0.1} = 10$ rad/sec

Poles : Double pole at $\omega_1 = 0$

$$\omega_2 = \frac{1}{0.2} = 5 \text{ rad/sec};$$

$$\omega_3 = \frac{1}{0.02} = 50 \text{ rad/sec.}$$

Step 3

Consider the term $\frac{20}{(j\omega)^2}$

Since $K = 20$

$$20 \log K = 20 \log 20$$

$$= 26 \text{ db}$$

Mark the point *26 db* at ω = *1 rad/sec*. Since there is a double pole at origin, draw a line with *– 40 db/dec* passing through the point *26 db* at ω = *1 rad/sec*. As the next corner frequency is due to a pole at *5 rad/sec* continue this line upto ω = *5 rad/sec*.

Step 4

From $\omega = 5$ *rad/sec* draw a line with slope – *60 db/dec* since there is a pole at *5 rad/sec.*

Continue the line with – *60 db/dec* until the next corner frequency is encountered, which in the present case is a zero with $\omega_a = 10$ *rad/sec.*

This zero contributes a + *20 db/dec* and hence after $\omega = 10$ *rad/sec*, the slope of the line would be

$$(-60 + 20) = -40 \text{ db/dec.}$$

Step 5

Draw a line with slope – *40 db/dec* at this point until the next corner frequency, which in this case is a pole at $\omega_3 = 50$ *rad/sec.* After this frequency, the slope of the magnitude plot will be

$$(-40 - 20) = -60 \text{ db/dec.}$$

Since there are no other poles or zeros, this line will continue as high frequency asymptote.

Step 6

Error table is constructed in Table. 6.3

Table 6.3. Error table for Ex. 6.2

Frequency	Error due to pole and zero factors, in db			Total error in db
	$\frac{1}{1+0.2j\omega}$	$0.1\,j\omega + 1$	$\frac{1}{0.02j\omega+1}$	
2.5	– 1	–	–	– 1
5	– 3	+ 1		– 2
10	– 1	+ 3		+ 2
20		+ 1		1
25			– 1	– 1
50			– 3	– 3
100			– 1	– 1

The corrections are marked on the magnitude plot. A smooth curve is drawn through these points and approaching the low frequency and high frequency asymptotes as shown in Fig. 6.10.

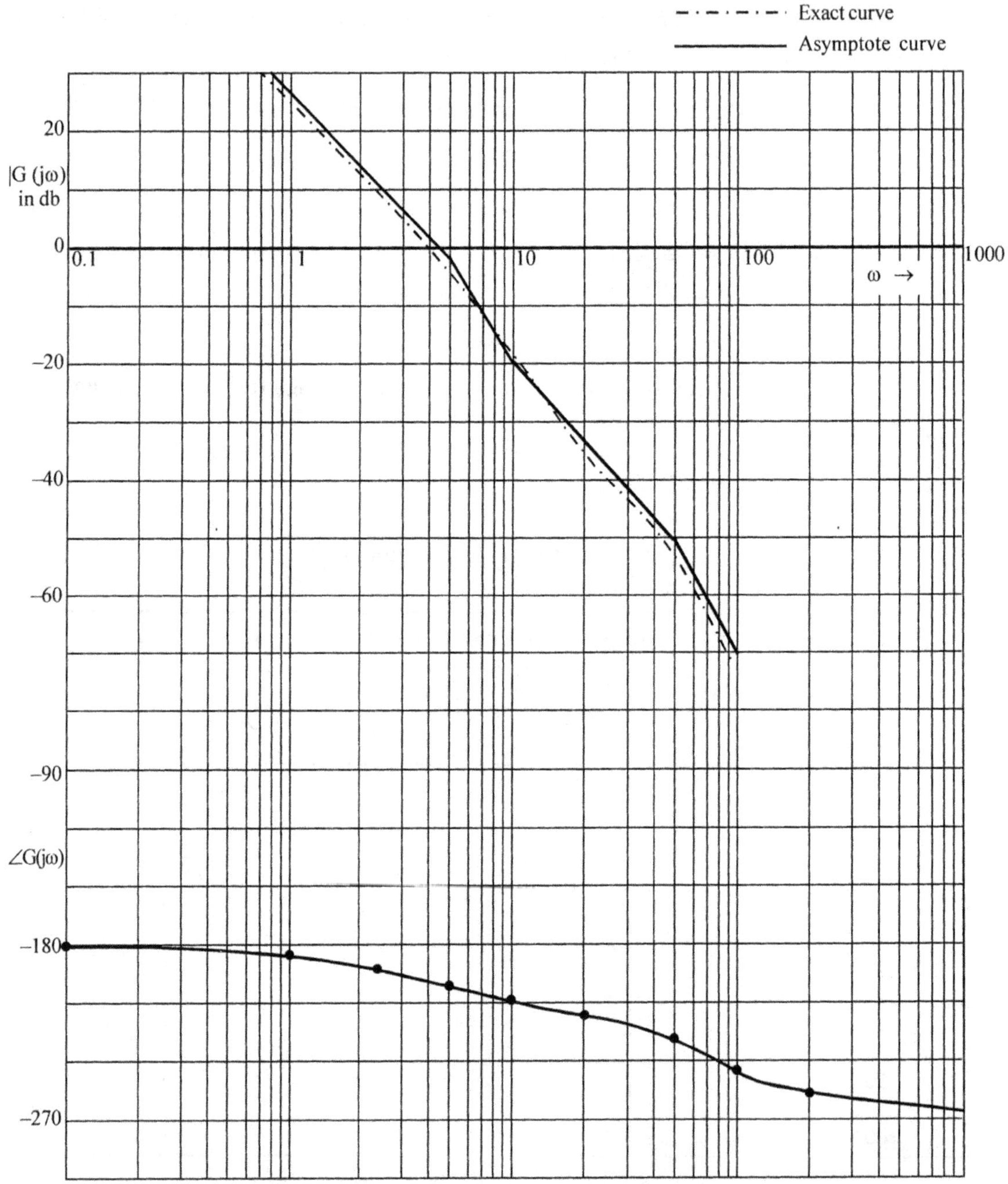

Fig. 6.10 Bode plot for Ex. 6.2

Step 7

The phase angle is calculated at different frequencies as shown in Table 6.4.

Table 6.4 Phase angle of $G(s) = \frac{20(0.1s+1)}{s^2(0.2s+1)(0.02s+1)}$

Frequency	Angles due to pole and zero factors				Total angle
	$\frac{1}{(j\omega)^2}$ $\phi_1 = -180$	$\frac{1}{0.2j\omega+1}$ $\phi_2 = -\tan^{-1} 0.2\omega$	$0.1j\omega+1$ $\phi_3 = \tan^{-1} 0.1\omega$	$\frac{1}{0.02j\omega+1}$ $\phi_4 = -\tan^{-1} 0.02\omega$	$\phi_T = \theta_1 + \theta_2 + \theta_3 + \theta_4$
0.1	– 180	– 1.14	0.57	– 0.11	– 180.68
1.0	– 180	– 11.3	5.71	– 1.14	– 186.73
2.5	– 180	– 26.56	14.0	– 2.86	– 195.42
5.0	– 180	– 45.0	26.56	– 5.71	– 204.16
10.0	– 180	– 63.43	45.0	– 11.3	– 209.73
20.0	– 180	– 76.0	63.43	– 21.8	– 214.37
50.0	– 180	– 84.3	78.69	– 45.0	– 230.63
100	– 180	– 87.0	84.3	– 63.43	– 246.13
200	– 180	– 88.6	87.0	– 75.96	– 257.56
1000	– 180	– 89.7	89.4	– 87.13	– 267.43

The phase angle plot is obtained as shown in Fig. 6.10.

Example 6.3

Obtain Bode plots for the system

$$G(s) = \frac{100}{s(s^2+12s+100)}$$

Solution :

Step 1

Put the transfer function in time constant form. For the complex poles, $\omega_n = 10$ and $2\delta\omega_n = 12$

$$\therefore \delta = \frac{12}{2\times 10} = 0.6$$

$$G(s) = \frac{100}{100s(1+0.12s+0.01s^2)}$$

$$G(j\omega) = \frac{1}{j\omega(1+0.12j\omega-0.01\omega^2)}$$

Step 2

The corner frequencies are

Zeros : None

Poles : $\omega = 0$, $\omega_n = 10$ rad/sec.

Step 3

Draw the low frequency asymptote corresponding to $\frac{1}{j\omega}$

At $\omega = 1$ *rad/sec,*

$$20 \log K = 20 \log 1 = 0 \text{ db.}$$

Draw a line with *–20 db/dec* passing through the point *0 db* at $\omega = 1$ *rad/sec*. Continue this line until the corner frequency of complex pole at $\omega_n = 10$ *rad/sec* is encountered.

Step 4

At $\omega_n = 10$ *rad/sec* since we have a set of complex conjugate poles the slope of the plot will change by *– 40db/dec*. Draw a line with *– 60 db/dec* at $\omega = 10$ *rad/sec.*

Step 5

Since there are no other poles and zeros, this magnitude plot is complete.

Step 6

Draw table of errors. Since we have complex poles with $\omega_n = 10$ *rad/sec* and $\delta = 0.8$, we have to calculate the error at different frequencies around $\omega_n = 10$ *rad/sec* and obtain the actual plot.

Table 6.5 Error table for $G(s) = \frac{1}{s(s^2 + 12s + 100)}$

Frequency	**error** $= -10 \log [(1 - 0.01\,\omega^2)^2 + 0.0144\omega^2]$ **for** $\omega < 10$ $= -10 \log [(1 - 0.01\,\omega^2)^2 + 0.0144\omega^2] + 40 \log \frac{\omega}{10}$ **for** $\omega > 10$
1	0.024
5	0.35
10	– 1.58
20	0.35
50	0.1
100	0.024

The actual plot is obtained in the usual way as shown in Fig. 6.11.

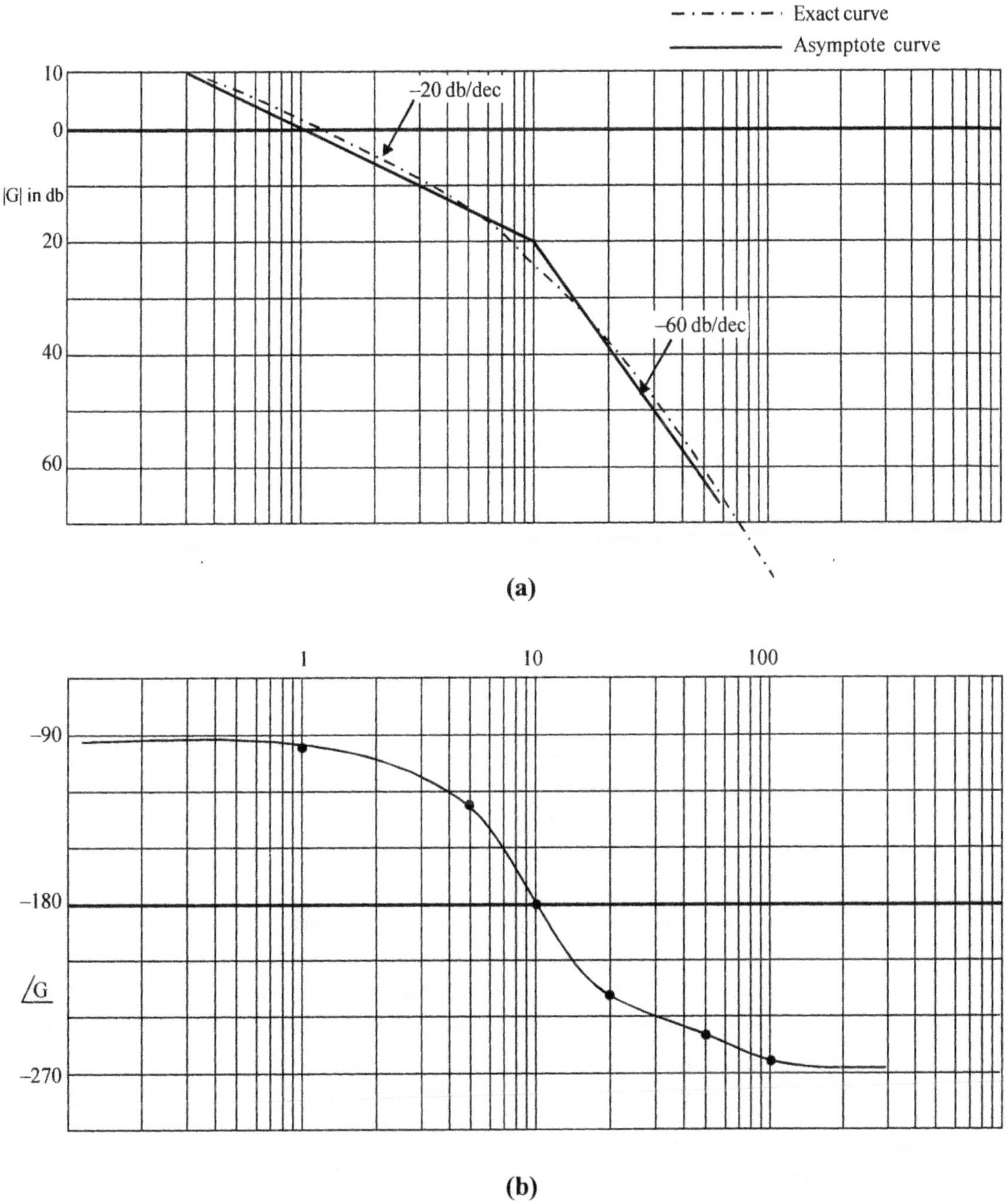

Fig. 6.11 Bode plot for Ex. 6.3 (a) Magnitude plot (b) Phase angle plot

Step 7

The phase angle plot is obtained by actually calculating the phase angle at various frequencies as in Table. 6.6.

Table. 6.6. Phase angle of $G(s) = \dfrac{100}{s(s^2 + 12s + 100)}$

Frequency	Angle due to poles and zeros		
	$\dfrac{1}{j\omega}$ $\phi_1 = -90^0$	$\dfrac{1}{1 + 0.12j\omega + \dfrac{(j\omega)^2}{100}}$ $\phi_2 = -\tan^{-1}\dfrac{0.12\omega}{1 - 0.01\omega^2}$	Total angle $\phi_T = \phi_1 + \phi_2$
1	– 90	– 6.91	– 96.91
5	– 90	– 38.66	– 128.66
10	– 90	– 90.0	– 180.0
20	– 90	– 141.34	– 231.34
50	– 90	– 165.96	– 255.96
100	– 90	– 173.0	– 263.0

Using the values in Table. 6.6 we can obtain the phase angle plot as shown in Fig. 6.11.

6.5.2 Polar Plots

Let us now consider another graphical representation of sinusoidal transfer function $G(j\omega)$. For a given value of ω, $G(j\omega)$ is a complex number and it has magnitude and angle. Thus

$$G(j\omega) = |G(j\omega)| \angle G(j\omega)$$

$$= M \angle\theta \qquad .....(6.32)$$

As ω is changed, both magnitude M and phase angle θ of $G(j\omega)$ can be represented as a phasor with magnitude M and angle θ for a given frequency. As ω is changed from 0 to ∞ this vector changes in magnitude and phase angle and the tip of this phasor traces a curve. This curve is known as polar plot of the given transfer function. This plot is useful in determining the stability of the system in frequency domain, using Nyquist stability criterion, to be discussed in chapter 7. It not only gives absolute stability of the system but also the relative stability.

To draw the exact plot the magnitude and angle are calculated for $\omega = 0$ to ∞ and polar graph sheets are used to plot these values. Polar graph sheet contains concentric circles uniformly spaced and a number of radial lines from the centre of these circles. At any given frequency, magnitude of transfer function can be marked using the circles and angle of the transfer function can be marked using the radial lines.

A sample plot is shown in Fig. 6.12.

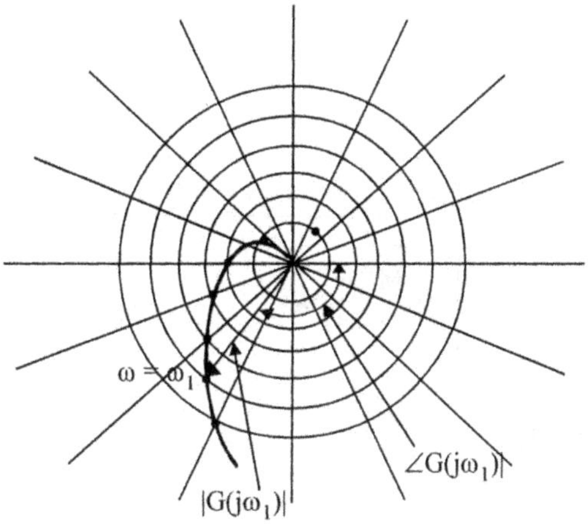

Fig. 6.12 Sample of polar plot

Usually a rough sketch is adequate to determine the stability and other aspects of a system. Let us consider some examples of obtaining the polar plots.

Example 6.4

Draw the polar plot of $G(s) = \dfrac{1}{1 + T_1 s}$

Solution :

The sinusoidal transfer function is given by

$$G(j\omega) = \frac{1}{1 + j\omega T_1}$$

$$= \frac{1}{\sqrt{1 + \omega^2 T_1^{\,2}}} \angle - \tan^{-1} \omega\, T_1$$

$$= M \angle\theta$$

At $\omega = 0$ $\quad M = 1 \quad \theta = 0$

This is represented by point A in Fig. 6. 13.

At $\omega = \infty$ $\quad M = 0 \quad \theta = -90$

This is represented by point 0 in Fig. 6.13.

For any $0 \leq \omega \leq \infty$, $M \leq 1$ and $0 \leq \theta \leq -90$. In fact the locus of the magnitude of $|G(j\omega)|$ can be shown to be a semi circle. The complete polar plot is shown in Fig. 6.13.

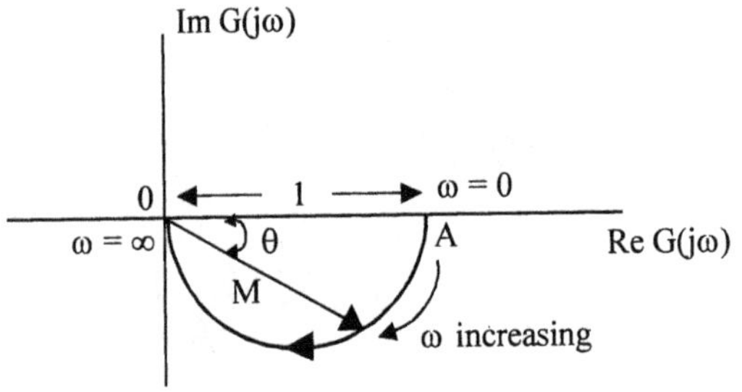

Fig. 6.13 Polar plot of $G(s) = \dfrac{1}{1+T_1 s}$

Example 6.5

Draw the polar plot of $G(s) = \dfrac{1}{s(1+T_1 s)}$

Solution :

$$G(j\omega) = \frac{1}{j\omega(1+j\omega T_1)}$$

$$= \frac{-T_1}{1+\omega^2 T_1^{\,2}} - j\frac{1}{\omega(1+\omega^2 T_1^{\,2})}$$

At $\omega = 0$ $\quad G(j\omega) = -T_1 - j\infty$

$$= \infty \angle -90^0$$

At $\omega = \infty$ $\quad G(j\omega) = -0 - j0 = 0 \angle -180^0$

The polar plot is sketched in Fig. 6.14.

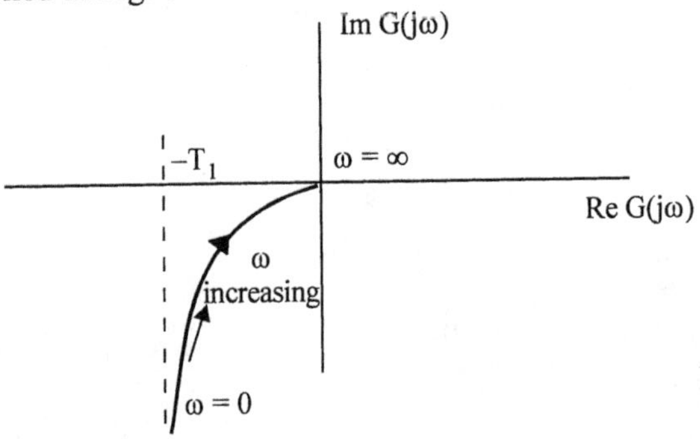

Fig. 6.14 Polar plot of $G(s) = \dfrac{1}{s(1+T_1 s)}$

Comparing the transfer functions of Exs. 6.4 and 6.5 we see that a pole at origin is added to the transfer function of Ex. 6.4. The effect of addition of a pole at origin to a transfer function can be seen by comparing the polar plots in Fig. 6.13 and 6.14. The plot in Fig. 6.13 is rotated by 90^0 in clock wise direction both at $\omega = 0$ and $\omega = 90^0$. At $\omega = 0$ the angle is -90^0 instead of 0 and at $\omega = \infty$ the angle is -180^0 instead of -90^0. We say that the whole plot is rotated by 90^0 in clockwise direction when a pole at origin is added.

A sketch of the polar plot of a given transfer function can be drawn by finding its behaviour at $\omega = 0$ and $\omega = \infty$.

Example 6.6

Draw the polar plot of $G(s) = \dfrac{1}{(1+T_1 s)(1+T_2 s)}$

Solution :

In this example, a non zero pole is added to the transfer function of example 6.4.

Let us examine the effect of this on the polar plot.

At $\omega = 0$ $\quad |G(j\omega)| = M = \left.\dfrac{1}{(1+j\omega T_1)(1+j\omega T_2)}\right|_{\omega=0} = 1$

At $\omega = \infty$ $\quad |G(j\omega)| = M = 0\angle -180^0$

(Since $\omega = \infty$, the real part can be neglected and hence the magnitude is 0 and angle is -180^0)

The polar plot is sketched in Fig. 6.15.

Thus we see that the nature of the plot is unaffected at $\omega = 0$ but the plot rotates by 90^0 in clockwise direction at $\omega = \infty$.

Similarly if a zero is added at some frequency, the polar plot will be rotated by 90 in anticlockwise direction at $\omega = \infty$.

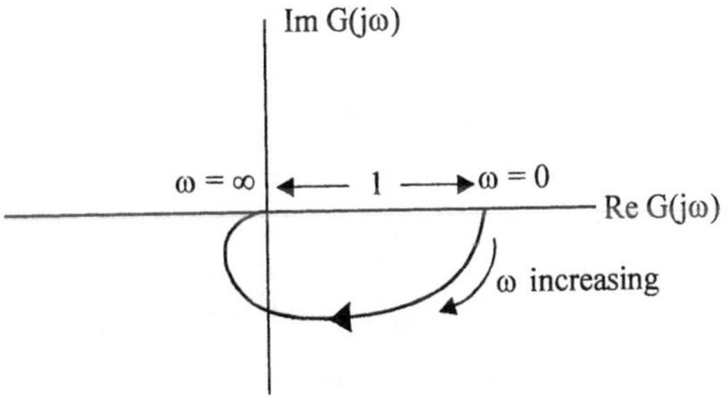

Fig. 6.15 Polar plot of $G(s) = \dfrac{1}{(1+T_1 s)(1+T_2 s)}$

Some examples of polar plots are shown in Fig. 6.16, which clearly demonstrate the aspect of adding poles at origin and poles at other frequencies.

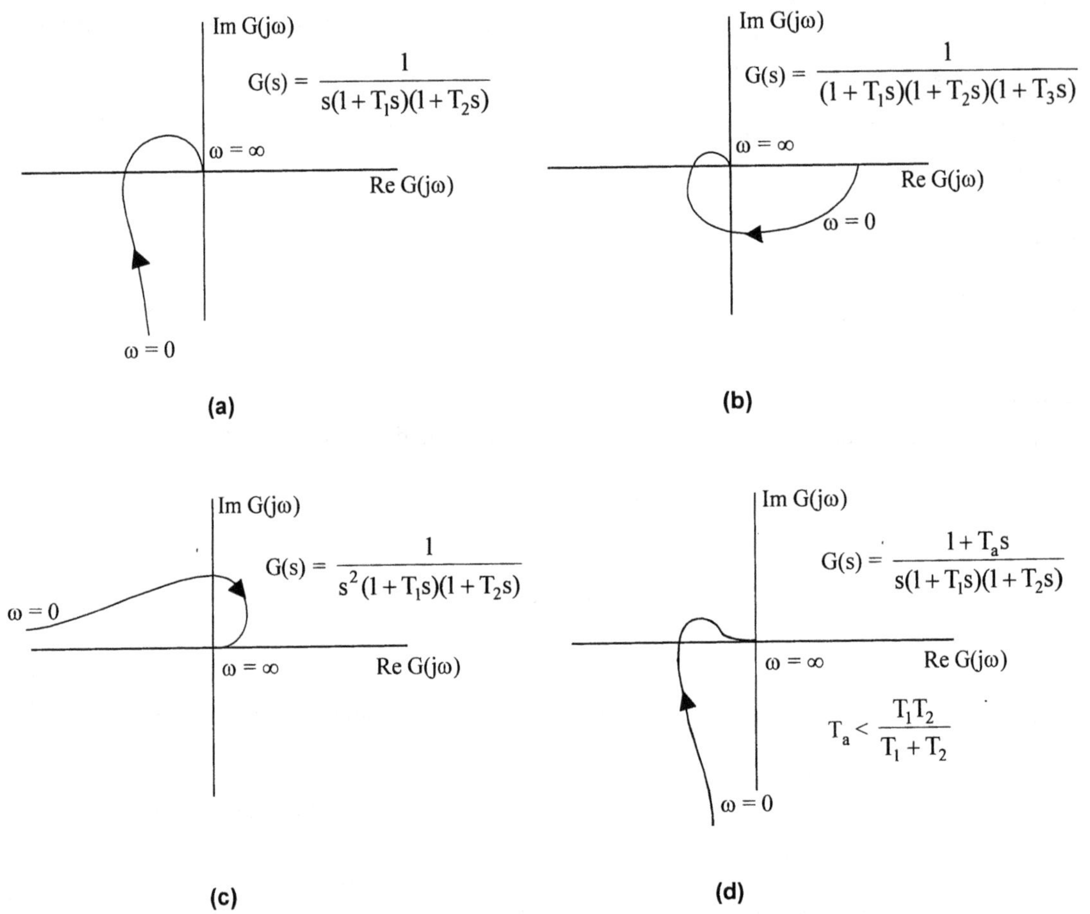

Fig. 6.16 Polar plot of some transfer functions.

6.5.3 Log Magnitude Vs Phase Plots

Another graphical representation of a transfer function is by plotting log magnitude of the transfer function in *db* versus the phase angle at various frequencies. Usually the Bode plots are first obtained and the magnitude in db and phase angle at a given frequency are read from them. Magnitude in db is plotted against phase to obtain the log magnitude Vs phase angle plot as shown in Fig. 6.17.

The advantages of these plots are that the relative stability can be determined with case and compensators can be designed easily.

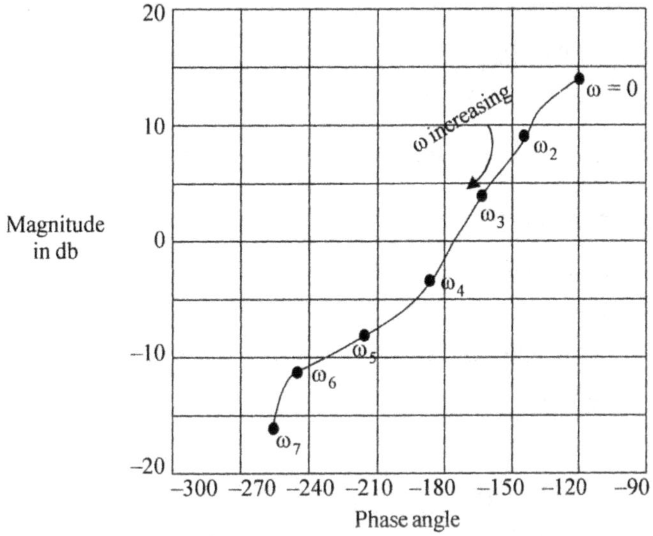

Fig. 6.17 Log magnitude Vs phase angle.

6.6 All Pass Systems

A system which passes all frequencies without any attenuation is known as an all pass system or all pass filter. Such systems are characterised by the transfer functions which have a zero in mirror image position with respect to the imaginary axis of *s*-plane for every pole in the left half of *s*-plane. We have seen in chapter 4, that for stable systems the poles must lie in the left half of *s*-plane only. If for every left half of *s*-plane pole there is a corresponding right half plane zero at mirror image position, such a system is known as an all pass system. Consider the system,

$$G(s) = \frac{1 - Ts}{1 + Ts}$$

The pole zero locations are shown in Fig. 6.18.

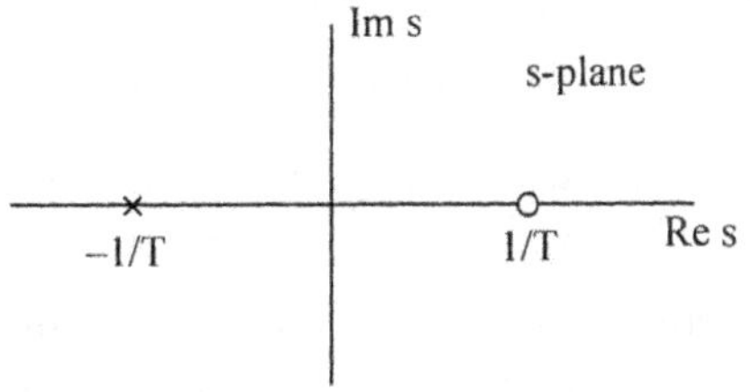

Fig. 6.18 Pole zero pattern of all pass system.

For $s = j\omega$, magnitude of G(s) is

$$|G(j\omega)| = \left|\frac{1 - j\omega T}{1 + j\omega T}\right| = \frac{(1 + \omega^2 T^2)^{\frac{1}{2}}}{(1 + \omega^2 T^2)^{\frac{1}{2}}} = 1$$

The magnitude is independent of ω and hence it passes all frequencies with a gain of unity.

The angle of G (jω) is given by

$$\angle G\,(j\omega) = -\,2\tan^{-1}\omega T$$

As ω is increased from 0 to ∞, the angle changes from 0 to -180^0.

6.7 Minimum Phase Systems

A system with all zeros and poles in the left half of *s*-plane is known as a minimum phase function. For a stable system all poles have to be necessarily in the left half of *s*-planes. However, there is no restriction on the zeros of the transfer function. Consider a system with one zero in the right half of *s*-plane and two poles in the left half of *s*-plane.

$$G(s) = \frac{1 - T_a s}{(1 + T_1 s)(1 + T_2 s)}$$

This can be written as

$$G(s) = \frac{1 + T_a s}{(1 + T_1 s)(1 + T_2 s)}\left(\frac{1 - T_a s}{1 + T_a s}\right)$$

$$= G_1(s).\ G_a(s)$$

Here $G(s)$ is considered as a product of two transfer functions $G_1(s)$ and $G_a(s)$. $G_1(s)$ has all poles and zero in the left half of *s*-plane. $G_a(s)$ is an all pass system. Since the contribution of $G_a(s)$ to magnitude of $G(s)$ is unity, the magnitude plot of $G(s)$ and $G_1(s)$ are identical. But the all pass system contributes angles *0* to $-$ *180* as ω is changed from *0* to -180^0, the angle of $G(j\omega)$ will be more than the angle of $G_1(j\omega)$ for all frequencies. Thus $G(j\omega)$ and $G_1(j\omega)$ have same magnitude plots but different phase angle plots. Thus it is evident that the phase angle of a transfer function could be changed without affecting its magnitude characteristic, by adding an all pass system.

Hence for a given magnitude characteristic $G_1(j\omega)$ has the least phase angle compared to $G(j\omega)$. Thus $G_1(j\omega)$ which has all poles and zeros in left half plane is known as minimum phase system. $G(j\omega)$ which has one or more zeros in right half plane, but has the same magnitude plot is a non minimum phase system. A minimum phase function has a unique magnitude, phase angle relationship. For a non minimum phase function for a given magnitude characteristic, the phase angle is more lagging than the minimum phase function.

These larger phase lags are usually detrimental to the system and hence they are avoided in control systems. The phase angle characteristics of minimum, all pass and non minimum phase functions $G_1(s)$, $G_a(s)$ and $G(s)$ are shown in Fig. 6.19.

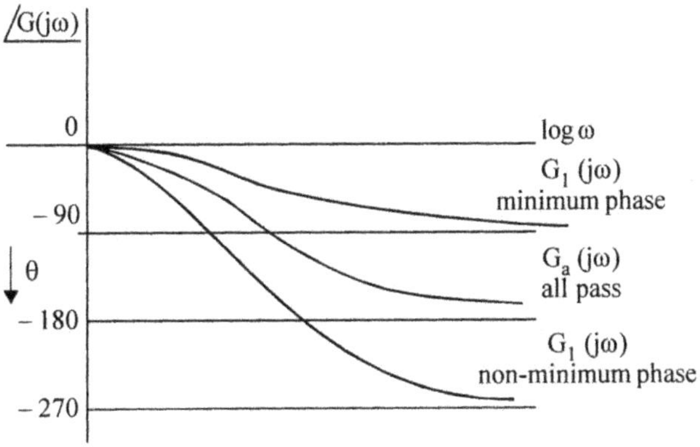

Fig. 6.19 Phase angle characteristics of minimum phase, all pass and non minimum phase systems

For a given non minimum phase transfer function, it is always possible to get the corresponding minimum phase function which has exactly the same magnitude characteristic but least phase angle characteristic. To do this, for every right half plane zero of the given transfer function, associate an all pass function and add a zero in the left half plane at corresponding position so that the magnitude curve remains the same.

Consider

$$G(s) = \frac{(1-2s)(1-5s)}{s(1+3s)(1+10s)}$$

$$= \frac{(1+2s)(1+5s)}{s(1+3s)(1+10s)} \cdot \left(\frac{1-2s}{1+2s}\right) \frac{(1-5s)}{(1+5s)}$$

$$= G_1(s).\ G_{a1}(s)\ G_{a2}(s)$$

$G_1(s)$ is required minimum phase function $G_{a1}(s)$ and G_{a2} are all pass functions. $G_1(s)$ has same magnitude curve as G(s) but has the least phase angle curve.

Problems

6.1 Find the frequency response specifications M_r, ω_r, ω_b for the systems with the following closed loop transfer functions.

(a) $\dfrac{16}{s^2 + 4.8s + 16}$

(b) $\dfrac{32}{s^2 + 8s + 32}$

(c) $\dfrac{100}{s^2 + 16s + 100}$

(d) $\dfrac{64}{s^2 + 6.4s + 64}$

6.2 For the following open loop transfer functions with unity feedback, draw the Bode plot and determine the frequency at which the plot crosses the 0 db line.

(a) $\dfrac{2000}{s(s+2)(s+100)}$

(b) $\dfrac{15(s+5)}{s(s^2+16s+100)}$

(c) $\dfrac{10s}{(0.1s+1)(0.01s+1)}$

6.3 Sketch the Bode flot for the following systems and determine the value of K for which the magnitude plot crosses the 0 db line at $\omega = 15$ rad/sec.

(a) $\dfrac{K(s+2)}{s(s+4)(s+10)}$

(b) $\dfrac{K}{s(1+s)(1+0.1s)(1+0.01s)}$

6.4 The asymptotic approximation of the gain plot is given in the following figures. Obtain the open loop transfer function in each case.

(a)

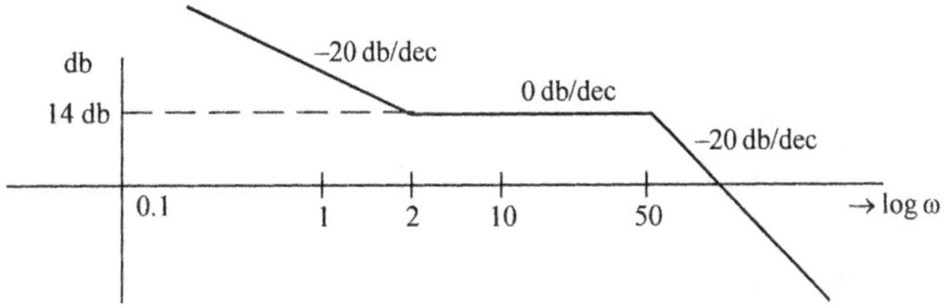

(b)

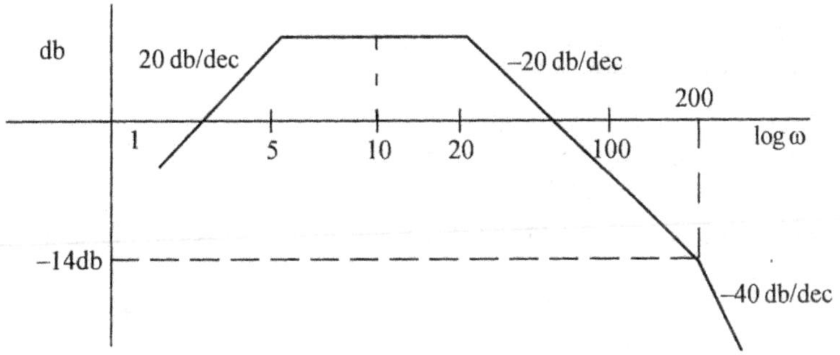

6.5 The following data refers to the frequency response test conducted on an open loop system. Plot the Bode magnitude plot and approximate by asymptotes to obtain the transfer function.

ω rad/sec	0.2	0.5	1	1.5	2	2.5	4	5	8	10	20	50
Gain	9.85	6.68	1.57	0.914	0.62	0.46	0.25	0.19	0.1	0.072	0.022	0.004

6.6 For the following unity feedback system obtain the values of K and τ to get a peak resonance of 1.26 at a resonance frequency of 10.5 rad/sec.

$$G(s) = \frac{K}{s(\tau s + 1)}$$

Also find the peak overshoot and settling time for a unit step input.

6.7 Draw the frequency response curve of the following closed loop transfer function and obtain M_r and ω_r.

$$\frac{C(s)}{R(s)} = \frac{540}{(s+15)(s^2+4s+36)}$$

Where will the poles of equivalent second order system be located.

6.8 Sketch the polar plots of the following open loop transfer functions. Find the frequency at which the plot crosses the negative real axis and the magnitude of G(jω) at this frequency.

(a) $\frac{10}{(s+1)(s+3)}$

(b) $\frac{10}{s(s+1)(s+5)}$

(c) $\frac{10}{(s+1)(s+3)(s+5)}$

(d) $\frac{(1+0.2s)}{s^2(1+0.01s)(1+0.05s)}$

6.9 Sketch the polar plot of the function,

$$G(s) = \frac{10}{s(s+5)(s+10)}$$

For what value of ω, |G(jω)| = 1. At this frequency what is the phase angle of G(jω).

7 Nyquist Stability Creterion and Closed Loop Frequency Response

7.1 Introduction

In chapter 4, it was shown that the location of the roots of the characteristic equation determines whether a system is stable or not. If all the roots lie in the left half of s-plane, the system is absolutely stable. If simple roots are present on the imaginary axis of the s-plane, sustained oscillations in the system will result. Using Routh Hurwitz criterion, a simple way of determining the roots in the right half of s-plane or on the imaginary axis was discussed. But Routh-Hurwitz criterion gives only absolute stability but does not tell us about how much stable the system is i.e., it can not throw any light on relative stability of the system. Root locus technique discussed in chapter 5, can be used to determine the location of closed loop poles on the s-plane from the location of poles and zeros of the open loop system when one parameter of the system, usually the gain, is varied. This also does not give a measure of the relative stability of the system.

A frequency domain technique is developed in this chapter, which gives a simple way of determining the absolute stability of the system, and also defines and determines the relative stability of a system.

This frequency domain criterion is known as Nyquist Stability Criterion. This method relates the location of the closed loop poles of the system with the frequency response of the open loop system. It is a graphical technique and does not require the exact determination of the closed loop poles. Open loop frequency response can be obtained by subjecting the system to a sinusoidal input of constant amplitude and variable frequency and measuring the amplitude and phase angle of the output.

The development of Nyquist Criterion is based on a theorem due to Cauchy, 'the principle of argument' in complex variable theory.

7.2 Principle of Argument

Consider a function of complex variable '*s*', denoted by *F*(*s*), which can be described as a quotient of two polynomials. Assuming that the two polynomials can be factored, we have

$$F(s) = \frac{(s+z_1)(s+z_2).....(s+z_m)}{(s+p_1)(s+p_2).....(s+p_n)} \quad .....(7.1)$$

Since $s = \sigma + j\omega$ is a complex variable, for any given value of *s*, *F*(*s*) is also complex and can be represented by $F(s) = u + jv$. For every point *s* in the *s*-plane at which *F*(*s*) and all its derivatives exist, ie, for points at which *F*(*s*) is analytic, there is a corresponding point in the *F*(*s*) plane. It means that *F*(*s*) in eqn. (7.1) maps points in *s*-plane at which *F*(*s*) is analytic into points in *F*(*s*) plane. In eqn. (7.1) $s = -p_1$, $s = -p_2$ are the poles of the function *F*(*s*) and therefore the function goes to infinity at these points. These points are also called singular points of the function *F*(*s*).

Now, consider a contour τ_s in *s*-plane as shown in Fig. (7.1) (a). Assume that this contour does not pass through any singular points of *F*(*s*). Therefore for every point on this contour, we can find a corresponding point in *F*(*s*) plane, or corresponding to the contour τ_s in *s*-plane there is a contour τ_f in *F*(*s*) plane as shown in Fig. 7.1 (b).

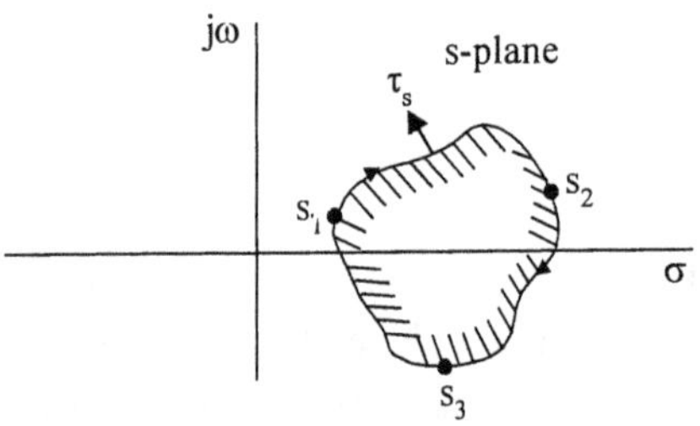

Fig. 7.1 (a) Arbitrary contour τ_s in s-plane not passing through singular points of F(s)

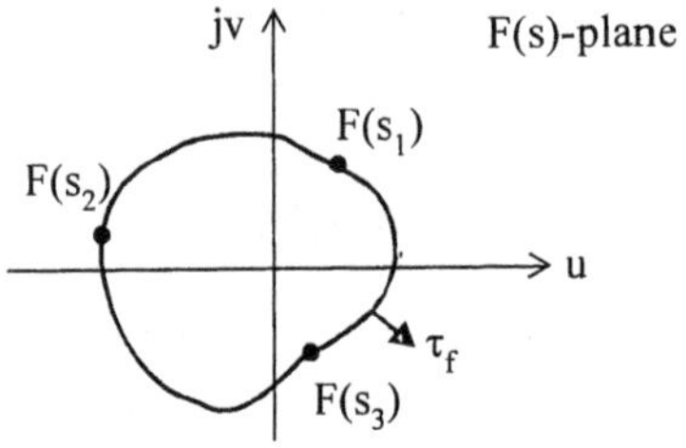

Fig. 7.1 (b) Corresponding F(s)-plane contour τ_f

Let us consider a closed contour and define the region to the right of the contour, when it is traversed in clockwise direction, to be enclosed by it. Thus the shaded region in Fig. 7.1(a) is considered to be enclosed by the closed contour τ_s. Let us investigate some of the properties of the mapping of this contour on to *F*(*s*)-plane when τ_s encloses (a) a zero of *F*(*s*) (b) a pole of *F*(*s*).

Case a : When τ_s encloses a zero of *F*(*s*)

Let $s = -z_1$ be encloses by the contour τ_s as shown in Fig. (7.2) (a).

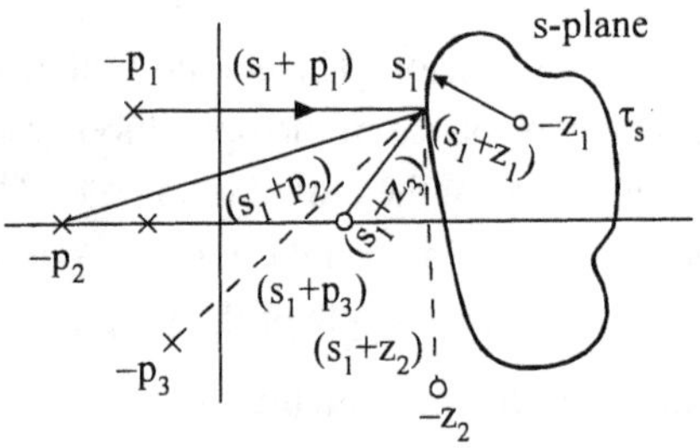

Fig. 7.2 (a) Contour t_s encloses one zero $s = z_1$ of F(s)

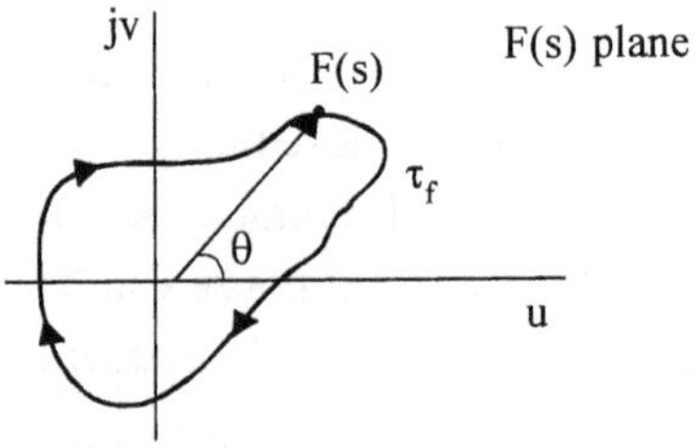

Fig. 7.2 (b) Corresponding F(s) plane contour τ_f

For any point s = s_1, we have

$$F(s_1) = \frac{(s_1+z_1)(s_1+z_2).....(s_1+z_m)}{(s_1+p_1)(s_1+p_2).....(s_1+p_n)}$$

$$= \frac{\alpha_1\,\alpha_2.....\alpha_m}{\beta_1\,\beta_2.....\beta_m} \underline{/\theta_1+\theta_2+....\theta_m-\phi_1-\phi_2....\phi_n} \quad .....(7.2)$$

where $\alpha_1 = |s_1 + z_1|$, $\alpha_2 = |s_1 + z_2|$, ... $\alpha_m = |s_1 + z_m|$

$\beta_1 = |s_1 + p_1|$, $\beta_2 = |s_1 + p_2|$, ... $\beta_n = |s_1 + p_n|$

and $\theta_1 = \underline{/(s_1 + z_1)}$, $\theta_2 = \underline{/(s_1 + z_2)}$, ... $\theta_m = \underline{/(s_1 + z_m)}$

$\phi_1 = \underline{/(s_1 + p_1)}$, $\phi_2 = \underline{/(s_1 + p_2)}$, ... $\phi_n = \underline{/(s_1 + p_n)}$

In the development of Nyquist criterion the magnitude of $F(s)$ is not important, as we will see later. Let us concentrate on the angle of $F(s)$.

In Fig. 7.2(a), as the point s moves on the contour τ_s in clockwise direction, and returns to the starting point, let us compute the total angle described by $F(s)$ vector as shown in Fig. 7.2(b).

The vector $(s + z_1)$ contributes a total angle of -2π to the angle of $F(s)$ as shown in Fig. 7.3(a) since the vector $(s + z_1)$ makes one complete rotation. This is because the point $s = -z_1$ lies inside the contour.

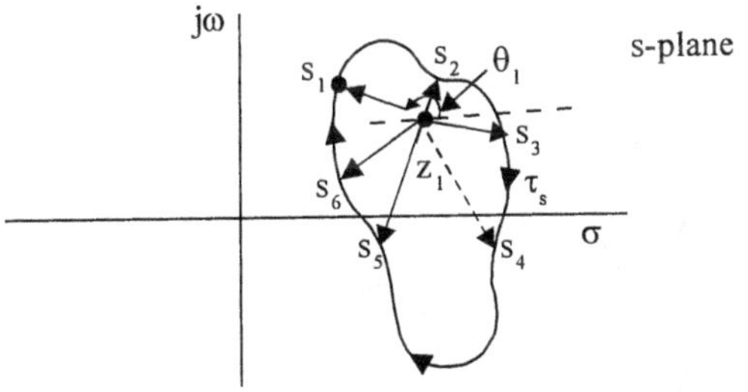

Fig. 7.3 (a) Angle contributed by $(s + z_1)$ to F(s)

The vector $(s + z_2)$, contributes zero net angle for one complete traversal of the point s on the contour τ_s in s-plane as shown in Fig. 7.3(b). This is because the point $s = -z_2$ lies outside the contour τ_s.

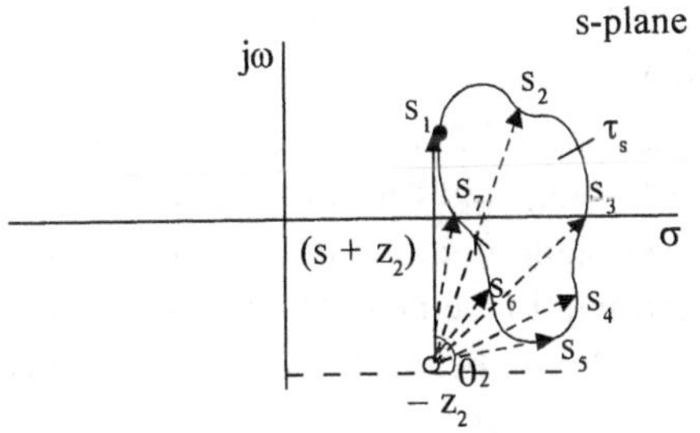

Fig. 7.3 (b) Angle contributed by $(s + z_2)$ to F(s)

Similarly all the zeros and poles which are not enclosed by the contour τ_s, contribute net zero angles to $F(s)$ for one complete traversal of a point s on the contour τ_s. Thus the total angle contributed by all the poles and zeros of $F(s)$ is equal to the angle contributed by the zero $s = -z_1$ which is enclosed by the contour. The $F(s)$ vector describes an angle of -2π and therefore the tip of $F(s)$ vector describes a closed contour about the origin of $F(s)$ plane in the clockwise direction. Similarly if k zeros are enclosed by the s-plane contour, the $F(s)$ contour will encircle the origin k times in the clockwise direction. Two cases for $k = 2$ and $k = 0$ are shown in the Fig. 7.4 (a) and (b).

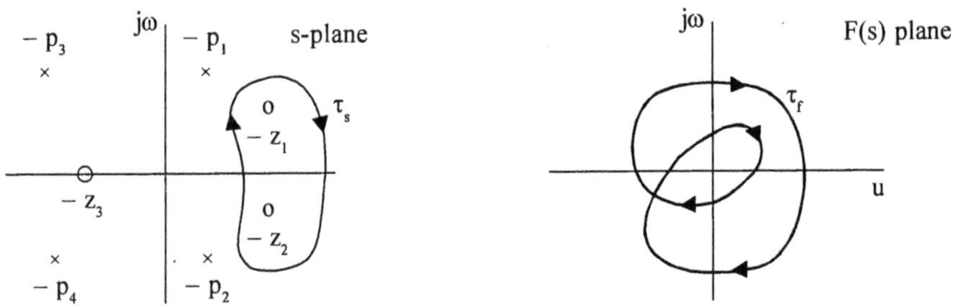

Fig. 7.4 (a) s-plane contour and F(s) plane contour for k = 2.

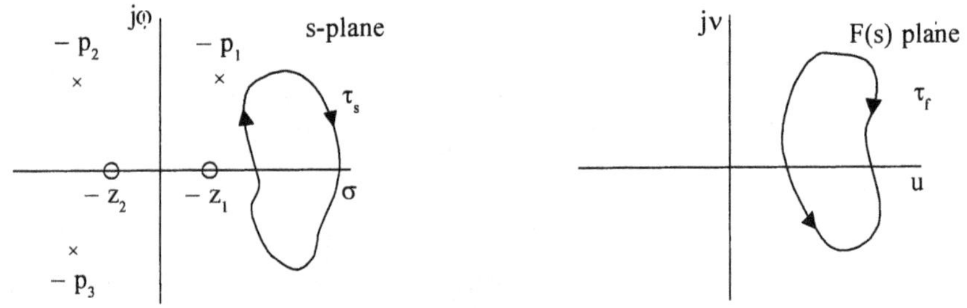

Fig. 7.4 (b) s-plane contour and F(s) plane contour for k = 0.

Case b : When τ_s encloses a pole of F(s)

When a pole of $F(s)$, $s = -p_1$ is enclosed by the contour τ_s, a net angle of 2π is contributed by the vector $(s + p_1)$ to $F(s)$ as the factor $(s + p_1)$ is in the denominator of $F(s)$. Thus the $F(s)$ plane contour will encircle the origin once in the anticlockwise direction.

If s-plane contour τ_s encloses P poles and Z zeros, the $F(s)$ plane contour encircles the origin P times in the anticlockwise direction and Z times in the clockwise direction. In otherwords, it encircles the origin of $F(s)$ plane (P – Z) times in the anticlockwise direction. The magnitude and hence the actual shape of the $F(s)$ plane contour is not important, but the number of times the contour encircles the origin is important in the development of Nyquist stability criterion, as will be discussed in the next section. This relation between the number of poles and zeros enclosed by the closed s-plane contour τ_s, and the number of encirclements of $F(s)$ plane contour τ_f is known as the principle of Argument.

7.2.1 Nyquist Criterion

For a feedback control system with loop transfer function given by :

$$G(s)\,H(s) = \frac{K(s+z_1)(s+z_2).....(s+z_m)}{(s+p_1)(s+p_2).....(s+p_n)} \quad m \le n$$

$$= K\frac{\pi_{i=1}^{m}(s+z_i)}{\pi_{j=1}^{n}(s+p_j)} \qquad .....(7.3)$$

The characteristic equation is given by

$$D(s) = 1 + G(s)\,H(s) = 1 + \frac{K\,\pi_{i=1}^{m}(s+z_i)}{\pi_{j=1}^{n}(s+p_j)} = 0$$

$$= \frac{\pi_{j=1}^{n}(s+p_j) + K\,\pi_{i=1}^{m}(s+z_i)}{\pi_{j=1}^{n}(s+p_j)} = 0 \qquad .....(7.4)$$

The numerator of eqn. (7.4) is a polynomial of degree n and hence it can be factored and written as

$$D(s) = \frac{(s+z_1')(s+z_2').....(s+z_n')}{\pi_{j=1}^{n}(s+p_j)} = 0 \qquad .....(7.5)$$

Thus it can be observed that :

1. The poles of the open loop system G(s) H(s) and poles of D(s) are the same (eqns. (7.3) and (7.5))
2. The roots of the characteristic equation D(s) = 0 are the zeros of D(s) given by $-z_1', -z_2',....-z_n'$ in eqn. (7.5).
3. The closed loop system will be stable if all the poles of the closed loop system, ie, all the roots of the characteristic equation lie in the left half of s-plane. In otherwords, no pole of the closed loop system should be in the right half of s-plane.
4. From eqn. (7.4) and (7.5), it is clear that even if some poles of open loop transfer function $(-p_1, -p_2, ... -p_n)$ lie in right half of s plane, the closed loop poles, or the zeros of D(s) = 0 ie $s = -z_1', -z_2'$ etc many all lie in the left half of s-plane. Thus even if the open loop system is unstable, the closed loop system may be stable.

In order to determine the stability of a closed loop system, we have to find if any of the zeros of characteristic equation D(s) = 0 in eqn. (7.5) lie in the right half of *s*-plane. If we consider an *s*-plane contour enclosing the entire right half of *s*-plane, plot the D(s) contour and find the number of encirclements of the origin, we can find the number of poles and zeros of D(s) in the right half of *s*-plane. Since the poles of D(s) are the same as open loop poles, the number of right half plane poles are known. Thus we can find the number of zeros of D(s) ie the number of closed loop poles in the right half of *s*-plane. If this number is zero, then the closed loop system is stable, otherwise the system is unstable.

7.3 Development of Nyquist Criterion

7.3.1 Nyquist Contour

Let us consider a closed contour, τ_N which encloses the entire right half of *s*-plane as shown in Fig. 7.5. This contour is known as a Nyquist Contour. It consists of the entire $j\omega$-axis and a semicircle of infinite radius.

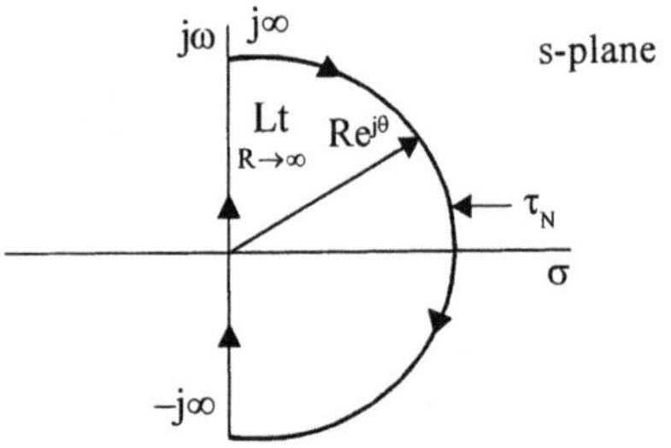

Fig. 7.5 Nyquist Contour

On the jω-axis,

$$s = j\omega \text{ and } \omega \text{ varies from } -\infty \text{ to } +\infty.$$

On the infinite semicircle,

$$s = \underset{R\to\infty}{\text{lt}} Re^{j\theta}, \ \theta \text{ varies from } +\frac{\pi}{2} \text{ to } 0 \text{ to } -\frac{\pi}{2}$$

Thus the Nyquist Contour encloses the entire right half of *s*-plane and is traversed in the clockwise direction.

7.3.2 Nyquist Stability Criterion

If $$D(s) = \frac{\prod_{i=1}^{n}(s+z_i)}{\prod_{j=1}^{m}(s+p_j)} \qquad .....(7.6)$$

is plotted for values of *s* on the Nyquist contour, the D(s) plane contour will encircle the origin N times in the counter clockwise direction, where

$$N = P - Z \qquad .....(7.7)$$

and P = number of poles of D(s) or the number of open loop poles in the right half of *s*-plane (R.H.S)

Z = Number of zeros of D(s) or the number of closed loop poles in the RHS.

If the closed loop system is stable,

$$Z = 0$$

Thus, for a stable closed loop system,

$$N = P \quad .....(7.8)$$

i.e., the number of counter clockwise encirclements of origin by the D(s) contour must be equal to the number of open loop poles in the right half of *s*-plane. Further, if the open loop system is stable, there are no poles of G(s) H(s) in the RHS and hence,

$$P = 0$$

∴ For stable closed loop system,

$$N = 0 \quad .....(7.9)$$

i.e., the number of encirclements of the origin by the D(s) contour must be zero.

Also observe that $G(s)\,H(s) = [1 + G(s)\,H(s)] - 1$

Thus G(s) H(s) contour and D(s) = 1 + G(s) H(s) differ by 1. If 1 is substracted from D(s) = 1 + G(s) H (s) for every value of *s* on the Nyquist Contour, G(s) H(s) contour will be obtained and the origin of D(s) plane corresponds to the point (– 1, 0) of G(s) H(s) plane, this is shown graphically in Fig. 7.6.

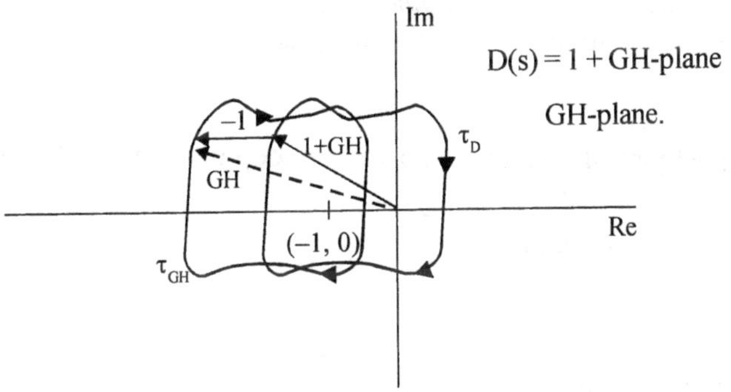

Fig. 7.6 D(s) = 1 + GH plane and GH plane Contours

If G(s) H(s) is plotted instead of 1 + G(s) H(s), the G(s) H(s) plane contour corresponding to the Nyquist Contour should encircle the (–1, j0) point P time in the counter clockwise direction, where P is the number of open loop poles in the RHS. The Nyquist Criterion for stability can now be stated as follows :

If the τ_{GH} Contour of the open loop transfer function G(s) H(s) corresponding to the Nyquist Contour in the *s*-plane encircles the (–1, j0) point in the counter clockwise direction, as many times as the number of poles of G(s) H(s) in the right half of *s*-plane, the closed loop system is stable. In the more common special case, where the open loop system is also stable, the number of these encirclements must be zero.

7.3.3 Nyquist Contour When Open Loop Poles Occur on jω-axis

If G(s) H(s) has poles on the jω-axis, 1 + G(s) H(s) also has these poles on the jω-axis. As the Nyquist Contour defined in section 7.2.2 passes through these jω-axis poles, this Contour is not suitable for the study of stability. No singulasitics of 1 + G(s) H(s) should lie on the *s*-plane Contour τ_s. In such cases a small semicircle is taken around these poles on the jω-axis towards the RHS so that these poles are bypassed. This is shown in Fig. 7.7.

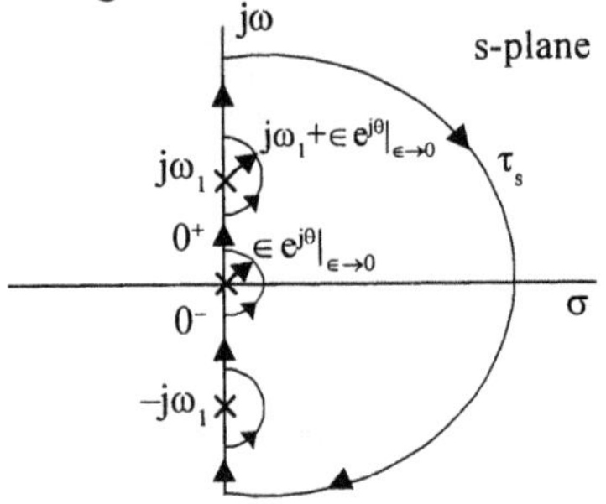

Fig. 7.7 Nyquist Contour when jω-axis poles are present

If an open loop pole is at $s = \pm j\omega_1$, near this point, *s* is taken to vary as given be eqn. (7.10).

$$s = j\omega_1 + \underset{\epsilon\to 0}{\text{lt}}\ \epsilon\, e^{j\theta},\ \theta \text{ changes from } -\frac{\pi}{2} \to 0 \to \frac{\pi}{2} \qquad \text{.....(7.10)}$$

This describes the semicircle around the pole $s = j\omega_1$ in the anti clockwise direction. Let us now consider some examples illustrating how the Nyquist plots are constructed and the stability deduced.

Example 7.1

Let us obtain the Nyquist plot of a system whose open loop transfer function G(s) H(s) is given by

$$G(s)\,H(s) = \frac{10}{(s+2)(s+4)}$$

Solution :

The Nyquist path is shown in Fig. 7.8(a).

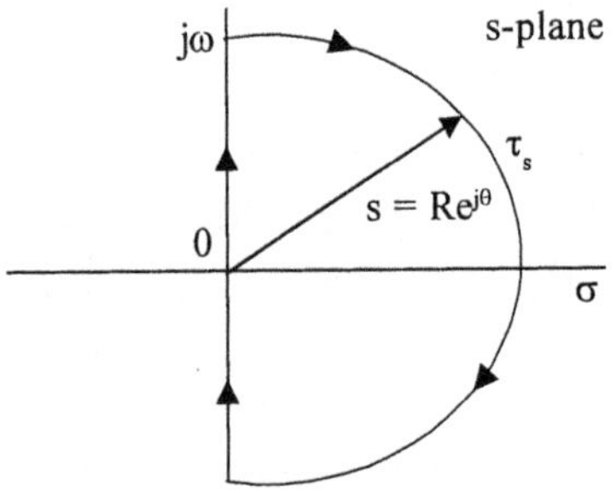

Fig. 7.8 (a) Nyquist path

Let us plot G(s) H(s) over the contour τ_s.

1. On the jω-axis if ω changes from 0 to +∞

$$s = j\omega$$

$$|G(s)\,H(s)|_{s=j\omega} = \frac{10}{(j\omega+2)(j\omega+4)}$$

For $\omega = 0$ $\quad |G(s)\,H(s)| = \frac{10}{8} = 1.25; \quad \angle G(s)\,H(s) = 0$

For $\omega = \infty$ $\quad |G(s)\,H(s)| = 0; \quad \angle G(s)\,H(s) = -180^0$

As the exact shape of G(s) H(s) is not required we can draw this part of the GH Contour by noting that the angle of GH changes from 0 to – 180 as ω changes from 0 to ∞.

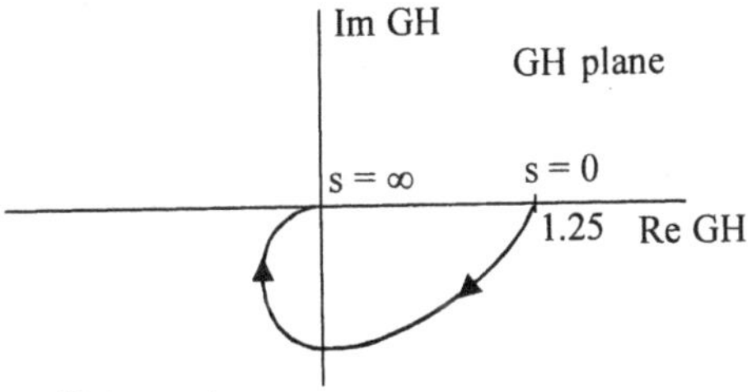

Fig. 7.8 (b) Part of GH plot for s = j0 to s = j∞

2. For the infinite semi circle described by $s = j\infty$ to 0 to $s = -j\infty$. For this part of the contour,

$$s = \lim_{R\to\infty} R\,e^{j\theta} \quad \theta = \frac{\pi}{2} \to 0 \to -\frac{\pi}{2}$$

$$G(s)\,H(s) = \frac{10}{(Re^{j\theta}+2)(Re^{j\theta}+4)}$$

As $R \to \infty$, $Re^{j\theta} >> 2$ and $Re^{j\theta} >> 4$

$$\therefore \quad G(s)\,H(s) = \lim_{R\to\infty} \frac{10}{R^2\,e^{j2\theta}} = \lim_{r\to 0} r\,e^{-j2\theta}$$

$$\text{where } r = \frac{10}{R^2} \text{ and as } R \to \infty,\ r \to 0$$

As $R \to \infty$, $|G(s)\,H(s)| \to 0$ and as θ changes from $\frac{\pi}{2}$ to 0 to $-\frac{\pi}{2}$

$$\angle G(s)\,H(s) = -\pi \text{ to } 0 \text{ to } \pi$$

This part of the G(s) H(s) plot is shown in Fig. 7.8(c).

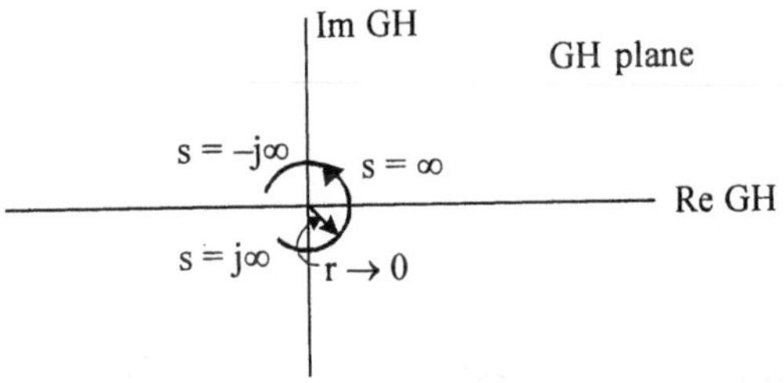

Fig. 7.8 (c) Nyquist plot corresponding to the Nyquist path of infinite semicircle.

3. For the Nyquist path on the jω-axis

$$s = j\omega \text{ and } \omega = -\infty \text{ to } \omega = 0,$$

$$G(s)\,H(s) = \frac{10}{(j\omega + 2)(j\omega + 4)}$$

At $s = -j\infty$ $\quad G(s)\,H(s) = \dfrac{10}{(-j\infty)(-j\infty)} = 0\ \angle{+180}$

At $s = j0$ $\quad G(s)\,H(s) = \dfrac{10}{(j0 + 2)(j0 + 4)} = 1.25\angle 0$

This part of the Nyquist plot is shown in Fig. 7.9 (d). The angle changes from +180 to 0 as ω changes from −∞ to 0.

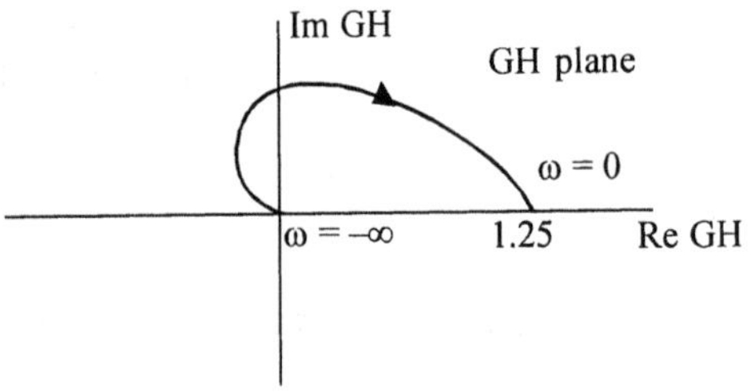

Fig. 7.9 (d) Nyquist plot for s = −j∞ to s = j0

The complete Nyquist plot for the contour τ_s is shown in Fig. 7.9(e).

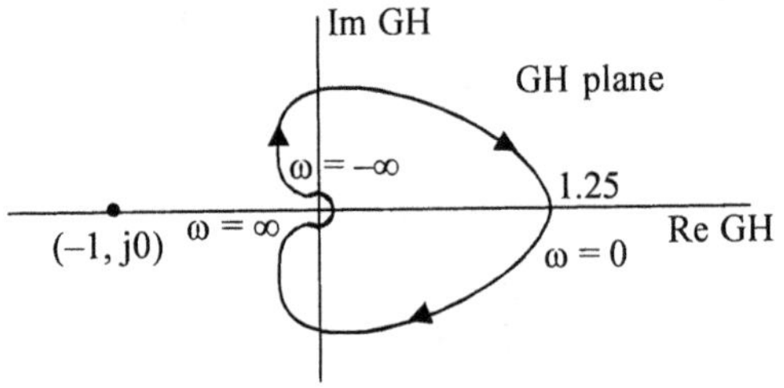

Fig. 7.9 (e) Complete Nyquist plot of $G(s)\,H(s) = \dfrac{10}{(s + 2)(s + 4)}$

The (−1, j0) point is also shown in the Fig. 7.9 (e), It is obvious that the Nyquist plot does not encircle the point (−1, j0). Therefore N = 0 Since there are no poles of G(s) H(s) in the right half of *s*-plane,

$$P = 0$$

$$\therefore \quad N = P - Z$$

$$0 = 0 - Z$$

or $\quad Z = 0$

Thus there are no zeros of 1 + G(s) H(s), ie., poles of the closed loop system in the right half of *s*-plane. Hence the system is stable.

Recall that G(s) H(s) is given by,

$$G(s)\,H(s) = \frac{\underset{i=1}{\overset{m}{\pi}}(s+z_i)}{\underset{j=1}{\overset{n}{\pi}}(s+z_j)} \text{ and } m \le n.$$

For many practical systems $m < n$. For such systems the infinite semicircle always maps into origin and the Nyquist plot goes round the origin with radius tending to zero. Since we are interested in the encirclement of the point (–1, j0), this part of the Nyquist plot is not required for determining the stability. Hence the mapping of infinite semicircle can be considered to be mapped on to the origin of GH plane. Moreover, since

$$G^*(j\omega)\,H^*(j\omega) = G(-j\omega)\,H(-j\omega)$$

the GH plot is symmetrical about the real axis. Hence if GH plot is obtained for values of s on the positive imaginary axis, the plot for values of s on the negative imaginary axis will be its mirror image. Therefore it is sufficient to plot the GH plot for $s = j0$ to $s = j\infty$. The infinite semi circle maps into origin. The plot for $s = -j\infty$ to $s = 0$ is the mirror image of its plot for $s = j0$ to $s = j\infty$.

Example 7.2

Determine the stability of the system

$$G(s)\,H(s) = \frac{10}{s(s+1)(s+4)}$$

Solution :

The Nyquist path is shown in Fig. 7.10(a).

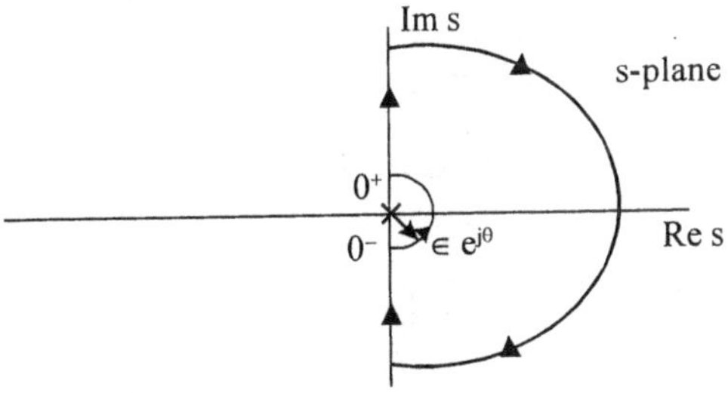

Fig. 7.10 (a) Nyquist path

1. For points on positive jω axis

$$s = j\omega \qquad 0^+ < \omega < \infty$$

$$G(s)\,H(s) = \frac{10}{j\omega(j\omega+1)(j\omega+4)}$$

For $\quad s = j0^+$

$$G(s)\,H(s) = \frac{10}{j0(j0+1)(j0+4)} = \infty \angle -90$$

For $s = j\infty$

$$G(s)\,H(s) = \frac{10}{j\infty(j\infty+1)(j\infty+4)} = 0\ \angle{-270}$$

This part of locus of G(s) H(s) corresponds to the curve ABO in Fig. 7.10(b).

2. The infinite semicircle maps on to the origin.
3. The locus of G(s) H(s) for $s = -j\infty$ to $s = j0$ on the negative imaginary axis corresponds to curve OBC in Fig. 7.10(b).

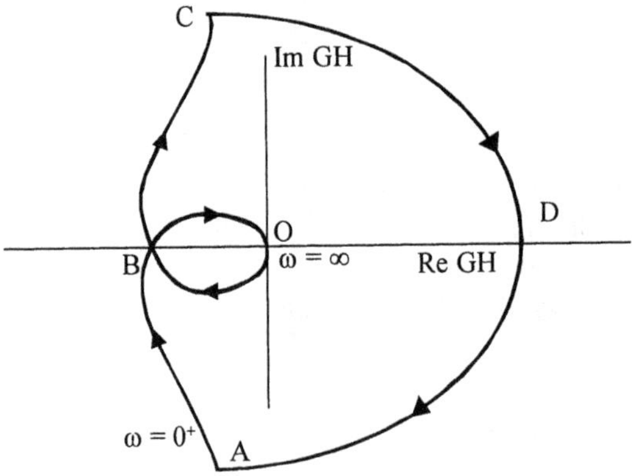

Fig. 7.10 (b) Complete Nyquist plot of $G(s)\,H(s) = \dfrac{10}{s(s+1)(s+4)}$

4. For the semicircle $s = \in e^{j\theta}$ around the pole at the origin,

$$G(s)\,H(s) = \frac{10}{\in e^{j\theta}(\in e^{j\theta}+1)(\in e^{j\theta}+4)}$$

As $\lim_{\in\to 0} \in e^{j\theta} << 1$ and $\in e^{j\theta} << 4$

$$\therefore \quad G(s)\,H(s) = \lim_{\in\to 0} \frac{10}{\in e^{j\theta}}$$

As θ varies from $-\dfrac{\pi}{2}$ to 0 to $\dfrac{\pi}{2}$

G(s) H(s) varies from $\infty\ \angle\dfrac{\pi}{2}$ to $\infty\ \angle 0$ to $\infty\ \angle -\dfrac{\pi}{2}$

This part of the GH plot corresponds to the infinite semicircle CDA.

5. To determine stability, we need to know whether the Nyquist plot encircles the (–1, j0) point or not. For this, we should determine where the locus crosses the negative real axis ie., the point B. Since the point B lies on the locus corresponding to $s = j\omega$ as ω changes from 0^+ to ∞, let us find the value of ω at which the imaginary part of G(s) H(s) is equal to zero.

$$G(s)\,H(s)\big|_{s=j\omega} = \frac{10}{j\omega(j\omega+1)(j\omega+4)}$$

Let us rationalise this and obtain its real and imaginary parts.

$$G(j\omega)H(j\omega) = \frac{10(1-j\omega)(4-j\omega)}{j\omega(1+\omega^2)(16+\omega^2)}$$

$$= \frac{10(4-\omega^2-j5\omega)}{j\omega(1+\omega^2)(16+\omega^2)}$$

$$= \frac{-50}{(1+\omega^2)(16+\omega^2)} - j\frac{10(4-\omega^2)}{\omega(1+\omega^2)(16+\omega^2)}$$

$$\text{Im}\, G(j\omega)H(j\omega) = 0$$

$$\therefore \quad \frac{10(4-\omega^2)}{\omega(1+\omega^2)(16+\omega^2)} = 0 \qquad .....(7.9)$$

$$\therefore \quad \omega^2 = 4$$

or $\omega = \pm 2$ rad/sec

At $\omega = 2$ rad/sec, we have,

$$G(j2)H(j2) = \frac{-50}{(1+4)(16+4)} = -0.5$$

Thus the Nyquist locus crosses the negative real axis at –0.5 for $\omega = \pm 2$ rad/sec. Thus the Nyquist plot is as shown in Fig. 7.10 (c) with the point (–1, j0) also indicated.

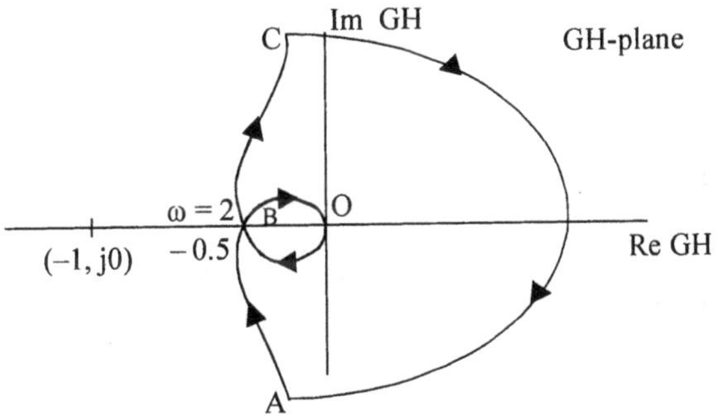

Fig. 7.10 (c) Nyquist plot with the critical point (–1, j0)

It is clear from the Fig. 7.10(c), the (–1, j0) point is not encircled by the GH locus. Hence the system is stable.

Example 7.3

Let us consider an example in which the open loop system has a pole in the right half of s-plane. Consider the system

$$G(s)H(s) = \frac{K}{(s+2)(s-1)}$$

Solution :

The Nyquist path is as shown in Fig. 7.11(a).

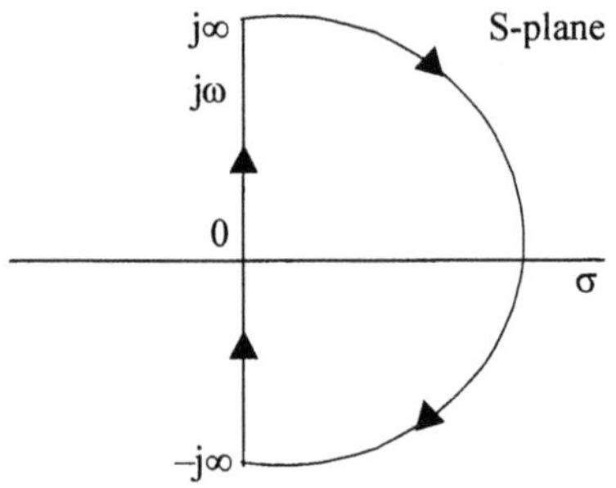

Fig. 7.11 (a) Nyquist path.

1. As s changes on jω axis,

$$s = j\omega \qquad 0 < \omega < \infty$$

$$G(s)\,H(s) = \frac{K}{(j\omega+2)(j\omega-1)}$$

At ω = 0, $$G(j0)\,H(j0) = -\frac{K}{2} = \frac{K}{2}\angle 180$$

At ω = ∞, $$G(j\infty)\,H(j\infty) = \frac{K}{(j\infty+2)(j\infty-1)} = 0\,\angle 180$$

2. As s moves on infinite semicircle G(s) H(s) maps on to the origin of GH-plane.
3. For $s = j\omega$ and $-\infty < \omega < 0$ the GH plot is mirror image of the plot for $0 < \omega < \infty$

The complete plot is as shown in Fig. 7.11(b).

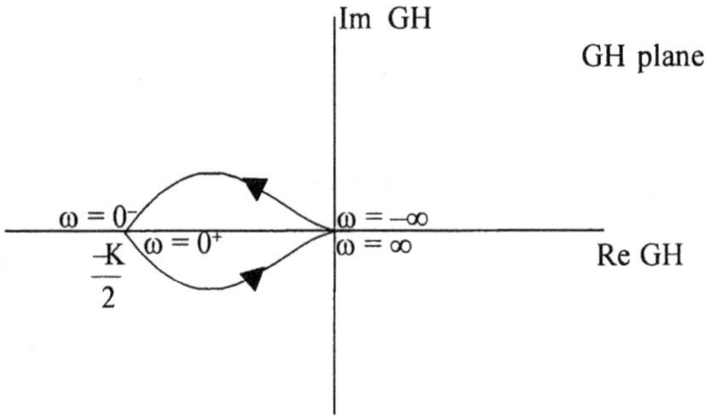

Fig. 7.11 (b) Nyquist plot of $GH = \dfrac{K}{(s+2)(s-1)}$.

The Nyquist plot encircles the (–1, j0) point once in anti clockwise direction, if $\frac{K}{2} > 1$ or $K > 2$, as shown in the Fig. 7.11(c).

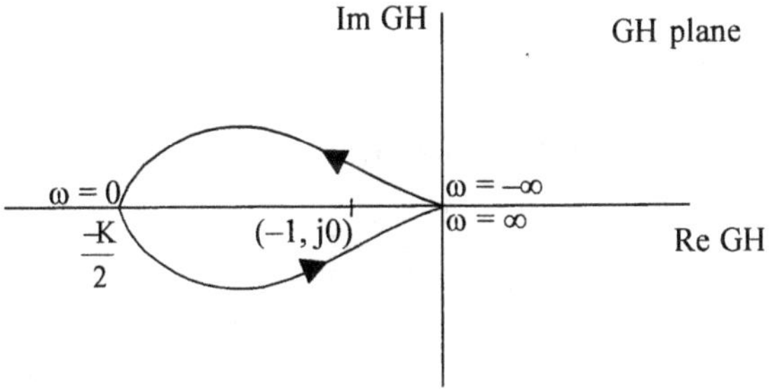

Fig. 7.11 (c) Nyquist plot for K > 2

For K > 2, we have

$$N = 1$$

Since there is one pole of G(s) H(s) in the right half of s-plane

$$P = 1$$

From $$N = P - Z$$

$$1 = 1 - Z$$

or $$Z = 0$$

Therefore there are no closed loop poles in the RHS and hence the system is stable for K > 2. If K < 2, the Nyquist plot is shown in Fig. 7.11(d).

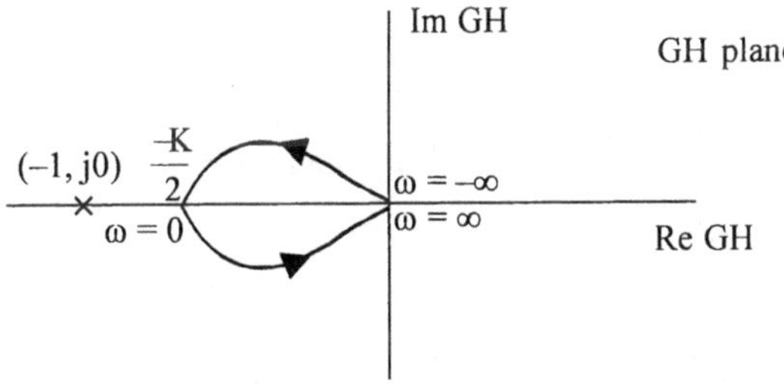

Fig. 7.11 (d) Nyquist plot for K < 2

Clearly, the Nyquist plot does not encircle (–1, j0) point and hence

$$N = 0$$

$$P = 1$$

$$N = P - Z$$

∴ $$Z = P = 1$$

There is one zero of the characteristic equation or equivalently, one pole of closed loop system in the RHS and therefore, the closed loop system is unstable for K < 2.

This examples illustrates the fact that, even if the open loop system is unstable the closed loop system may be stable. If some of the open loop poles are in the RHS, for stability the Nyquist plot should encircle the origin, in counter clockwise direction, as many times as there are RHP poles of the openloop system.

Example 7.4

Obtain the range of values of K for which the system with open loop transfer function

$$G(s)\ H(s) = \frac{K(s+1)}{s^2(s+2)(s+4)}$$

is stable.

Solution :

(i) For the Nyquist path shown in Fig. 7.12 (a) and for values of s on section I of the path

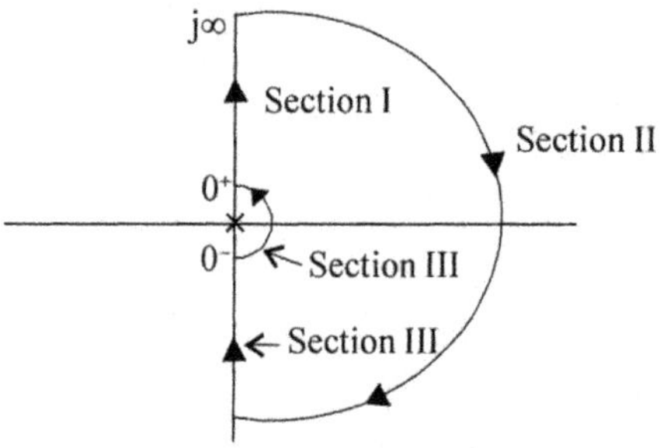

Fig. 7.12 (a) Nyquist path for example 7.4

$$s = j\omega \qquad 0^+ < \omega < \infty$$

$$G\ (j\omega)\ H\ (j\omega) = \frac{K(j\omega+1)}{(j\omega)^2(j\omega+2)(j\omega+4)}$$

for $\omega = 0^+$

$$G\ (j0^+)\ H\ (j0^+) = \frac{K}{(j0)^2}$$

$$= \infty\ \angle{-180^0}$$

for $\omega = \infty$

$$G\ (j\infty)\ H\ (j\infty) = \frac{K(j\infty+1)}{(j\infty)^2(j\infty+2)(j\infty+4)}$$

$$= \frac{K.j\infty}{(j\infty)^2(j\infty)(j\infty)}$$

$$= 0\ \angle{-270}$$

The segment of Nyquist plat near $s = j0^+$ and $s = j\infty$ is shown in Fig. 7.12 (b).

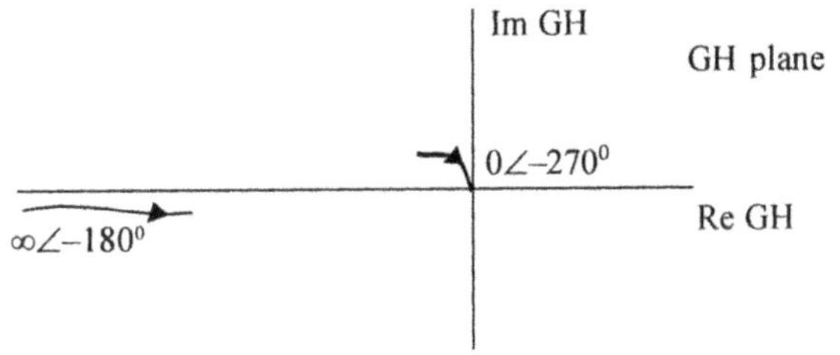

Fig. 7.12 (b) Part of Nyquist plot at s = j0⁺ and s = j∞

It is clear from the plot shown in Fig. (7.12) (b) that the plot has to cross the negative real axis for some value of s = jω. Let us find this point. At this point the imaginary part of G(jω) H(jω) should be zero or the angle of G(jω) H(jω) = $\pm 180^0$

Equating $\quad \angle G(j\omega)\,H(j\omega) = -\,180^0$

$$\tan^{-1}\omega - 180 - \tan^{-1}\frac{\omega}{2} - \tan^{-1}\frac{\omega}{4} = -\,180$$

$$\tan^{-1}\omega = \tan^{-1}\frac{\omega}{2} + \tan^{-1}\frac{\omega}{4}$$

Taking tangent of the angles on both sides, we have,

$$\tan(\tan^{-1}\omega) = \tan\left(\tan^{-1}\frac{\omega}{2} + \tan^{-1}\frac{\omega}{4}\right)$$

$$\omega = \frac{\frac{\omega}{2} + \frac{\omega}{4}}{1 - \frac{\omega^2}{8}}$$

Simplifying and solving for ω, we have,

$$\omega = \sqrt{2} \text{ rad/sec}$$

At this value of ω,

$$|G(j\omega)\,H(j\omega)| = -\frac{K(1+\omega^2)^{\frac{1}{2}}}{\omega^2(\omega^2+4)^{\frac{1}{2}}(\omega^2+4)^{\frac{1}{2}}}$$

$$= -\frac{K(1+2)^{\frac{1}{2}}}{2(2+4)^{\frac{1}{2}}(2+16)^{\frac{1}{2}}}$$

$$= -\,0.0722K$$

Thus the GH plot crosses the negative real axis at $-0.0722K$ for $s = j\sqrt{2}$.

(ii) For selection II

The GH plot maps on to the origin of GH plane.

(iii) For selection III

The GH plot is a mirror image of plot for $\omega = 0^+$ to ∞.

(iv) For selection IV

$$s = \underset{\epsilon \to 0}{\text{lt}} \ \epsilon\, e^{j\theta} \qquad \theta \to -\frac{\pi}{2} \to 0 \to \frac{\pi}{2}$$

$$G(s)\,H(s) = \frac{K(\epsilon e^{j\theta} + 1)}{\epsilon^2 e^{j2\theta}(\epsilon e^{j\theta} + 1)(\epsilon e^{j\theta} + 4)}$$

as $\epsilon \to 0$ $\qquad G(s)\,H(s) = \dfrac{K}{4 \underset{\epsilon \to 0}{\text{lt}} \epsilon^2 e^{j2\theta}}$

as θ changes from $-\frac{\pi}{2} \to 0 \to \frac{\pi}{2}$ on the semicircle around the origin.

G(s) H(s) changes from $\infty \angle \pi \to \infty \angle 0 \to \infty \angle -\pi$.

The complete plot is shown in Fig. 7.12 (c),

for $\qquad 0.0722\,K < 1 \quad \text{or } K < \dfrac{1}{0.0722} = 13.85$

and $\qquad 0.0722\,K > 1 \quad \text{or } K > \dfrac{1}{0.0722} = 13.85$

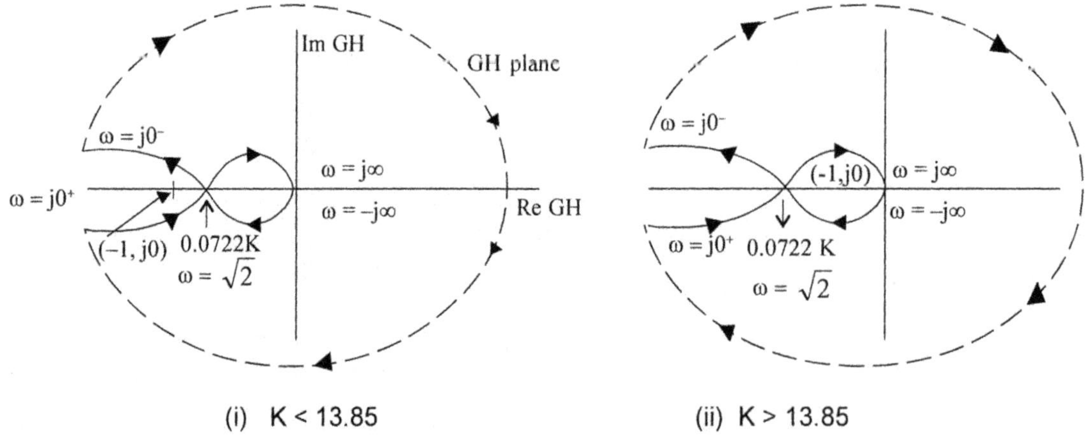

Fig. 7.12 (c) Nyquist plot for example 7.4.

From plot (i) of Fig. 7.12 (c), the point (–1, j0) is not encircled by the GH plot and

hence, $N = 0$

since $P = 0$

and $N = P - Z$

$Z = 0$

Thus the system is stable for $K < 13.85$

From plot (ii) of Fig. 7.12 (c), the point (–1, j0) is encircled twice in clockwise direction.

$\therefore$ $N = -2$

$\because$ $P = 0$

and $N = P - Z$

$-2 = -Z$

$Z = 2$

There are two zeros of the characteristic equation in the RHP and hence the system is unstable for $K > 13.85$. The range of values of K for which the system is stable, is given by,

$$0 < K < 13.85$$

Example 7.5

Determine the stability of the system

$$G(s)\,H(s) = \frac{K(s+2)^2}{s^3}$$

Solution :

Consider the Nyquist Centour shown in Fig. 7.13 (a).

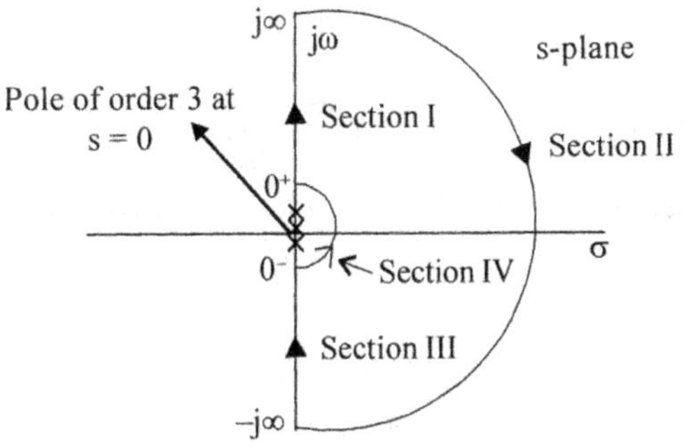

Fig. 7.13 (a) Nyquist path for example 7.5

The open loop system has a triple pole at the origin and Nyquist path is taken with an indentation around the origin with a semi circle to bypass these poles.

Section I

$$s = j\omega \qquad \omega \to 0^+ \text{ to } \infty$$

$$G(j\omega)\,H(j\omega) = \frac{K(j\omega+2)^2}{(j\omega)^3}$$

for $\omega = 0^+$

$$G(j0^+)\,H(j0^+) = \frac{K(j0^+ + 2)^2}{(j0^+)^3} = \infty \angle{-270}$$

for $\omega = \infty$

$$G(j\infty)\,H(j\infty) = \frac{K(j\infty+2)^2}{(j\infty)^3} = 0 \angle{-90}$$

As the plot goes from -270^0 to -90^0, it has to cross the negative real axis for some value ω. At this value of ω the imaginary part of $G(j\omega)\,H(j\omega)$ should be zero.

$$G(j\omega)\,H(j\omega) = \frac{K(j\omega+2)^2}{(j\omega)^3}$$

$$= \frac{K(-\omega^2+4+4j\omega)}{-j\omega^3}$$

$$= \frac{jK(4-\omega^2)}{\omega^3} - \frac{4K}{\omega^2}$$

Equating the imaginary part to zero, we have

$$\frac{K(4-\omega^2)}{\omega^3} = 0$$

or $\omega^2 = 4$

$$\omega = \pm 2 \text{ rad/sec}$$

and $|G(j\omega)\,H(j\omega)| = -4\,\frac{K}{4} = -K$

The Nyquist plot for section I of the Nyquist plot is shown in Fig. 7.13 (b).

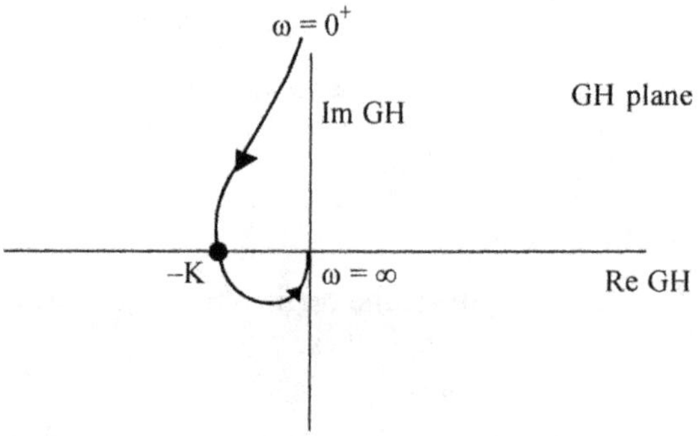

Fig. 7.13 (b) Nyquist plot for section I of Nyquist path.

Section II

For values of *s* on this section the plot of GH will map on to the origin.

Section III

For this section the plot is mirror image of plot for Section I.

Section IV

On this section,

$$s = \lim_{\in \to 0} \in e^{j\theta} \qquad \theta \to -\frac{\pi}{2} \to 0 \to \frac{\pi}{2}$$

$$G(s)\,H(s)\Big|_{s=\in e^{j\theta}} = \frac{K(\in e^{j\theta}+2)^2}{(\in e^{j\theta})^3}$$

As $\in \to 0$

$$G(s)\,H(s) = \frac{4K}{\in^3 e^{j3\theta}} = \infty\, e^{-j3\theta}$$

For different values of θ, $\angle G(s)\,H(s)\Big|_{s=\in e^{j\theta}}$ is tabulated.

Table 7.1

θ	∠G(s) H(s)
– 90	+ 270
– 60	+ 180
– 30	+ 90
0	0
30	– 90
+ 60	– 180
90	– 270

The plot of G(s) H(s) for this section is shown in Fig. 7.13 (c).

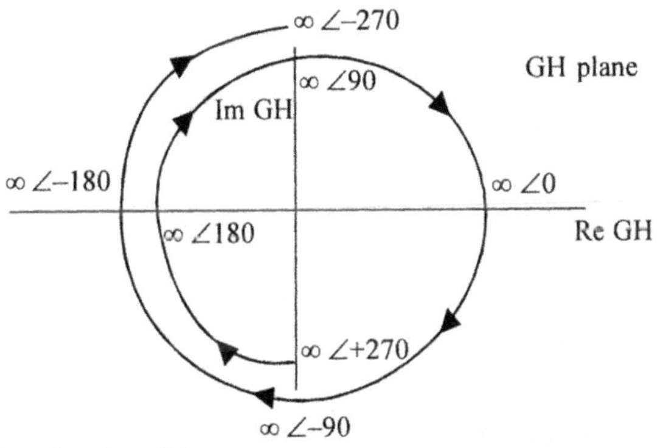

Fig. 7.13 (c) Plot of GH for Section IV.

The complete Nyquist plot is shown in Fig. 7.13 (d).

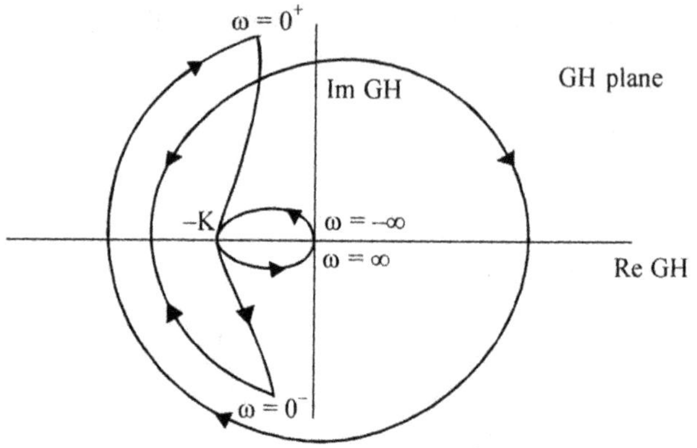

Fig. 7.13 (d) Complete Nyquist plot of G(s) H(s) = $\frac{K(s+2)^2}{s^3}$

For K > 1, the point (–1, j0) is traversed once in anticlockwise direction and once in clockwise direction as can be seen for Fig. 7.13 (d). The net encirclements are zero and since there are no poles in the right half of s-plane, the system is stable.

For K > 1, the point (–1, j0) is traversed twice in clockwise direction and therefore,

$$N = -2$$

Since $P = 0$

and $N = P - Z$

$$-2 = 0 - Z$$

$$Z = 2$$

The system is unstable.

Example 7.6

Comment on the stability of the system.

$$G(s)\,H(s) = \frac{K(s+10)(s+2)}{(s+0.5)(s-2)}$$

Solution :

The Nyquist path consists of the entire jω axis and the infinite semicircle enclosing the right half of s-plane.

For $s = j\omega$ and $\omega \to 0$ to ∞

$$G(s)\,H(s) = \frac{K(j\omega+10)(j\omega+2)}{(j\omega+0.5)(j\omega-2)}$$

for $\omega = 0$

$$G(s)\,H(s) = \frac{20k}{-1} = -\,20k = 20K\angle 180$$

for $\omega = \infty$

$$G(s)\,H(s) = \lim_{\omega \to 0} \frac{K(j\omega)^2\left(1+\dfrac{10}{j\omega}\right)\left(1+\dfrac{2}{j\omega}\right)}{(j\omega)^2\left(1+\dfrac{0.5}{j\omega}\right)\left(1-\dfrac{2}{j\omega}\right)}$$

$$= K$$

To find the possible crossing of negative real axis,

$$\text{Im}\, G(j\omega)\, H(j\omega) = 0$$

$$\text{Im} \frac{K(j\omega+10)(j\omega+2)(-j\omega+0.5)(-j\omega-2)}{(\omega^2+0.25)(\omega^2+4)} = 0$$

$$\text{Im}\,(-\omega^2 + 20 + 12j\omega)\,(-\omega^2 - 1 + 1.5j\omega) = 0$$

$$-1.5\omega^2 + 30 - 12 - 12\omega^2 = 0$$

$$\omega^2 = \frac{18}{13.5} = \frac{4}{3}$$

$$\omega = 1.1547 \text{ rad/sec}$$

$$\text{Re}\,[G(j\omega)\, H(j\omega)]_{\omega = 1.1547} = K \left.\frac{-(20-\omega^2)(1+\omega^2)-18\omega^2}{(\omega^2+.25)(\omega^2+4)}\right|_{\omega = 1.1547}$$

$$= -8K$$

Hence the Nyquist plot crosses the negative real axis at – 8K for $\omega = 1.1547$ rad/sec.

The infinite semicircle of Nyquist path maps into the origin of GH plane. The negative imaginary axis maps into a mirror image of the Nyquist plot of the positive $j\omega$ axis. Hence the complete Nyquist plot is shown in Fig. 7.14.

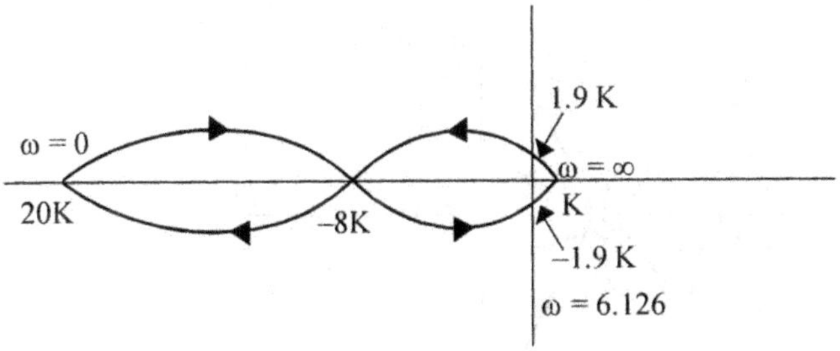

Fig. 7.14 Complete Nyquist plot of system in example 7.6.

By equating the real part of $G(j\omega)\, H(j\omega)$ to zero, we can get the crossing of $j\omega$-axis also. The plot crosses the $j\omega$-axis at $|G(j\omega)\, H(j\omega)| = -1.9$ K for $\omega = 6.126$ rad/sec. This is also indicated in the Fig. 7.14. From Fig. 7.14 it is clear that if $8K > 1$ or $K > 0.125$, $(-1, j0)$ point is encircled once in anticlockwise direction and hence

$$N = 1$$

Since $P = 1$

and $N = P - Z$

$Z = 0$

∴ The system is stable for $K > 0.125$.

If $K < 0.125$, the $(-1, j0)$ point is encircled once in the clockwise direction and hence $N = -1$

Since $P = 1$

and $N = P - Z$

$Z = 2$

There are two closed loop poles in the RHP and hence the system is unstable.

7.3.4 Nyquist Stability Criterion for Systems which are Open Loop Stable

When the open loop transfer function does not contain any poles in the RHP we have $P = 0$ and for stability $N = 0$ ie., there should not be any encirclements of $(-1, j0)$ point by the Nyquist plot. The Nyquist criterion can be simplified for all such cases and we can avoid drawing the entire Nyquist plot. It is sufficient to draw the plot for points on the positive $j\omega$ axis of s-plane and conclude about the stability. In otherwords, the polar plot discussed in section 6.5.2 is sufficient to determine the stability of open loop stable control systems.

Consider the Nyquist plot shown in Fig. 7.10(c) for the system,

$$G(s)\,H(s) = \frac{10}{s(s+1)(s+4)}$$

The system is open loop stable. The Nyquist plot is repeated in Fig. 7.15(a) for convenience

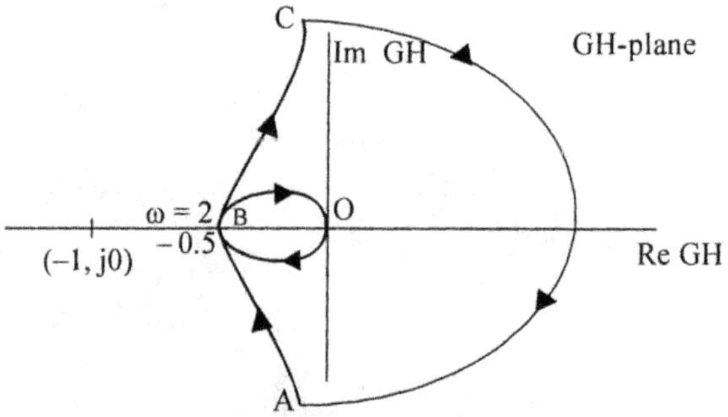

Fig. 7.15 (a) Nyquist plot of $G(s)\,H(s) = \dfrac{10}{s(s+1)(s+4)}$

It is clear that the $(-1, j0)$ point is not *encircled* by the Nyquist plot and hence the system is stable.

Let us now consider the portion of the plot for $\omega = 0^+$ to $\omega = \infty$ and the point $(-1, j0)$ as shown in Fig. 7.15 (b).

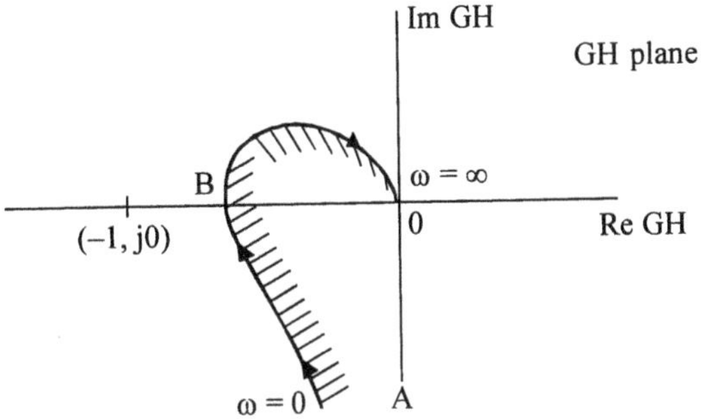

Fig. 7.15 (b) Polar plot of the system $G(s)\,H(s) = \dfrac{10}{s(s+1)(s+4)}$

Let us define the term, 'enclosed'.

A point or region is said to be enclosed by a closed path if it is found to lie to the right of the path when the path is traversed in a prescribed direction. For example, shaded regions shown in Fig. 7.15 (c) and 7.15 (d) are said to be enclosed by the closed contour. The point A is said to be enclosed by the contour τ in Fig. 7.15 (c) but the point B in Fig. 7.15 (d) is not enclosed by τ. The point C outside the path in Fig. 7.15 (c) is enclosed.

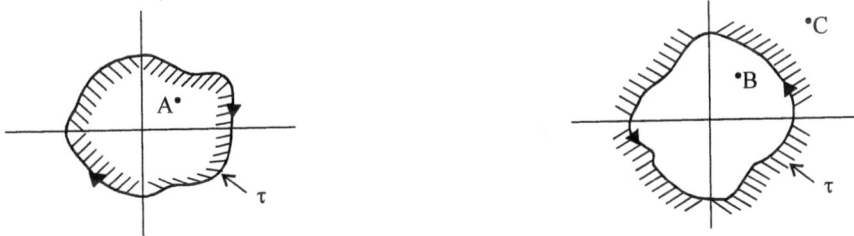

Fig. 7.15 (c) Point A enclosed by τ. **(d)** Point B not enclosed but point C enclosed by τ.

The point A in Fig. 7.15(c) is encircled by the contour τ in clockwise direction, and the point is also enclosed by the contour. In Fig. 7.15(d), the point B is encircled by the contour τ in counter clockwise direction but it is not enclosed, whereas the point C is not encircled by the contour τ, but it is enclosed by the contour τ.

Now, coming to the discussion about the stability of open loop systems, it is clear that the Nyquist Contour should not encircle the $(-1, j0)$ point, for the system to be stable. In Fig. 7.15(a) it can be seen that for stable system $(-1, j0)$ point should not be enclosed by the Nyquist contour. If we consider the plot for $\omega = 0$ to $\omega = \infty$ as shown in Fig. 7.15(b), if the point $(-1, j0)$ is not enclosed by the plot, then the system is stable.

Thus the Nyquist Criterion in the case of open loop stable systems can be stated as follows :

The Nyquist plot for $\omega = 0$ to $\omega = \infty$ (ie., the polar plot of G(s) H(s)) should not enclose the point $(-1, j0)$ point when the plot is traversed in the indicated direction from $\omega = 0$ to $\omega = \infty$. The point $(-1, j0)$ is known as the critical point.

For an unstable system the point (–1, j0) will be enclosed by the polar plot as shown in Fig. 7.16.

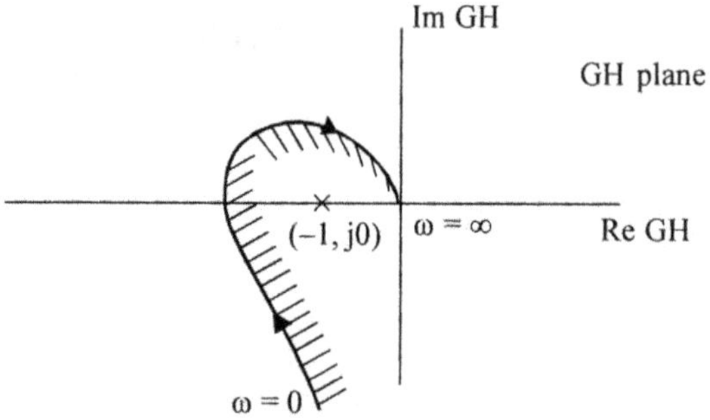

Fig. 7.16 Nyquist plot for an unstable system.

The polar plot, which can be obtained by conducting a frequency response test on the given system, can be used to determine the stability of the system.

7.4 Relative Stability

So far we have considered only the absolute stability of the system. The Nyquist criterion tells us whether the system is stable or not by the location of the critical point with respect to the Nyquist plot. But often we are interested in knowing how stable the system is, if it is already stable. This aspect of stability is known as relative stability of the system.

Consider a third order system given by

$$G(s)\,H(s) = \frac{K}{s(s+1)(s+4)}$$

Consider the plot for four different values of K.

(i) K = 30

The polar plot of the system and its time response for a unit step input are as shown in Fig. 7.17 (a) and (b) respectively.

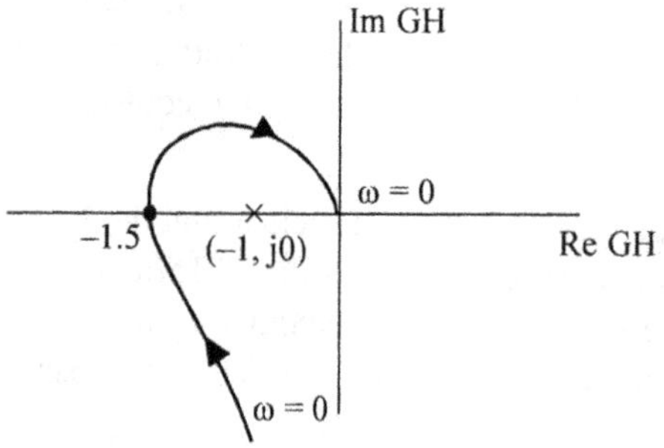

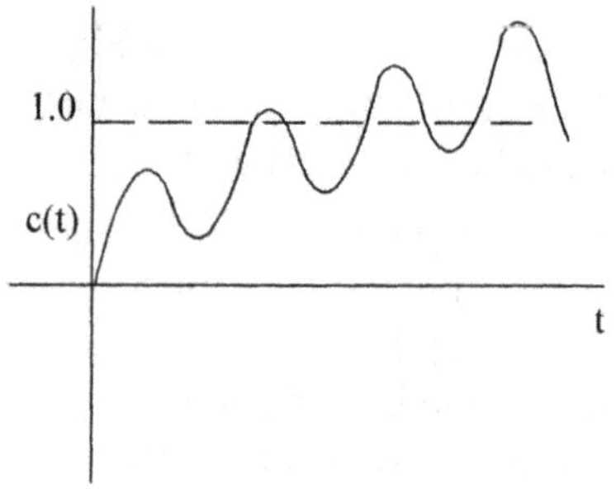

Fig. 7.17 **(a)** Polar plot of GH for K = 30 **(b)** Step response for K = 30

The polar plot shows that the (–1, j0) point is enclosed and hence the system is unstable. The step response of this system increases with time.

(ii) K = 20

The polar plot for this case and its step response are shown in Fig. 7.18(a) and (b)

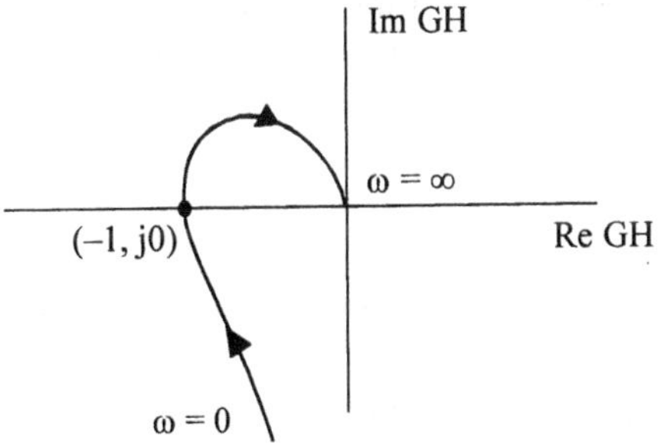

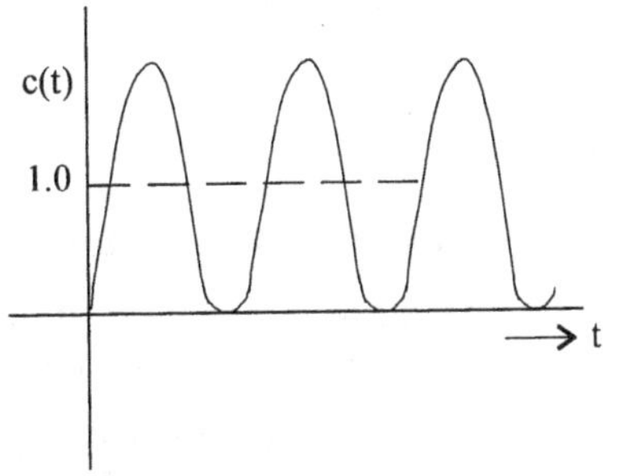

Fig. 7.18 (a) Polar plot of with K = 20 **(d)** Step response for K = 20

The polar plot passes through the critical point and the step response is purely oscillatory, which corresponds to closed loop poles on the jω-axis. Thus, if the polar plot passes through (–1, j0), the poles of the closed system lie on the jω axis and the system is marginally or limitedly stable.

(iii) K = 15

The polar plot and the step response are shown in Fig. 7.19 (a) and (b) respectively.

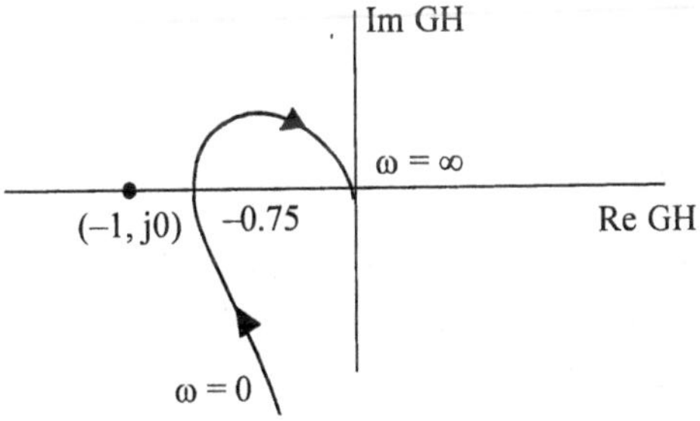

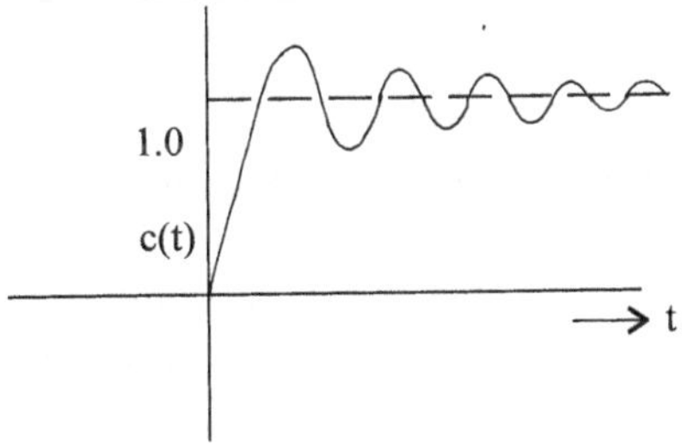

Fig. 7.19 (a) Polar plot of with K = 15 **(b)** Step response for K = 15

The system is stable for K = 15 but the point (–1, j0) is very near to the GH plot. The step response is highly oscillatory and has large overshoot.

(iv) K = 10

The polar plot and the step response are as sketched in Fig. 7.20 (a) and (b) respectively.

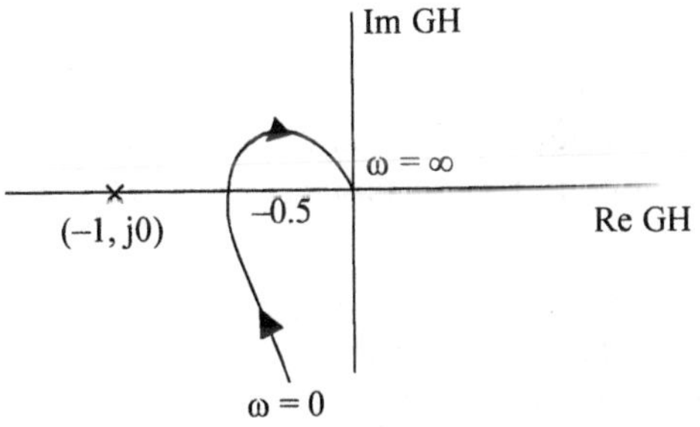

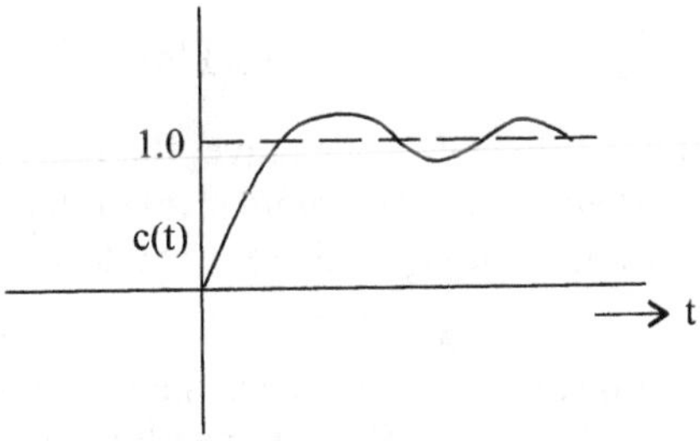

Fig. 7.20 (a) Polar plot of with K = 10 **(b)** Step response for K = 10

The GH plot is far away from the critical point and the step response has less oscillations and low overshoot.

From the above discussion it is clear that the system is more stable if the critical point (–1, j0) is far away from the GH plot. Thus, the distance between the critical point and the GH plot can be used to quantitatively obtain the relative stability of a system. It is to be noted that the relative stability studies are made on open loop stable systems only.

7.4.1 Measures of Relative Stability : Gain Margin

We have seen that the nearness of the critical point (–1, j0) to the GH plot can be used to quantize the relative stability of a system. One such measure is the "Gain Margin".

Consider the polar plot shown in Fig. 7.21. The point where the G(s) H(s) plot crosses the negative real axis is known as the phase cross over frequency.

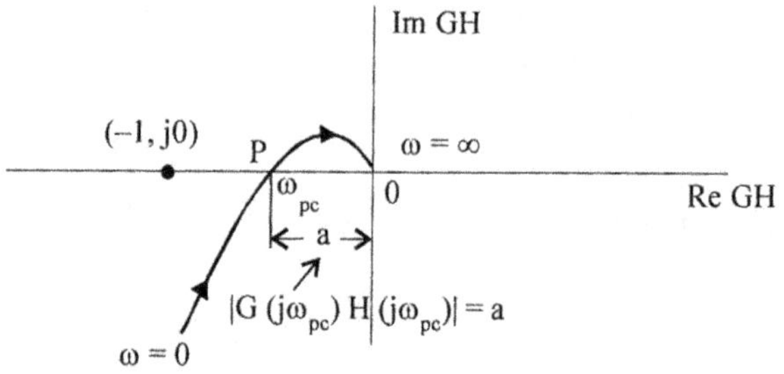

Fig. 7.21 Polar plot and the phase cross over frequency.

At $\omega = \omega_{pc}$, the G(s) H(s) plot crosses the negative real axis, i.e, the phase angle changes from -180^0 to $+180^0$. This frequency is therefore called as phase crossover frequency ω_{pc}. The magnitude of G(s) H(s) at $s = j\omega_{pc}$ is given by OP and is equal to $|G(j\omega_{pc}) H(j\omega_{pc})|$. The gain margin is defined as

$$\text{Gain Margin} \quad (GM) = \frac{1}{|G(j\omega_{pc})H(j\omega_{pc})|} = \frac{1}{a} \qquad .....(7.10)$$

Usually gain margin is specified in terms of decibels, db, as

$$\text{Gain margin in db} = 20\log_{10}\frac{1}{|G(j\omega_{pc})H(j\omega_{pc})|}$$

$$= 20\log_{10}\frac{1}{a} \qquad .....(7.11)$$

If a > 1, the GM is negative, the polar plot encloses the critical point and the system is unstable. The system is more and more stable if the phase cross over point P is nearer and nearer to the origin. The gain margin is positive for stable systems.

The magnitude of $G(j\omega) H(j\omega)$ at any frequency ω, in particular at $\omega = \omega_{pc}$, will increase, as the loop gain K is increased. Thus the phase cross over point P moves nearer to the critical point, as the gain K is increased. For a particular value of K the value $|G(j\omega_{pc}) H(j\omega_{pc})|$ becomes unity and the system will be on the verge of instability.

Thus the gain margin can also be defined as the amount of increase in the gain that can be permitted before the system becomes unstable.

From eqn. (7.10), if the gain is increased by $\frac{1}{a}$, $|G(j\omega_{pc})\,H(j\omega_{pc})| = 1$ and the system will be on the verge of instability. Now consider two systems with polar plots as shown in Fig. 7.22.

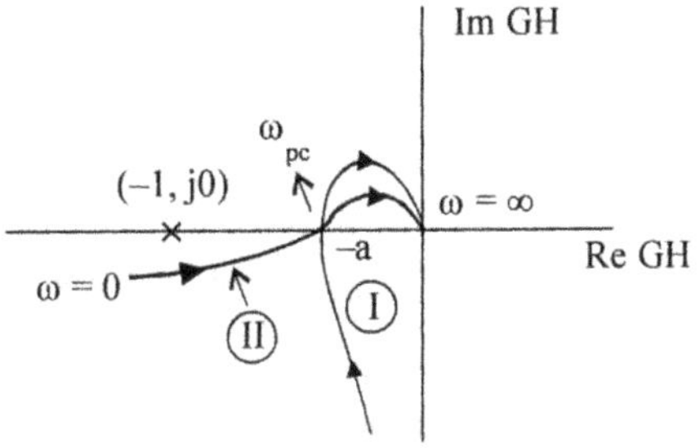

Fig. 7.22 Polar plots of two systems with same Gain Margin.

The two systems have the same cross over point *a* and hence the two systems have the same gain margin. But the system II is nearer to the critical point than the system I at some other frequency and therefore the system I is relatively more stable than the system II, even though they both have the same gain margin. Hence we need another factor to judge the relative stability of a system and is defined as phase margin.

7.4.2 Phase Margin

Consider the polar plot shown in Fig. 7.23. Draw a circle with origin as centre and unit radius.

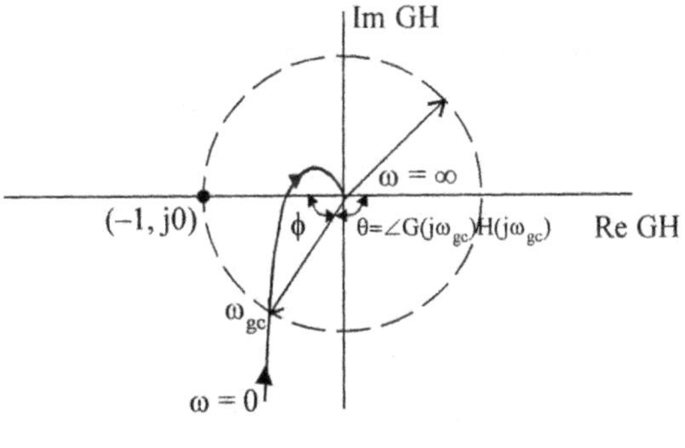

Fig. 7.23 Definition of gain cross over frequency and phase margin.

The frequency at which the polar plot crosses the unit circle, is known as Gain cross over frequency, ω_{gc}. For frequencies greater than ω_{gc}, the magnitude of G(s) H(s) becomes less than unity. If now an additional phase lag of $\phi = 180 - \angle G(j\omega_{gc})\,H(j\omega_{gc}) = 180 - \theta$ is added without changing the magnitude at this frequency, the polar plot of the system will cross (−1, j0) point. If the polar plot is rotated by an angle equal to ϕ in the clockwise direction, the system becomes unstable. Therefore, the phase margin can be defined as follows.

"Phase margin is the amount of phase lag that can be introduced into the system at the gain cross over frequency to bring the system to the verge of instability".

The phase margin is measured positively from the -180^0 line in the counter clockwise direction. The phase margin is positive for stable systems.

The value of the phase margin is given by

$$\text{PM } \phi = \angle G(j\omega_{gc})\, H(j\omega_{gc}) + 180$$

where the angle of $G(j\omega_{gc})\, H(j\omega_{gc})$ is measured negatively.

Gain Margin and phase margin are the two relative stability measures which are usually specified as frequency response specifications. These specifications are valid only for open loop stable systems. A large value of gain margin or phase margin denotes a highly stable system but usually a very slow system. On the other hand a small value of either of these indicates a highly oscillatory system. Systems with a gain margin of about 5 to 10 db and a phase margin of around 30 to 40^0 are considered to be reasonably stable systems.

For many practical systems a good gain margin ensures a good phase margin also. But there may be systems with good gain margin but low phase margin and vice vasa as shown in Fig. 7.24 (a) and (b).

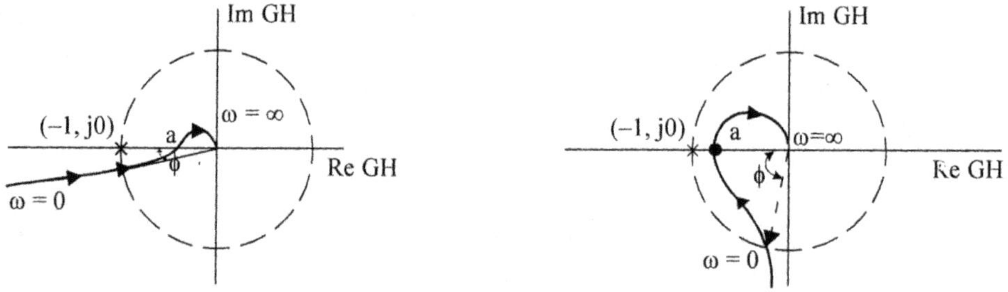

Fig. 7.24 (a) System with good gain margin but low phase margin.
(b) System with good phase margin but poor gain margin

For a second order system, the polar plot does not cross the negative real axis at all, for any value of gain K, as shown in Fig. 7.25. Hence the gain margin of such a system is infinity.

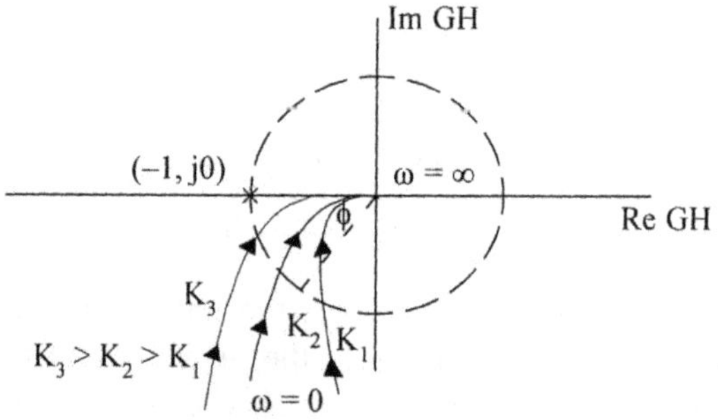

Fig. 7.25 Polar plot of second order system for different values of gain

But the phase margin reduces as the value of K is increased. Thus for a second order system the appropriate relative stability measure is the phase margin rather than the gain margin. Hence usually phase margin is specified as one of the frequency domain specifications in the design of a control system rather than the gain margin.

Example 7.7

Consider a unity feedback system with

$$G(s) = \frac{K}{s(0.5s+1)(0.05s+1)}$$

Let us find the gain margin and phase margin for K = 1.

(a) Gain Margin

The polar plot of G(s) is shown in Fig. 7.26.

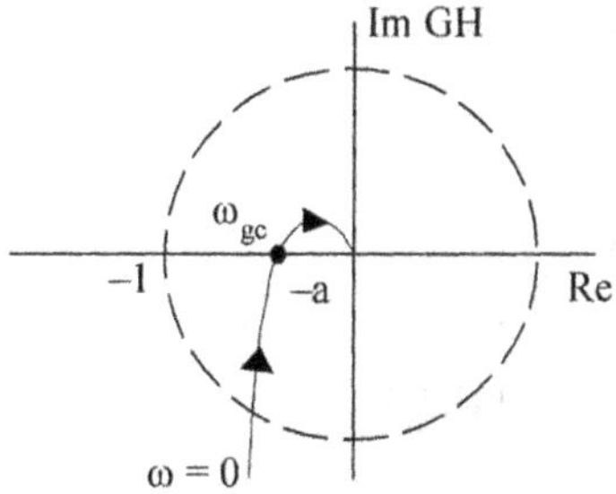

Fig. 7.26 Polar plot of G(s) H(s)

To obtain the phase cross over frequency we will equate the imaginary part of G(jω) to zero.

$$G(j\omega) = \frac{1}{j\omega(0.5j\omega+1)(0.05j\omega+1)}$$

$$= \frac{1}{-0.55\omega^2 + j\omega\left(1-0.025\omega^2\right)}$$

$$\text{Im } G(j\omega) = 1 - 0.025\,\omega^2 = 0$$

$$\omega_{pc}^2 = \frac{1}{0.025} = 40$$

$$\omega_{pc} = \pm\sqrt{40}$$

$$\omega_{pc} = 6.325 \text{ rad/sec.}$$

At this frequency, the real part of G(jω) is

$$\text{Re } G(j\omega_{pc}) = -\frac{1}{0.55\times 40}$$

$$= -0.0454$$

The GH plot crosses negative real axis at –0.0454 for ω_{pc} = 6.325 rad/sec.

Thus $a = 0.0454$

and $$GM = 20\log\frac{1}{a}$$

$$= 20\log\frac{1}{0.0454}$$

$$= 26.86 \text{ db}$$

The gain can be increased by a factor of $\frac{1}{a} = \frac{1}{0.0454} = 22$ for the system to remain stable.

(b) Phase Margin

Let us first find the gain cross over frequency. At this frequency, the gain is unity.

$$|G(j\omega)\ H(j\omega)| = 1$$

$$\frac{1}{\omega\sqrt{1+0.25\omega^2}\ \sqrt{1+0.0025\omega^2}} = 1$$

ie., $\omega^2\ (1 + 0.25\omega^2)\ (1 + 0.0025\omega^2) = 1$

Solving for ω^2 we get

$$\omega^2 = 0.91$$

and $\omega_{gc} = 0.954$ rad/sec

$$\angle G(j\omega_{gc}) = -90 - \tan^{-1} 0.5\omega_{gc} - \tan^{-1} 0.05\ \omega_{gc}$$

$$= -118.232^0$$

Thus phase margin is

$$\phi = -118.232 + 180$$

$$= 61.768^0$$

7.4.3 Adjustment of Gain K for Desired Values of Gain and Phase Margins Using Polar Plots

We can also calculate the required gain K to obtain a specified gain margin and phase margin for a given system.

Consider the Ex 7.7 again.

(a) Let the gain margin desired be 15db.

$$GM = 20 \log \frac{1}{a} = 15$$

$$\therefore\ a = 0.1778$$

For K = 1 the intercept was 0.0454. To get an intercept of 0.1778, we increase the gain by a facts 0.1778/0.0454.

$$\therefore\ K = 3.92$$

(b) Let the phase margin desired be 45^0. Then from definition,

$$\angle G(j\omega_{gc}) + 180 = PM = \phi$$

$$-90 - \tan^{-1} 0.5\omega - \tan^{-1} 0.05\omega + 180 = 45$$

Taking tangent of the angles on both sides, we have,

$$\frac{0.55\omega}{1-0.025\omega^2} = \tan 45 = 1$$

$$0.025\omega^2 + 0.55\omega - 1 = 0$$

$\therefore$ $\omega_{gc} = 1.69$ rad/sec

At this frequency $|G(j\omega_{gc})| = 1$

$$\therefore \quad \frac{K}{1.69\sqrt{1+0.25\times1.69^2}\sqrt{1+0.0025\times1.69^2}} = 1$$

$$\therefore \quad K = 2.22$$

Thus if K is increased to 2.22, the phase margin will be 45^0.

Note : Gain and phase margins can be adjusted to the desired values analytically, only for systems upto second order or third order, if one of the three open loop poles occur at the origin and there are no open loop zeros. Otherwise graphical procedure must be used to solve the problem.

7.4.4 Gain Margin and Phase Margin Using Bode Plot

Bode plot also can be used to determine the gain and phase margins of a system and adjust the gain for a given gain margin or phase margin. Let us first consider how phase and gain margins could be determined using the Bode plot.

Example 7.8

Determine gain and phase margin for the unity feedback system using Bode plot.

$$G(s) = \frac{10}{s(0.5s+1)(0.05s+1)}$$

Solution :

The Bode magnitude and phase angle plots are shown in Fig. 7.27.

When the phase angle curve crosses the -180^0 line, the frequency is read from the Bode phase angle curve. This is phase crossover frequency ω_{pc}. At this frequency the magnitude in db is read from the magnitude curve. This is the gain margin of the system. If at ω_{pc}, the gain is positive, it means that $|G(j\omega)\ H(j\omega)|$ is greater than 1 and hence the system is unstable. If at the phase cross over frequency the gain of the system is zero, the system is oscillatory. Finally if at ω_{pc} the gain is negative, $|G(j\omega_{pc})| < 1$ and hence the system is stable.

Similarly, the frequency at which the gain curve crosses 0 db line, is the gain cross over frequency, ω_{gc}. At this frequency the angle of $G(j\omega)\ H(j\omega)$ can be read off from the phase angle curve. Than

Phase margin $\quad \phi = \angle G(j\omega_{gc})\ H(j\omega_{gc}) + 180$

The angle of $G(j\omega_{gc})\ H(j\omega_{gc})$ read above the -180^0 line gives the phase margin for a stable system. If the angle of $G(j\omega_{gc})\ H(j\omega_{gc})$ lies below the -180^0, the phase margin is taken as negative and the system will be unstable.

In the example 7.8.

$$\omega_{pc} = 6.4 \text{ rad/sec}$$

and $\quad GM = 8\text{db}$

$$\omega_{gc} = 4 \text{ rad/sec}$$

and $\quad PM = \phi = 15^0$

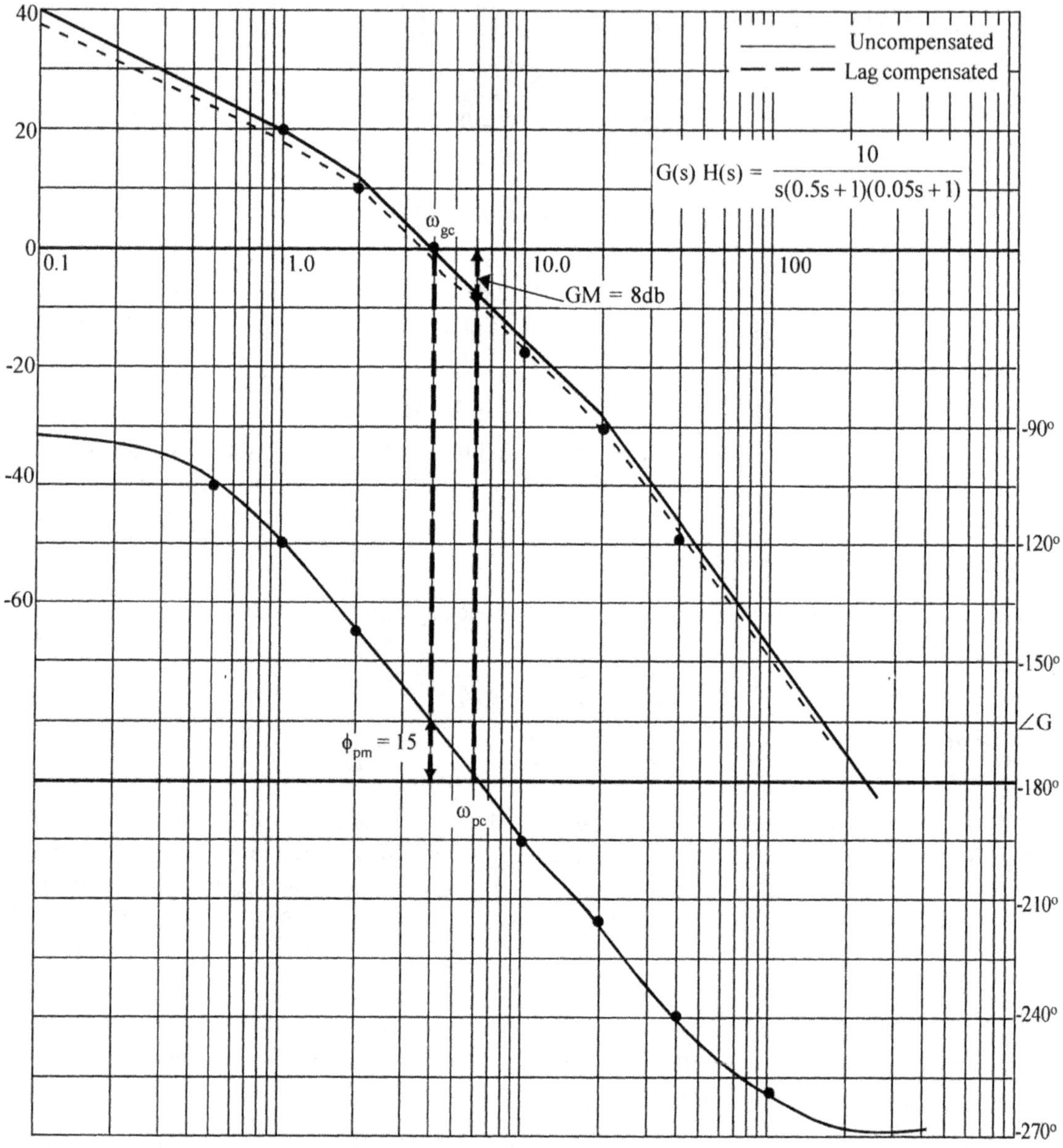

Fig. 7.27 Bode plot for the system in Ex. 7.8

The system gain could be adjusted to a desired value using Bode plot.

Let us consider an example to illustrate this procedure.

Example 7.9

Determine the gain and phase margin for the unity feedback system with K = 1.

$$G(s) = \frac{K}{s\,(0.5s+1)(0.05s+1)}$$

Determine the value of K for obtaining (a) gain margin of 20 db (b) phase margin of 40^0

Solution :

With K = 1, the magnitude and phase plots are drawn in Fig. 7.28.

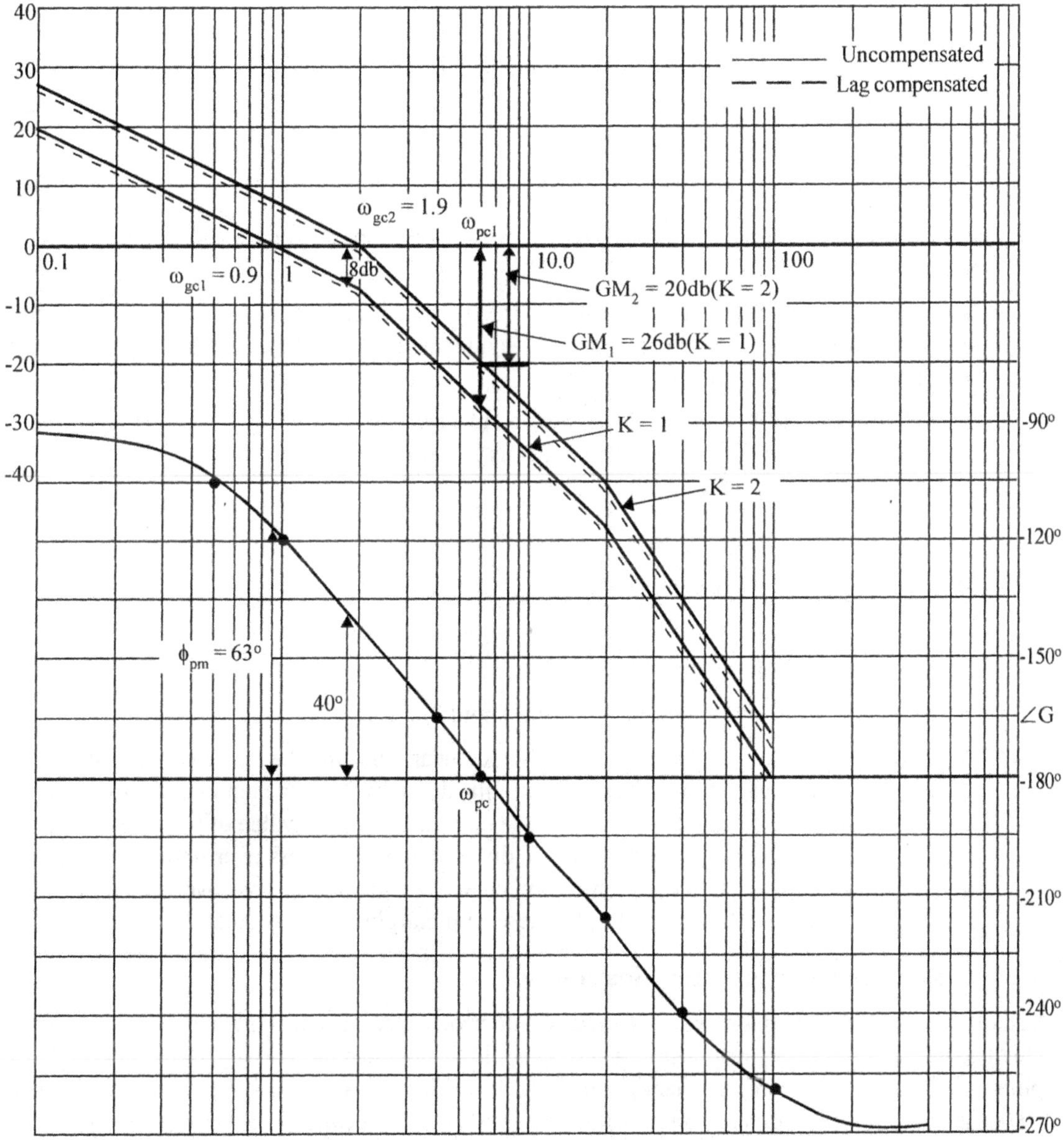

Fig. 7.28 Bode plot for the system in Ex. 7.9

From the figure

$$\omega_{gc} = 0.9 \text{ rad/sec.}$$

$$\phi_{PM} = 63^0$$

$$\omega_{pc} = 6 \text{ rad/sec}$$

$$GM = 26 \text{ db}$$

(i) To obtain a gain margin of 20db, the gain of the system should be increased by (26 – 20) = 6db without changing the phase cross over frequency, ie, the gain K must be increased such that

$$20 \log K = 6$$

or

$$K = 1.995 \simeq 2$$

By making K = 2, the entire gain curve is shifted up by 6db. The phase curve is unaltered. The gain curve for K = 2 is shown in Fig. 7.28.

(ii) To obtain a phase margin of 40^0, locate the frequency at which the phase curve has an angle of $(-180 + 40) = -140^0$. Find the gain at this frequency. Increase the value of K so that the gain curve crosses 0 db line at this frequency.

From Fig. 7.28, -140^0 phase angle is obtained at a frequency of 1.9 rad/sec. This should be the new gain cross over frequency when the K is increased.

$$\omega_{gc} \text{ (new)} = 1.9 \text{ rad/sec.}$$

At this frequency, the gain is read off from the magnitude plot for K = 1.

Gain at ω_{gc} (new) = 8db

$$\therefore \quad 20 \log K = 8$$

$$K = 2.51$$

Thus to get a phase margin of 40^0, the gain should be increased to 2.51.

Note : Increasing the gain from K = 1 to K = K_1 means shifting the entire gain curve up by 20 log K_1 db. It is equivalent to shifting the 0 db line down by the same amount.Thus if Bode plot for K = 1 is known.the gain margin and phase margin for a given value of K = K_1 can be found out shifting the 0 db line down by 20 log K db. The new gain cross over frequency and hence the new phase margin can be read from the curves. At the phase cross over frequency which remains same for K = 1 and K = K_1, the new gain margin can be found.

7.5 Closed Loop Frequency Response

Usually we are interested in the time response of a control system. But this is often difficult to obtain. On the otherhand frequency response of a system could be obtained more easily. Bode plots and polar plots could be used to represent them graphically. In chapter 6, we have shown that the frequency response and time response are correlated to one another for a second order system. If time domain specifications are given, frequency domain specifications can be easily obtained. The design of a controller or compensator is carried out in frequency domain. The frequency response of the compensated system can now be translated back to estimate the time response. Let us consider the frequency domain specifications of a control system.

1. ***Resonance peak, M_r :*** This is the maximum of the magnitude of the closed loop frequency response. From chapter 6, eqn. (6.8)

$$M_r = \frac{1}{2\delta\sqrt{1-\delta^2}}$$

It is seen that, M_r solely depends on the damping factor δ. A small value of δ corresponds to large overshoot in time response and large resonance peak in frequency response.

2. ***Resonance frequency, ω_r :*** This is the frequency at which resonance peak occurs. From eqn. (6.7)

$$\omega_r = \omega_n \sqrt{1-2\delta^2}$$

Resonance frequency is dependent on ω_n and δ and hence it is indicative of frequency of oscillations and speed of response of the time response.

3. ***Bandwidth, ω_d :*** It is defined as the range of frequencies for which the magnitude of the frequency response is more than –3db. The frequency at which the gain is –3db, is known as the cut off frequency. Signals with frequencies above this frequency are attenuated. Bandwidth is a measure of the ability of the system to reproduce the input signals in the output. It also throws light on noise rejection capabilities of the system. More significantly it indicates the rise time of time response of the system for a given damping factor. Fast response or small rise time is obtained for systems with large bandwidth.

4. ***Cutoff rate :*** The rate of change of magnitude curve of the Bode plot at the cut off frequency is known as the cut off rate. This is indicative of the ability of the system to distinguish between signal and noise. However, sharp cut off often results in large resonance peak and hence less stable system.

5. ***Gain margin and phase margin :*** These are the relative stability measures and are indicative of the nearness of the closed loop poles to the jω axis. Usually phase margin, rather than gain margin, is specified as one of the frequency domain specifications. The phase margin is directly related to the damping factor, δ for a second order system. Let us derive the relationship between phase margin and the damping factor.

Consider a standard second order system with

$$G(s)\,H(s) = \frac{\omega_n^2}{s\,(s+2\delta\omega_n)}$$

$$G(j\omega)\,H(j\omega) = \frac{\omega_n^2}{j\omega\,(j\omega+2\delta\omega_n)} \qquad .....(7.12)$$

At the gain cross over frequency ω_{gc}, $|G(j\omega)\,H(j\omega)| = 1$

$$\therefore \qquad \frac{\omega_n^2}{\omega_{gc}\sqrt{\omega_{gc}^2 + 4\delta^2\omega_n^2}} = 1$$

or
$$\omega_{gc}^4 + 4\delta^2\,\omega_{gc}^2\,\omega_n^2 - \omega_n^4 = 0$$

This is a quadratic in ω_{gc}^2 and the solution yields,

$$\omega_{gc}^2 = \omega_n^2\,\sqrt{(4\delta^4 + 1)} - 2\delta^2 \qquad \text{.....(7.13)}$$

Phase margin of the system is given by,

$$\phi_{pm} = -\,90 - \tan^{-1}\frac{\omega_{gc}}{2\delta\omega_n} + 180 \qquad \text{.....(7.14)}$$

$$= 90 - \tan^{-1}\frac{\omega_{gc}}{2\delta\omega_n}$$

$$\tan\phi_{pm} = \tan\left(90 - \tan^{-1}\frac{\omega_{gc}}{2\delta\omega_n}\right)$$

$$= \cot\tan^{-1}\frac{\omega_{gc}}{2\delta\omega_n} = \frac{2\delta\omega_n}{\omega_{gc}}$$

$$\therefore \qquad \phi_{pm} = \tan^{-1}\frac{2\delta\omega_n}{\omega_{gc}} \qquad \text{.....(7.15)}$$

From eqn. (7.13), we have

$$\frac{\omega_n}{\omega_{gc}} = \frac{1}{\sqrt{\sqrt{4\delta^4 + 1} - 2\delta^2}} \qquad \text{.....(7.16)}$$

Substituting eqn. (7.16) in eqn. (7.15), we have,

$$\phi_{pm} = \tan^{-1}\left[\frac{2\delta}{\sqrt{\sqrt{4\delta^4 + 1} - 2\delta^2}}\right] \qquad \text{.....(7.17)}$$

Eqn. (7.17) gives a relationship between the phase margin and the damping factor for a second order system for $\delta < 1$. Eqn. (7.17) is plotted for δ ranging for 0 to 0.8 in Fig. 7.29.

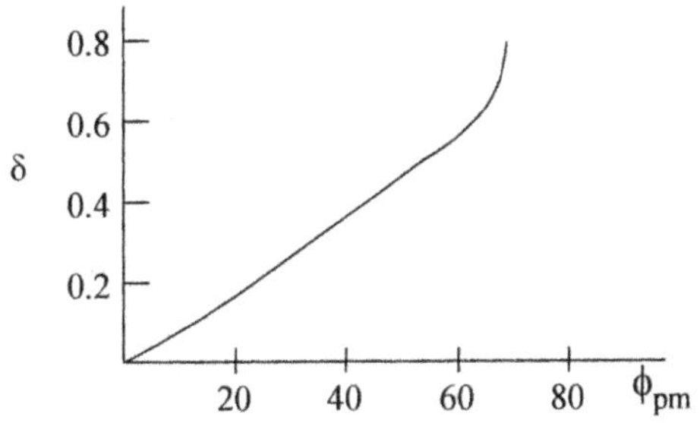

Fig. 7.29 Curve showing the variation of ϕ_{pm} with δ for a scond order system.

As seen from Fig. 7.29, for values of $\delta < 0.7$ the curve is seen to be linear and hence can be approximate by a straight line,

$$\delta = 0.01\phi_{pm} \qquad .....(7.18)$$

Here ϕ_{pm} is to be taken in degrees.

Frequency response can be obtained for a given system to estimate the resonant peak and resonance frequency which are the two important frequency domain specifications. But this is often difficult.

Thus graphical techniques are developed to determine M_r and ω_r from the open loop frequency response.

7.5.1 Constant M Circles

Consider the polar plot of the open loop transfer function of a unity feedback system. A point on the polar plot is given by :

$$G(s)|_{s = j\omega} = G(j\omega) = x + jy.$$

The closed loop frequency response is given by

$$T(j\omega) = \frac{C(j\omega)}{R(j\omega)} = \frac{G(j\omega)}{1+G(j\omega)} = \frac{x+jy}{1+x+jy} \qquad .....(7.19)$$

$$\therefore \qquad |T(j\omega)|^2 = \frac{x^2+y^2}{(1+x)^2+y^2}$$

Let $\qquad |T(j\omega)| = M$

$$\therefore \qquad M^2 = \frac{x^2+y^2}{(1+x)^2+y^2}$$

$$M^2(1+x)^2 + M^2y^2 = x^2 + y^2$$

Rearranging, we have

$$x^2 (M^2 - 1) + 2xM^2 + y^2 (M^2 - 1) = -M^2 \qquad .....(7.20)$$

$$\underline{x^2 + \frac{2M^2}{M^2 - 1} x} + y^2 = -\frac{M^2}{M^2 - 1} \qquad ......(7.21)$$

Making a perfect square of the terms underlined in eqn. 7.21, we have,

$$\underline{\left(x + \frac{M^2}{M^2 - 1}\right)^2} + y^2 = -\frac{M^2}{M^2 - 1} + \frac{M^4}{(M^2 - 1)^2}$$

$$= \frac{-M^2(M^2 - 1) + M^4}{(M^2 - 1)^2}$$

$$= \left(\frac{M}{M^2 - 1}\right)^2 \qquad .....(7.22)$$

Eqn. (7.22) represents a circle with a radius of $\frac{M}{M^2 - 1}$ and centre at $\left(-\frac{M^2}{M^2 - 1}, 0\right)$.

For various assumed values of M, a family of circles can be drawn which represent the eqn. (7.22). These circles are called constant M-circles.

Properties of M-circles :

1. For M = 1, the centre of the circle is at $\left(\underset{M \to 1}{\text{lt}} \frac{-M^2}{M^2 - 1}, 0\right)$ ie., $(-\infty, 0)$.

   The radius is also infinity

   Substituting M = 1 in eqn. 7.20, we have

   $$2x = -1$$

   or $$x = -\frac{1}{2}$$

   This M = 1 is a straight line parallel to y axis at $x = -\frac{1}{2}$.

2. For M > 1, centre of the circle is on the negative real axis and as $M \to \infty$, the centre approaches (–1, j0) point and the radius approaches zero; ie (–1, j0) point represents a circle for $M = \infty$.

3. For $0 < M < 1$, $-\frac{M^2}{M^2-1}$ is positive and hence the centre is on the positive real axis.
4. For $M = 0$, the centre is at (0, 0) and radius is 0; ie., origin represents the circle for $M = 0$.
5. As M is made smaller and smaller than unity, the centre moves from $+\infty$ towards the origin on the positive real axis.

The M circles are sketched in Fig. 7.30.

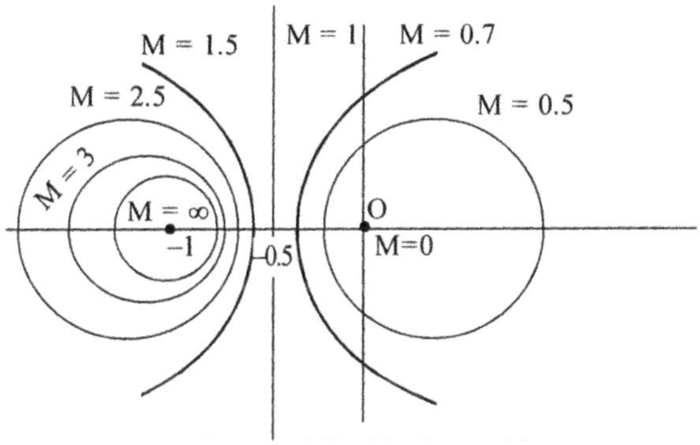

Fig. 7.30 Constant M-circles.

7.5.2 Constant N circles

Constant N circles are obtained for the points on the open loop polar plot which result in constant phase angle for the closed loop system. Consider the phase angle of the closed loop transfer function.

$$\angle T(j\omega) = \theta = \angle \frac{x + jy}{1 + x + jy}$$

$$= \tan^{-1}\frac{y}{x} - \tan^{-1}\frac{y}{1+x} \qquad .....(7.23)$$

Taking tangent of the angles on both sides of eqn. 7.23, we have

$$\tan\theta = \frac{\frac{y}{x} - \frac{y}{1+x}}{1 + \frac{y^2}{x(1+x)}} = \frac{y}{x^2 + y^2 + x} \qquad .....(7.24)$$

Let $\tan\theta = N$

Then $$\frac{y}{x^2 + y^2 + x} = N$$

Rearranging, we get,

$$N(x^2 + x) + Ny^2 - y = 0$$

$$N\left(x+\frac{1}{2}\right)^2 + N\left(y-\frac{1}{2N}\right)^2 = \frac{N}{4}+\frac{1}{4N}$$

$$\left(x+\frac{1}{2}\right)^2 + \left(y-\frac{1}{2N}\right)^2 = \frac{1}{4}\left(\frac{N^2+1}{N^2}\right) \qquad .....(7.25)$$

Eqn. (7.25) represents the equation of a family of circles for different values of N with

centre at $$\left(-\frac{1}{2}, \frac{1}{2N}\right) \qquad .....(7.26)$$

and Radius $$= \frac{\sqrt{N^2+1}}{2N} \qquad .....(7.27)$$

These circles are known as constant N circles.

Properties of N circles :

1. All the circles pass through the origin (0, 0) and the point (–1, 0) as these points satisfy eqn. (7.25) irrespective of value of N.
2. From eqn. (7.26) we see that the centres of these circles lie on $x = -\frac{1}{2}$ line.
3. As $N \to 0^+$ centre tends to $\left(-\frac{1}{2}, +\infty\right)$ and as $N \to 0^-$ centre tends to $\left(-\frac{1}{2}, -\infty\right)$ and radius becomes infinity, ie, x-axis represents the N = 0 line.

   ie., $\tan\theta = N = 0$

   $\theta = 0 \text{ or } \pm 180$
4. For both positive and negative values of N, the radius remains the same. For positive N, the centre lies on the line $x = -\frac{1}{2}$ and above the x-axis. For negative N, the centre lies on $x = -\frac{1}{2}$ and below the x-axis. The locii are symmetrical about x-axis. Positive N corresponds to $+\theta$ and negative N corresponds to $-\theta$.

The constant N-circles are shown in Fig. 7.31. Instead of marking the values of N on the various circles, value of $\phi = \tan^{-1}N$ are marked so that the phase angle can be read from the curves.

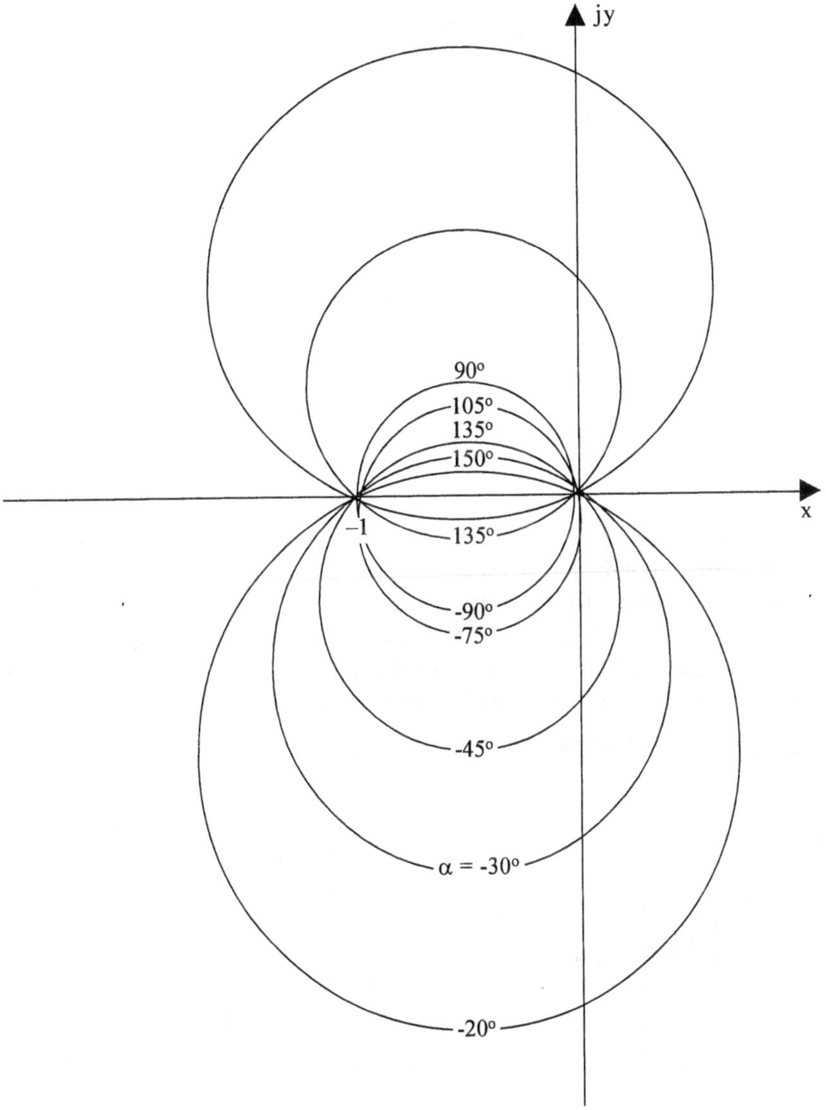

Fig. 7.31 Constant N circles

Printed charts, in which M – N locii are plotted, are available. The polar plot of G(jω) can be drawn on this graph and at any value of ω, G(jω) point is located. At that point, the magnitude M and phase angle ϕ of the closed loop transfer function can be read off. The complete closed loop frequency response can be plotted in this way for any given G(s).

From the closed loop frequency response, resonance peak, resonance frequency, bandwidth etc. can be easily calculated.

If H(s) $\neq$ 1, ie for non unity feedback systems, we can modify the procedure used for finding the closed loop frequency response as discussed below.

Let
$$T(j\omega) = \frac{C(j\omega)}{R(j\omega)} = \frac{G(j\omega)}{1+G(j\omega)H(j\omega)}$$

$$= \frac{G(j\omega)H(j\omega)}{1+G(j\omega)H(j\omega)} \cdot \frac{1}{H(j\omega)}$$

$$= T_1(j\omega)\frac{1}{H(j\omega)}$$

Where,
$$T_1(j\omega) = \frac{G_1(j\omega)}{1+G_1(j\omega)} \text{ and } G_1(j\omega) = G(j\omega)H(j\omega)$$

We can use M – N circle locii for determining T_1 (jω) and T(jω) can be obtained by multiplying T_1 (jω) by $\frac{1}{H(j\omega)}$.

7.5.3 Nichols Charts

Though the constant M and N circles plotted on polar coordinates are useful in the design of control systems, it is more convenient to plot them on gain phase plane (gain in decibels Vs phase in degrees). It is easier to construct a Bode plot of G(s) rather than polar plot. At any frequency |G(jω)| and ∠G(jω) can be obtained from the Bode plot and transferred to the gain-phase plot. From this it is easier to read resonance peak and other frequency domain specifications which may be used in the design. Let us consider plotting of M-circle on to the gain phase plot.

Consider an M circle for M = M_1 as shown in Fig. 7.32 (a).

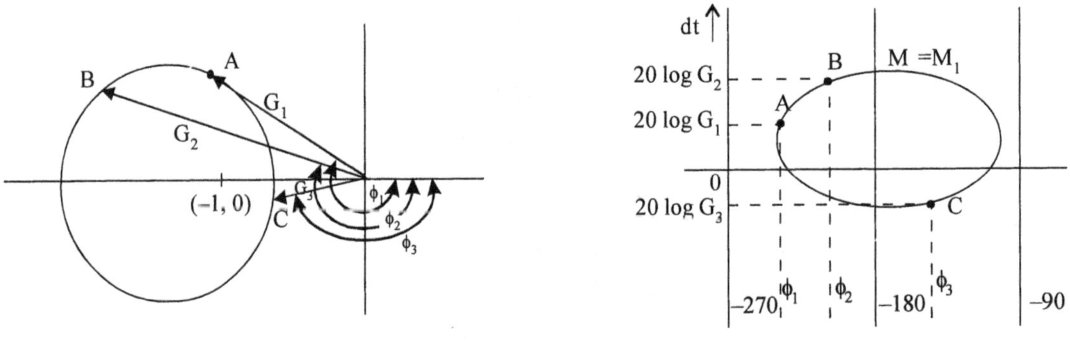

Fig. 7.32 (a) M circle for M = M_1 in GH plane **(b)** M circle in gain phase plane.

Consider a point A on this M circle. Join A to the origin. The magnitude of this vector OA in db and the phase angle measured negatively from the real axis are transferred to the gain-phase plane as shown in Fig. 7.33 (b). This graph contains angles on the x-axis for 0 to -360^0 on a linear scale. The y-axis represents gain in decibels. Similarly points B at C can also be plotted. This gives rise to a contour gain phase plot. The constant N locii are also plotted on the same plane in a similar manner. The critical point (–1, j0) in GH plane corresponds to 0 db, -180^0, point in the gain phase plot. These M and N locii were first conceived and plotted by NB Nichols and are called the Nichols charts.

The Nichol's chart for 0 to -180^0 and -180^0 to -360^0 are mirror images of each other. Nichols chart is shown in Fig. 7.33.

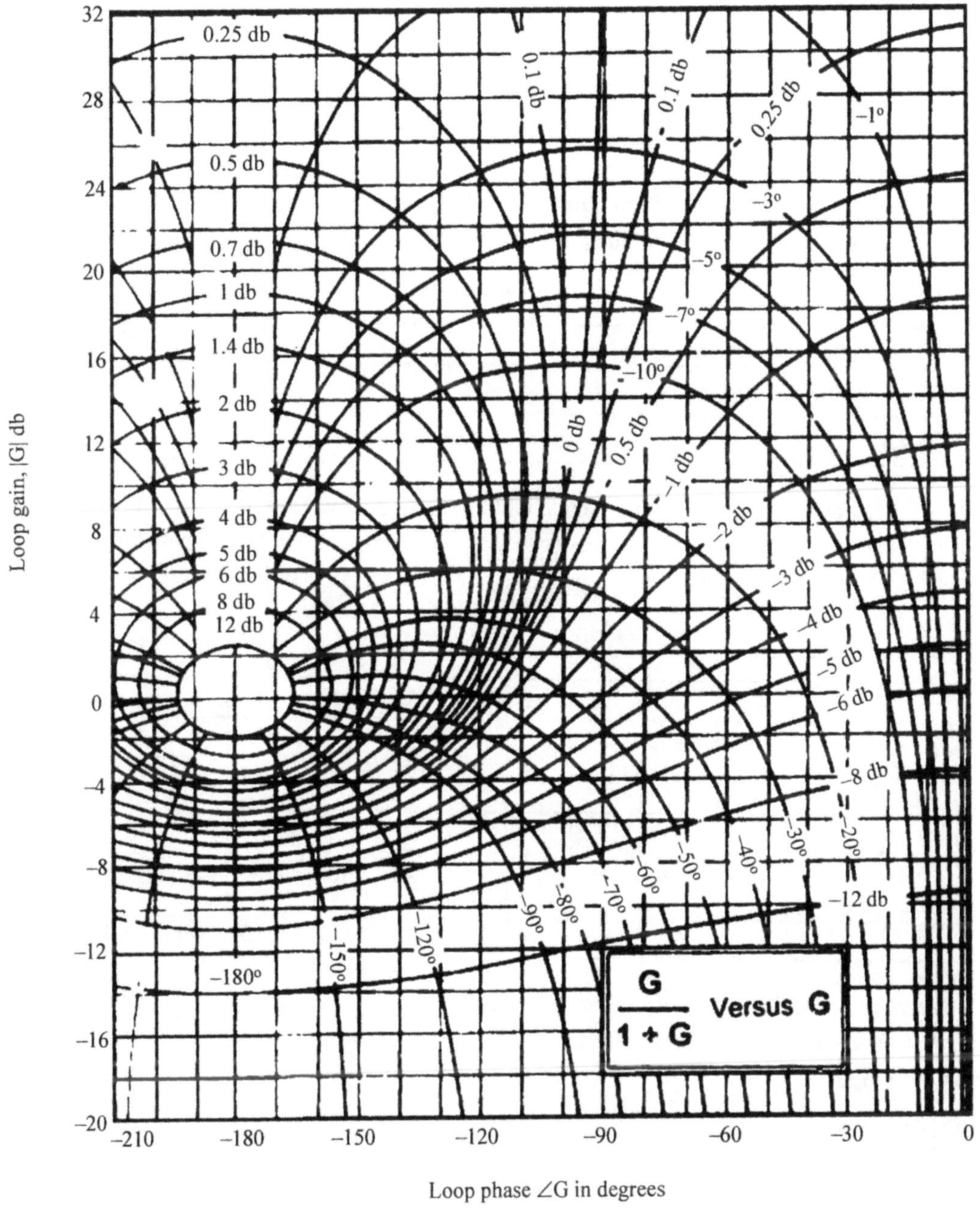

Fig. 7.33 Nichol's chart

Example 7.10

Obtain the closed loop frequency response of a unity feedback system whose open loop transfer function is given by,

$$G(s) = \frac{10}{s(0.5s+1)(0.05s+1)}$$

Use Nichol's chart.

Solution :

The Bode plot of the given open loop transfer function is already drawn in Fig. 7.27. The magnitude of $G(j\omega)$ and angle of $G(j\omega)$ are obtained at different frequencies and tabulatd in Table 7.2.

Table 7.2

ω in rad/sec	0.5	1.0	2	3	4	6	8	10	20
$\lvert G(j\omega)\rvert$ in db	26	20	14	6.5	2	–5	–10.5	–15	–26
$\angle G(j\omega)$ in deg	–105	–120	–140	–156	–165	–180	–189	–195	–219

These readings are transferred on to the Nichol's chart as shown in Fig. 7.34. The magnitude and angle of the closed loop frequency response are read from the Nichol's chart and tabulated in Table 7.3. From the magnitude values, the value of M is calculated and tabulated.

Table 7.3

ω in rad/sec	0.5	1	2	3	4	4.3	4.8	5.2	6	8	10
$\left\lvert\frac{G}{1+G}\right\rvert$ in db	0.1	0.4	1.4	4	11	14	15	8	2	–8	–14
$\angle\frac{G}{1+G}$	3	–5.5	–9	–18	–35	–80	–100	–160	–180	–190	–195
M	1	1.05	1.17	1.38	3.55	5.0	5.62	2.5	1.26	0.4	0.2

The value of M and angle of closed loop frequency response are plotted as a function of ω in Fig. 7.35. The value of peak resonance M_r, the resonance frequency ω_r, and the bandwidth are read from the graph.

From the graph,

$$M_r = 5.6$$

$$\omega_r = 4.8 \text{ rad/sec}$$

$$\omega_b = 6.8 \text{ rad/sec}$$

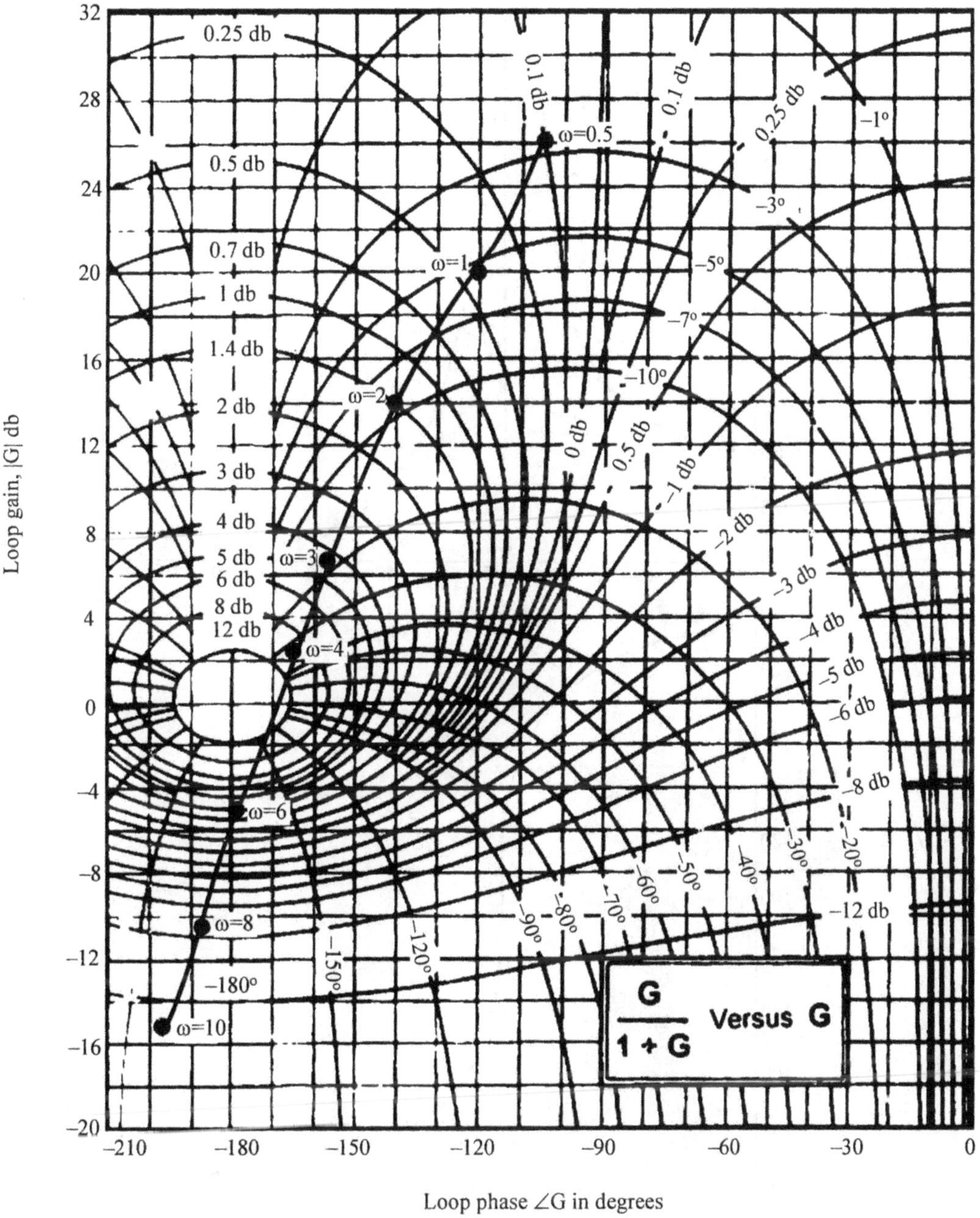

Fig. 7.34 Open loop frequency response superposed on Nichol's chart for Ex. 7.10

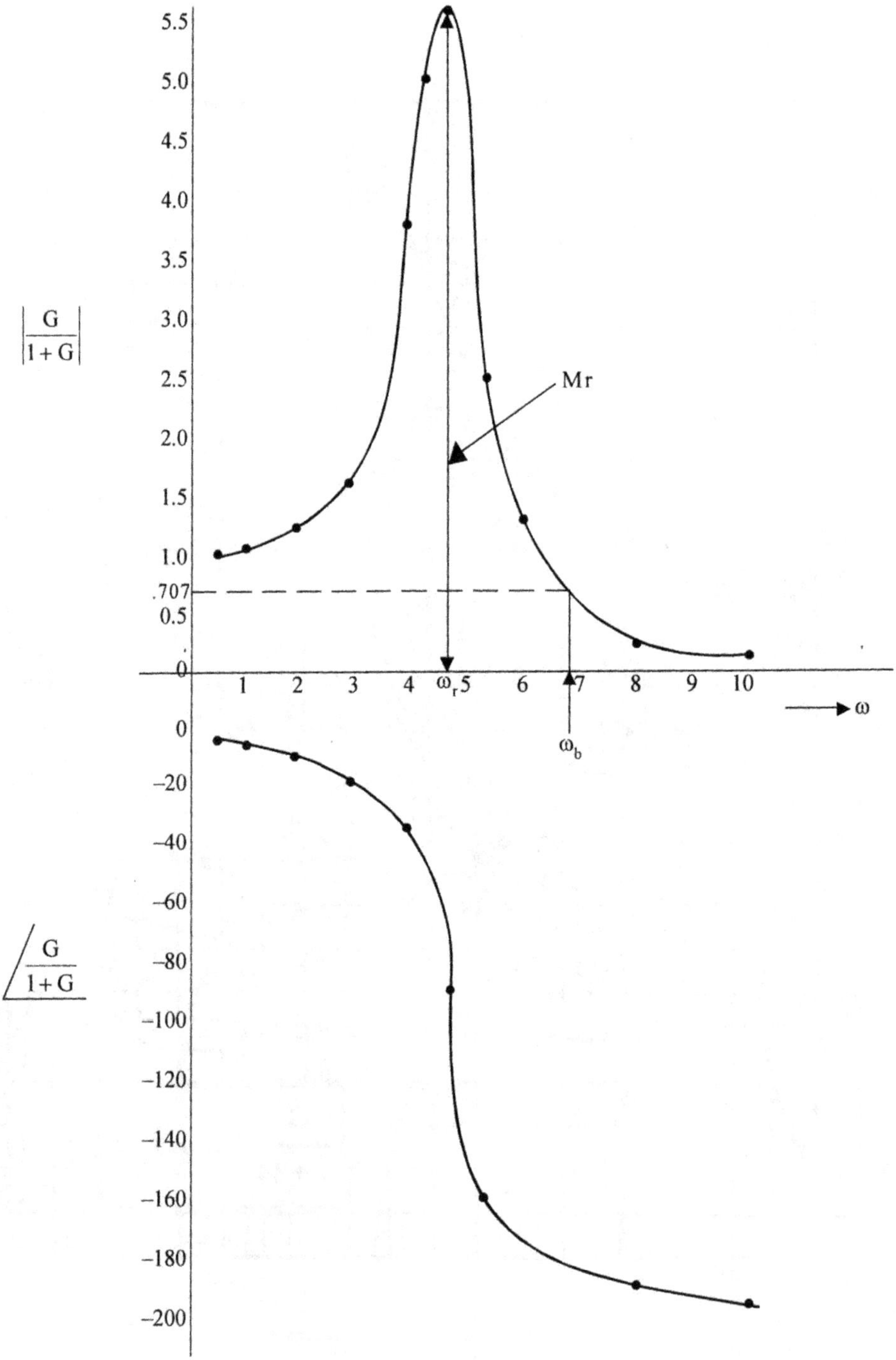

Fig. 7.35 Closed loop frequency response of the system of Ex. 7.10.

Problems

7.1 Draw the complete Nyquist plots for the following open loop transfer functions. If the system is unstable, how many poles of the closed loop system are in the right half of *s*-plane.

(a) $G(s)\,H(s) = \dfrac{2(s+0.25)}{s^2(s+1)(s+0.5)}$

(b) $G(s)\,H(s) = \dfrac{1}{s^2+50}$

(c) $G(s)\,H(s) = \dfrac{2(1+0.5s)(s+1)}{(1+10s)(1-s)}$

7.2 Obtain the polar plot of the following system.

$$G(s)\,H(s) = \frac{K}{(T_1 s+1)}$$

If to this system, poles at origin are added as shown below, sketch the polar plots in each case and comment about the behaviour at $\omega = 0$ and $\omega = \infty$ in each case.

(a) $\dfrac{K}{s(T_1 s+1)}$

(b) $\dfrac{K}{s^2(T_1 s+1)}$

(c) $\dfrac{K}{s^3(T_1 s+1)}$

7.3 Consider again the polar plot of

$$G(s)\,H(s) = \frac{K}{T_1 s+1}$$

If now, non zero real poles are added as shown below sketch the polar plots in each case. Comment on their behaviour at $\omega = 0$ and $\omega = \infty$

(a) $\dfrac{K}{(1+sT_1)(1+sT_2)}$

(b) $\dfrac{K}{(1+sT_1)(1+sT_2)(1+sT_3)}$

(c) $\dfrac{K}{(1+sT_1)(1+sT_2)(1+sT_3)(1+sT_4)}$

7.4 Consider the following polar plot. If now a pole at origin and a pole at $s = -\frac{1}{T_2}$ are added, sketch the polar plot.

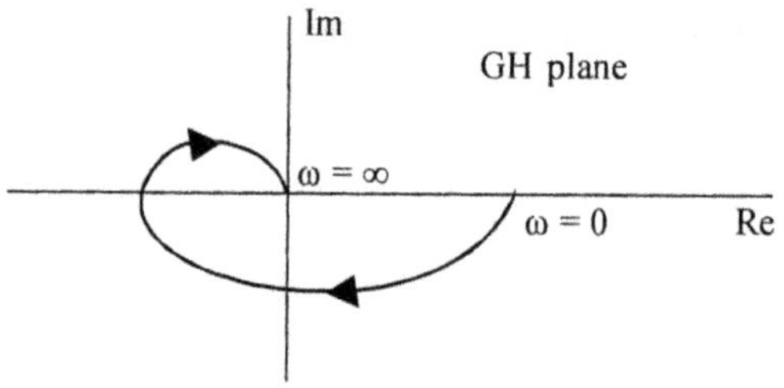

Fig. P 7.4

7.5 Consider the polar plot of,

$$G(s)\,H(s) = \frac{K}{1 + sT_1}$$

Zeros are added as shown below. Sketch the polar plots in each case and comment on the effect of adding a zero on the polar plot.

(a) $K\,\dfrac{1 + sT_a}{1 + sT_1} \qquad T_a < T_1$

(b) $K\,\dfrac{1 + sT_a}{1 + sT_1} \qquad T_a > T_1$

7.6 Sketch the Nyquist plot of,

$$G(s)\,H(s) = \frac{10}{(2s + 1)}$$

A pole at $s = 0$ and a zero at $s = -1$ are added to this open loop transfer function. Sketch the resulting Nyquist plot. Compare the two plots in terms of behaviour at $\omega = 0$ and $\omega = \infty$

7.7 Draw the Nyquist plot for the system,

$$G(s)\,H(s) = \frac{K}{s(1 + 2s)(1 + 5s)}$$

Find the critical value of K for stability. If now a derivative control is used with $G_c(s) = (1 + 0.5s)$, will the system become more stable ? What is the new value of K for which the system will be stable.

7.8 Comment on the stability of the systems whose Nyquist plots are shown below. Find the number of closed loop poles in the RHP in each case.

(a)

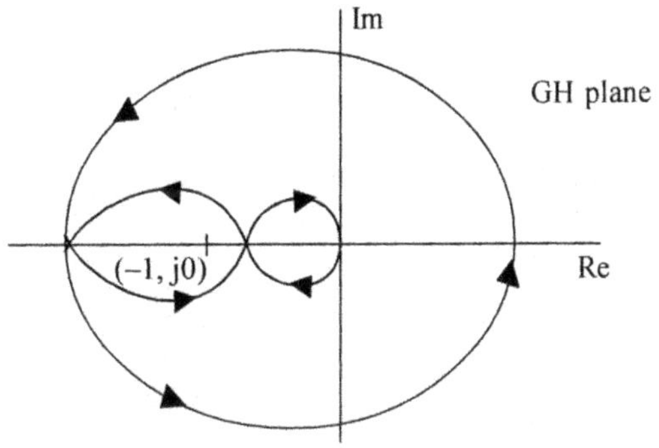

Open loop poles are all in LHP

(b)

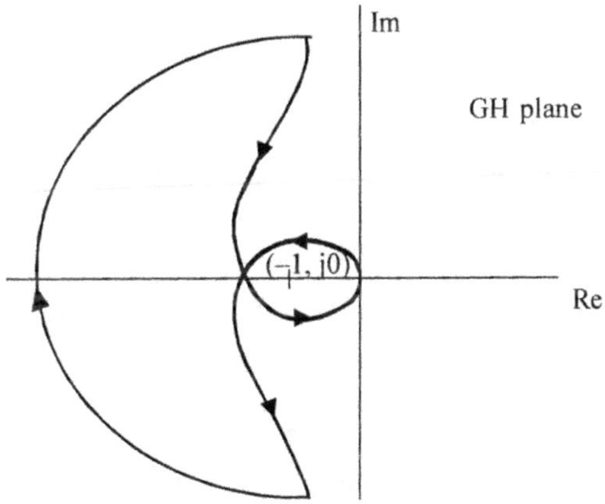

One open loop pole in RHP

7.9 Find the gain margin and phase margin for the following system with transfer function G(s) H(s) given by,

$$\frac{5}{s(1+0.1s)(1+0.2s)}$$

7.10 Consider the system,

$$G(s)\ H(s) = \frac{K}{s(1+0.2s)(1+0.05s)}$$

using Nyquist plot

(a) Find the gain margin and phase margin for K = 1

(b) What value of K will result in a gain margin of 15 db

(c) What value of K will result in a phase margin of 45°

7.11 Solve problem 7.10 using Bode plot.

7.12 Obtain the closed loop frequency response of the system

$$G(s)\,H(s) = \frac{10}{s(0.1s+1)(0.05s+1)}$$

using Nichol's chart. Find the peak resonance M_r, resonance frequency, ω_r and bandwidth ω_b.

8 Design in Frequency Domain

8.1 Preliminaries of Design

A control system is usually required to meet three time response specifications, namely, steady state accuracy, damping factor and settling time. The steadystate accuracy is specified in terms of the permissible steadystate error to a step, velocity or accleration input. Peak overshoot to step input is indicative of the damping factor and the speed of response is indicated either by rise time or the settling time. Only one of these two quantities have to be specified since both these quantities depend on δ and ω_n.

Steady state accuracy is usually satisfied by a proper choice of the error constants K_p, K_v and K_a depending on the type of the system. A damping factor of about 0.28 to 0.7, corresponding to a peak overshoot of 40% to 5% to a step input is usually satisfactory. A standard second order system is described by the transfer function,

$$G(s) = \frac{K_v}{s(\tau s + 1)}$$

where τ is the time constant of the system or plant and K_v is the gain of the amplifier. It may not be possible to change the time constant of the plant in a practical situation. For example, the selection of an actuator may depend on the type of power supply and space available. It may also be restricted by the economic considerations. Generally the speed of a servo motor, which is used as an actuator is higher than that required by the load. A gear train may have to be introduced to suit the desired speed. All these different components constitute a plant and it may not exactly suit the design specifications. Hence we have only one quantity, K_v, which can be varied to satisfy one of the specifications, usually the steady state error requirement.

Thus, from the above discussion, it can be concluded that additional subsystem called 'Compensator or Controller' has to designed to adjust the parameters of the overall system to satisfy the design criterion. This compensator may be used in series with the plant in the forward path as shown in Fig. 8.1 (a) or in the feedback path shown in Fig. 8.1 (b). The compensation in the first case is known as series or cascade compensation and the later is known as feedback compensation. The compensator may be a passive network or an active network. They may be electrical, mechanical, hydraulic, preumatic or any other type of devices. We will consider electrical networks as compensators in this chapter.

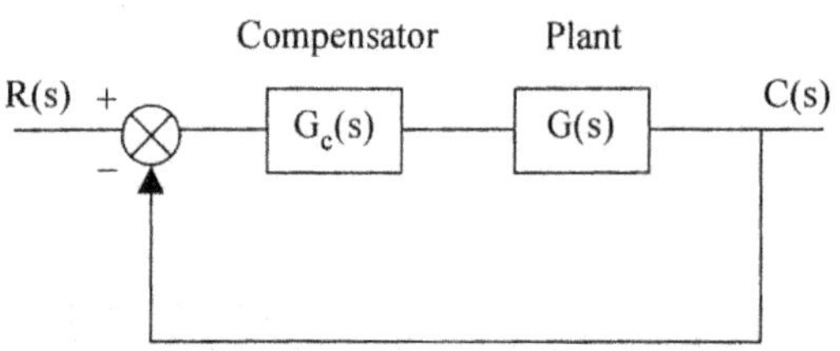

Fig. 8.1 (a) Series or cascade compensation configuration.

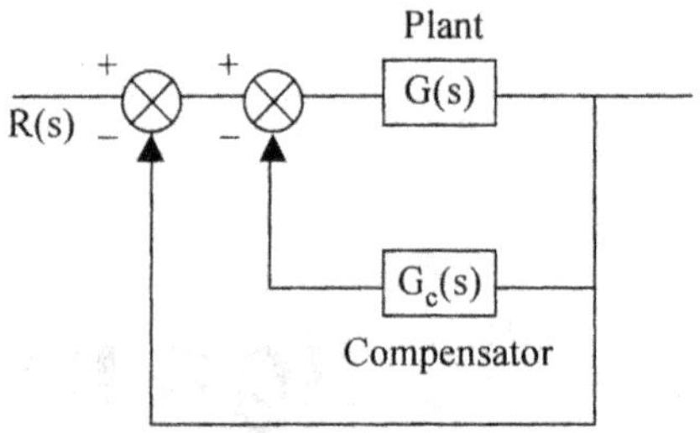

Fig. 8.1 (b) Feedback compensation configuration.

The design of a compensator can be carried out in either time domain, using root locus technique or in frequency domain using Bode plots or Nichol's charts. The desired performance specifications of transient response can be given either in time domain, namely, peak overshoot, settling time, rise time etc. or in frequency domain, viz, peak resonance, resonance frequency, Bandwidth, gain margin, phase margin etc. The steadystate performance measure is specified in terms of the error constants or steadystate error to a step, velocity or accleration inputs. A suitable combination of time domain and frequency domain specifications may also be given. Since the time response and frequency response of a second order system are well correlated, the performance measures from one domain could easily be converted to those in the other domain.

In this chapter we will discuss the design of compensators for d.c control systems in frequency domain using Bode plots. D.C control systems, as discussed in chapter 3, have unmodulated signals and compensators must be designed to operate on d.c. input signals. These signals usually have a range of frequencies 0 to 20 Hz. A.C control systems, on the other hand, have suppressed carrier frequency signals and operate on frequency ranges of 60 to 400 Hz or more. The technique developed here will not be suitable in the design of compensators for a.c control systems.

8.2 A Design Example

Let us consider the design of a control system whose open loop transfer function is given by,

$$G(s) = \frac{K_v}{s(0.5s+1)(0.05s+1)}$$

Let it be required that the steady state error e_{ss}, for a velocity input be less then 0.2. This can be obtained by a suitable choice of K_v. Since

$$e_{ss} = \frac{1}{K_v}$$

$$K_v = \frac{1}{0.2} = 5$$

The Bode plot for the system is shown in Fig. 8.2.

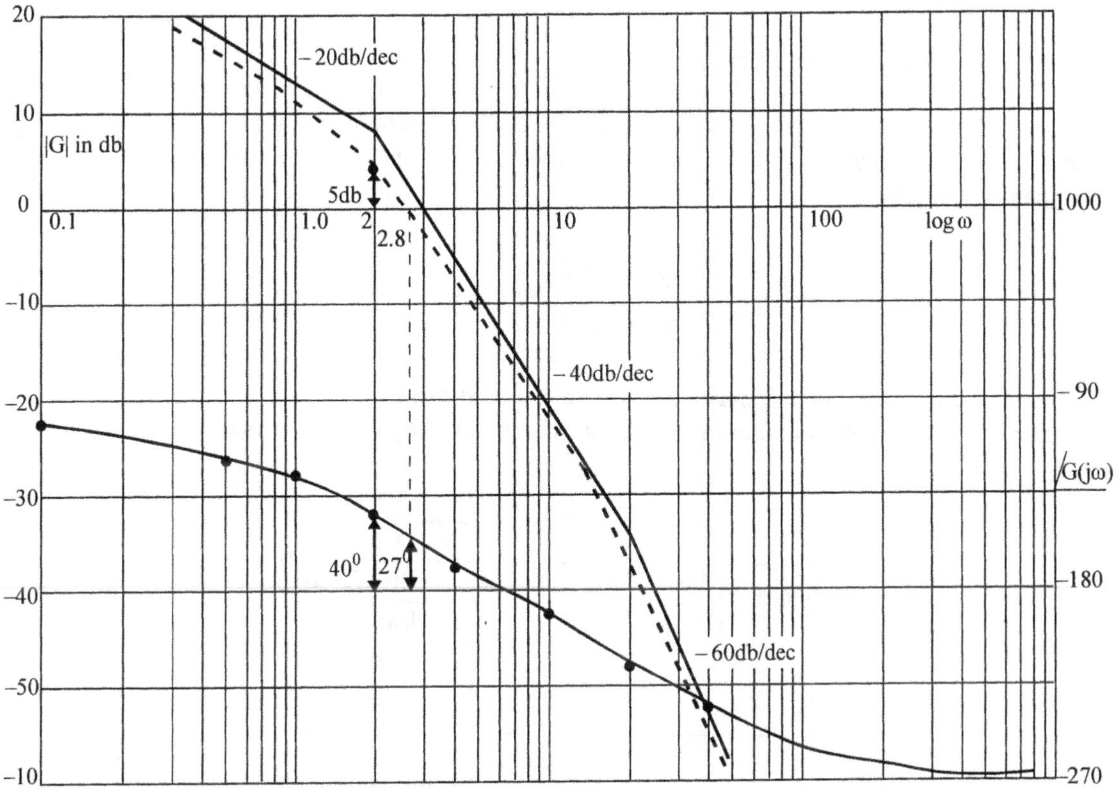

Fig. 8.2 Bode plot of $G(s) = \dfrac{K_v}{s(0.5s+1)(0.05s+1)}$ with $K_v = 5$.

From the figure the gain cross over frequency is obtained as ω_{gc} = 2.8 rad/sec and the phase margin is 27°.

Let it be required that the phase margin be 40°. Let a compensating zero be added in cascade with the forward path transfer function such that it contributes and angle of $Q_c = 40^o - 27^o = 13^o$ at the gain cross over frequency.

Thus $$G_c(s) = (\tau s + 1) \qquad .....(8.1)$$

Select τ such that

$$\angle G_c(j\omega_{gc}) = \tan^{-1} \omega_{gc\tau} = 13^o$$

or $$\tau = 0.082$$

i.e, a zero is added at

$$Z_c = \frac{1}{\tau} = 12.2$$

This adds an angle of 13° at the gain cross over frequency and hence the required phase margin is obtained. Adding a zero at s = –12.2 does not change the gain plot appreciably near the gain cross over frequency.

Since an isolated zero is not physically realisable, we must add a pole in addition to the compensating zero. This pole is added far away from the jω-axis so that its effect on transient response is negligible. The positions of pole and zero of the compensator may be adjusted to get the required phase margin, yet not affecting much the gain plot at the gain cross over frequency. The transfer function of the compensator is given by,

$$G_c(s) = \frac{s+z_c}{s+p_c} = \frac{s+\frac{1}{\tau}}{s+\frac{1}{\alpha\tau}}; \qquad \alpha = \frac{z_c}{p_c} < 1 \text{ and } \tau > 0 \qquad .....(8.2)$$

As the zero is nearer to the jω-axis than the pole, it contributes a net positive angle, and hence the compensator is termed as a *lead compensator*. The block diagram of the compensated system is given in Fig. 8.3.

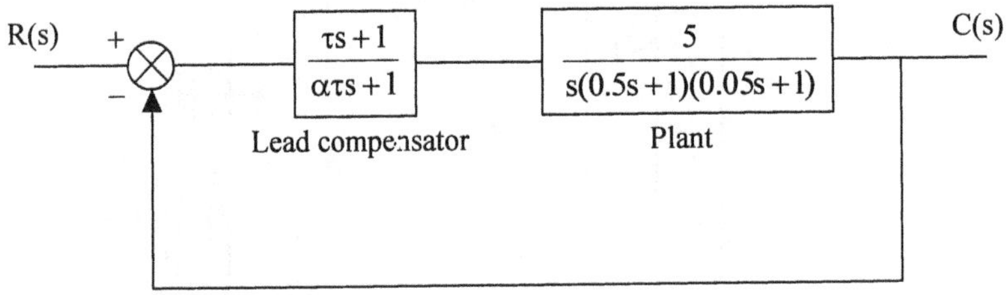

Fig. 8.3 Block diagram of lead compensated system.

The required phase margin can also be obtained by changing the gain cross over frequency so that the required phase margin could be obtained at this new gain cross over frequency. From Fig. 8.2, the required phase margin is obtained at a frequency of ω_{gc} = 2 rad/sec. The gain curve can be made to pass through 0 db line at ω_{gc} = 2 rad/sec by reducing the system gain or reducing K_v. At ω_{gc} = 2 rad/sec the gain can be read from the gain curve, which is 5db. If K_v is reduced such that,

$$20 \log K_v = 5$$

or $$K_v = 1.78$$

the gain curve crosses at 2 rad/sec.

The transient response specifications are met but the error constant is reduced to 1.78. It does not satisfy the steadystate error requirement of $e_{ss} < 0.2$.

Consider now improving the steady state performance by increasing the *type* of the system by including a compensating pole at the origin. The error constant becomes ∞. However, adding a pole at the origin makes the uncompensated type-1 system to become compensated type-2 system which is inherently unstable under closed loop conditions. This is evident from the root locus plot of the system as shown in Fig. 8.4.

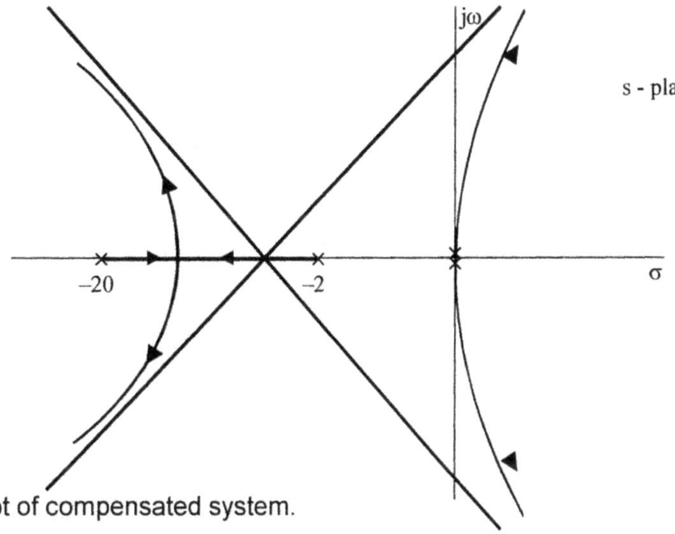

Fig. 8.4 Root locus plot of compensated system.

In order to obviate this difficulty, a compensating zero is added very close to the pole at the origin and in the left half of the s-plane. This will ensure that the compensated system is of type-2 and the effect of the pole on the transient response is nullified by the presence of a zero very near to it. Thus the compensator will have a transfer function,

$$G_c(s) = \frac{s + z_c}{s} \qquad .....(8.3)$$

On physical realisability considerations, the pole at the origin is shifted slightly to the left of the origin and the transfer function becomes,

$$G_c(s) = \frac{s + z_c}{s + p_c}; \quad \frac{z_c}{p_c} = \beta > 1 \qquad .....(8.4)$$

$\beta > 1$ ensures that the zero is to the left of the pole. Since the pole is nearer to the origin, this transfer function introduces negative angle, i.e, lagging angle and hence this type of compensator is termed as a lag compensator.

From the above discussion it is clear that a lead compensator improves the transient performance while preserving the steadystate performance and a lag compensator improves the steadystate performance while preserving the transient performance.

If both transient and steadystate performance have to be improved, a lag lead compensator may be used. A type-0 or type-1 system may be compensated by using any the above compensators but a type-2 systems which is inherently unstable, can be compensated by using lead compensator only as this alone can increase the margin of stability.

8.3 Realisation of Compensators and their Characteristics

The three types of compensators, viz, lead, lag and lag-lead compensators can be realised by electrical, mechanical, pneumatic, hydrautic and other components. We will discuss the realisation of these compensators by electrical RC networks. We will also obtain their frequency response characteristics which are useful in their design.

8.3.1 Lead Compensator

We have seen in section 8.2 that the pole zero configuration of a lead compensator is as given in Fig. 8.5.

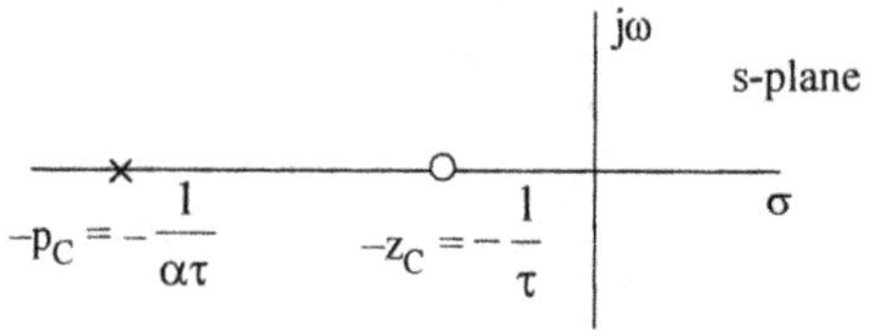

Fig. 8.5 Pole zero configuration of a lead compensator.

The transfer function of a lead compensator is given by,

$$G_c(s) = \frac{s+\frac{1}{\tau}}{s+\frac{1}{\alpha\tau}} ; \alpha < 1, \tau > 0 \quad .....(8.5)$$

$$= \alpha\left(\frac{\tau s+1}{\alpha\tau s+1}\right) \quad .....(8.6)$$

This transfer function can be realised by an RC network shown in Fig. 8.6.

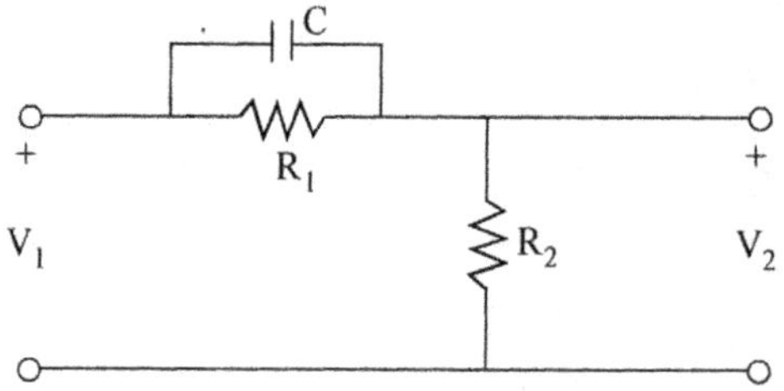

Fig. 8.6 RC network realisation of a lead compensator.

The transfer function of the network is given by,

$$\frac{V_2(s)}{V_1(s)} = \frac{R_2}{R_2 + \frac{R_1 . \frac{1}{Cs}}{R_1 + \frac{1}{Cs}}} = \frac{R_2(R_1Cs+1)}{R_1R_2Cs + R_1 + R_2}$$

$$= \frac{R_2}{R_1+R_2} \cdot \frac{R_1Cs+1}{\frac{R_1R_2C}{R_1+R_2}s+1} \quad .....(8.7)$$

Comparing eqn. (8.7) with eqn. (8.6) we have

$$\tau = R_1 C \quad .....(8.8)$$

$$\alpha = \frac{R_2}{R_1+R_2} < 1 \quad .....(8.9)$$

Since $\alpha < 1$, it is a lead compensator. Given the time constant τ and the d.c gain α of the lead network, the three components of RC network, namely, R_1, R_2 and C have to be determined.

Since eqns. (8.8) and (8.9) are the two relations to be used, one component may be chosen arbitraritly. This may be decided by the required impedance level of the network.

To obtain the frequency response of the network, we have,

$$G_c(j\omega) = \alpha \frac{(1+j\omega\tau)}{1+j\alpha\omega\tau}; \qquad \alpha < 1 \qquad .....(8.10)$$

For $\omega = 0$,

$G_c(j\omega) = \alpha$ and $\alpha < 1$

which means that the network produces an attenuation of α, and this d.c attenuation has to be cancelled by using an amplifier of gain $A = \frac{1}{\alpha}$. Thus the compensating network takes the form as shown in Fig. 8.7.

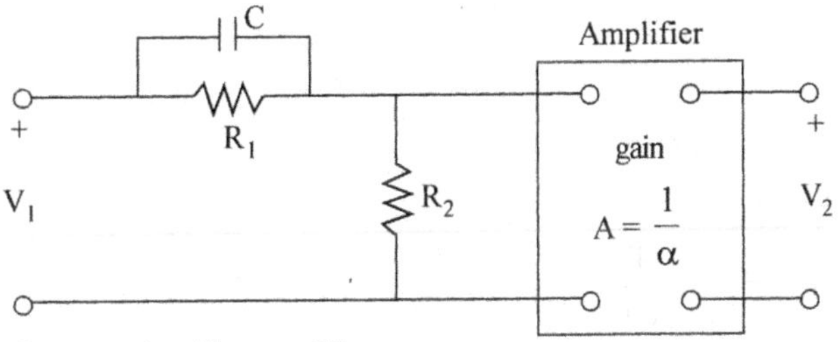

Fig. 8.7 Phase lead network with amplifier.

Thus, the sinusoidal transfer function of the lead network is given by

$$G_c(j\omega) = \frac{1+j\omega\tau}{1+j\alpha\omega\tau}; \qquad \alpha < 1 \qquad .....(8.11)$$

Under steady state conditions, the output of this network leads the input and hence this network is known as a lead network.

The Bode plot of eqn. (8.11) is shown in Fig. 8.8.

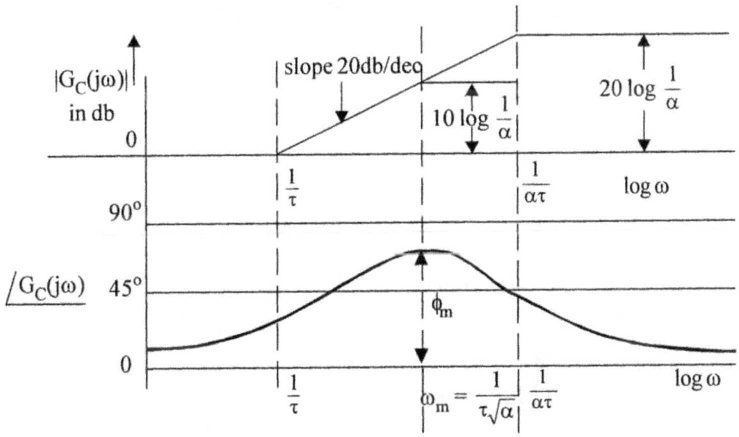

Fig. 8.8 Bode plot of lead network.

From eqn. (8.11), $\angle G(j\omega)$ is given by

$$\phi = \tan^{-1} \omega\tau - \tan^{-1} \alpha\omega\tau \qquad .....(8.12)$$

or

$$\tan \phi = \frac{\omega\tau(1-\alpha)}{1+\alpha\omega^2 \tau^2} \qquad .....(8.13)$$

The angle ϕ is a function of ω and it will be a maximum when $\tan \phi$ is a maximum with respect to ω.

$$\frac{d(\tan\phi)}{d\omega} = \frac{(1+\alpha\omega^2\tau^2)\,\tau\,(1-\alpha) - \omega\tau\,(1-\alpha)\,2\alpha\omega\tau^2}{(1+\alpha\omega^2\tau^2)^2} = 0$$

$\therefore$

$$1 + \alpha\omega^2 \tau^2 - 2\alpha\omega^2 \tau^2 = 0$$

$$\alpha\omega^2 \tau^2 = 1$$

or

$$\omega^2 = \frac{1}{\alpha\tau^2}$$

$\therefore$

$$\omega_m = \frac{1}{\tau\sqrt{\alpha}} \qquad .....(8.14)$$

Eqn. 8.14 can be written as

$$\omega_m^2 = \frac{1}{\tau} \cdot \frac{1}{\alpha\tau}$$

$\therefore$ ω_m is the geometric mean of the two corner frequencies $\frac{1}{\tau}$ and $\frac{1}{\alpha\tau}$

Substituting the value of $\omega = \omega_m$ in eqn. 8.13 we have

$$\tan \phi_m = \frac{\frac{1}{\sqrt{\alpha}}(1-\alpha)}{1+1}$$

$$= \frac{1-\alpha}{2\sqrt{\alpha}}$$

From the right angled triangle

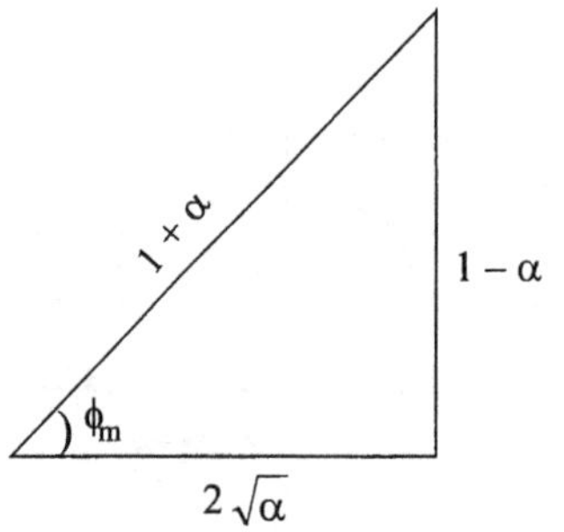

we have $$\sin \phi_m = \frac{1-\alpha}{1+\alpha} \qquad .....(8.15)$$

Solving for α, we get,

$$\alpha = \frac{1-\sin\phi_m}{1+\sin\phi_m} \qquad .....(8.16)$$

Also, the gain at ω_m is obtained as,

$$|G_c(j\omega_m)| = 20 \log \left(\frac{1+\omega_m^2\tau^2}{1+\alpha^2\,\omega_m^2\tau^2}\right)^{\frac{1}{2}}$$

Substituting $\omega_m = \dfrac{1}{\tau\sqrt{\alpha}}$, we have,

$$|G_c(j\omega_m)| = 10 \log \left(\frac{1+\frac{1}{\alpha}}{1+\alpha}\right)$$

$$= 10 \log \frac{1}{\alpha} \qquad .....(8.17)$$

The lead compensator is required to provide the necessary angle at the gain crossover frequency, to obtain the required phase margin. Knowing the value of required angle, ϕ_m, to be provided by the network, α parameter can be obtained using eqn. 8.17. The value of α required to get a maximum phase lead, ϕ_m, is tabulated in Table 8.1.

Table 8.1

ϕ_m	α
0	1
20	0.49
40	0.217
60	0.072
80	0.0076

From the table it can be seen that for obtaining larger phase angles, α has to be reduced to very low value. For very low values of α, the pole of the lead network is located for away from the jω-axis and from the Bode magnitude diagram it can be seen that the network has a large gain of $20 \log \frac{1}{\alpha}$ db at high frequencies. Thus the high frequency noise signals are amplified and signal to noise ratio becomes very poor. Hence it is not desirable to have low values of α. A value of $\alpha = 0.1$ is considered to be suitable. When large phase angles are desired, two lead networks in cascade may be used.

8.3.2 Lag Compensator

The transfer function of a lag compensator is given by,

$$G_c(s) = \frac{s+\frac{1}{\tau}}{s+\frac{1}{\beta\tau}},\ \beta > 1,\ \tau > 0 \qquad .....(8.18)$$

$\beta > 1$ ensures that the pole is to the right of the zero. The pole zero configuration is given in Fig. 8.9.

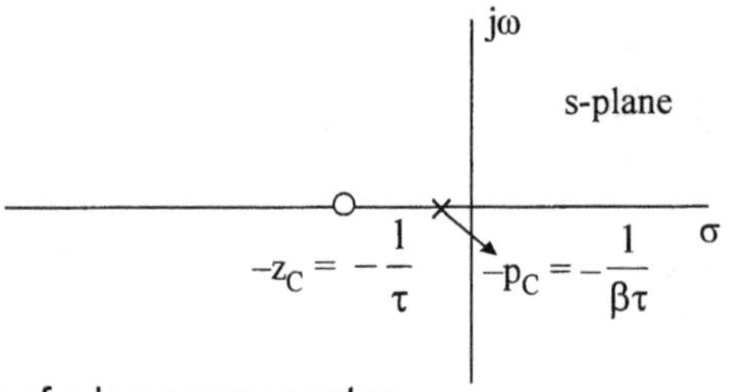

Fig. 8.9 Pole zero pattern of a lag compensator.

The RC network realisation is shown in Fig. 8.10.

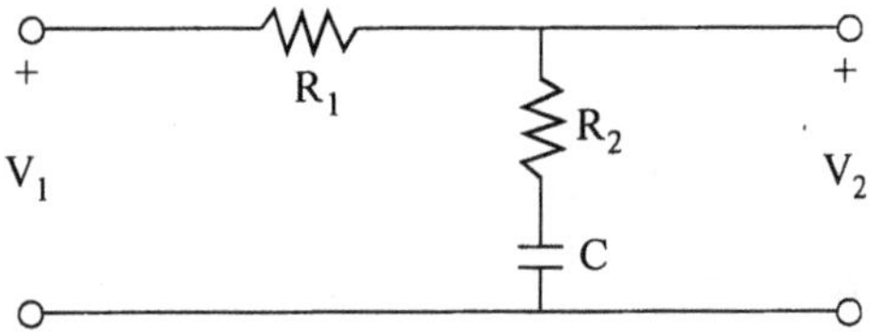

Fig. 8.10 An RC lag network.

$$G_c(s) = \frac{V_2(s)}{V_1(s)} = \frac{R_2+\frac{1}{Cs}}{R_1+R_2+\frac{1}{Cs}} = \frac{R_2Cs+1}{(R_1+R_2)Cs+1} \qquad .....(8.19)$$

$$= \frac{R_2Cs+1}{\frac{R_1+R_2}{R_2}R_2Cs+1} \qquad .....(8.20)$$

$$= \frac{\tau s+1}{\beta\tau s+1} \qquad .....(8.21)$$

where $\tau = R_2C$ and $\beta = \frac{R_1+R_2}{R_2} > 1$ .....(8.22)

The sinusoidal transfer function is given by

$$G_c(j\omega) = \frac{j\tau\omega+1}{j\beta\tau\omega+1} \qquad .....(8.23)$$

The d.c gain of this network is unity and hence no d.c amplifier is required, as in the case of a lead network.

The Bode diagram is given in Fig. 8.11.

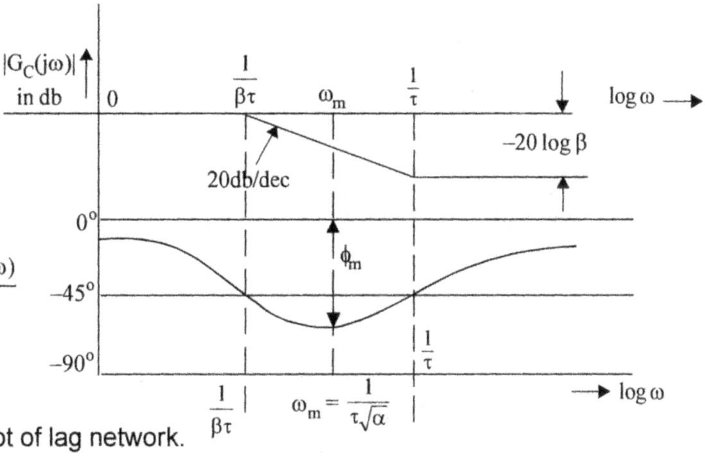

Fig. 8.11 Bode plot of lag network.

Under steady state conditions the output of the network lags the input and therefore this network is known as a lag network. The maximum phase lag ϕ_m is again obtained from eqn. (8.15) and frequency at which it occurs, from eqn. (8.14), replacing α by β. At high frequencies, the network produces an attenuation of 20 log β and hence signal to noise ratio is improved by this network.

8.3.3 Lag-Lead Compensator

The general form of the transfer function of lag lead compensator is,

$$G_c(s) = \underbrace{\frac{s+\frac{1}{\tau_1}}{s+\frac{1}{\beta\tau_1}}}_{\text{lag}} \; \underbrace{\frac{s+\frac{1}{\tau_2}}{s+\frac{1}{\alpha\tau_2}}}_{\text{lead}}; \qquad \beta > 1,\ \alpha < 1;\quad \tau_1, \tau_2 > 0 \qquad .....(8.24)$$

Transfer function given by eqn. (8.24) can be realised by an RC network as shown in Fig. 8.12.

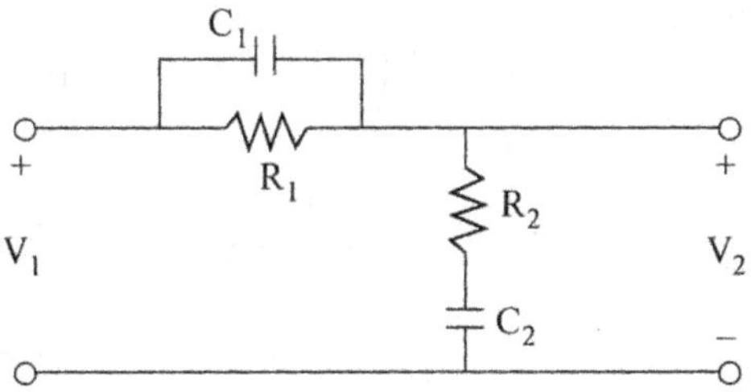

Fig. 8.12 Lag lead RC network.

The transfer function of this network can be derived as,

$$G_c(s) = \frac{V_2(s)}{V_c(s)} = \frac{\left(s+\frac{1}{R_1C_1}\right)\left(s+\frac{1}{R_2C_2}\right)}{s^2+\left(\frac{1}{R_1C_1}+\frac{1}{R_2C_2}+\frac{1}{R_2C_1}\right)s+\frac{1}{R_1R_2C_1C_2}} \qquad .....(8.25)$$

Comparing eqns. (8.25) and (8.24), we have

$$\tau_1 = \frac{1}{R_1C_1} \; ; \tau_2 = \frac{1}{R_2C_2} \quad .....(8.26)$$

$$R_1 R_2 C_1 C_2 = \alpha\beta \, \tau_1 \, \tau_2 \quad .....(8.27)$$

and
$$\frac{1}{R_1C_1} + \frac{1}{R_2C_2} + \frac{1}{R_2C_1} = \frac{1}{\beta\tau_1} + \frac{1}{\alpha\tau_2} \quad .....(8.28)$$

From eqns. (8.26) and (8.27) we have

$$\alpha\beta = 1 \quad .....(8.29)$$

This means that α and β cannot be chosen independently. In view of eqn. (8.29) we can write $G_c(s)$ as

$$G_c(s) = \frac{\left(s + \frac{1}{\tau_1}\right)\left(s + \frac{1}{\tau_2}\right)}{\left(s + \frac{1}{\beta\tau_1}\right)\left(s + \frac{\beta}{\tau_2}\right)} \; ; \beta > 1 \quad .....(8.30)$$

where
$$\tau_1 = R_1 C_1 \qquad \tau_2 = R_2 C_2 \quad .....(8.31)$$

and
$$\frac{1}{R_1C_1} + \frac{1}{R_2C_2} + \frac{1}{R_2C_1} = \frac{1}{\beta\tau_1} + \frac{\beta}{\tau_2} \quad .....(8.32)$$

The pole zero configuration of a lag lead network is given in Fig. 8.13.

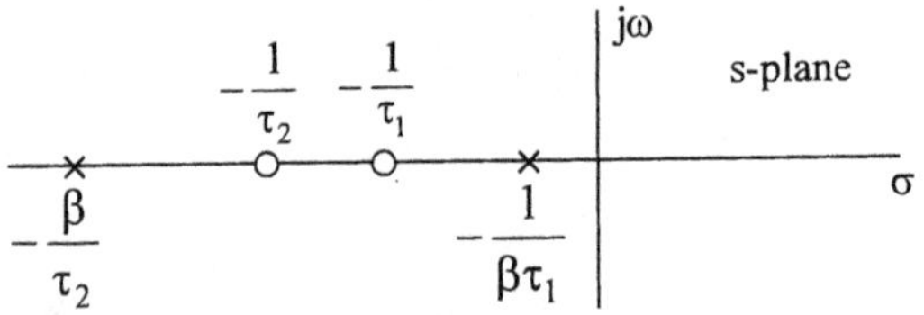

Fig. 8.13 Pole zero configuration of lag lead network.

The Bode plot is shown in Fig. 8.14.

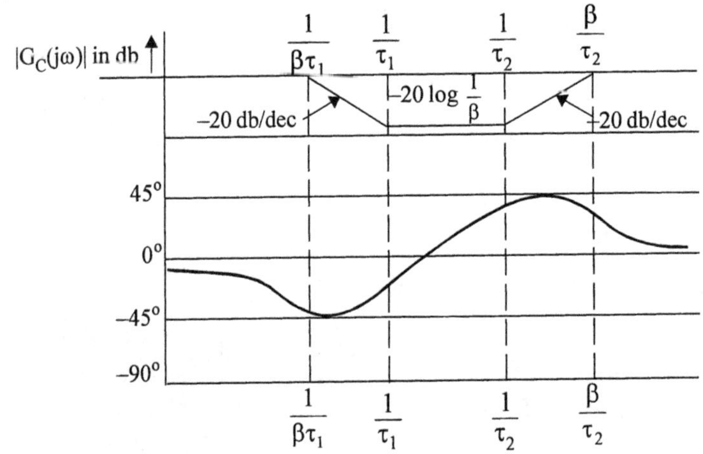

Fig. 8.14 Bode plot of Lag-lead network

8.4 Cascade Compensation in Frequency Domain Using Bode Plots

In order to design a compensator using Bode plot, the specifications are to be given in frequency domain. If specifications are given in time domain, they can be easily converted into frequency domain using time and frequency domain correlations discussed in chapter 6. The frequency domain specifications are usually given by :

1. Phase margin ϕ_m or peak resonance M_r which indicates relative stability.
2. Bandwidth ω_b or resonance frequency ω_r which indicates rise time or settling time
3. Error constants which indicates steadystate errors.

The design is carried out in the frequency domain using Bode plots. The time response of the compensated system must be obtained to check the time response specifications. This is necessary because the time and frequency domain specifications are correlated under the assumption that the compensated system is of second order, or it has a pair of dominant closed loop poles. Based on the time response characteristics, the design may be altered sustably. The frequency domain method of design is easy to apply but the time response specifications cannot be directly controlled in this method.

8.4.1 Design of Lead Compensator

The procedure for the design of a lead compensator is developed in the following.

1. The system gain K is adjusted to satisfy the steady state error criterion, as specified by the appropriate error constants K_p, K_v or K_a or the steady state error e_{ss}.
2. The Bode plot of the system with the desired K is plotted. The phase margin of the uncompensated system is read from the graph (ϕ_1). If this is satisfactory no compensator is necessary. If the phase margin falls short of the desired phase margin, additional phase lead ϕ_m, must be provided by a lead compensator at the gain cross over frequency.
3. Since maximum phase lead, ϕ_m, of a RC lead network occurs at a frequency ω_m, which is the geometric mean of the two corner frequencies ω_1 and ω_2 of the network, it is desirable to have this maximum phase lead occur at the gain cross over frequency of the uncompensated system. Thus the two corner frequencies of the lead compensator must be located on either side of the gain crossover frequency ω_{gc1}. But the gain curve of the compensator also effects the gain curve of the uncompensated system. It lifts up the gain curve around the gain cross over frequency and hence the compensated gain curve crosses 0 db line at a slightly higher frequency ω_{gc2}. This is shown in Fig. 8.15 for a type one, second order system.

$$G(s) = \frac{K_v}{s(\tau s + 1)}$$

In Fig. 8.15 the frequency, ω_m, at which maximum phase lead occurs in a phase lead network, is made to coincide with the gain cross over frequency, ω_{gc1}, of the uncompensated system. The effect of this is two fold.

1. The gain cross over frequency moves to ω_{gc2} which is higher than ω_{gc1}. Hence the phase margin obtained after the compensation, is not equal to $\phi_1 + \phi_m$ (Fig. 8.15) as desired.

2. The new phase margin has to be read at the new gain cross over frequency, ω_{gc2}, and is equal to $\phi_2 + \phi_3$ as shown in Fig. 8.15. Obviously $\phi_2 < \phi_1$, and $\phi_3 < \phi_m$ and hence the phase margin obtained will fall short of the desired value.

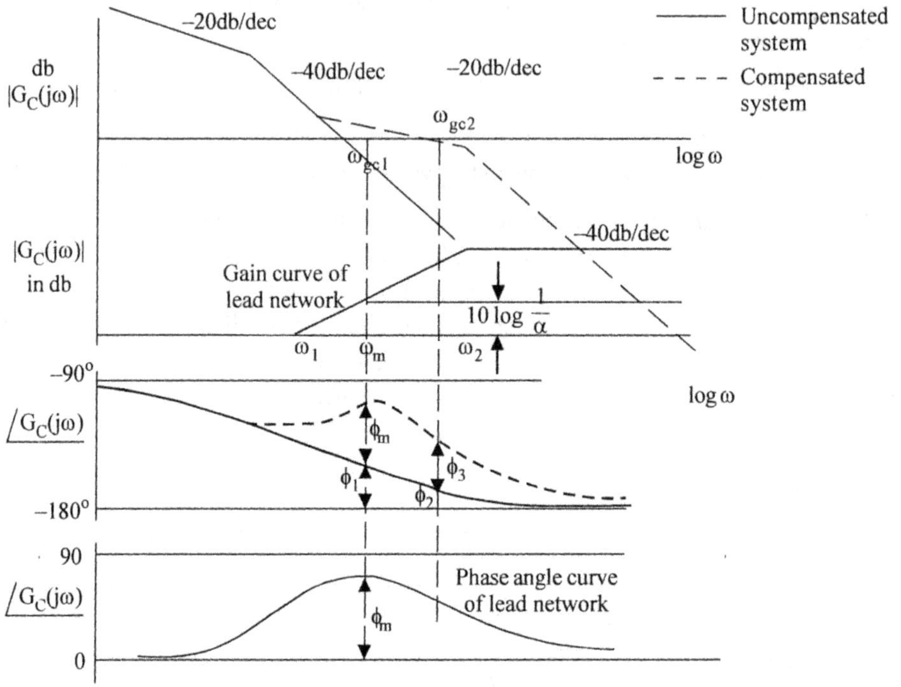

Fig. 8.15 Bode plots for (a) un compensated system (b) compensated system (c) lead network.

3. It is clear from the above discussion that if ω_m of lead network is made to coincide with ω_{gc1}, desired phase margin is not obtained for the compensated network. Thus it is desirable to make ω_m coincide with ω_{gc2} so that maximum phase lead occurs at the new gain crossover frequency. But this new gain cross over frequency is not known before hand.
4. The above difficulty is overcome as follows. The phase lead network provides a gain of $10 \log \frac{1}{\alpha}$ at ω_m as given by eqn. (8.17). If ω_m is to be made to coincide with the new gain cross over frequency ω_{gc2}, ω_{gc2} must be chosen such that the uncompensated system has a gain of $-10 \log \frac{1}{\alpha}$ at this frequency. Thus if $\omega_m = \omega_{gc2}$, the gain provided by the lead network will be $10 \log \frac{1}{\alpha}$ and hence the gain curve of the uncompensated system will be lifted up by $10 \log \frac{1}{\alpha}$ at this frequency and hence the gain crossover of compensated system occurs at this frequency as shown in Fig. 8.16. α The parameter of the lead network can be calculated by knowing the required phase lead to be provided by the network using eqn. (8.16).

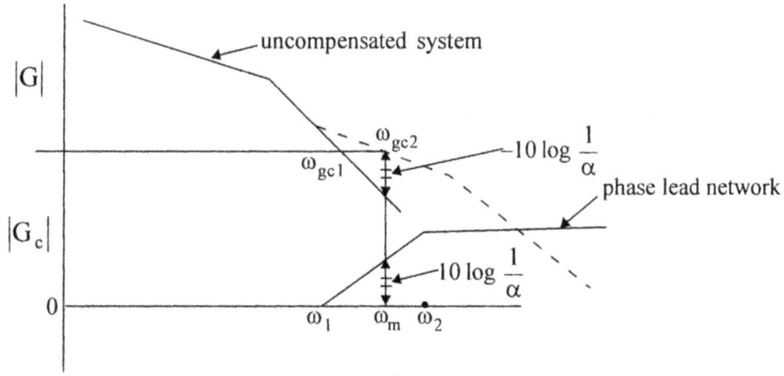

Fig. 8.16 Locating ω_m at ω_{gc2} where $|G(j\omega_{gc2})| = -10 \log \frac{1}{\alpha}$.

5. The maximum phase angle to be provided by the lead network is obtained as follows :

   At the gain crossover frequency of the uncompensated system, the system has a phase margin of ϕ_1 which is less than the desired phase margin ϕ_d. Hence an additional phase lead of $\phi_d - \phi_1$ has to be provided by the lead network. But the system will not be providing a phase lead of ϕ_1 after compensation because the gain crossover is going to shift to a new higher value ω_{gc2}, at which the phase angle provided by the system will be ϕ_2, as shown in Fig. 8.15, which is less than ϕ_1. Also ω_{gc2}, and hence ϕ_2 is not known to start with. In order to compensate this unknown shortfall, an additional phase angle ϕ_ϵ ranging from 5 to 20° is to be provided. Thus the maximum phase lead ϕ_m to be provided by the lead network is taken to be,

$$\phi_m = \phi_d - \phi_1 + \phi_\epsilon \qquad \text{.....(8.33)}$$

   ϕ_ϵ may be around 5° if the slope of the uncompensated system at ω_{gc1} is –40 db/dec or less and may go to 15° to 20° if it is –60 db/dec.

6. The RC phase lead network is designed as follows

   Maximum phase lead to be provided is given by,

$$\phi_m = \phi_d - \phi_1 + \phi_\epsilon$$

   Then α is calculated from, $\alpha = \dfrac{1-\sin\phi_m}{1+\sin\phi_m}$

   We have to make $\omega_m = \omega_{gc2}$, where ω_{gc2} is the frequency at which,

$$|G(j\,\omega_{gc2})| = -10 \log \frac{1}{\alpha}.$$

   Knowing α, we can locate ω_{gc2} from the magnitude plot.

   From these values, the corner frequencies of lead network are obtained as,

$$\omega_1 = \frac{1}{\tau} = \omega_m \sqrt{\alpha} \qquad \text{.....(8.34)}$$

$$\omega_2 = \frac{1}{\alpha\tau} = \frac{\omega_m}{\sqrt{\alpha}} \qquad \text{.....(8.35)}$$

The transfer function of the lead network is

$$G_c(s) = \frac{s + \frac{1}{\tau}}{s + \frac{1}{\alpha\tau}}$$

7. It is observed that the gain cross over frequency is increased after compensation. Gain cross over frequency can be taken as a measure of the bandwidth of the system. Since the bandwidth is increased by the lead compensation, speed of response is improved. The actual bandwidth can be obtained by transferring the data from Bode plot to Nichol's chart.

8. Since the gain at higher frequencies is increased by 20 log $\frac{1}{\alpha}$, the noise frequencies are amplified and hence the signal to noise ratio deteorates.

9. The values of R_1, R_2 and C can be calculated using,

$$\tau = R_1 C \qquad .....(8.36)$$

$$\alpha = \frac{R_2}{R_1 + R_2} \qquad .....(8.37)$$

Since one of the three elements R_1, R_2 and C can be chosen arbitrarily, as there are only two equations, impedance level required or the cost of the compensator may be used to select this element.

The design procedure may be summarised as follows.

(i) Select the gain K to satisfy steadystate error requirements.

(ii) Using this value of K draw the Bode magnitude and phase plots of the uncompensated system. Obtain the phase margin of the uncompensated system. Let this be ϕ_1,

(iii) Determine the phase lead required.

$$\phi_m = \phi_d - \phi_1 + \phi_\epsilon$$

where

ϕ_d – Desired phase margin

ϕ_1 – Available phase margin

ϕ_ϵ – Additional phase angle required to compensate the reduction in phase angle due to increase in gain crossover frequency after compensation.

(iv) Determine α of the RC network as

$$\alpha = \frac{1 - \sin\phi_m}{1 + \sin\phi_m}$$

If ϕ_m required is more than 60° it is recommended to use two RC networks in cascade, each providing an angle of $\frac{\phi_m}{2}$.

(v) Find the gain of the phase lead network at ω_m, which is $10 \log \frac{1}{\alpha}$.

Locate the frequency at which the gain of the uncompensated system is $-10 \log \frac{1}{\alpha}$. Let this frequency be ω_{gc2}. Make $\omega_{gc2} = \omega_m$. Read the phase margin ϕ_2, provided by the system at this frequency ω_m. If $\phi_1 - \phi_2$ is $> \phi_\in$, increase $\phi_\in$ and recalculate ϕ_m, α and ω_m until $\phi_1 - \phi_2 \leq \phi_\in$.

(vi) Compute the two corner frequencies of the lead network from eqn.s (8.34), (8.35)

$$\omega_1 = \frac{1}{\tau} = \omega_m \sqrt{\alpha}$$

$$\omega_2 = \frac{1}{\alpha\tau} = \frac{\omega_m}{\sqrt{\alpha}}$$

The transfer function of lead network

$$G_c(s) = \frac{s + \frac{1}{\tau}}{s + \frac{1}{\alpha\tau}}$$

(vii) Draw the Bode plot of the compensated system and check the phase margin. If it falls short of the desired phase margin increase $\phi_\in$ and redesign.

(viii) Obtain the elements of RC phase lead network. Using egs (8.36) and (8.37).

Let us now illustrate the procedure by some examples.

Example 8.1

Consider a system

$$G_c(s) = \frac{K_v}{s(s+1)}$$

The specifications are :

e_{ss} for a velocity input should be less than 0.1.

Phase margin should be greater than 40°.

Solution :

1. Choose K_v such that $e_{ss} < 0.1$

Since $$e_{ss} = \frac{1}{K_v}$$

$$K_v = \frac{1}{e_{ss}} = \frac{1}{0.1} = 10$$

2. The Bode plot is drawn for $K_v = 10$

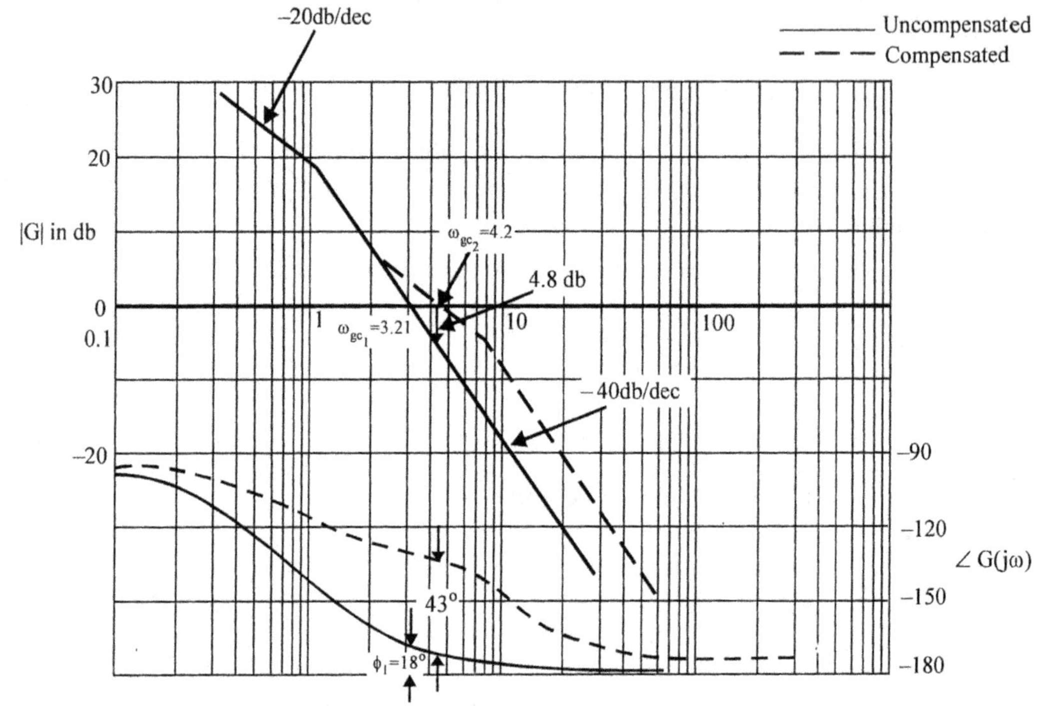

Fig. 8.17 Design of phase lead compensator for Ex. 8.1.

The gain crossover frequency is,

$$\omega_{gc1} = 3.2 \text{ rad/sec}$$

and phase margin of uncompensated system is,

$$\phi_1 = 18^\circ$$

3. Since the desired phase margin is $\phi_d = 40^\circ$, the phase lead to be provided by the phase lead network is

$$\begin{aligned}\phi_m &= \phi_d - \phi_1 + \phi_\epsilon \\ &= 40 - 18 + 8 \qquad (\phi_\epsilon \text{ is assumed to be } 8^\circ) \\ &= 30^\circ\end{aligned}$$

4.

$$\begin{aligned}\alpha &= \frac{1-\sin\phi_m}{1+\sin\phi_m} \\ &= \frac{1-\sin 30}{1+\sin 30} \\ &= \frac{1}{3}\end{aligned}$$

5. Gain of the phase lead network at ω_m is,

$$|G_c(j\omega_m)| = 10 \log \frac{1}{\alpha}$$
$$= 10 \log 3$$
$$= 4.77 \text{ db}$$

Frequency at which $|G(j\omega_{gc2})| = -4.77$ db is read from the magnitude plot of $G(j\omega)$

$$\omega_{gc2} = \omega_m = 4.2 \text{ rad/sec}$$

At $\omega_m = 4.2$ rad/sec, the phase margin provided by the system, ϕ_2, is 14°. Hence $\phi_1 - \phi_2 = 18.14 = 4^o$ which is less than $\phi_\epsilon = 8$. If desired, ϕ_ϵ can be reduced to 5° or 6°.

6. The corner frequencies of the lead networks are :

$$\omega_1 = \frac{1}{\tau} = \omega_m \sqrt{\alpha}$$
$$= \frac{4.2}{\sqrt{3}}$$
$$= 2.425 \text{ rad/sec}$$

$$\omega_2 = \frac{1}{\alpha\tau} = \frac{\omega_m}{\sqrt{\alpha}}$$
$$= 4.2\sqrt{3}$$
$$= 7.27 \text{ rad/sec}$$

∴ The transfer function of the lead compensator is,

$$G_c(s) = \frac{s + 2.425}{s + 7.27}$$
$$= \frac{1}{3}\frac{0.412s + 1}{0.1376s + 1}$$

7. The Bode plot for the compensated system is drawn. From Fig. (8.17) the phase margin is

$$\phi_{pm} = 43^o$$

This satisfies the design specification.

8. The RC phase lead network is designed with,

$$R_1C = \tau = 0.412$$
$$\alpha = \frac{R_2}{R_1 + R_2} = \frac{1}{3}$$

Assuming $C = 1\ \mu F$

$$R_1 = 412\ K\Omega$$
$$R_2 = 206\ K\Omega$$

The lead network is shown in Fig. 8.18.

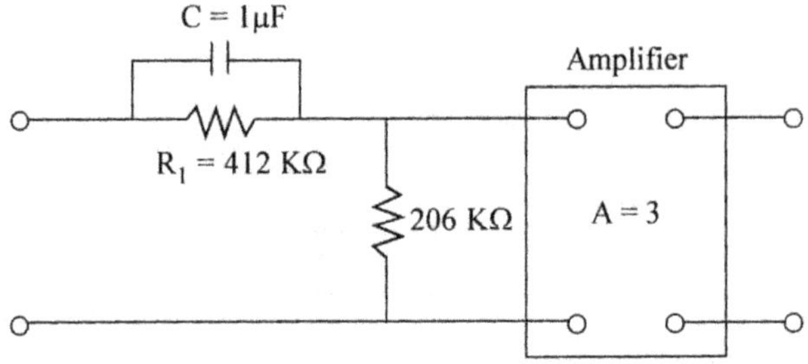

Fig. 8.18 RC phase lead network for Ex. 8.1.

An amplifier with a gain of $\frac{1}{\alpha} = 3$ is used to nullify the d.c attenuation introduced by the lead network.

The open loop transfer function of the compensated system is

$$G_c(s)\ G(s) = \frac{10(0.412s+1)}{s(0.1376s+1)(s+1)}$$

Example 8.2

Let the system to be compensated be,

$$G(s) = \frac{K}{s(1+0.1s)(1+0.25s)}$$

The specifications to be met are

$$K_v = 10 \text{ and } \phi_{pm} \geq 40^o$$

Solution :

1. Since G(s) is in time constant form

$$K = K_v = 10$$

2. The Bode plot for this value of K is drawn.

   The gain crossover frequency ω_{gc1} = 6.4 rad/sec.

   The phase margin of the system is $\phi_{pm} = \phi_1 = 0^o$.

   This indicates that the system is highly oscillatory.

3. Since the desired phase margin $\phi_d = 40^o$.

$$\begin{aligned}\phi_m &= \phi_d - \phi_1 + \phi_\epsilon \\ &= 40 - 0 + 15^o \\ &= 55^o\end{aligned}$$

The phase curve of the system has a steep slope around the gain cross over frequency as seen from Fig. 8.18. Hence ϕ_ϵ is chosen to be 15°.

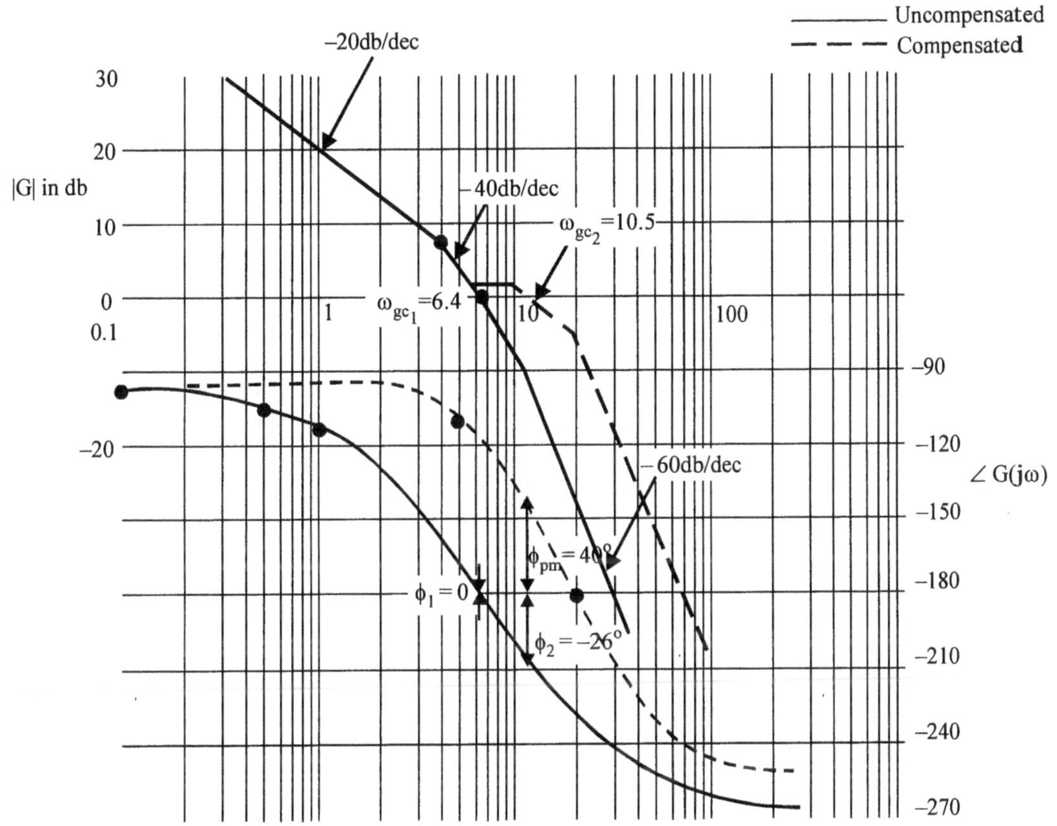

Fig. 8.19 Bode plot or the Ex. 8.2.

4.
$$\alpha = \frac{1-\sin\phi_m}{1+\sin\phi_m}$$
$$= \frac{1-\sin 55}{1+\sin 55}$$
$$= 0.0994 \simeq 0.1$$

5. Gain of the phase lead network at ω_m is

$$|G_c(j\omega_m)| = 10 \log \frac{1}{0.1} = 10\text{db}$$

Frequency at which the magnitude of the uncompensated system is –10 db is read from the graph as $\omega_{gc2} = \omega_m = 10$ rad/sec.

At $\omega_{gc2} = 10$ rad/sec, the phase margin ϕ_2, provided by the uncompensated system, from the graph, is –23°.

$$\phi_1 - \phi_2 = 0 - (-23)$$
$$= 23$$
$$> \phi_\epsilon$$

Hence $\phi_\epsilon = 15$ is not sufficient and hence a higher value has to be chosen. This is because the uncompensated system is oscillatory and the phase angle is changing steeply around the gain cross over frequency.

Since ϕ_ϵ chosen is already high, let us design for a higher value of $\phi_m = 66^\circ$ and design two phase lead networks each providing 33°.

$$\alpha = \frac{1-\sin 33}{1+\sin 33}$$

Total magnitude provided by the two phase lead networks at ω_m is,

$$2\left(10 \log \frac{1}{\alpha}\right) = 10.45 \text{ db.}$$

From the graph magnitude of -10.45 db for uncompensated system occurs at $\omega = \omega_{gc2} = 10.5$ rad/sec. The phase margin ϕ_2 at this frequency is -26°. Hence the total phase margin obtained is equal to $(66 - 26) = 40^\circ$ which satisfies the specifications. Thus, we will have two identical sections of phase lead networks each having the following parameters.

$$\alpha = 0.3$$

$$\phi_m = 33^\circ$$

$$\omega_m = 10.5 \text{ rad/sec}$$

6. Each phase lead network has the following corner frequencies

$$\omega_1 = \frac{1}{\tau} = \omega_m \sqrt{\alpha}$$

$$= 10.5\sqrt{0.3}$$

$$= 5.75 \text{ rad/sec}$$

$$\omega_2 = \frac{1}{\alpha\tau} = \frac{\omega_m}{\sqrt{\alpha}}$$

$$= 19.17 \text{ rad/sec}$$

The transfer function of each of the lead compensators, is

$$G_{c1}(s) = \frac{s+5.75}{s+19.17}$$

$$= 0.3\,\frac{0.174s+1}{0.052s+1}$$

7. The Bode plot for the compensated system is drawn, from which the phase margin is approximately 40°. This satisfies the design specifications.
8. The single section of phase lead network is designed as follows.

Choose $C = 1\ \mu F$

$$R_1 C = \frac{1}{\tau} = 5.75$$

$$R_1 = \frac{5.75}{1\times 10^{-6}} = 5.75\ M\Omega$$

$$\alpha = \frac{R_2}{R_1 + R_2} = 0.3$$

$$R_2 = 2.46 \text{ M}\Omega$$

The complete phase lead network is as shown in Fig. 8.20. The amplifier should provide a gain of $\left(\frac{1}{0.3}\right)^2$ and is normally placed in between the two phase lead networks.

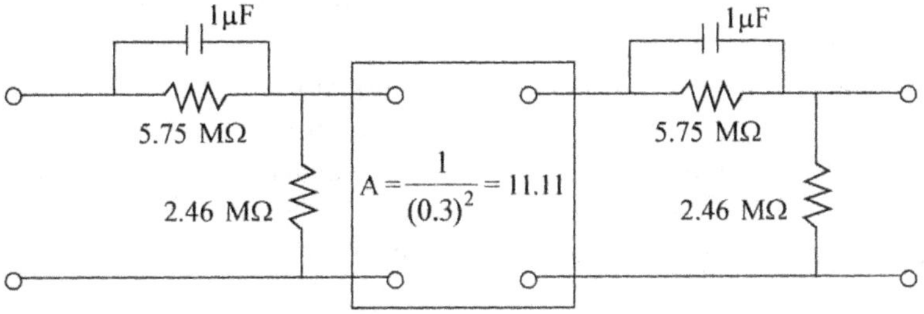

Fig. 8.20 Phase lead network for Ex. 8.2.

The open loop transfer function of the compensated system is

$$G(s) = \left(\frac{0.174s + 1}{0.052s + 1}\right)^2 \frac{10}{s(1 + 0.1s)(1 + 0.25s)}$$

The main effects of phase lead compensation may be summarised as follows :

Frequency response

1. Phase lead is provided around the resonant frequency.
2. For a specified gain constant K, the slope of the magnitude curve is reduced at the gain cross over frequency. The relative stability is therefore improved. The resonance peak is reduced.
3. The gain cross over frequency and hence the bandwidth of the system is increased.

Time response

1. Overshoot to step input is reduced.
2. The rise time is small.

When large phase leads are required, two or more lead networks are used so that α is greater than 0.1 for each network and gain crossover frequency is not unduly increased. Inherently unstable systems, like type 2 systems can be effectively compensated using phase lead networks.

For systems with low damping ratios, the phase shift may decrease rapidly near the gain cross over frequency and phase lead compensation may be ineffective. This rapid change in phase angle may be due to,

(a) Two simple corner frequencies placed close to each other near the gain cross over frequency.
(b) A double pole placed near the gain cross over frequency.
(c) A complex conjugate pole near the gain cross over frequency.

In these cases also two or more phase lead networks may be used to achieve desired specifications.

8.4.2 Design of Lag Compensator

The procedure for design of a lag compensator is developed in the following :

1. The gain constant is set as per the steadystate error requirements.
2. Bode plot is drawn. If the phase margin obtained is adequate no other compensation is required. If phase margin is not adequate, locate a frequency, ω_{gc2} at which the desired phase margin is available.
3. The gain of the uncompensated system at this frequency is measured. This gain will be positive for type 1 or type 0 systems.
4. The attenuation characteristic of phase lag network at high frequencies is used to reduce the gain of the uncompensated system to zero at ω_{gc2}, so that this frequency becomes the gain crossover frequency of the compensated system, as shown in Fig. 8.21. The phase lag introduced by the lag network is detrimental to the system. Hence the frequency at which maximum phase lag occurs for the phase lag network must be located for away from the gain crossover frequency.

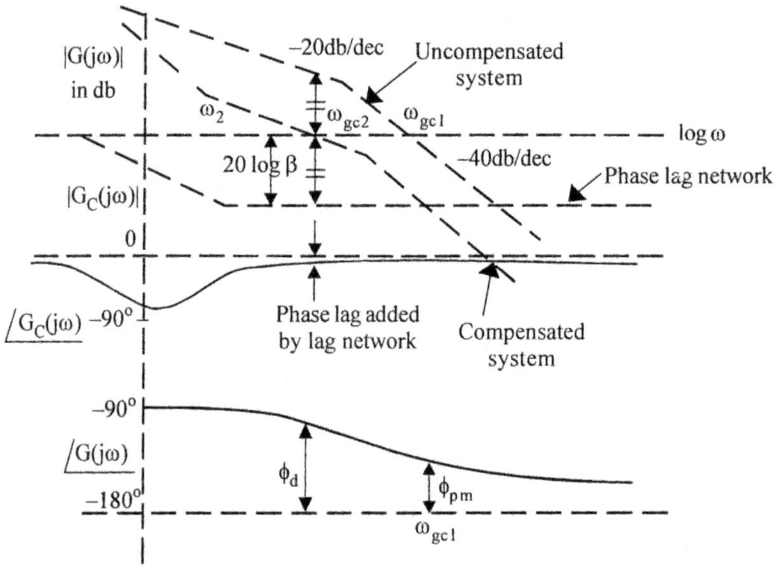

Fig. 8.21 Bode plots of G(jω) and G_c(jω).

5. The corner frequencies of the lag network are located far to the left of the new gain cross over frequency, ω_{gc2}, so that the phase lag provided by the lag network does not appreciably affect the phase of the system at this frequency. However a small negative angle will be added to the system phase at ω_{gc2}, as shown in Fig. 8.21. To compensate this, the phase margin to be provided is taken to be slightly higher than the desired phase margin. Thus the phase margin is taken to be

$$\phi_m = \phi_d + \phi_\epsilon$$

ϕ_ϵ is taken to be in the range of 5° to 15°. If the corner frequencies of lag network are closer to the new gain cross over frequency, higher value of ϕ_ϵ is chosen. The upper corner frequency ω_2 of the lag network is placed, usually, one octave to 1 decade below the new gain cross over frequency ω_{gc2}.

The design steps are summerised as follows :

1. Choose the gain K to satisfy steadystate error specifications.
2. Draw the Bode plot with this value of K. If the phase margin of this system is inadequate, design a lag compensator.
3. The phase margin to be provided is given by,

$$\phi_{pm} = \phi_d + \phi_\epsilon \qquad .....(8.38)$$

where ϕ_ϵ ranges between 5° to 15°.

4. Locate the frequency ω_{gc2} at which the phase margin ϕ is available on the gain plot of G(s). Find the magnitude of G(s) at this frequency.
5. The attenuation to be provided by the lag network is given by,

$$20 \log \beta = |G\,(j\omega_{gc2})|_{db} \qquad .....(8.39)$$

Find the value of β from this equation.

6. The upper corner frequency, ω_2, of the lag compensator is chosen to be between $\frac{\omega_{gc2}}{2}$ to $\frac{\omega_{gc2}}{10}$. Thus

$$\omega_2 = \frac{1}{\tau}$$

7. With τ and β known, the phase lag network can be designed. The transfer function of the Lag network is,

$$G_c(s) = \frac{\tau s + 1}{\beta \tau s + 1}$$

8. Draw the Bode plot of the compensated system to check the specifications.

   If necessary, choose a different value of τ and redesign to satisfy the specifications.

Example 8.3

Design a lag compensator for the unity feedback system with,

$$G(s) = \frac{K}{s(s+2)}$$

to satisfy the following specifications,

$$K_v = 10$$

$$\phi_{pm} \geq 32^o$$

Solution :

1. $$G(s) = \frac{K}{2s(0.5s+1)}$$

since $$K_v = 10$$

$$\frac{K}{2} = K_v$$

or $$K = 2K_v = 20$$

2. Bode plot is drawn for K = 20

   From the graph, gain cross over frequency ω_{gc1} = 4.4 rad/sec

   Phase margin = 25°

   Phase margin falls short of the desired phase margin. Hence a lag compensator is designed.

3. Phase margin to be provided is

$$\phi_{pm} = \phi_d + \phi_\epsilon$$

$$= 32 + 8^o$$

$$= 40^o$$

4. Frequency at which the phase margin of 40° is available is, ω_{gc2} = 2.5 rad/sec.

5. The gain of the system at this frequency is 11 db.

   ∴ The attenuation to be provided by the lag network is –11 db.

$$20 \log \beta = 11$$

∴ $$\beta = 3.55$$

6. The upper corner frequency is chosen to be,

$$\omega_2 = \frac{\omega_{gc2}}{6} = \frac{2.5}{6} \simeq 0.4 \text{ rad/sec}$$

$$\omega_2 = \frac{1}{\tau}$$

$$\tau = \frac{1}{0.4} = 2.5 \text{ sec}$$

7. $$\omega_1 = \frac{1}{\beta\tau} = \frac{0.4}{3.55} = 0.11 \text{ rad/sec}$$

   The transfer function of the lag network is,

$$G_c(s) = \frac{2.5s+1}{9.1s+1}$$

8. Bode plot of the compensated system is drawn as shown in Fig. 8.22.

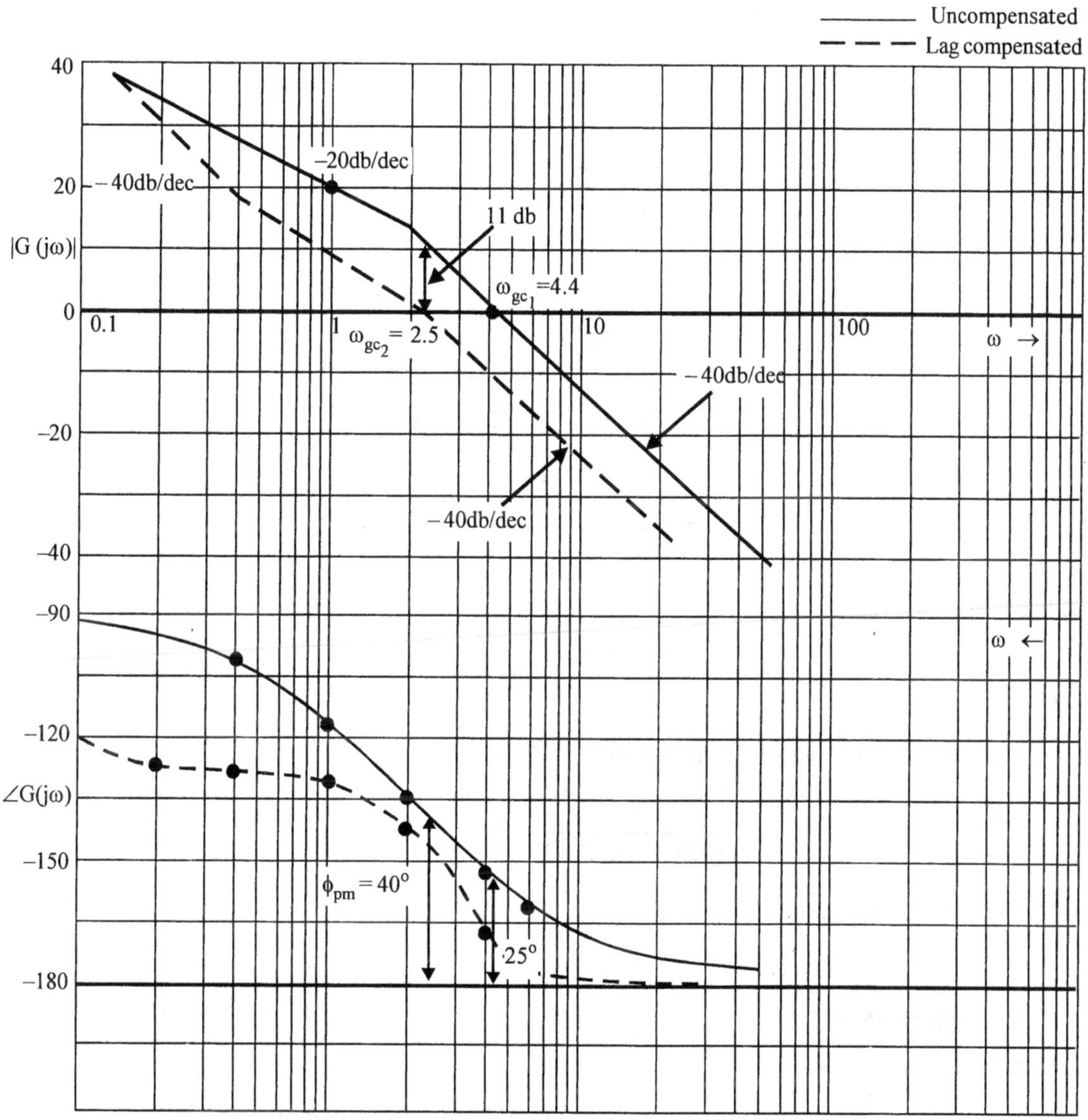

Fig. 8.22 Design of lag compensator for Ex. 8.3.

The gain crossover frequency is 2.4 rad/sec

The phase margin is 32°.

Since the specifications are satisfied, the lag compensation is adequate.

$$\tau = R_2 C = 2.5$$

If $C = 1\ \mu F$

$R_2 = 2.5\ M\Omega$

and $$\frac{R_1 + R_2}{R_2} = \beta = 3.55$$

$R_1 = 6.375\ M\Omega.$

The lag network is

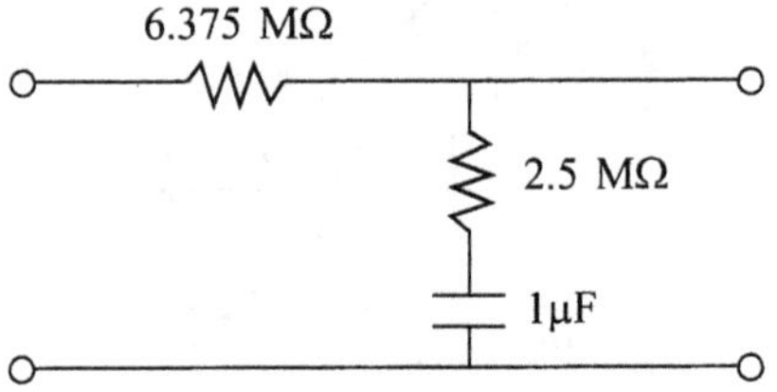

Fig. 8.23 RC lag network for Ex 8.3.

Example 8.4

Design a phase lag compensator for the system

$$G(s) = \frac{K}{s(0.5s+1)(0.2s+1)}$$

with the following specifications

$$K_v \geq 5$$

The damping factor, $\delta = 0.4$

Solution :

From eqn. (6.17) the frequency domain specification ϕ_{pm} corresponding to $\delta = 0.4$ is,

$$\phi_d = \phi_{pm} = 43^o$$

1. The steady state error requirement is met by,

$$K = K_v = 5$$

2. The Bode plot is drawn with $K = 5$ in Fig. 8.24.

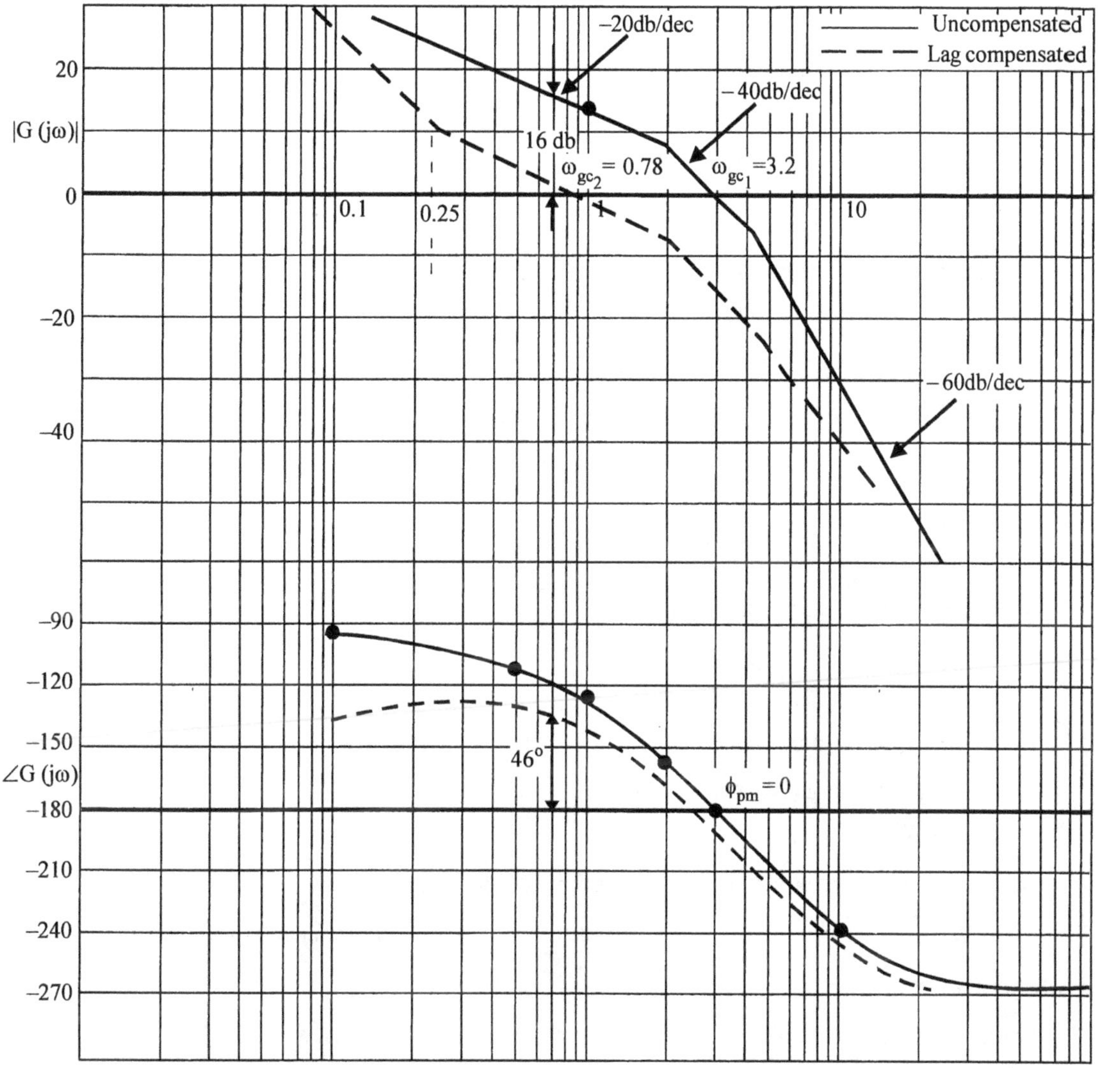

Fig. 8.24 Design of a lag compensator for Ex. 8.4.

From the graph $\omega_{gc1} = 3.2$ rad/sec

$$\phi_{pm} = 0^o$$

3. Phase margin to be provided is,

$$\phi_{pm} = \phi_d + \phi_\in$$

$$= 43 + 15 = 58^o$$

$\phi_\in$ is chosen to be 15° because the gain crossover frequency is very small and the corner frequencies of lag network will be still smaller. A lag network at very low frequencies requires a large value of capacitor and the cost of the compensator will increase.

4. A phase margin of 58^0 occurs at $\omega_{gc2} = 0.78$ rad/sec.
5. The gain of the system at this frequency is 16 db.

   $\therefore$ Attenuation to be provided by the lag network is – 16 db.

$$\therefore 20 \log \beta = 16$$

$$\beta = 6.3$$

6. The upper corner frequency is chosen to be,

$$\omega_2 = \frac{\omega_{gc2}}{3} = \frac{0.78}{3} \simeq 0.25$$

$$\tau = \frac{1}{\omega_2} = \frac{1}{0.25} = 4 \text{ sec}$$

$$\omega_1 = \frac{1}{\beta\tau} = \frac{0.25}{6.3} = 0.04 \text{ rad/sec}$$

7. The transfer function of the lag network is

$$G_c(s) = \frac{4s+1}{25s+1}$$

8. Bode plot of the compensated system is drawn as shown in Fig. 8.24.

   From the graph the gain cross over frequency is 0.7 rad/sec.

   The phase margin is 46°.

   As the specifications are satisfied, the design is complete.

   The effects of lag compensation on the response may be summarised as follows.

Frequency response

1. For a given relative stability, the velocity error constant is increased.
2. The gain cross over frequency is decreased, which in turn means lesser bandwidth.
3. For a given gain K, the magnitude curve is attenuated at lower frequencies. Thus the phase margin is improved. Resonance peak is also reduced.
4. The bandwidth is reduced and hence its noise characteristics are better.

Time response

1. Time response is slower since undamped natural frequency is reduced.

   The lag compensator can be designed only if the required phase margin is available at any frequency. A type-2 system is absolutely unstable and a lag compensator can not be designed for this system.

8.4.3 Design of Lag-Lead Compensator

A lead compensator increases the gain cross over frequency and hence the bandwidth is also increased. For higher order systems with large error constants, the phase lead required may be large, which in turn results in larger bandwidth. This may be undesirable due to noise considerations. On the other hand a lag compensator results in a lower bandwidth and generally a sluggish system. To satisfy the additional requirement on bandwidth, a lag or lead compensator alone may not be satisfactory.

Hence a lag-lead compensator is used. The procedure for the design of lag-lead compensator is as follows :

1. The required error constant is satisfied by choosing a proper gain. The phase margin and the bandwidth are obtained from the Nichols chart. If the phase margin falls short and the bandwidth is smaller than the desired value, a lead compensator is designed.
2. If the bandwidth is larger than the desired value, lag compensator is attempted, if the phase margin desired is available at any frequency. If this results in a lower bandwidth than the desired bandwidth, a lag lead compensator is to be designed.
3. To design a lag-lead compensator, we start with the design of a lag compensator. A lag compensator is designed to partially satisfy the phase margin requirement. It means that the gain cross over frequency, ω_{gc2}, is chosen to be higher than that to be used, if a full lag compensation is designed. This ensures that the bandwidth is not reduced excessively by the lag compensator.
4. The β and ω_{gc2} are known and the corner frequencies of the lag compensator can be obtained.

   Since $\alpha = \dfrac{1}{\beta}$, the frequency at which 20 log α is available on the magnitude plot is the new gain crossover frequency ω_{gc3}. The maximum phase lead available is,

   $$\phi_m = \sin^{-1}\left(\frac{1-\alpha}{1+\alpha}\right)$$

   and $$\omega_m = \omega_{gc3}$$

   With these parameters, a lead compensator is designed. Here, for the lead network,

   $$\omega_1 = \frac{1}{\tau} = \omega_m \sqrt{\alpha}$$

   $$\omega_2 = \frac{1}{\alpha\tau} = \frac{\omega_m}{\sqrt{\alpha}}$$

   The log magnitude vs phase angle curve of the lag-lead compensated network is drawn on the Nichols chart, from which the bandwidth can be obtained. The procedure is illustrated by an example.

Example 8.5

Consider the system,

$$G(s) = \frac{K}{s(s+5)(s+10)}$$

Design a compensator to satisfy the following specifications.

e_{ss} for a velocity input ≤ 0.045

$$\phi_{pm} \geq 45^o$$

Bandwidth $\quad \omega_b = 11$ rad/sec

Solution :

1. $$e_{ss} = \frac{1}{K_v}$$

$$K_v = \frac{1}{e_{ss}} = \frac{1}{0.04} = 25$$

$$\frac{K}{50} = 25$$

or $\quad K = 1250.$

2. Bode plot and log magnitude Vs phase angle plot on Nichols chart are drawn in Fig. 8.25 and Fig. 8.26 respectively.

   From the graph in Fig. 8.25 the phase cross over frequency is 10.5 rad/sec.

   The phase margin is -22°

   From the graph of Fig. 8.26, the Bandwidth is 13.5 rad/sec.

   Thus, since the Bandwidth is already large, a lead compensator will further increase it. A lag compensator provides the required phase margin at a frequency of about 2.3 rad/sec, which makes the bandwidth much smaller than desirable. Hence a lag-lead compensator only can satisfy all the specifications.

3. Let us partially compensate first by a lag compensator. Choose the gain crossover frequency to be 4 rad/sec. At this frequency the gain of the uncompensated system is 16 db.

   $$\therefore \quad 20 \log \beta = 16$$

   $$\beta = 6.3 \simeq 8 \text{ (say)}$$

   Choose $\omega_1 = 1$ rad/sec, then $\omega_2 = \frac{1}{8} = 0.125$ rad/sec. The transfer function of the lag compensator is $\quad G_{c1}(s) = \frac{s+1}{8s+1}$

   The Bode plot of lag compensated system is also drawn in Fig. 8.25.

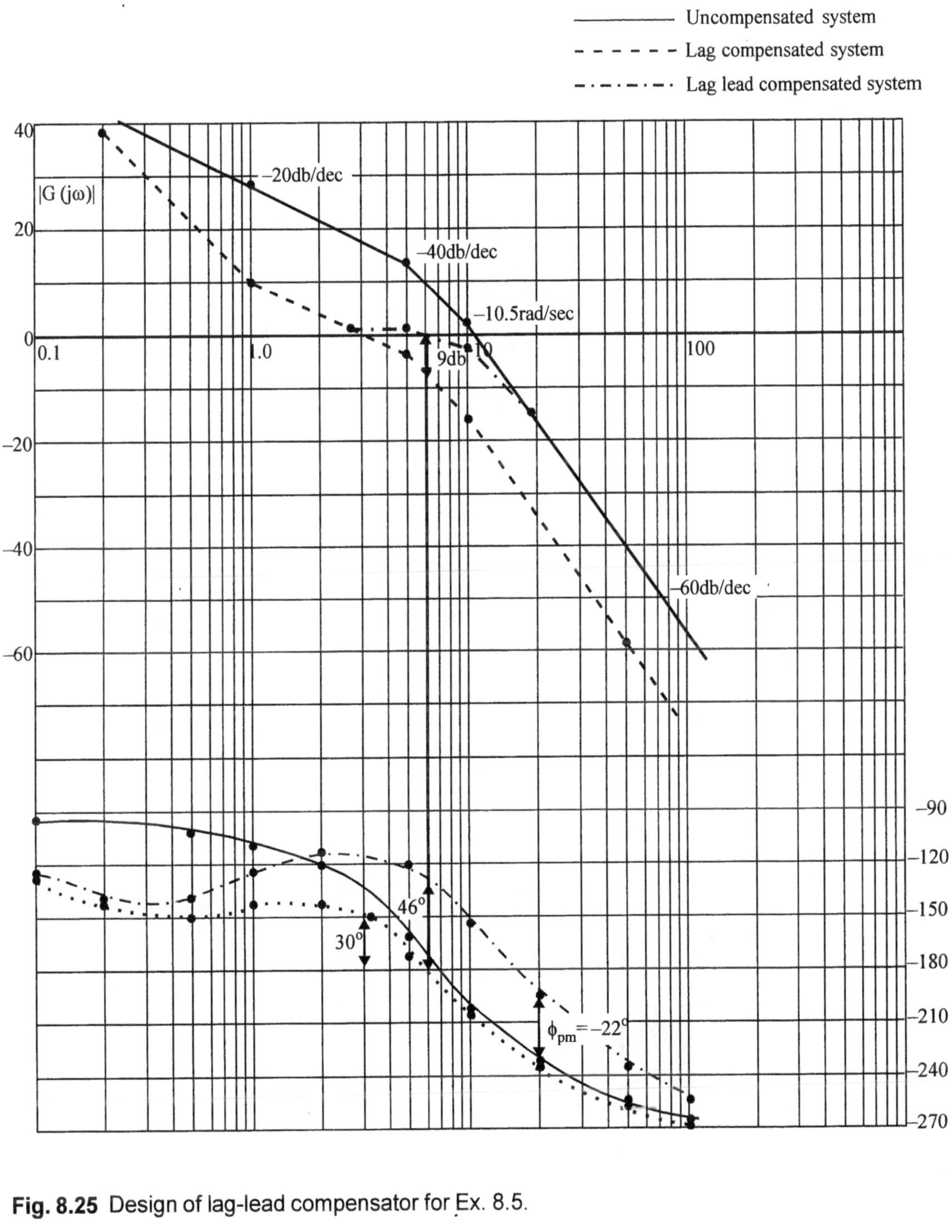

Fig. 8.25 Design of lag-lead compensator for Ex. 8.5.

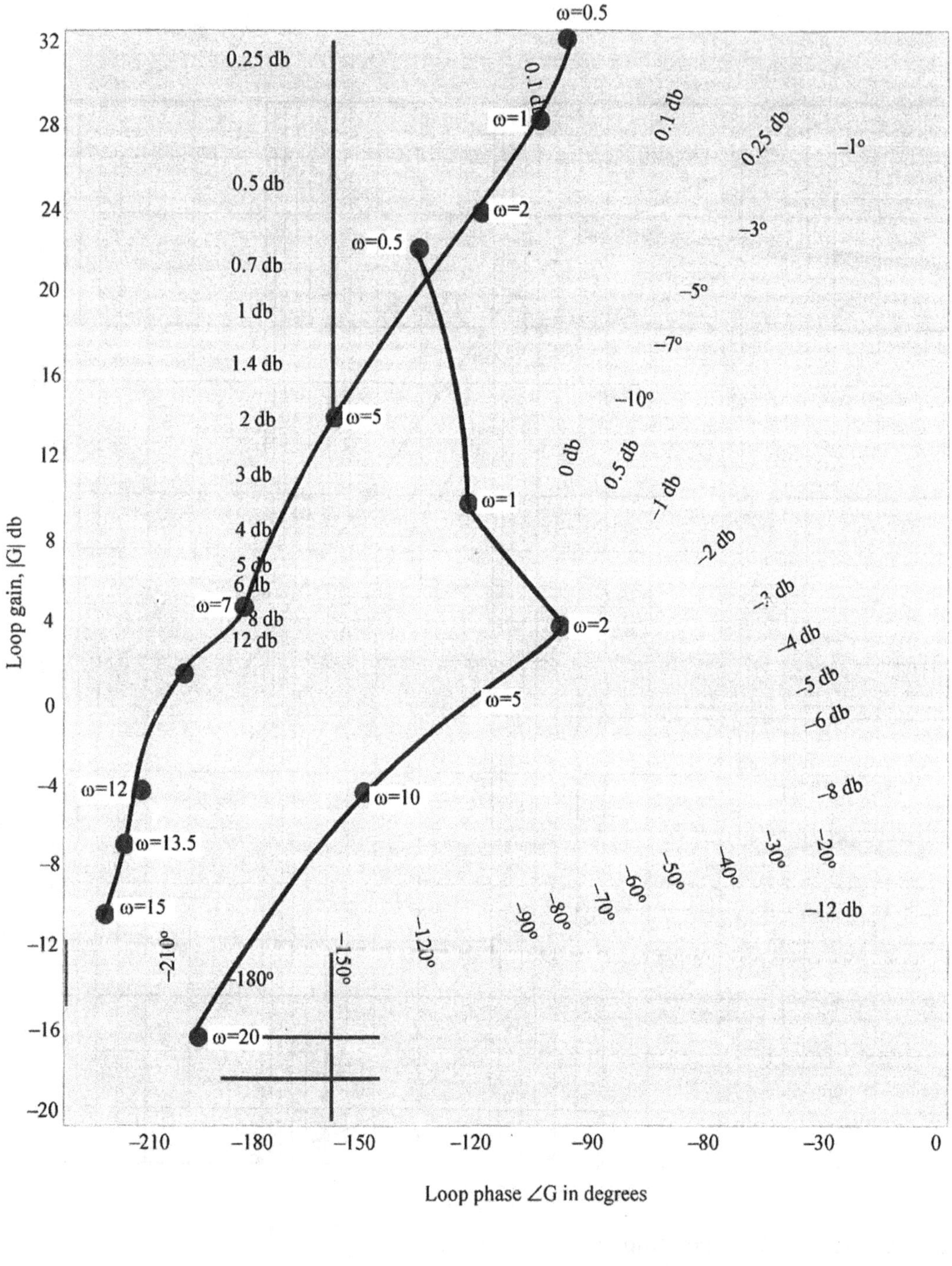

Fig. 8.26 Gain-phase plots of uncompensated and leg-lead compensated systems superposed on Nichol's chart for Ex 8.5.

4. Let us now design a lead compensator. Since $\beta = 8$,

$$\alpha = \frac{1}{\beta} = \frac{1}{8} = 0.125$$

The maximum phase lead provided by the lead network is,

$$\phi_m = \sin^{-1}\left(\frac{1-\alpha}{1+\alpha}\right)$$

$$= \sin^{-1}\frac{7}{9}$$

$$= 51^{\circ}$$

The gain to be provided by the lead network is

$$10 \log \alpha = 10 \log \frac{1}{8} = 9 \text{ db.}$$

This gain occurs at a frequency, $\omega_m = 6.8$ rad/sec on the lag compensated gain plot.

Thus $\omega_m = 6.8$ rad/sec

The corner frequencies of the lead network are,

$$\omega_1 = \omega_m\sqrt{\alpha} = \frac{6.8}{\sqrt{8}} = 2.4 \text{ rad/sec}$$

$$\tau_1 = 0.416 \text{ sec}$$

$$\omega_2 = \frac{\omega_m}{\sqrt{\alpha}} = 6.8\sqrt{8} = 19.23 \text{ rad/sec}$$

$$\tau_2 = 0.052 \text{ sec}$$

The transfer function of the lead compensator is

$$G_{c2}(s) = \frac{0.416s+1}{0.052s+1}$$

The open loop transfer function of the compensated system becomes,

$$G(s)\,G_{c1}(s)\,G_{c2}(s) = \frac{1250}{s(s+5)(s+10)} \cdot \frac{s+1}{8s+1} \cdot \frac{0.416s+1}{0.052s+1}$$

The Bode plot of the lag-lead compensated network is drawn, from which the phase margin is obtained as 46°, which is acceptable. The $|G|$ in db and angle of G are read from Bode plot for the compensated system and are tabulated in Table 8.2.

Table 8.2 $|G|$ in db and $\angle G$ at various frequencies

ω	0.5	1.0	2.0	5.0	10	20
$\|G\|$ db	22	10	4	2	– 4	– 16
$\angle G$	– 138	– 125	– 112	– 121	– 154	– 195

These values are transferred to Nichol's chart. The intersection of – 3 db line with the curve is obtained at a frequency where |G| = 7 db. This magnitude of 7 db occurs at a frequency of 11 rad/sec as read from the Bode magnitude plot of the compensated system. This frequency is the bandwidth of the compensated system.

Thus $\omega_b = 11$ rad/sec.

which is satisfactory.

This completes the design of a lag-lead compensator for the system.

Problems

8.1 The open loop transfer function of a unity feedback system is given by,

$$G(s) = \frac{5}{s(s+1)(0.5s+1)}$$

What is the phase margin of this system. If a lag compensator given by,

$$G_c(s) = \frac{10s+1}{100s+1}$$

is added in cascade with the forward path transfer function, determine,

(i) Phase margin

(ii) Gain cross over frequency

(iii) Steady state error to a unity velocity input

(iv) Gain margin

8.2 A unity feedback system has a open loop transfer function,

$$G(s) = \frac{2}{s(s+0.5)}$$

what is the steadystate error for a unit velocity input of the system. Design a lag compensator so that the steadystate error remains the same but a phase margin of 45° is achieved. Also obtain the RC network to realise this compensator.

8.3 A unity feedback system has an open loop transfer function,

$$G(s) = \frac{K}{s(s+4)(s+10)}$$

It is desired to have a resonance peak of $M_r = 1.232$. What value of K will give this resonance peak. Design a cascade compensator to obtain a resonance peak of 1.13, with the same value of velocity error constant.

8.4 For a unity feedback system with,

$$G(s) = \frac{K}{s(s+2)}$$

design a cascade lead compensator so that the steady state error for a unit velocity input is 0.05 and the phase margin is 50°. What is the gain margin for these compensated system.

8.5 Design a lag lead compensator for the unity feedback system with,

$$G(s) = \frac{K}{s(s+1)(s+2)}$$

and satisfying the specifications,

$K_v = 10\ sec^{-1}$

$\phi_{pm} = 50^o$

B.W > 2 rad/sec

8.6 Design a suitable compensator for a unity feedback system,

$$G(s) = \frac{K}{s(s+1)(s+4)}$$

to satisfy the following specifications.

damping ratio $\delta = 0.5$

Velocity error constant $K_v = 5\ sec^{-1}$

Find the settling time of the compensated system for a unit step input.

8.7 Design a suitable compesator for the unity feedback system,

$$G(s) = \frac{K}{s(0.1s+1)(0.2s+1)}$$

The specifications are,

$K_v = 25\ sec^{-1}$

Phase margin $\phi_{Pm} > 45^o$

Bandwidth $\omega_b = 12$ rad/sec

8.8 Consider the unity feedback system,

$$G(s) = \frac{K}{s^2(s+5)}$$

Design a suitable compensator to satisfy the specifications,

$\phi_{Pm} = 50^o$

$K_a = 0.2$

$\omega_b = 0.8$ rad/sec

8.9 For the unity feedback system with,

$$G(s) = \frac{K}{s(s+1)(s+6)}$$

design a lag compensator to get a phase margin of 40° and a velocity error constant of $5\ sec^{-1}$. Find the bandwidth of the compensated system and also the settling time for a unit step input.

8.10 For the system in Problem 8.9 design a lead compensator to achieve the same specifications. Find the bandwidth and the settling time of the compensated system.

9 State Space Analysis of Control Systems

9.1 Introduction

Mathematical modelling of a system plays an important role in the analysis and design of control systems. Transfer function is one such model, which we have used for analysis and design of control systems in the previous chapters. This is a useful representation if the system is linear, time invariant and has a single input and single output (SISO). It is also defined for systems with zero initial conditions only. The tools developed, viz, root locus technique, Bode plot, Nyquist plot, Nichol's chart etc are powerful in the analysis and design of control systems. But transfer function representation is not useful for,

1. Systems with initial conditions
2. Nonlinear systems
3. Time varying systems and
4. Multiple input multiple output (MIMO) systems

Further, the output for a given input can only be found. It does not throw any light on the variation of internal variables. Sometimes this information is necessary because, some internal variables may go out of bounds, eventhough the output remains within the desired limits. The methods discussed so far are known as classical methods. In this method, the output only is fedback to obtain the desired performance of the system. This may not result in the best or optimum performance of the system. It may be desirable to feedback additional internal variables to achieve better results. The design procedures in classical theory are mostly trial and error procedures.

A need for a represenation which overcomes all the above draw backs was felt and the state space representation of the system was evolved. This representation forms the basis for the development of modern control systems. This representation contains the information about some of the internal variables along with the output variable and is amenable for analysis and design using digital computer. It is suitable for representing linear, nonlinear, time invariant, time varying, SISO and MIMO systems.

9.2 State Variables

Consider an RLC network excited by an input v(t) as shown in Fig. 9.1.

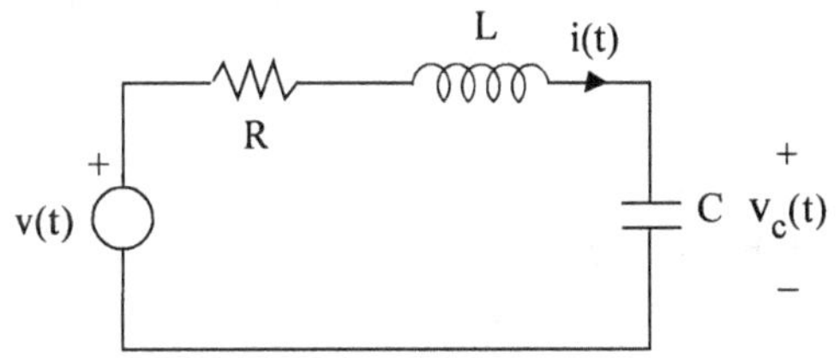

Fig. 9.1 An RLC circuit

The dynamic behaviour of this circuit can be understood by considering the loop equation,

$$Ri + L\frac{di}{dt} + \frac{1}{C}\int_{-\infty}^{t} i\,dt = v(t) \qquad .....(9.1)$$

Differentiating eqn. (9.1) we get a second order differential equation,

$$L\frac{d^2i}{dt^2} + R\frac{di}{dt} + \frac{i}{C} = \frac{dv}{dt} \qquad .....(9.2)$$

The solution of eqn. (9.2) requires two initial conditions, namely, i(o) and $\frac{di}{dt}$ (o). i(o) is the current throught the inductor at t = 0 and from eqn. (9.1) with t = 0, we have,

$$\frac{di}{dt}(0) = \frac{v(0) - Ri(0) - \frac{1}{C}\int_{-\infty}^{0} i\,dt}{L}$$

The quantities $\frac{1}{C}\int_{-\infty}^{0} i\,dt$ is the voltage across the capacitor at t = 0 and hence $\frac{di}{dt}$ (0) is dependent on the inital voltage across the capacitor in addition to v(0) and i(0). Thus, if we know v(t) for $t \geq 0$, i(0), the current through the inductor at t = 0 and $v_c(0)$, the voltage across the capacitor at t = 0, the dynamic response of the system can be easily evaluated. Inductor and the capacitor are the two energy storing elements in the network which are responsible for the behaviour of the network alongwith the input. Thus we can treat the current through the inductor and voltage across the capacitor, as the characterising variables of the network. If these variable are known at any time $t = t_0$, the network response can be easily found out. Hence these two variables describe the state of the network at any time t and these are the minimum number of variables that should be known at $t = t_0$ to obtain the dynamic response of the network.

Now we can define the state and state variables as :

The minimum number of variables required to be known at time $t = t_o$ alongwith the input for $t \geq 0$, to completely determine the dynamic response of a system for $t > t_o$, are known as the state variables of the system. The state of the system at any time 't' is given by the values of these variables at time 't'.

If the dynamic behaviour of a system can be described by an nth order differential equation, we require n initial conditions of the system and hence, a minimum of n state variables are required to be known at $t = t_0$ to completely determine the behaviour of the system to a given input. It is a standard practice to denote these n state variables by $x_1(t)$, $x_2(t)$ ..... $x_n(t)$ and m inputs by $u_1(t)$, $u_2(t)$, ... $u_m(t)$ and p outputs by $y_1(t)$, $y_2(t)$ ..... $y_p(t)$.

The system is described by n first order differential equations in these state variables :

$$\frac{dx_1}{dt} = \dot{x}_1 = f_1 (x_1, x_2 \dots x_n; u_1, u_2 \dots u_m, t) \qquad .....(9.3)$$

$$\vdots \qquad \vdots \qquad \vdots$$

$$\frac{dx_n}{dt} = \dot{x}_n = f_n (x_1, x_2 \dots x_n; u_1, u_2 \dots u_m, t)$$

The functions $f_1, f_2 \dots f_n$ may be time varying or time invariant and linear or nonlinear in nature. Using vector notation to represent the states, their derivatives, and inputs as :

$$X(t) = \begin{bmatrix} x_1(t) \\ x_2(t) \\ \vdots \\ x_n(t) \end{bmatrix}; \; \dot{X}(t) = \begin{bmatrix} \dot{x}_1(t) \\ \dot{x}_2(t) \\ \vdots \\ \dot{x}_n(t) \end{bmatrix}; \; U(t) = \begin{bmatrix} u_1(t) \\ u_2(t) \\ \vdots \\ u_m(t) \end{bmatrix} \qquad .....(9.4)$$

where X(t) is known as state vector and U(t) is known as input vector. We can wrie eqns. (9.3) in a compact form as,

$$\dot{X}(t) = f\,[X(t), U(t), t] \qquad .....(9.5)$$

where

$$f\,[X(t), U(t), t] = \begin{bmatrix} f_1 & [X(t), u(t), t] \\ f_2 & [X(t), u(t), t] \\ \vdots & \\ f_n & [X(t), u(t), t] \end{bmatrix}$$

If the functions f are independent of t, the system is a time invariant system and eqn. (9.5) is written as,

$$\dot{X}(t) = f\,[X(t), U(t)] \qquad .....(9.6)$$

The outputs $y_1, y_2 \dots y_p$ may be dependent on the state vector X(t) and input vector U(t) and may be written as,

$$Y(t) = g\,[X(t), U(t)] \qquad .....(9.7)$$

where $$Y(t) = \begin{bmatrix} y_1(t) \\ y_2(t) \\ \vdots \\ y_p(t) \end{bmatrix}$$

is known as output vector.

For a single input single output systems (SISO) U(t) and Y(t) are scalars. Once the system state is known at any time t, the output can be easily found out since eqn. (9.7) is only algebraic equation and not a dynamic relation.

9.3 State Equations for Linear Systems

For linear systems, in eqn. (9.3), the derivatives of state variables can be expressed as linear combinations of the state variables and inputs.

$$\dot{x}_1 = a_{11} x_1 + a_{12} x_2 + ... + a_{1n} x_n + b_{11} u_1 + b_{12} u_2 + ... + b_{1m} u_m$$

$$\dot{x}_2 = a_{21} x_1 + a_{22} x_2 + ... + a_{2n} x_n + b_{21} u_1 + b_{22} u_2 + ... + b_{2m} u_m$$

$$\vdots$$

$$\dot{x}_n = a_{n1} x_1 + a_{n2} x_2 + ... + a_{nn} x_n + b_{n1} u_1 + b_{n2} u_2 + ... + b_{nm} u_m \quad .....(9.8)$$

where a_{ij}^s and b_{ij}^s are constants.

Eqns. (9.8) can be written in a matrix form as,

$$\dot{X}(t) = A X(t) + BU(t) \quad .....(9.9)$$

where X(t) is a n × 1 state vecotr

A is a n × n constant system matrix

B is a n × m constant input matrix

U(t) is a m × 1 input vector.

$$A = \begin{bmatrix} a_{11} & a_{12} & \cdots & a_{1n} \\ a_{21} & a_{22} & \cdots & a_{2n} \\ \vdots & \vdots & & \vdots \\ a_{n1} & a_{n2} & \cdots & a_{nn} \end{bmatrix}; \quad B = \begin{bmatrix} b_{11} & b_{12} & \cdots & b_{1m} \\ b_{21} & b_{22} & \cdots & b_{2m} \\ \vdots & \vdots & & \\ b_{n1} & b_{n2} & \cdots & b_{nm} \end{bmatrix}$$

Similarly, the outputs can also be expressed as linear combinations of state variables and inputs as,

$$y_1(t) = c_{11} x_1 + c_{12} x_2 ... + c_{1n} x_n + d_{11} u_1 + d_{12} u_2 + ... + d_{1m} u_m$$

$$\vdots$$

$$y_p(t) = c_{p1} x_1 + c_{p2} x_2 ... + c_{pn} x_n + d_{p1} u_1 + d_{p2} u_2 + ... + d_{pm} u_m \quad .....(9.10)$$

Eqns. (9 -10) can be written compactly as,

$$Y(t) = C\,X(t) + D\,U(t) \qquad .....(9.11)$$

Where Y (t) is a p × 1 output vector

C is a p × n output matrix

D is a p × m transmission matrix

$$C = \begin{bmatrix} c_{11} & c_{12} & \cdots & c_{1n} \\ \vdots & & & \vdots \\ c_{p1} & c_{p2} & \cdots & c_{pn} \end{bmatrix}; \; D = \begin{bmatrix} d_{11} & d_{12} & \cdots & d_{1n} \\ \vdots & & & \vdots \\ d_{p1} & d_{p2} & \cdots & d_{pn} \end{bmatrix}$$

The complete state model of the linear system is given by eqn. (9.9) are (9.11).

$$\dot{X}(t) = A\,X(t) + BU(t) \qquad .....(9.12)$$

$$Y(t) = C\,X(t) + DU(t) \qquad .....(9.13)$$

Eqn. (9.12) is known as the state equation,

and eqn. (9.13) is known as the output equation.

For a single input single output system

$$\dot{X}(t) = A\,X(t) + bu \qquad .....(9.14)$$

$$y(t) = C\,X(t) + du \qquad .....(9.15)$$

Where b and d are (n × 1) and (p × 1) vectors respectively. u(t) is the single input and y(t) is the single output. In most of the control systems, the output is not directly coupled to the input and hence y(t) is not dependent on u(t). Hence eqn. (9.15) is written as,

$$y(t) = C\,X(t) \qquad .....(9.16)$$

For time invariant systems, the matrices A, B, C and D are constant matrices. For time varying systems, the elements of A, B, C and D matrices are functions of time. In this book we will be concerned with linear, time invariant, single input single output systems only.

State variable representation of a system is not unique. For a given system we may define different sets of variables to describe the behaviour of the system. In all such different representation the number of state variables required are the same, and this number is known as the order of the system. Let us consider the RLC circuit again, shown in Fig. 9.1. If the current, i, through the inductor and voltage, v_c, across the capacitor are taken as state variables, we have,

$$C\frac{dv_c}{dt} = i(t) \qquad .....(9.17)$$

$$L\frac{di}{dt} = v_L(t) = v - i\,R - v_c \qquad .....(9.18)$$

Defining

$$v_c(t) \triangleq x_1(t)$$

$$i(t) \triangleq x_2(t)$$

$$v(t) \triangleq u(t)$$

We have $$\dot{x}_1(t) = \frac{1}{C} x_2(t) \qquad .....(9.19)$$

$$\dot{x}_2(t) = -\frac{1}{L} x_1(t) - \frac{R}{L} x_2(t) + \frac{1}{L} u(t) \qquad .....(9.20)$$

In matrix form, eqns. (9.19), (9.20) can be written as,

$$\begin{bmatrix} \dot{x}_1(t) \\ \dot{x}_2(t) \end{bmatrix} = \begin{bmatrix} 0 & \frac{1}{C} \\ -\frac{1}{L} & -\frac{R}{L} \end{bmatrix} \begin{bmatrix} x_1(t) \\ x_2(t) \end{bmatrix} + \begin{bmatrix} 0 \\ \frac{1}{L} \end{bmatrix} u(t) \qquad .....(9.21)$$

or $$\dot{X}(t) = A\,X(t) + b\,u(t)$$

where $$A = \begin{bmatrix} 0 & \frac{1}{C} \\ -\frac{1}{L} & -\frac{R}{L} \end{bmatrix}; \; b = \begin{bmatrix} 0 \\ \frac{1}{L} \end{bmatrix} \qquad .....(9.22)$$

Now, consider the eqn. (9.1) for the RLC circuit,

$$Ri + L\frac{di}{dt} + \frac{1}{C}\int_{-\infty}^{t} i\,dt = v \qquad .....(9.23)$$

The charge q is given by,

$$q(t) = \int_{-\infty}^{t} i\,dt \qquad .....(9.24)$$

and $$i = \frac{dq}{dt} \qquad .....(9.25)$$

In terms of the variable q, eqn. (9.23) can be written as

$$L\frac{d^2q}{dt^2} + R\frac{di}{dt} + \frac{q}{c} = v \qquad .....(9.26)$$

Now if we define the state variables as :

$$x_1(t) = q(t) \qquad .....(9.27)$$

and $$x_2(t) = \frac{dq}{dt} = i(t) = \dot{x}_1(t) \qquad .....(9.28)$$

Thus $$\dot{x}_1(t) = x_2 \qquad .....(9.29)$$

$$\dot{x}_2(t) = \frac{d^2q}{dt^2} = \frac{1}{L}v - \frac{R}{L}\frac{dq}{dt} - \frac{q}{LC}$$

$$= -\frac{1}{LC}x_1 - \frac{R}{L}x_2 + \frac{1}{L}U(t) \qquad .....(9.30)$$

Eqns. (9.29) and (9.30) can be put in matrix form as,

$$\begin{bmatrix} \dot{x}_1(t) \\ \dot{x}_2(t) \end{bmatrix} = \begin{bmatrix} 0 & 1 \\ -\dfrac{1}{LC} & -\dfrac{R}{L} \end{bmatrix} \begin{bmatrix} x_1 \\ x_2 \end{bmatrix} + \begin{bmatrix} 0 \\ \dfrac{1}{L} \end{bmatrix} u(t) \qquad .....(9.31)$$

Eqns (9.21) and (9.31) give two different representations of the same system in state variable form. Thus, the state variable representation of any system is not unique.

9.4 Canonical Forms of State Models of Linear Systems

Since the state variable representation of a system is not unique, we will have infinite ways of choosing the state variables. These different state variables are uniquely related to each other. If a new set of state variables, are chosen as a linear combination of the given state variables X, we have

$$X = PZ \qquad .....(9.32)$$

where P is a nonsingular n × n constant matrix, so that

$$Z = P^{-1} X. \qquad .....(9.33)$$

From eqn. (9.32),

$$\dot{X} = P\,\dot{Z} = AX + Bu$$

$$= APZ + Bu$$

$$\therefore \qquad \dot{Z} = \bar{A}\,Z + \bar{B}\,u. \qquad .....(9.34)$$

Eqn. (9.34) is the representation of the same system in terms of new state variables Z and

$$\bar{A} = P^{-1} AP$$

and $$\bar{B} = P^{-1} B.$$

Since P is a non singular matrix, P^{-1} exists. Now let us consider some standard or canonical forms of state models for a given system.

9.4.1 Phase Variable Form

When one of the variables in the physical system and its derivates are chosen as state variables, the state model obtained is known to be in the phase variable form. The state variables are themselves known as phase variables. Usually the output of the system and its derivates are chosen as state variables. We will derive the state model when the system is described either in differential equation form or in transfer function form.

A general n^{th} order differential equation is given by

$$\overset{(n)}{y} + a_1 \overset{(n-1)}{y} + ... + a_{n-1}\dot{y} + a_n y = b_o \overset{(m)}{u} + b_1 \overset{(m-1)}{u} + ... + b_{m-1}\dot{u} + b_m u \qquad .....(9.35)$$

Where a_i's and b_j's are constants and m and n are integers with $n \geq m$.

$$\overset{(n)}{y} \triangleq \frac{d^n y}{dt^n} \text{ and } \overset{(m)}{u} \triangleq \frac{d^m u}{dt^m}$$

The *n* initial conditions are y(o), $\dot{y}$ (o), ..., $\overset{(n-1)}{y}$ (o). If, for the present, we assume the initial conditions to be zero, we can obtain the transfer function of the system from eqn. (9.35) as,

$$T(s) = \frac{Y(s)}{U(s)} = \frac{b_o s^m + b_1 s^{m-1} + ... + b_{m-1} s + b_m}{s^n + a_1 s^{n-1} + ... + a_{n-1} s + a_n} \qquad .....(9.36)$$

The initial conditions can be taken care of, after obtaining the state space representation.

Case (a) m = 0

It the derivatives of the inputs are not present in eqn. (9.35) and it can be written as,

$$\overset{(n)}{y} + a_1 \overset{(n-1)}{y} + ... + a_{n-1} \dot{y} + a_n y = b_o u \qquad .....(9.37)$$

Let us define the state variables as,

$$x_1 = y$$
$$x_2 = \dot{x}_1 = \dot{y}$$
$$x_3 = \dot{x}_2 = \ddot{y}$$
$$\vdots$$
$$x_n = \dot{x}_{n-1} = \overset{(n-1)}{y} \qquad .....(9.38)$$

and $\dot{x}_n = \overset{(n)}{y}$

From eqn. (9.37) we have,

$$\dot{x}_n = b_o u - a_1 \overset{(n-1)}{y} - a_2 \overset{n-2}{y} \; ... - a_{n-1} \dot{y} - a_n y$$
$$= b_o u - a_1 x_n - a_2 x_{n-1} \; ... - a_{n-1} x_2 - a_n x_1$$

The above equations can be written in a matrix form as,

$$\begin{bmatrix} \dot{x}_1 \\ \dot{x}_2 \\ \vdots \\ \dot{x}_n \end{bmatrix} = \begin{bmatrix} 0 & 1 & 0 & \cdots & 0 \\ 0 & 0 & 1 & \cdots & 0 \\ \vdots & \vdots & \vdots & & \vdots \\ 0 & 0 & 0 & \cdots & 1 \\ -a_n & -a_{n-1} & -a_{n-2} & \cdots & -a_1 \end{bmatrix} \begin{bmatrix} x_1 \\ x_2 \\ \vdots \\ x_n \end{bmatrix} + \begin{bmatrix} 0 \\ 0 \\ \vdots \\ 0 \\ b_o \end{bmatrix} u \qquad .....(9.39)$$

In vector matrix notation we have,

$$\dot{X} = AX + bu \qquad .....(9.40)$$

In this equation, we can observe that the system matrix A is in a special form. the diagonal above the main diagonal of the matrix contains all 1^s and the last row contains the negatives of the coefficients, a^s_j, of the differential equation. All other elements are zeros. Such a form of the matrix is known as companion form or Bush's form. Similarly, the vector *b* contains all elements to be zero except the last one. Hence, in view of these observations, the state space equations given by eqn. (9.39) can be written down directly from the differential eqn. (9.37).

The output y is equal to x_1 and hence the output equation is given by,

$$y = CX \qquad .....(9.41)$$

where $c = [1 \ 0 \ 0 \ ..... \ 0]$

If the transfer function is known instead of the differential equation, we can easily obtain a differential form as shown below.

For m = 0, we have from eqn. (9.36)

$$T(s) = \frac{Y(s)}{U(s)} = \frac{b_0}{s^n + a_1 s^{n-1} + ... + a_n} \qquad .....(9.42)$$

$$(s^n + a_1 s^{n-1} + ... + a_n) Y(s) = b_o U(s)$$

or

$$\overset{(n)}{y}(t) + a_1 \overset{(n-1)}{y} + ... + a_n y = b_o u \qquad .....(9.43)$$

Eqn. (9.43) is the same as eqn. (9.37) and hence the state space model is again given by eqns. (9.39) and (9.41). A block diagram of the state model in eqn. (9.39) is shown in Fig. (9.2). Each block in the forward path represents an integration and the output of each integrator is takne a state variable.

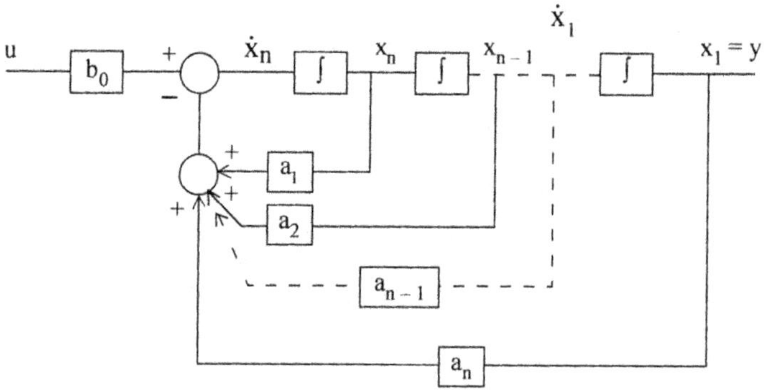

Fig. 9.2 Block diagram representation of eqn. (9.39).

We can also represent eqn. (9.39) in signal low graph representation as shown in Fig. (9.3).

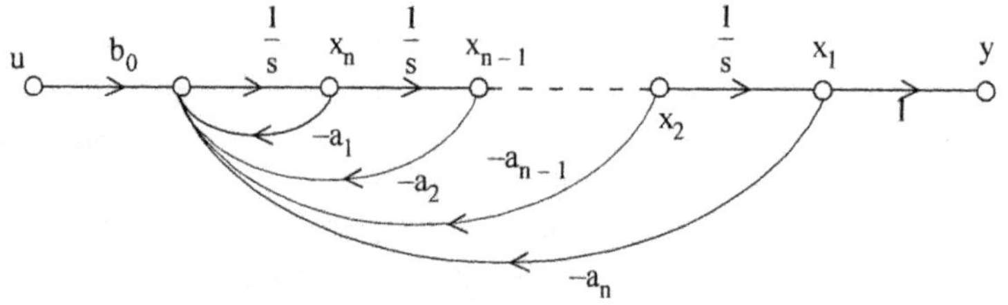

Fig. 9.3 Signal flow graph representation of eqn. (9.39).

If we have non zero initial conditions, the initial conditions can be related to the initial conditions on the state variables. Thus, from eqn. (9.38).

$$x_1(0) = y(0)$$

$$x_2(0) = \dot{y}(0)$$

$$\vdots$$

$$x_n(0) = \overset{(n-1)}{y}(0) \qquad \text{.....(9.44)}$$

Example 9.1

Obtain the phase variable state model for the system

$$\ddot{y} + 2\ddot{y} + 3\dot{y} + y = u$$

Solution

Here n = 3 and $a_1 = 2$, $a_2 = 3$, $a_3 = 1$ and b = 1. Hence the state model can be directly written down as,

$$\begin{bmatrix} \dot{x}_1 \\ \dot{x}_2 \\ \dot{x}_3 \end{bmatrix} = \begin{bmatrix} 0 & 1 & 0 \\ 0 & 0 & 1 \\ -1 & -3 & -2 \end{bmatrix} \begin{bmatrix} x_1 \\ x_2 \\ x_3 \end{bmatrix} + \begin{bmatrix} 0 \\ 0 \\ 1 \end{bmatrix} u$$

$$X(0) = [y(0),\ \dot{y}(0),\ \ddot{y}(0)]^T$$

and $y = [1 \;\; 0 \;\; 0]X$

Example 9.2

Obtain the companion form of state model for the system whose transfer function is given by,

$$T(s) = \frac{Y(s)}{U(s)} = \frac{2}{s^3 + s^2 + 2s + 3}$$

Solution :

Case a :

The state model in companion form can be directly written down as,

$$\dot{X} = \begin{bmatrix} 0 & 1 & 0 \\ 0 & 0 & 1 \\ -3 & -2 & -1 \end{bmatrix} X + \begin{bmatrix} 0 \\ 0 \\ 2 \end{bmatrix} u$$

$$y = [1 \;\; 0 \;\; 0]X$$

Case b : $m \neq 0$

Let us consider a general case where m = n.

The differential equation is given by

$$\overset{(n)}{y} + a_1 \overset{(n-1)}{y} + a_2 \overset{n-2}{y} + \ldots + a_{n-1} \overset{n-1}{y} + a_n y = b_0 \overset{(n)}{u} + b_1 \overset{(n-1)}{u} + \ldots + b_n u \qquad \text{.....(9.45)}$$

The transfer function of the system represented by eqn. (9.45) is,

$$T(s) = \frac{Y(s)}{U(s)} = \frac{b_0 s^n + b_1 s^{n-1} + ... + b_{n-1}s + b_n}{s^n + a_1 s^{n-1} + ... + a_{n-1} s + a_n} \qquad .....(9.46)$$

Eqn. (9.46) can be written as

$$T(s) = \frac{Y_1(s)}{U(s)} \cdot \frac{Y(s)}{Y_1(s)} = \frac{1}{s^n + a_1 s^{n-1} + ... + a_{n-1}s + a_n} \cdot (b_0 s^n + b_1 s^{n-1} + .. + b_n)$$

where

$$\frac{Y_1(s)}{U(s)} = \frac{1}{s^n + a_1 s^{n-1} + ... + a_{n-1}s + a_n} \qquad .....(9.47)$$

and

$$\frac{Y(s)}{Y_1(s)} = b_0 s^n + b_1 s^{n-1} + ... + b_n \qquad .....(9.48)$$

Eqn. (9.47) is same as eqn. (9.42) with b =1.

Hence its state space representation is given by eqn. 9.39.

$$\begin{bmatrix} \dot{x}_1 \\ \dot{x}_2 \\ \vdots \\ \dot{x}_n \end{bmatrix} = \begin{bmatrix} 0 & 1 & 0 & \cdots & 0 & 0 \\ 0 & 0 & 1 & \cdots & 0 & 0 \\ \vdots & & & & & \\ 0 & 0 & 0 & \cdots & 0 & 1 \\ -a_n & -a_{n-1} & -a_{n-2} & \cdots & -a_2 & -a_1 \end{bmatrix} \begin{bmatrix} x_1 \\ x_2 \\ \vdots \\ x_n \end{bmatrix} + \begin{bmatrix} 0 \\ 0 \\ \vdots \\ 0 \\ 1 \end{bmatrix} u \qquad .....(9.49)$$

and

$$y_1 = x_1 \qquad .....(9.50)$$

The signal flow graph of this system is the same as in Fig. (9.3) with $y = y_1$ and $b_0 = 1$.

From eqn. (9.48), we have

$$Y(s) = (b_0 s^n + b_1 s^{n-1} + ... + b_n) Y_1(s)$$

or

$$y = b_0 \overset{(n)}{y_1} + b_1 \overset{(n-1)}{y_1} + ... + b_{n-1} \dot{y}_1 + b_n y_1$$

From eqn. (9.49) and (9.50), we have

$$y = b_0 \dot{x}_n + b_1 x_n + ... + b_{n-1} x_2 + b_n x_1$$

Substistuting for $\dot{x}_n$ from eqn. (9.49), we have

$$y = b_0 (-a_n x_1 - a_{n-1} x_2 + ... - a_2 x_{n-1} - a_1 x_n + u) +$$
$$b_1 x_n + ... + b_{n-1} x_2 + b_n x_1$$

$$\therefore \quad y = x_1 (b_n - b_0 a_n) + x_2 (b_{n-1} - b_0 a_{n-1})$$
$$+ ... + x_n (b_1 - b_0 a_1) + b_0 u \qquad .....(9.51)$$

$$y = [(b_n - b_0 a_n), (b_{n-1} - b_0 a_{n-1}), \ldots (b_1 - b_0 a_1)] \begin{bmatrix} x_1 \\ x_2 \\ \vdots \\ x_n \end{bmatrix} + b_0 u \qquad .....(9.52)$$

The signal flow graph of the system of eqn. (9.45) is obtained by modifying signal flow graph shown in Fig. (9.3) as,

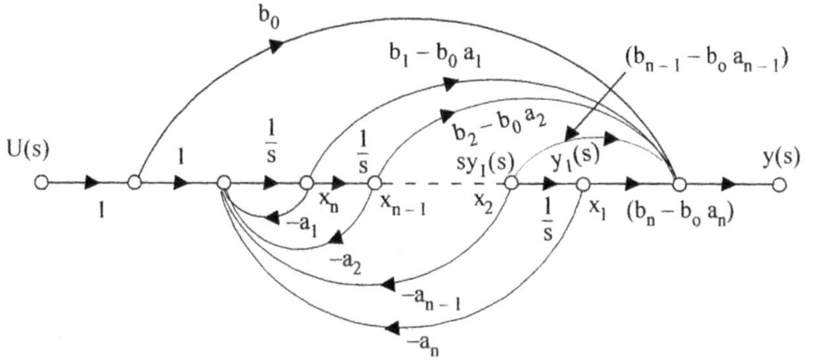

Fig. 9.4 Signal flow graph of system of eqn. (9.45)

For the more common case, where $m \leq n - 1$ in eqn. (9.35), $b_0 = 0$ and eqn. (9.52) can be written as

$$y = [b_n \;\; b_{n-1} \;\ldots\; b_1]\, X \qquad .....(9.53)$$

Where b_n^s are the coefficients of the numerator polynomical of eqn. (9.46). In this case, the state space representation of eqn. (9.35) can be written down by inspection.

Example 9.3

Obtain the state space representation of the system whose differential equation is given by,

$$\dddot{y} + 2\ddot{y} + 3\dot{y} + 6y = \ddot{u} - \dot{u} + 2u$$

Also draw the signal flow graph for the system.

Solution

In the given differential equation,

$$a_1 = 2,\; a_2 = 3,\; a_3 = 6$$

and $\quad b_0 = 0,\; b_1 = 1,\; b_2 = -1$ and $b_3 = 2$

Substituting in eqns. (9.49) and (9.52), we have

$$\dot{X} = \begin{bmatrix} 0 & 1 & 0 \\ 0 & 0 & 1 \\ -6 & -3 & -2 \end{bmatrix} X + \begin{bmatrix} 0 \\ 0 \\ 1 \end{bmatrix} u$$

$$y = [2 \;\; -1 \;\; 1]\, X$$

The signal flow graph of the system is given in Fig. (9.5).

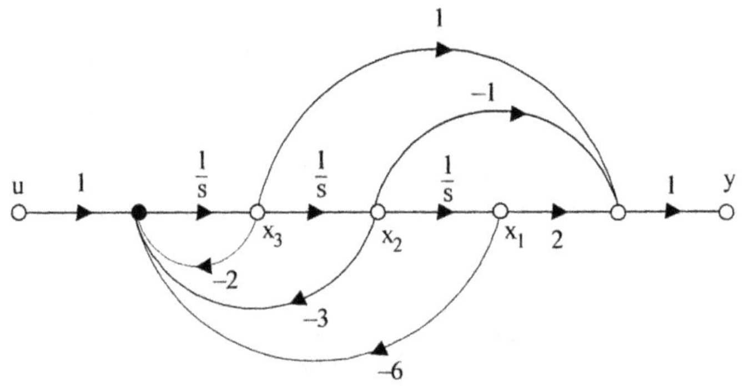

Fig. 9.5 Signal flow graph for Ex. (9.3)

Phase variable representation is a simple method of obtaining the state space representation of a system. This plays a very important role in the design of a control systems in state space. But this representation is not useful in practice, as these phase variables often do not represent physical variables and hence are not available for measurement or control. They are given by the output and its derivatives. Higher order derivatives of the output are difficult to obtain in practice. Hence let us consider another more useful representation of state model of a system.

9.4.2 Diagonal Form

This form is also known as canonical variable form or normal form. The system matrix A in this case is obtained as a diagonal matrix. Let us consider the transfer function of the system given by

$$T(s) = \frac{Y(s)}{U(s)} = \frac{b_0 s^n + b_1 s^{n-1} + \ldots + b_{n-1} s + b_n}{s^n + a_1 s^{n-1} + \ldots + a_{n-1} s + a_n} \qquad \text{.....(9.54)}$$

Case a : All the poles of T(s) are distinct and given by $-p_1, -p_2, \ldots - p_n$. Expanding T(s) in partial fractions, we have,

$$T(s) = \frac{Y(s)}{U(s)} = b_0 + \sum_{i=1}^{n} \frac{k_i}{s + p_i} \qquad \text{.....(9.55)}$$

Eqn. (9.55) can be represented by a block diagram in Fig. (9.6) (a) and signal flow graph in Fig. (9.6) (b).

Defining the output of each integrator as a state variable, as shown in Figs. (9.6) (a), (b) we have,

$$\dot{x}_1 = u - p_1 x_1$$

$$\dot{x}_2 = u - p_2 x_2 \qquad \text{.....(9.56)}$$

$$\vdots$$

$$\dot{x}_n = u - p_n x_n$$

$$y = k_1 x_1 + k_2 x_2 + \ldots + k_n x_n + b_0 u \qquad \text{.....(9.57)}$$

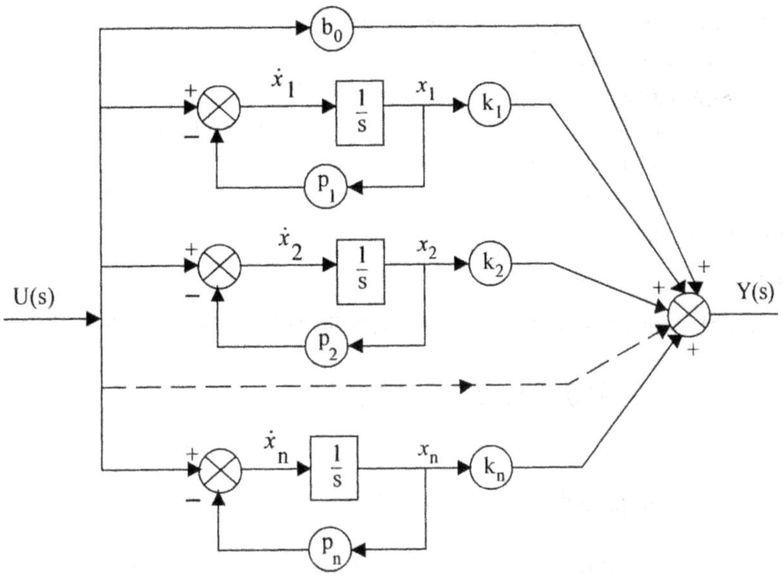

Fig. 9.6 (a) Block diagram representation of eqn. (9.55).

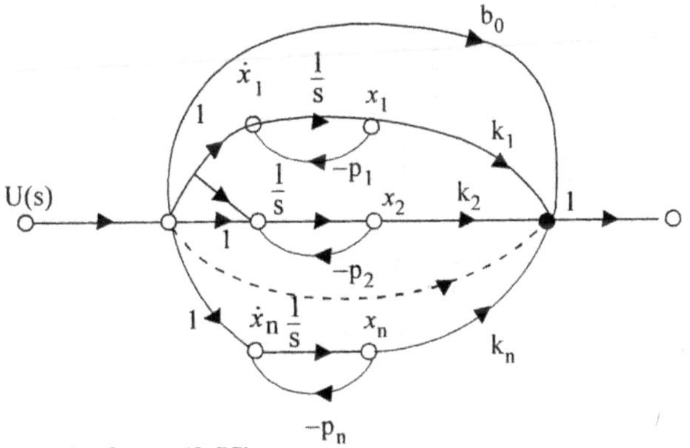

Fig. 9.6 (b) Signal flow graph of eqn. (9.55).

Expressing eqn. (9.56) and (9.57) in matrix form.

$$\begin{bmatrix} \dot{x}_1 \\ \dot{x}_2 \\ \vdots \\ \dot{x}_n \end{bmatrix} = \begin{bmatrix} -p_1 & 0 & 0 & \cdots & 0 \\ 0 & -p_2 & 0 & \cdots & 0 \\ \vdots & & & & \\ 0 & 0 & 0 & \cdots & -p_n \end{bmatrix} \begin{bmatrix} x_1 \\ x_2 \\ \vdots \\ x_n \end{bmatrix} + \begin{bmatrix} 1 \\ 1 \\ \vdots \\ 1 \end{bmatrix} u \qquad \text{.....(9.58)}$$

$$y = [k_1 \;\; k_2 \; \ldots \; k_n] \begin{bmatrix} x_1 \\ x_2 \\ \vdots \\ x_n \end{bmatrix} + b_0 u \qquad \text{.....(9.59)}$$

In eqn. (9.58) we observe that the system matrix A is in diagonal form with poles of T(s) as its diagonal elements. Also observe that the column vector *b* has all its elements as 1^s. The *n* equations represented by eqn. (9.58) are independent of each other and can be solved independently. These equations are said to be decoupled. This feature is an important property of this normal form which is useful in the analysis and design of control systems in state variable form. It is also pertinent to mention that the canonical variables are also not physical variables and hence not available for measurement or control.

Example 9.4

Obtain the normal form of state model for the system whose transfer function is given by

$$T(s) = \frac{Y(s)}{U(s)} = \frac{s+1}{s(s+2)(s+4)} \qquad .....(9.60)$$

Solution

T(s) can be expanded in partial fractions as,

$$T(s) = \frac{1}{8s} + \frac{1}{4(s+2)} - \frac{3}{8(s+4)} \qquad .....(9.61)$$

The state space representation is given by,

$$\dot{X} = \begin{bmatrix} 0 & 0 & 0 \\ 0 & -2 & 0 \\ 0 & 0 & -4 \end{bmatrix} X + \begin{bmatrix} 1 \\ 1 \\ 1 \end{bmatrix} u$$

$$Y = \begin{bmatrix} \frac{1}{8} & \frac{1}{4} & -\frac{3}{8} \end{bmatrix} X$$

Case b : Some poles of T(s) in eqn. (9.54) are repeated.

Let us illustrate this case by an example.

$$T(s) = \frac{Y(s)}{U(s)} = \frac{s}{(s+1)^2(s+2)} \qquad .....(9.62)$$

The partial fraction expansion is given by,

$$T(s) = \frac{2}{s+1} - \frac{1}{(s+1)^2} - \frac{2}{s+2} \qquad .....(9.63)$$

The simulation of this transfer function by block diagram is shown in Fig. (9.7).

Defining the output of each integrator in Fig. (9.7) as a state variable, we have

$$\dot{x}_1 = x_2 - x_1$$

$$\dot{x}_2 = u - x_2$$

$$\dot{x}_3 = u - 2x_3 \qquad .....(9.64)$$

and $$y = -x_1 + 2x_2 - 2x_3 \qquad .....(9.65)$$

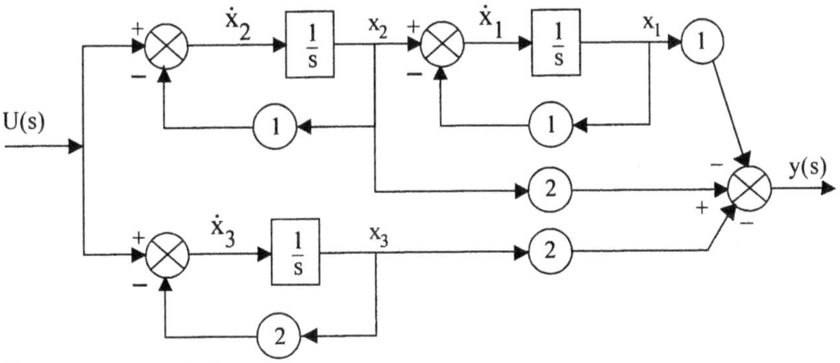

Fig. 9.7 Block diagram representation.

Eqns. (9.64) and (9.65) can be put in matrix form as,

$$\dot{X} = \begin{bmatrix} -1 & 1 & 0 \\ 0 & -1 & 0 \\ 0 & 0 & -2 \end{bmatrix} X + \begin{bmatrix} 0 \\ 1 \\ 1 \end{bmatrix} u \quad .....(9.66)$$

$$y = [-1 \ \ 2 \ \ -2]\, X \quad .....(9.67)$$

The above procedure can be generalised to a system with more repeated poles. Let λ_1 be repeated twice, λ_2 be repeated thrice and λ_3 and λ_4 be distinct in a system with n = 7. The state model for this system will be,

$$\begin{bmatrix} \dot{x}_1 \\ \dot{x}_2 \\ \dot{x}_3 \\ \dot{x}_4 \\ \dot{x}_5 \\ \dot{x}_6 \\ \dot{x}_7 \end{bmatrix} = \begin{bmatrix} \lambda_1 & 1 & 0 & 0 & 0 & 0 & 0 \\ 0 & \lambda_1 & 0 & 0 & 0 & 0 & 0 \\ 0 & 0 & \lambda_2 & 1 & 0 & 0 & 0 \\ 0 & 0 & 0 & \lambda_2 & 1 & 0 & 0 \\ 0 & 0 & 0 & 0 & \lambda_2 & 0 & 0 \\ 0 & 0 & 0 & 0 & 0 & \lambda_3 & 0 \\ 0 & 0 & 0 & 0 & 0 & 0 & \lambda_4 \end{bmatrix} \begin{bmatrix} x_1 \\ x_2 \\ x_3 \\ x_4 \\ x_5 \\ x_6 \\ x_7 \end{bmatrix} + \begin{bmatrix} 0 \\ 1 \\ 0 \\ 0 \\ 1 \\ 1 \\ 1 \end{bmatrix} u \quad .....(9.68)$$

Matrix eqn. (9.68) is partitioned as shown by the dotted lines and it may be represented by eqn. (9.69).

$$\begin{bmatrix} \dot{x}_1 \\ \dot{x}_2 \\ \dot{x}_3 \\ \dot{x}_4 \end{bmatrix} = \begin{bmatrix} J_1 & 0 & 0 & 0 \\ 0 & J_2 & 0 & 0 \\ 0 & 0 & J_3 & 0 \\ 0 & 0 & 0 & J_4 \end{bmatrix} \begin{bmatrix} x_1 \\ x_2 \\ x_3 \\ x_4 \end{bmatrix} + \begin{bmatrix} b_1 \\ b_2 \\ b_3 \\ b_4 \end{bmatrix} u \quad .....(9.69)$$

Eqn. (9.69) is in diagonal form. The sub matrices J_1 and J_2 contain the repeated ples λ_1 and λ_2 respectively on their diagonals and the super diagonal elements are all ones. J_1 and J_2 are known as Jordon blocks. The matrix A itself is known to be of Jordon form. J_1 is a Jordon block of order 2 and J_2 is a Jordon block of order 3. J_3 and J_4 corresponding to non repeated ples λ_3 and λ_4 are said to be of order 1. The column vectors b_1 and b_2 have all zero elements except the last element which is a 1. b_3 and b_4 are both unity and correspond to non repeated roots λ_3 and λ_4.

Eqn. (9.68)is said to be the Jordon form representation of a system.

9.5 Transfer Function from State Model

From a given transfer function we have obtained different state space models. Now, let us consider how a transfer function can be obtained from a given state space model. If the state model is in phase variable form, the transfer function can be directly written down by inspection in view of eqns. (9.39) and (9.42) or eqns. (9.46), (9.49) and (9.52).

Alternatively, the signal flow graph can be obtained from the state model and Mason's gain formula can be used to get the transfer function.

Example 9.5

Obtain the transfer function for the system,

$$\begin{bmatrix} \dot{x}_1 \\ \dot{x}_2 \\ \dot{x}_3 \end{bmatrix} = \begin{bmatrix} 0 & 1 & 0 \\ 0 & 0 & 1 \\ -1 & -2 & -3 \end{bmatrix} \begin{bmatrix} x_1 \\ x_2 \\ x_3 \end{bmatrix} + \begin{bmatrix} 0 \\ 0 \\ 1 \end{bmatrix} u$$

$$y = [1 \ \ 0 \ \ 0]\, X$$

The transfer function is of the form, for n = 3

$$T(s) = \frac{b_0 s^3 + b_1 s^2 + b_2 s + b_3}{s^3 + a_1 s^2 + a_2 s + a_3}$$

From the matrix A, b, and C we have

$$a_1 = 3,\ a_2 = 2,\ a_3 = 1$$

$$b_0 = 0,\ b_1 = 0,\ b_2 = 0,\ b_3 = 1$$

Hence $$T(s) = \frac{1}{s^3 + 3s^2 + 2s + 1}$$

Example 9.6

Obtain the transfer function of the system

$$\dot{X} = \begin{bmatrix} 0 & 1 & 0 \\ 0 & 0 & 1 \\ -2 & -4 & -6 \end{bmatrix} X + \begin{bmatrix} 0 \\ 0 \\ 1 \end{bmatrix} u$$

$$Y = [1 \ \ -2 \ \ 3]\, X + 2u$$

Here $a_1 = 6$ $\quad a_2 = 4$ $\quad a_3 = 2$

$b_0 = 2$ $\quad b_3 - b_0\, a_3 = 1$ $\quad b_2 - b_0\, a_2 = -2$

$b_1 - b_0\, a_1 = 3$

$b_3 = 1 + 2(2) = 5$

$b_2 = -2 + 2(4) = 6$

$b_1 = 3 + 2(6) = 15$

$$\therefore \qquad T(s) = \frac{2s^3 + 15s^2 + 6s + 5}{s^3 + 6s^2 + 4s + 2}$$

If the state model is in Jordon form, we can obtain the transfer function as shown in Ex 9.7.

Example 9.7

Obtain the transfer function for the system

$$\dot{X} = \begin{bmatrix} -1 & 1 & 0 \\ 0 & -1 & 0 \\ 0 & 0 & -2 \end{bmatrix} X + \begin{bmatrix} 0 \\ 0 \\ 1 \end{bmatrix} u$$

$$Y = [-1 \;\; 3 \;\; 3]\, X$$

Solution

From the system matrix A, the three poles of the transfer function are –1, –1 and –2. The residues at these poles are –1, 3 and 3 (from the matrix C). The transfer function is, therefore, given by

$$T(s) = \frac{-1}{s+1} + \frac{3}{(s+1)^2} + \frac{3}{s+2}$$

$$= \frac{2s^2 + 6s + 7}{(s+1)^2(s+2)}$$

If the matrix A is not in either companion form or Jordon form, the transfer function can be derived as follows.

$$\dot{X} = A X + bu \qquad .....(9.70)$$

$$Y = CX + du \qquad .....(9.71)$$

Taking Laplace transform of eqn. (9.70), assuming zero initial conditions, we have

$$s X(s) = A X(s) + b U(s)$$

$$(sI - A) X(s) = b U(s)$$

$$X(s) = (sI - A)^{-1} b U(s) \qquad .....(9.72)$$

Taking Laplace transform of eqn. (9.71) and substituting for X(s) from eqn. (9.72), we get

$$Y(s) = C (sI - A)^{-1} b U(s) + d U(s) \qquad .....(9.73)$$

$$= [C (sI - A)^{-1} b + d] U(s)$$

$$\therefore \qquad T(s) = C (sI - A)^{-1} b + d$$

If *d* is equal to zero, for a commonly occuring case,

$$T(s) = C (sI - A)^{-1} b \qquad .....(9.74)$$

Since a transfer function representation is unique for a given system, eqn. (9.74) is independent of the form of A. For a given system, different system matrices may be obtained, but the transfer function will be unique.

Since $$(sI - A)^{-1} = \frac{Adj(sI - A)}{|sI - A|}$$

the roots of $$|sI - A| = 0 \qquad .....(9.75)$$

are the poles of the transfer function T(s) and eqn. (9.75) is known as the characteristic equation of the matrix A.

Example 9.8

Obtain the transfer function for the system,

$$\dot{X} = \begin{bmatrix} -1 & 0 & -1 \\ 0 & -1 & 1 \\ 1 & -2 & -3 \end{bmatrix} X + \begin{bmatrix} 0 \\ 1 \\ 1 \end{bmatrix} u$$

$$Y = [1 \quad 0 \quad 1] X$$

Solution

$$(sI - A) = \begin{bmatrix} s+1 & 0 & 1 \\ 0 & s+1 & -1 \\ -1 & 2 & s+3 \end{bmatrix}$$

$$(sI - A)^{-1} = \frac{1}{s^3 + 5s^2 + 10s + 6} \begin{bmatrix} s^2 + 4s + 5 & 2 & -(s+1) \\ 1 & s^2 + 4s + 4 & s+1 \\ s+1 & -2(s+1) & (s+1)^2 \end{bmatrix}$$

The transfer function is given by,

$$T(s) = \frac{1}{\Delta(s)} [1 \quad 0 \quad 1] \begin{bmatrix} s^2 + 4s + 5 & 2 & -(s+1) \\ 1 & s^2 + 4s + 4 & s+1 \\ s+1 & -2(s+1) & (s+1)^2 \end{bmatrix} \begin{bmatrix} 0 \\ 1 \\ 1 \end{bmatrix}$$

Where $\Delta(s) = s^3 + 5s^2 + 10s + 6$

$$T(s) = \frac{1}{\Delta(s)} [1 \quad 0 \quad 1] \begin{bmatrix} 1 - s \\ s^2 + 5s + 5 \\ (s+1)(s-1) \end{bmatrix}$$

$$= \frac{s(s-1)}{s^3 + 5s^2 + 10s + 6}$$

9.6 Diagonalisation

The state model in the diagonal or Jordon form is very useful in understanding the system properties and evaluating its response for any given input. But a state space model obtained by considering the real physical variables as state variables is seldom in the canonical form. These physical variables can be readily measured and can be used for controlling the response of the system. Hence a state model obtained, based on physical variables, is often converted to canonical form and the properties of the system are studied. Hence we shall consider the techniques used for converting a general state space model to a Jordon form model. Consider a system with the state space model as,

$$\dot{X} = AX + bu \quad .....(9.76)$$

$$y = CX + du \quad .....(9.77)$$

Let us define a new set of state variables 'Z', related to the state variables X by a non singular matrix P, such that

$$X = PZ \quad .....(9.78)$$

$$\dot{X} = P\dot{Z} \quad .....(9.79)$$

Substituting eqns. (9.78) and (9.79) in eqn. (9.76), we have

$$P\dot{Z} = APZ + bu$$

or $$\dot{Z} = P^{-1}APZ + P^{-1}bu \quad .....(9.80)$$

also $$y = CPZ + du \quad .....(9.81)$$

Let us select the matrix P such that $P^{-1}AP = J$ where J is a Jordon matrix.

Thus $$\dot{Z} = JZ + \bar{b}u \quad .....(9.82)$$

$$y = \bar{c}Z + du \quad .....(9.83)$$

where $$\bar{b} = P^{-1}b$$

and $$\bar{c} = CP$$

Now how to choose the matrix P such that $P^{-1}AP$ is a Jordon matrix ? In order to answer this question, we consider the eigenvalues and eigenvectors of the matrix A.

9.6.1 Eigenvalues and Eigenvectors

Consider the equation

$$AX = Y \quad .....(9.84)$$

Here an $n \times n$ matrix A transforms an $n \times 1$ vector X to another $n \times 1$ vector Y. The vector X has a direction in the state space. Let us investigate, whether there exists a vector X, which gets transformed to another vector Y, in the same direction as X, when operated by the matrix A.

This means $Y = \lambda X$ where λ is a scalar.

$\therefore$ $$AX = \lambda X \quad .....(9.85)$$

or $$[A - \lambda I]X = 0 \quad .....(9.86)$$

Eqn. (9.86) is a set of homogeneous equations and it will have a nontrivial ($X \neq 0$) solution if and only if,

$$|A - \lambda I| = 0 \qquad .....(9.87)$$

Eqn. (9.87) results in a polynomial in λ given by,

$$\lambda^n + a_1 \lambda^{n-1} + a_2 \lambda^{n-2} + ... + a_{n-1}\lambda + a_n = 0 \qquad .....(9.88)$$

There are n values of λ which satisfy eqn. (9.88). These values are called as eigen values of the matrix A. Eqn. (9.88) is known as the characteristic equation of matrix A. Since eqns. (9.87) and (9.75) are similar, the eigen values are also the poles of the transfer function.

For any eigenvalue $\lambda = \lambda_i$, from eqn. (9.86) we have,

$$[A - \lambda_i I] X = 0 \qquad .....(9.89)$$

For this value of λ_i, we know that a nonzero vector X_i exists satisfying the eqn. (9.89). This vector X_i is known as the eigenvector corresponding to the eigenvalue λ_i. Since there are n eigenvalues for a nth order matrix A, there will be n eigenvectors for a given matrix A. Since $|A - \lambda I| = 0$, the rank of the matrix $(A - \lambda I) < n$.

If the rank of the matrix A is $(n - 1)$, there will be one eigenvector corresponding to each λ_i. Let these eigen vectors corresponding to $\lambda_1, \lambda_2, ... \lambda_n$ be $m_1, m_2, ... m_n$ respectively.

Then, from eqn. (9.85),

$$\begin{aligned} Am_1 &= \lambda_1 m_1 \\ Am_2 &= \lambda_2 m_2 \\ &\vdots \\ Am_n &= \lambda_n m_n \end{aligned} \qquad .....(9.90)$$

Eqn. (9.90) can be written in matrix form as,

$$A [m_1, \ m_2, \ ... \ m_n] = [\lambda_1 m_1, \ \lambda_2 m_2, \ ... \ \lambda_n m_n] \qquad .....(9.91)$$

We can write eqn. (9.91) as,

$$A [m_1, \ m_2, \ ... \ m_n] = [m_1, \ m_2, \ ... \ m_n] \begin{bmatrix} \lambda_1 & 0 & \cdots & 0 \\ 0 & \lambda_2 & \cdots & 0 \\ \vdots & & & \\ 0 & 0 & \cdots & \lambda_n \end{bmatrix} \qquad .....(9.92)$$

Defining, $\qquad M = [m_1, \ m_2, \ ... \ m_n]$

where M is called as the modal matrix, which is the matrix formed by the eigen vectors, we have,

$$AM = MJ \qquad .....(9.93)$$

Where J is the diagonal matrix formed by the eigen values as its diagonal elements.

From eqn. (9.93) we have

$$J = M^{-1} AM \qquad .....(9.94)$$

This is the relation to be used for diagonalising any matrix A. If the matrix P in eqn. 9.80 is chosen to be M, the model matrix of A, we get the diagonal form.

If all the eigenvalues of A are distinct, it is always possible to find n linearly independent eigenvectors and the J matrix will be diagonal. On the other hand if some of the eigenvalues are repeated we may not be able to get n linearly independent eigenvectors. Corresponding to the repeated eigen values, we will have to obtain, what are known as generalised eigenvectors. These vectors also will be linearly independent and the resulting J matrix will be in the Jordon form as given in eqn. (9.68). Let us illustrate these different cases by some examples.

Example 9.9

This examples illustrates the case when all the eigen values are distinct. Consider the matrix

$$A = \begin{bmatrix} 1 & 0 & 2 \\ 0 & 2 & 1 \\ 0 & 0 & 3 \end{bmatrix}$$

Solution

Since A is in triangular form the eigen values are the values on diagonal.

$$\lambda_1 = 1, \ \lambda_2 = 2 \text{ and } \lambda_3 = 3$$

Consider $|A - \lambda I|$

$$|A - \lambda I| = \begin{vmatrix} 1-\lambda & 0 & 2 \\ 0 & 2-\lambda & 1 \\ 0 & 0 & 3-\lambda \end{vmatrix}$$

It can be shown that the eigen vector can be obtained as any non zero column of adj $(A - \lambda I)$. Thus,

$$\text{adj } |A - \lambda I| = \begin{bmatrix} (2-\lambda)\ (3-\lambda) & 0 & -2\ (2-\lambda) \\ 0 & (1-\lambda)\ (3-\lambda) & -(1-\lambda) \\ 0 & 0 & (1-\lambda)\ (2-\lambda) \end{bmatrix}$$

Let us find the eigen vector corresponding to $\lambda = 1$.

$$\text{adj } (A - I) = \begin{bmatrix} 2 & 0 & -2 \\ 0 & 0 & 0 \\ 0 & 0 & 0 \end{bmatrix}$$

The two non zero column of adj $(A - I)$ are linearly dependent and hence any one of them can be taken as an eigen vector.

$$\therefore \qquad m_1 = \begin{bmatrix} 2 \\ 0 \\ 0 \end{bmatrix} \text{ or } \begin{bmatrix} 1 \\ 0 \\ 0 \end{bmatrix}$$

(Any constant multiple of the vector also qualifies as an eigen vector)

Similarly for $\lambda = 2$ and $\lambda = 3$ we have

$$m_2 = \begin{bmatrix} 0 \\ 1 \\ 0 \end{bmatrix} \qquad m_3 = \begin{bmatrix} 1 \\ 1 \\ 1 \end{bmatrix}$$

$$\therefore \qquad M = \begin{bmatrix} 1 & 0 & 1 \\ 0 & 1 & 1 \\ 0 & 0 & 1 \end{bmatrix}$$

$$M^{-1} = \begin{bmatrix} 1 & 0 & -1 \\ 0 & 1 & -1 \\ 0 & 0 & 1 \end{bmatrix}$$

$$M^{-1}AM = \begin{bmatrix} 1 & 0 & -1 \\ 0 & 1 & -1 \\ 0 & 0 & -1 \end{bmatrix} \begin{bmatrix} 1 & 0 & 2 \\ 0 & 2 & 1 \\ 0 & 0 & 3 \end{bmatrix} \begin{bmatrix} 1 & 0 & 1 \\ 0 & 1 & 1 \\ 0 & 0 & 1 \end{bmatrix}$$

Carrying out the multiplication, we have,

$$J = \begin{bmatrix} 1 & 0 & 0 \\ 0 & 2 & 0 \\ 0 & 0 & 3 \end{bmatrix}$$

Example 9.10

Let us consider the case when roots are repeated. Consider

$$A = \begin{bmatrix} 1 & 1 & 2 \\ 0 & 2 & 1 \\ 0 & 0 & 2 \end{bmatrix}$$

Solution :

The eigen values of A are

$$\lambda_1 = 1 \text{ and } \lambda_2 = \lambda_3 = 2$$

The adj $(A - \lambda I)$ is given by

$$\text{Adj } (A - \lambda I) = \begin{bmatrix} (2-\lambda)^2 & -(2-\lambda) & 1-2\,(2-\lambda) \\ 0 & (1-\lambda)\,(2-\lambda) & -(1-\lambda) \\ 0 & 0 & (1-\lambda)\,(2-\lambda) \end{bmatrix} \qquad \text{.....(9.95)}$$

For $\qquad \lambda = 1$

$$\text{Adj } (A - I) = \begin{bmatrix} 1 & -1 & -1 \\ 0 & 0 & 0 \\ 0 & 0 & 0 \end{bmatrix}$$

Any non zero column of Adj $(A - \lambda I)$ qualifies as an eigen vector.

$$\therefore \qquad m_1 = \begin{bmatrix} 1 \\ 0 \\ 0 \end{bmatrix}$$

Since $\lambda = 2$ is a repeated eigen value, it is possible to get either two linearly independent eigen vectors or one linearly independent eigen vector and one generalised eigen vector. In general, if λ_i is repeated r times, the number of linearly independent eigen vectors is equal to the nullity of $(A - \lambda_i I)$ or [n - rank of $(A - \lambda_i I)$] where n is the order of the matrix A.

In the present example,

$$[A - 2I] = \begin{bmatrix} -1 & 1 & 2 \\ 0 & 0 & 1 \\ 0 & 0 & 0 \end{bmatrix}$$

The rank $\qquad \rho\,[A - 2I] = 2$

$\therefore$ nullity $\qquad \eta\,[A - 2I] = 3 - 2 = 1$

We can find only one linearly independent vector for the repeated eigen value $\lambda = 2$. It can be obtained by any non zero column of adj $[A - \lambda I]$. From eqn. (9.95) with $\lambda = 2$, we have,

$$\text{Adj}\,[A - 2I] = \begin{bmatrix} 0 & 0 & 1 \\ 0 & 0 & 1 \\ 0 & 0 & 0 \end{bmatrix}$$

$$\therefore \qquad m_2 = \begin{bmatrix} 1 \\ 1 \\ 0 \end{bmatrix}$$

The third eigen vector can be obtained by considering the differential of column 3 of eqn. (9.95), with respect to λ and putting $\lambda = 2$.

$$m_3 = \frac{d}{d\lambda}\begin{bmatrix} 1-2\,(2-\lambda) \\ -(1-\lambda) \\ (1-\lambda)\,(2-\lambda) \end{bmatrix}_{\lambda=2}$$

$$= \begin{bmatrix} 2 \\ 1 \\ 2\lambda - 3 \end{bmatrix}_{\lambda=2}$$

$$= \begin{bmatrix} 2 \\ 1 \\ 1 \end{bmatrix}$$

The modal matrix is given by

$$M = \begin{bmatrix} 1 & 1 & 2 \\ 0 & 1 & 1 \\ 0 & 0 & 1 \end{bmatrix}$$

$$M^{-1} = \begin{bmatrix} 1 & -1 & -1 \\ 0 & 1 & -1 \\ 0 & 0 & 1 \end{bmatrix}$$

$$J = M^{-1}AM = \begin{bmatrix} 1 & -1 & -1 \\ 0 & 1 & -1 \\ 0 & 0 & 1 \end{bmatrix} \begin{bmatrix} 1 & 1 & 2 \\ 0 & 2 & 1 \\ 0 & 0 & 2 \end{bmatrix} \begin{bmatrix} 1 & 1 & 2 \\ 0 & 1 & 1 \\ 0 & 0 & 1 \end{bmatrix}$$

$$= \begin{bmatrix} 1 & 0 & 0 \\ 0 & 2 & 1 \\ 0 & 0 & 2 \end{bmatrix}$$

This is the required result.

Example 9.11

Consider the matrix,

$$A = \begin{bmatrix} 1 & 1 & 2 \\ 0 & 2 & 0 \\ 0 & 0 & 2 \end{bmatrix}$$

Obtain the Jordon form of the matrix.

Solution :

The eigen values of A are,

$$\lambda_1 = 1, \ \lambda_2 = \lambda_3 = 2$$

The adj $(A - \lambda i)$ is,

$$\begin{bmatrix} (2-\lambda)^2 & -(2-\lambda) & -2(2-\lambda) \\ 0 & (1-\lambda)\ (2-\lambda) & -(1-\lambda) \\ 0 & 0 & (1-\lambda)\ (2-\lambda) \end{bmatrix}$$

The eigen vector of $\lambda_1 = 1$ is any non zero column of Adj $(A - \lambda_i I)$.

$$\text{Adj}\ (A - I) = \begin{bmatrix} -1 & 1 & 2 \\ 0 & 0 & 0 \\ 0 & 0 & 0 \end{bmatrix}$$

The three columns of (A – I) are linearly dependent.

$$\therefore \qquad m_1 = \begin{bmatrix} 1 \\ 0 \\ 0 \end{bmatrix}$$

For the repeated eigen value $\lambda_2 = \lambda_3 = 2$

$$(A - 2I) = \begin{bmatrix} -1 & 1 & 2 \\ 0 & 0 & 0 \\ 0 & 0 & 0 \end{bmatrix}$$

The rank $\rho(A - 2I) = 1$

and nullity $\eta(A - 2I) = 3 - 1 = 2$

Hence we can find two linearly independent eigen vectors for the repeated eigen value $\lambda = 2$. The two eigen vectors can be found by considering the equation

$$(A - \lambda I)\, X = 0 \text{ with } \lambda = 2$$

$$\therefore \qquad \begin{bmatrix} -1 & 1 & 2 \\ 0 & 0 & 0 \\ 0 & 0 & 0 \end{bmatrix} \begin{bmatrix} x_1 \\ x_2 \\ x_3 \end{bmatrix} = \begin{bmatrix} 0 \\ 0 \\ 0 \end{bmatrix}$$

or $\qquad -x_1 + x_2 + 2x_3 = 0$

Choose x_1, x_2 and x_3 to satisfy this equations. Since there is only one equation and 3 variables, we can choose two variables arbitrarily.

We have $\qquad x_1 = x_2 + 2x_3$

One solution is obtained by taking $x_1 = 0$, $x_2 = 2$ and $x_3 = -1$. Another linearly independent solution is obtained by taking

$$x_1 = 1,\ x_2 = 1 \text{ and } x_3 = 0$$

$$\therefore \qquad m_2 = \begin{bmatrix} 0 \\ 2 \\ -1 \end{bmatrix} \text{ and } m_3 = \begin{bmatrix} 1 \\ 1 \\ 0 \end{bmatrix}$$

The modal matrix is,

$$M = \begin{bmatrix} 1 & 0 & 1 \\ 0 & 2 & 1 \\ 0 & -1 & 0 \end{bmatrix}$$

$$M^{-1} = \begin{bmatrix} 1 & -1 & -2 \\ 0 & 0 & -1 \\ 0 & 1 & 2 \end{bmatrix}$$

The Jordon matrix is given by,

$$J = M^{-1}AM$$

$$= \begin{bmatrix} 1 & -1 & -2 \\ 0 & 0 & -1 \\ 0 & 1 & 2 \end{bmatrix} \begin{bmatrix} 1 & 1 & 2 \\ 0 & 2 & 0 \\ 0 & 0 & 2 \end{bmatrix} \begin{bmatrix} 1 & 0 & 1 \\ 0 & 2 & 1 \\ 0 & -1 & 0 \end{bmatrix}$$

$$= \begin{bmatrix} 1 & 0 & 0 \\ 0 & 2 & 0 \\ 0 & 0 & 2 \end{bmatrix}$$

Here we observe that the Jordon matrix is purely a diagonal matrix eventhough a root is repeated.

The diagonolisation of the matrix A, when it is in the companion form deserves special attention.

(a) All the roots of A are distinct and are given by $\lambda_1, \lambda_2, \ldots \lambda_n$

In this case the modal matrix can be shown to be equal to,

$$M = \begin{bmatrix} 1 & 1 & \cdots & 1 \\ \lambda_1 & \lambda_2 & & \lambda_n \\ \lambda_1^2 & \lambda_2^2 & & \lambda_n^2 \\ \vdots & \vdots & & \vdots \\ \lambda_1^{n-1} & \lambda_2^{n-1} & & \lambda_n^{n-1} \end{bmatrix}$$

This matrix is a special matrix and is known as Vander Monde Matrix.

(b) When some roots of A are repeated. Let us consider an example in which the order of the matrix A is 6 and the roots are

$$\lambda_1, \lambda_1, \lambda_1, \lambda_2, \lambda_2, \lambda_3$$

When the matrix A is in companion form, for any repeated root of A, we can find only one linearly independent eigenvector and the other eigenvectors corresponding to this root are generalised eigenvectors. The modal matrix in this case can be obtained as,

$$M = \begin{bmatrix} 1 & 0 & 0 & 1 & 0 & 1 \\ \lambda_1 & 1 & 0 & \lambda_2 & 1 & \lambda_3 \\ \lambda_1^2 & 2\lambda_1 & 1 & \lambda_2^2 & 2\lambda_2 & \lambda_3^2 \\ \vdots & \vdots & \vdots & \vdots & \vdots & \vdots \\ \lambda_1^{n-1} & \frac{d}{d\lambda_1}\left(\lambda_1^{n-1}\right) & \frac{1}{2!}\frac{d^2}{d\lambda_1^2}\left(\lambda_1^{n-1}\right) & \lambda_2^{n-1} & \frac{d}{d\lambda_2}\left(\lambda_2^{n-1}\right) & \lambda_3^{n-1} \end{bmatrix}$$

If the root λ_j is repeated r times, the qth eigen vector $(q \le r)$ corresponding to this eigen value is given by

$$\frac{1}{(q-1)}\frac{d^{q-1}}{d\lambda_j^{q-1}} \quad [1 \quad \lambda_j \quad \lambda_j^2 \quad \ldots\ldots \quad \lambda_j^{n-1}]^T$$

The model matrix thus obtained is known as the modified Vander Monde matrix.

9.7 Solution of State Equation

The state equation is given by,

$$\dot{X}(t) = AX(t) + bu(t) \quad .....(9.96)$$

With the initial condition $X(o) = X_o$

We can write eqn. (9.96) as,

$$\dot{X}(t) - AX(t) = bu(t) \quad .....(9.97)$$

Multiplying both sides of Eqn. (9.97) by e^{-At}, we have

$$e^{-At}[\dot{X}(t) - AX(t)] = e^{-At} bu(t) \quad .....(9.98)$$

Consider $$\frac{d}{dt}[e^{-At} X(t)] = e^{-At}\dot{X}(t) - Ae^{-At} X(t)$$

$$= e^{-At}(\dot{X}(t) - AX(t)] \quad .....(9.99)$$

From eqn. (9.98) and (9.99), we can write,

$$\frac{d}{dt}[e^{-At} X(t)] = e^{-At} bu(t) \quad .....(9.100)$$

Integrating both sides of eqn. (9.100) between the limits 0 to t, we get

$$e^{-At} X(t) \Big|_0^t = \int_0^t e^{-A\tau} b\, u(\tau)\, d\tau$$

$$e^{-At} X(t) - X_0 = \int_0^t e^{-A\tau} b\, u(\tau)\, d\tau$$

$\therefore$ $$X(t) = e^{At} X_0 + e^{At}\int_0^t e^{-A\tau} b\, u(\tau)\, d\tau$$

or $$X(t) = e^{At} X_0 + \int_0^t e^{A(t-\tau)} b\, u(\tau)\, d\tau. \quad .....(9.101)$$

The first term on the right hand side of eqn. (9.101) is the homogeneous solution of eqn. (9.96) and the second term is the forced solution. The matrix exponential e^{At} is defined by the infinite series,

$$e^{At} = I + At + \frac{A^2 t^2}{2!} + ... + \frac{A^j t^j}{j!} + ...$$

Eqn. (9.101) can be generalised to any initial state at $t = t_0$ rather than $t = 0$, as

$$X(t) = e^{A(t-t_0)} X(t_0) + \int_{t_0}^t e^{A(t-\tau)} b\, u(\tau)\, d\tau. \quad .....(9.102)$$

Let us concentrate on the homogenous solution.

If u(t) = 0, we have

$$X(t) = e^{At} X_o \qquad .....(9.103)$$

This equation gives the relation relation between the initial state X_o and the state at any time *t*. The transition from the state X_o to X(t) is carried out by the matrix exponential e^{At}. Hence this matrix function is known as the state transition matrix (STM). If the STM is known for a given system the response to any input can be obtained by using eqn. (9.101). This is a very important concept in the state space analysis of any system.

9.7.1 Properties of State Transition Matrix

Let the state transition matrix be denoted by,

$$\phi(t) = e^{At}$$

Some useful properties of STM are listed below.

1. $\phi(0) = e^{Ao} = I;$ I is a unit matirx
2. $\phi^{-1}(t) = [e^{At}]^{-1} = e^{-At} = \phi(-t)$
3. $\phi(t_1 + t_2) = e^{A(t1 + t2)} = e^{At}{}_1 . e^{At}{}_2$
   $= \phi(t_1)\, \phi(t_2) = \phi(t_2)\, \phi(t_1)$

The solution of the state equation can be written in terms of the function $\phi(t)$ as,

$$X(t) = \phi(t) X_0 + \int_0^t \phi(t - \tau)\, b\, u(\tau)\, d\tau \qquad .....(9.104)$$

In the solution of state equation, the STM plays an important role and hence we must find ways of computing this matrix.

9.7.2 Methods of Computing State Transition Matrix

There are several methods of computing the matrix exponential e^{At}. Let us consider some of them.

(a) Laplace transform method

Let $\dot{X} = AX; \quad X(0) = X_0 \qquad .....(9.105)$

Taking Laplace transform of eqn. (9.105), we have,

$$(s X(s) - X_0) = A X(s)$$

This equation can be written as,

$$[sI - A] X(s) = X_0$$

$$\therefore \quad X(s) = (sI - A)^{-1} X_0 \qquad .....(9.106)$$

Taking inverve Laplace transform of eqn. (9.106) we get,

$$X(t) = \mathcal{L}^{-1} (sI - A)^{-1} X_0 \qquad .....(9.107)$$

Comparing eqn. (9.107) with eqn. (9.103), it is easy to see that,

$$e^{At} = \mathcal{L}^{-1} [(sI - A)^{-1}] \qquad .....(9.108)$$

Example 9.12

Find the homogenous solution of the system,

$$\dot{X} = \begin{bmatrix} 0 & 1 \\ -2 & -3 \end{bmatrix} X ; \qquad X_0 = \begin{bmatrix} 1 \\ 0 \end{bmatrix}$$

Solution

The solution of the given system is given by,

$$X(t) = e^{At}\, X_0$$

Let us compute the state transition matrix e^{At} using Laplace transform method.

$$e^{At} = \mathcal{L}^{-1}\,[(sI - A)^{-1}]$$

$$(sI - A) = \begin{bmatrix} s & 0 \\ 0 & s \end{bmatrix} - \begin{bmatrix} 0 & 1 \\ -2 & -3 \end{bmatrix}$$

$$= \begin{bmatrix} s & -1 \\ 2 & s+3 \end{bmatrix}$$

$$(sI - A)^{-1} = \frac{1}{s^2 + 3s + 2} \begin{bmatrix} s+3 & 1 \\ -2 & s \end{bmatrix}$$

$$= \begin{bmatrix} \dfrac{s+3}{(s+1)\,(s+2)} & \dfrac{1}{(s+1)\,(s+2)} \\ \dfrac{-2}{(s+1)\,(s+2)} & \dfrac{s}{(s+1)\,(s+2)} \end{bmatrix}$$

$$e^{At} = \mathcal{L}^{-1}\,[sI - A]^{-1}$$

$$= \begin{bmatrix} 2e^{-t} - e^{-2t} & e^{-t} - e^{-2t} \\ -2e^{-t} + 2e^{-2t} & -e^{-t} + 2e^{-2t} \end{bmatrix}$$

The homogeneous solution of the state equation is given by,

$$X(t) = e^{At}\, X_0$$

$$= \begin{bmatrix} 2e^{-t} - e^{-2t} & e^{-t} - e^{-2t} \\ -2e^{-t} + 2e^{-2t} & -e^{-t} + 2e^{-2t} \end{bmatrix} \begin{bmatrix} 1 \\ 0 \end{bmatrix}$$

$$= \begin{bmatrix} 2e^{-t} - e^{-2t} \\ -2e^{-t} - 2e^{-2t} \end{bmatrix}$$

The solution can be obtained in the Laplace transfer domain itself and the time domain solution can be obtained by finding the Laplace inverse. This obviates the need for finding Laplace inverse of all the elements of e^{At}, Thus

$$X(s) = (sI - A)^{-1} X_o$$

$$X(s) = \begin{bmatrix} \frac{s+3}{(s+1)(s+2)} & \frac{1}{(s+1)(s+2)} \\ \frac{-2}{(s+1)(s+2)} & \frac{s}{(s+1)(s+2)} \end{bmatrix} \begin{bmatrix} 1 \\ 0 \end{bmatrix}$$

$$= \begin{bmatrix} \frac{s+3}{(s+1)(s+2)} \\ \frac{-2}{(s+1)(s+2)} \end{bmatrix}$$

$$\therefore \quad X(t) = \begin{bmatrix} 2e^{-t} - e^{-2t} \\ -2e^{-t} + 2e^{-2t} \end{bmatrix}$$

(b) Cayley - Hamilton Technique

Any function of a matrix f(A) which can be expressed as an infinite series in powers of A can be obtained by considering a polynomial function $g(\lambda)$ of order $(n - 1)$, using Cayley Hamilton theorem, Here n is the order of the matrix A.

Cayley - Hamilton theorem states that any matrix satisfies its own characteristic equation.

The characteristic equation of a matrix A is given by,

$$q(\lambda) = |\lambda I - A| = \lambda^n + a_1 \lambda^{n-1} + ... + a_{n-1} \lambda + a_n = 0 \qquad .....(9.109)$$

The Cayley Hamilton theorem says that,

$$q(A) = A^n + a_1 A^{n-1} + ... + a_{n-1} A + a_n I = 0 \qquad .....(9.110)$$

Let $$f(A) = b_o I + b_1 A + b_2 A^2 + ... + b_n A^n + b_{n+1} A^{n+1} + ... \qquad .....(9.111)$$

Consider a scalar function,

$$f(\lambda) = b_o + b_1 \lambda + b_2 \lambda^2 + ... + b_n \lambda^n + b_{n+1} \lambda^{n+1} + ... \qquad .....(9.112)$$

Let $f(\lambda)$ be divided by the characteristic polynomial $q(\lambda)$ given by eqn. (9.109). Let the remainder polynomial be $g(\lambda)$. Since $q(\lambda)$ is of order n, the remainder polynomial $g(\lambda)$ will be of order $(n - 1)$ and let $Q(\lambda)$ be the quotient polynomial. Thus,

$$f(\lambda) = Q(\lambda)\, q(\lambda) + g(\lambda) \qquad .....(9.113)$$

Let $$g(\lambda) = \alpha_0 + \alpha_1 \lambda + \alpha_1 \lambda^2 + ... + \alpha_{n-1} \lambda^{n-1}$$

The function f(A) can be obtained from eqn. (9.113) by replacing λ with A

Thus $$f(A) = Q(A)\, q(A) + g(A) \qquad .....(9.114)$$

But $q(A) = 0$ by Cayley Hamilton theorem.

$\therefore$ $f(A) = g(A)$ .....(9.115)

$$= \alpha_0 + \alpha_1 A + \alpha_2 A^2 + ... + \alpha_{n-1} A^{n-1} \quad .....(9.116)$$

Since $q(\lambda_i) = 0$ for $i = 1, 2, ... n$

we have from eqn. (9.113),

$$f(\lambda_i) = g(\lambda_i) \text{ for } i = 1, 2, ... n \quad .....(9.117)$$

Eqn. (9.117) gives n equations for the n unknown coefficients $\alpha_0, \alpha_1, ... \alpha_{n-1}$ of $g(\lambda)$

We notice that g(A) is a polynomial of order (n – 1) only and hence f(A) which is of infinite order can be computed interms of a finite lower order polynomial.

To summarise, the procedure for finding a matrix function is :

(i) Find the eigen vlaues of A.

(ii) (a) If all the eigenvalues of A are distinct solve for the coefficients α_j of $g(\lambda)$ using

$$f(\lambda_i) = g(\lambda_i) \text{ for } i = 1, 2, ... n$$

(b) If some eigenvalues are repeated, the procedure is modified as shown below.

Let $\lambda = \lambda_k$ be repeated r times.

Then $$\left.\frac{d^j q(\lambda)}{d\lambda^j}\right|_{\lambda=\lambda_k} = 0 \text{ for } j = 0, 1, ... r - 1$$

Using this equation in eqn. (9.114), we get

$$\left.\frac{d^j f(\lambda)}{d\lambda^j}\right|_{\lambda=\lambda_k} = \left.\frac{d^j g(\lambda)}{d\lambda^j}\right|_{\lambda=\lambda_k} \quad \text{for } j = 0, 1, ...r - 1$$

This gives the required r equations corresponding to the repeated eigen value λ_k. Proceeding in a similar manner for other repeated eigenvalues and using eqn. (9.116) we get the necessary equations to evaluate,

$$\alpha_j, \text{ for } j = 0, 1, 2 ... n - 1$$

(iii) The matrix function f(A) can be computed using the relation,

$$f(A) = g(A)$$

The procedure is illustrated using a few examples.

Example 9.13

Find $$f(A) = A^4 + 2A^3$$

where $$A = \begin{bmatrix} 0 & 1 \\ -6 & -5 \end{bmatrix}$$

Solution :

The eigenvalues of A are obtained from,

$$q(\lambda) = \begin{vmatrix} -\lambda & 1 \\ -6 & -5-\lambda \end{vmatrix} = 0$$

$$\lambda(\lambda + 5) + 6 = 0$$

$$\lambda^2 + 5\lambda + 6 = 0$$

$$\text{or } \lambda = -3, -2$$

Since the order of A is 2,

$$g(\lambda) = \alpha_0 + \alpha_1 \lambda$$

Now $f(\lambda) = \lambda^4 + 2\lambda^3$

$$g(\lambda) = f(\lambda) \text{ for } \lambda = -3, -2$$

$$g(-3) = \alpha_0 - 3\,\alpha_1 = (-3)^4 + 2\,(-3)^3 = 27$$

$$g(-2) = \alpha_0 - 2\,\alpha_1 = (-2)^4 + 2\,(-2)^3 = 0$$

Solving for α_0 and α_1, we get

$$\alpha_0 = -54;\ \alpha_1 = -27$$

$\therefore$ $g(\lambda) = -54 - 27\lambda$

$$f(A) = g(A) = -54\,I - 27\,A$$

$$f(A) = \begin{bmatrix} -54 & 0 \\ 0 & -54 \end{bmatrix} - 27 \begin{bmatrix} 0 & 1 \\ -6 & -5 \end{bmatrix}$$

$$= \begin{bmatrix} -54 & -27 \\ 162 & 81 \end{bmatrix}$$

Example 9.14

For $A = \begin{bmatrix} 0 & 1 & 0 \\ 0 & 0 & 1 \\ -6 & -11 & -6 \end{bmatrix}$

Find $f(A) = e^{At}$

Solution :

The characteristic equation is given by

$$q(\lambda) = \begin{vmatrix} -\lambda & 1 & 0 \\ 0 & -\lambda & 1 \\ -6 & -11 & -6-\lambda \end{vmatrix} = 0$$

The eigenvalues are

$$\lambda = -1, -2, \text{ and } -3$$

$$f(\lambda) = e^{\lambda t}$$

Since A is of order 3, the remainder polynomial is

$$g(\lambda) = \alpha_0 + \alpha_1 \lambda + \alpha_2 \lambda^2$$

$$g(\lambda_i) = f(\lambda_i) \text{ for } i = 1, 2, 3$$

$$\alpha_0 - \alpha_1 + \alpha_2 = e^{-t}$$

$$\alpha_0 - 2\alpha_1 + 4\alpha_2 = e^{-2t}$$

$$\alpha_0 - 3\alpha_1 + 9\alpha_2 = e^{-3t}$$

Solving these equations for α_0, α_1 and α_2, we have,

$$\alpha_0 = 3e^{-t} - 3e^{-2t} + e^{-3t}$$

$$\alpha_1 = 2.5e^{-t} - 4e^{-2t} + \frac{3}{2} e^{-3t}$$

$$\alpha_2 = 0.5e^{-t} - e^{-2t} + 0.5\, e^{-3t}$$

$$\therefore \quad f(A) = e^{At} = g(A)$$

$$= \alpha_0 I + \alpha_1 A + \alpha_2 A^2$$

Substituting for A we get

$$e^{At} = \begin{bmatrix} \alpha_0 & \alpha_1 & \alpha_2 \\ -6\alpha_2 & \alpha_0 - 11\alpha_2 & \alpha_1 - 6\alpha_2 \\ -6\alpha_1 + 36\alpha_2 & -11\alpha_1 + 60\alpha_2 & \alpha_0 - 6\alpha_1 + 25\alpha_2 \end{bmatrix}$$

Substituting the values of α_0, α_1 and α_2, we get,

$$e^{At} = \begin{bmatrix} 3e^{-t} - 3e^{-2t} + e^{-3t} & 2.5e^{-t} - 4e^{-2t} + 1.5e^{-3t} & 0.5e^{-t} - e^{-2t} + 0.5e^{-3t} \\ -3e^{-t} + 6e^{-2t} + 3e^{-3t} & -2.5e^{-t} + 8e^{-2t} - 4.5e^{-3t} & -0.5e^{-t} + 2e^{-2t} - 1.5e^{-3t} \\ 3e^{-t} - 12e^{-2t} + 9e^{-3t} & 2.5e^{-t} - 16e^{-2t} + 13.5e^{-3t} & 0.5e^{-t} - 4e^{-2t} + 4.5e^{-3t} \end{bmatrix}$$

Example 9.15

Find e^{At} for $A = \begin{bmatrix} 0 & 1 \\ -4 & -4 \end{bmatrix}$

Solution :

The characteristic equation is

$$\lambda^2 + 4\lambda + 4 = 0$$

$$(\lambda + 2)^2 = 0 \text{ or } \lambda = -2, -2$$

We have $\quad f(\lambda) = e^{\lambda t}$ and $g(\lambda) = \alpha_0 + \alpha_1 \lambda$

Since the eigen value is repeated,

$$f(\lambda_i) = g(\lambda_i)$$

and $$\frac{df(\lambda_i)}{d\lambda_i} = \frac{dg(\lambda_i)}{d\lambda_i}$$

$$f(-2) = g(-2)$$

$$\alpha_0 - 2\alpha_1 = e^{-2t}$$

and $$t\, e^{\lambda t}\Big|_{\lambda=-2} = \alpha_1$$

or $$t\, e^{-2t} = \alpha_1$$

$\therefore$ $$\alpha_0 = e^{-2t} + 2t\, e^{-2t}$$

$\therefore$ $$f(A) = e^{At} = g(A) = \alpha_0 I + \alpha_1 A$$

$$= \begin{bmatrix} \alpha_0 & 0 \\ 0 & \alpha_0 \end{bmatrix} + \alpha_1 \begin{bmatrix} 0 & 1 \\ -4 & -4 \end{bmatrix}$$

$$= \begin{bmatrix} \alpha_0 & \alpha_1 \\ -4\alpha_1 & \alpha_0 - 4\alpha_1 \end{bmatrix}$$

$$e^{At} = \begin{bmatrix} e^{-2t} + 2t\, e^{-2t} & t\, e^{-2t} \\ -4t\, e^{-2t} & e^{-2t} - 2t\, e^{-2t} \end{bmatrix}$$

Example 9.16

Obtain the solution of the state equation,

$$\dot{X} = \begin{bmatrix} 0 & 1 \\ -1 & -2 \end{bmatrix} X + \begin{bmatrix} 0 \\ 1 \end{bmatrix} u$$

$X_0 = [0 \quad 1]^T$ and u is a unit step input.

Solution :

First the STM is computed. Using Laplace transform method

$$(sI - A)^{-1} = \begin{bmatrix} s & -1 \\ 1 & s+2 \end{bmatrix}^{-1}$$

$$= \frac{1}{s^2 + 2s + 1} \begin{bmatrix} s+2 & 1 \\ -1 & s \end{bmatrix}$$

Consider the state equation,

$$\dot{X} = AX + bu$$

Taking Laplace transform of this equation,

$$sX(s) - X_0 = AX(s) + b\,U(s)$$

or $$(sI - A)\,X(s) = X_0 + b\,U(s)$$

$\therefore$ $$X(s) = (sI - A)^{-1} X_0 + (sI - A)^{-1} b\,U(s)$$

Substituting the relevent values,

$$X(s) = \frac{1}{s^2 + 2s + 1}\begin{bmatrix} s+2 & 1 \\ -2 & s \end{bmatrix}\begin{bmatrix} 0 \\ 1 \end{bmatrix} + \begin{bmatrix} \dfrac{s+2}{s^2+2s+1} & \dfrac{1}{s^2+2s+1} \\ \dfrac{-1}{s^2+2s+1} & \dfrac{s}{s^2+2s+1} \end{bmatrix}\begin{bmatrix} 0 \\ 1 \end{bmatrix}\frac{1}{s}$$

$$= \begin{bmatrix} \dfrac{1}{(s+1)^2} \\ \dfrac{s}{(s+1)^2} \end{bmatrix} + \begin{bmatrix} \dfrac{1}{s(s+1)^2} \\ \dfrac{1}{(s+1)^2} \end{bmatrix}$$

$$= \begin{bmatrix} \dfrac{1}{s(s+1)} \\ \dfrac{1}{s+1} \end{bmatrix}$$

Taking inverse Laplace transform, we get,

$$X(t) = \begin{bmatrix} 1 - e^{-t} \\ e^{-t} \end{bmatrix}$$

Example 9.17

Find the state response of the system,

$$\dot{X} = \begin{bmatrix} 0 & 1 & 0 \\ 0 & 0 & 1 \\ -6 & -11 & -6 \end{bmatrix} X + \begin{bmatrix} 0 \\ 0 \\ 1 \end{bmatrix} u(t)$$

For $X_0 = [1 \;\; 0 \;\; 0]^T$ and a unit step input.

Solution :

We have already calculated the STM for the A matrix of this example, in Ex. 9.12. Using this result in the response of the system.

$$X(t) = e^{At} X_0 + \int_0^t e^{A(t-\tau)}\, b\, u(\tau)\, d\tau$$

From Ex. (9.14),
$$e^{At}X_0 = \begin{bmatrix} 3e^{-t} - 3e^{-2t} + e^{-3t} \\ -3e^{-t} + 6e^{-2t} + 3e^{-3t} \\ 3e^{-t} - 12e^{-2t} + 9e^{-3t} \end{bmatrix}$$

and
$$e^{A(t-\tau)}bu = \begin{bmatrix} 0.5e^{-(t-\tau)} - e^{-2(t-\tau)} + 0.5e^{-3(t-\tau)} \\ -0.5e^{-(t-\tau)} + 2e^{-2(t-\tau)} - 1.5e^{-3(t-\tau)} \\ 0.5e^{-(t-\tau)} - 4e^{-2(t-\tau)} + 4.5e^{-3(t-\tau)} \end{bmatrix}$$

Integrating between the limits 0 to t, we get

$$\int_0^t e^{A(t-\tau)}\, bu(\tau)d\tau = \begin{bmatrix} 0.5e^{-(t-\tau)} - \dfrac{e^{-2(t-\tau)}}{2} + 0.5\dfrac{e^{-3(t-\tau)}}{3} \\ -0.5e^{-(t-\tau)} + e^{-2(t-\tau)} - 0.5e^{-3(t-\tau)} \\ 0.5e^{-(t-\tau)} - 2e^{-2(t-\tau)} + 1.5e^{-3(t-\tau)} \end{bmatrix}_0^t$$

$$= \begin{bmatrix} \dfrac{1}{6} - 0.5e^{-t} + 0.5e^{-2t} - \dfrac{1}{6}e^{-3t} \\ 0.5e^{-t} - e^{-2t} + 0.5e^{-3t} \\ -0.5e^{-t} + 2e^{-2t} - 1.5e^{-3t} \end{bmatrix}$$

$$\therefore \quad X(t) = e^{At}x_0 + \int_0^t e^{A(t-\tau)}\, b\, u\,(t)\, d\tau$$

$$= \begin{bmatrix} 3e^{-t} - 3e^{-2t} + e^{-3t} \\ -3e^{-t} + 6e^{-2t} + 3e^{-3t} \\ 3e^{-t} - 12e^{-2t} + 9e^{-3t} \end{bmatrix} + \begin{bmatrix} \dfrac{1}{6} - 0.5e^{-t} + 0.5e^{-2t} - \dfrac{1}{6}e^{-3t} \\ 0.5e^{-t} - e^{-2t} + 0.5e^{-3t} \\ -0.5e^{-t} + 2e^{-2t} - 1.5e^{-3t} \end{bmatrix}$$

$$= \begin{bmatrix} \dfrac{1}{6} + 2.5e^{-t} - 2.5e^{-2t} + \dfrac{5}{6}e^{-3t} \\ -2.5e^{-t} + 5e^{-2t} + 3.5e^{-3t} \\ 2.5e^{-t} - 10e^{-2t} + 7.5e^{-3t} \end{bmatrix}$$

Example 9.18

Obtain the solution of the following state equation by first obtaining the canmical form.

$$\dot{X} = \begin{bmatrix} 1 & 0 & 2 \\ 0 & 2 & 1 \\ 0 & 0 & 3 \end{bmatrix} X + \begin{bmatrix} 0 \\ 0 \\ 1 \end{bmatrix} U$$

With the initial condition $X(0) = [1 \;\; 0 \; -1]^T$

Solution :

First let us obtain the canmical form of the given state equation. The system matrix A is given by,

$$A = \begin{bmatrix} 1 & 0 & 2 \\ 0 & 2 & 1 \\ 0 & 0 & 3 \end{bmatrix}$$

Let $\qquad X = PZ$

Where P is the modal matrix .

Modal matrix for this matrix A was obtained in Ex 9.9. Using the result, we have

$$P = \begin{bmatrix} 1 & 0 & 1 \\ 0 & 1 & 1 \\ 0 & 0 & 1 \end{bmatrix}$$

and

$$P^{-1} = \begin{bmatrix} 1 & 0 & -1 \\ 0 & 1 & -1 \\ 0 & 0 & 1 \end{bmatrix}$$

Using the transformation, we have

$$\dot{Z} = P^{-1} A P Z + P^{-1} bu$$

$$= \begin{bmatrix} 1 & 0 & 0 \\ 0 & 2 & 0 \\ 0 & 0 & 3 \end{bmatrix} Z + \begin{bmatrix} 1 & 0 & -1 \\ 0 & 1 & -1 \\ 0 & 0 & 1 \end{bmatrix} \begin{bmatrix} 0 \\ 0 \\ 1 \end{bmatrix} u$$

($\because$ P^{-1} AP gives the diagonal matrix with the eigen values 1, 2 and 3 on the diagonal)

$$\dot{Z} = \begin{bmatrix} 1 & 0 & 0 \\ 0 & 2 & 0 \\ 0 & 0 & 3 \end{bmatrix} Z + \begin{bmatrix} -1 \\ -1 \\ 1 \end{bmatrix} U$$

and $Z(0) = P^{-1} X(0)$

$$= \begin{bmatrix} 1 & 0 & -1 \\ 0 & 1 & -1 \\ 0 & 0 & 1 \end{bmatrix} \begin{bmatrix} 1 \\ 0 \\ 0 \end{bmatrix}$$

$$= \begin{bmatrix} 1 \\ 0 \\ 0 \end{bmatrix}$$

Solution of the above equation is given by,

$$Z(t) = e^{Jt} Z(o) + \int_0^t e^{J(t-\tau)} \bar{b}\, U(\tau) d\tau$$

where $J = \begin{bmatrix} 1 & 0 & 0 \\ 0 & 2 & 0 \\ 0 & 0 & 3 \end{bmatrix}$ and $\bar{b} = \begin{bmatrix} -1 \\ -1 \\ 1 \end{bmatrix}$

The state transmition matrix e^{Jt} can be easily obtained since J is in diagonal form. It is given by,

$$e^{Jt} = \begin{bmatrix} e^t & 0 & 0 \\ 0 & e^{2t} & 0 \\ 0 & 0 & e^{3t} \end{bmatrix}$$

Thus

$$Z(t) = \begin{bmatrix} e^t & 0 & 0 \\ 0 & e^{2t} & 0 \\ 0 & 0 & e^{3t} \end{bmatrix} \begin{bmatrix} 1 \\ 0 \\ 0 \end{bmatrix} + \int_0^t \begin{bmatrix} e^{t-\tau} & 0 & 0 \\ 0 & e^{2(t-\tau)} & 0 \\ 0 & 0 & e^{3(t-\tau)} \end{bmatrix} \begin{bmatrix} -1 \\ -1 \\ 1 \end{bmatrix} d\tau$$

$$Z(t) = \begin{bmatrix} e^t \\ 0 \\ 0 \end{bmatrix} + \int_0^t \begin{bmatrix} -e^{(t-\tau)} \\ -e^{2(t-\tau)} \\ e^{3(t-\tau)} \end{bmatrix} d\tau$$

$$Z(t) = \begin{bmatrix} e^t \\ 0 \\ 0 \end{bmatrix} + \begin{bmatrix} e^{(t-\tau)} \\ \dfrac{e^{2(t-\tau)}}{2} \\ \dfrac{-e^{3(t-\tau)}}{3} \end{bmatrix}_0^t$$

$$Z(t) = \begin{bmatrix} e^t \\ 0 \\ 0 \end{bmatrix} + \begin{bmatrix} 1-e^t \\ \dfrac{1}{2}-\dfrac{e^{2t}}{2} \\ -\dfrac{1}{3}+\dfrac{e^{3t}}{3} \end{bmatrix}$$

$$= \begin{bmatrix} 1 \\ \dfrac{1}{2}(1-e^{2t}) \\ \dfrac{1}{3}(-1+e^{3t}) \end{bmatrix}$$

$\because$ $X(t) = PZ(t)$,

We have $$X(t) = \begin{bmatrix} 1 & 0 & 1 \\ 0 & 1 & 1 \\ 0 & 0 & 1 \end{bmatrix} \begin{bmatrix} 1 \\ \dfrac{1}{2}(1-e^{2t}) \\ \dfrac{1}{3}(e^{3t}-1) \end{bmatrix}$$

$$X(t) = \begin{bmatrix} \dfrac{e^{3t}}{3}+\dfrac{2}{3} \\ \dfrac{1}{6}-\dfrac{1}{2}e^{2t}+\dfrac{1}{3}e^{3t} \\ \dfrac{1}{3}(e^{3t}-1) \end{bmatrix}$$

9.8 Qualitative Analysis of Control Systems

Sofar we have discussed,

(a) how to model a given physical system and

(b) solution of such a system for a given input, i.e., quantitative analysis of the system.

We will now discuss some fundamental concepts about the control of these systems, i.e., qualitative aspects of these systems, when a controller is to be designed to obtain a desired response.

There are two basic questions to be answered before we design a suitable controller for a given system, namely,

(i) can we transfer the system from any initial state to any desired final state in finite time by applying a suitable unconstrained control.

(ii) by measuring the output for a finite length of time, can we determine the initial state of the system.

These aspects were first introduced by Kalman and are defined as controllability and observability of the system respectively. These aspects play a very important role in the design of a controller for a given system. In this section we will only give elementary treatment of these aspects of the system.

9.8.1 State Controllability

Definition

A system is said to be state controllable at $t = t_0$ if it is possible to transfer the initial state $X(t_0)$ to any final state $X(t_f)$ in finite time, using an unconstrained control signal u(t). If every state is controllable, the system is said to be completely state controllable.

Kalman's test for controllability

Consider the dynamical system described by the state equation,

$$\dot{X} = AX + Bu \qquad .....(9.118)$$

where X is a n-dimensional state vector

u is an scalar control signal

A is an $n \times n$ system matrix

B is an $n \times 1$ input matrix

Without loss of generality, let us consider the initial time $t_0 = 0$ and the final desired state to be the origin of the state space.

The solution of the state eqn. (9.118) is given by,

$$X(t) = e^{At} X(0) + \int_0^t e^{A(t-\tau)} B\, u(\tau)\, d\tau$$

Since the final state at $t = t_f$ i.e., $X(t_f) = 0$, we have,

$$X(t_f) = 0 = e^{At_f}\ X(0) + \int_0^{t_f} e^{A(t_f-\tau)} B\ u(\tau)\ d\tau$$

or
$$X(0) = -e^{-At_f} \int_0^{t_f} e^{A(t_f-\tau)}\ B\ u(\tau)\ d\tau$$

$$X(0) = -\int_0^{t_f} e^{-A\tau} B\ u(\tau)\ d\tau \qquad .....(9.119)$$

If we are able to obtain a u(t) for any given X(0) from eqn. (9.119), the system is controllable. Using eqn. (9.116) with $f(A) = e^{-At}$, we can write $e^{-A\tau}$ as,

$$e^{-A\tau} = \sum_{k=0}^{n-1} \alpha_k(\tau)A^k \qquad .....(9.120)$$

Substituting eqn. (9.120) in eqn. (9.119), we have,

$$X(0) \quad = -\int_0^{t_f} \sum_{k=0}^{n-1} \alpha_k(\tau)A^k\ B\ u(\tau)\ d\tau$$

$$= -\sum_{k=0}^{n-1} A^k B \int_0^{t_f} \alpha_k\ u(\tau)\ d\tau \qquad .....(9.121)$$

Let
$$\int_0^{t_f} \alpha_k\ u(\tau)\ d\tau = \beta_k$$

Then
$$X(0) = -\sum_{k=0}^{n-1} A^k B\ \beta_k \qquad .....(9.122)$$

Eqn. (9.122) can be written in the matrix form,

$$X(0) = \left[B \vdots AB \vdots A^2B \vdots ..... \vdots A^{n-1}B\right] \begin{bmatrix} \beta_0 \\ \beta_1 \\ \beta_2 \\ \vdots \\ \beta_{n-1} \end{bmatrix} \qquad .....(9.123)$$

A control u(t) exists, which transfers any given initial state X(0) to the origin of state space, if the vector $[\beta_0, \beta_1, ... \beta_{n-1}]^T$ can be obtained from eqn. (9.123). It requires that the $n \times n$ matrix

$$\left[B \vdots AB \vdots A^2B \vdots ... \vdots A^{n-1}B\right]$$

be non singular, i.e., its rank be equal to n.

This result can be extended to a more general case where the input vector is of dimension r. In the equation

$$\dot{X} = AX + Bu \qquad .....(9.124)$$

u is an m-vector and

B is an $n \times m$ matrix

The controllability condition in this case can be stated as :

The $(n \times nm)$ matrix

$$Q_C = \left[B \vdots AB \vdots |A^{n-1}B\right] \qquad .....(9.125)$$

has a rank n.

The matrix Q_C is known as the controllability matrix. For complete state controllability, therefore, all the columns of Q_C must be linearly independent. This test for deciding the controllability is known as Kalman's test for controllability.

Gilbert's test for Controllability

Let us now consider an alternate test for examining the controllability of a given system. Let us suppose that the eigenvalues of A in eqn. (9.124) are distinct. It is possible to transform the A matrix into a diagonal matrix using a transformation matrix P. As already discussed in section 9.6, if

$$X = PZ$$

eqn. (9.124) can be transformed to

$$\dot{Z} = P^{-1} APZ + P^{-1} Bu \qquad .....(9.126)$$

where P matrix is the modal matrix and P^{-1} AP is a diagonal matrix with eigen values on the diagonal. Eqns. (9.126) are first order uncoupled differential equations. The kth equation is given by

$$\dot{Z}_k = \lambda_k Z_k + \alpha_{k1} u_1 + \alpha_{k2} u_2 + ... + \alpha_{km} u_m \qquad .....(9.127)$$

where λ_k is the kth eigenvalue of A

and $\alpha_{k1}, \alpha_{k2}, ... \alpha_{km}$, are the kth row elements of P^{-1} B

If all the elements of kth row of P^{-1} B, i.e.,

$$\alpha_{k1} = \alpha_{k2} = ... = \alpha_{km} = 0$$

then no control signal exists in eqn. (9.127) and hence the state variable Z_k cannot be controlled.

Hence the condition for complete state controllability is that none of the rows of $P^{-1}B$ in eqn. (9.126) should contain all zeros.

The above treatment can be extended to the case where the matrix A has repeated eigen values. In this case the matrix A can be transformed to Jordon's cononical form. If P is the modal matrix which transforms A into its Jordon form, let

$$X = PZ$$

and hence

$$\begin{aligned}\dot{Z} &= P^{-1} APZ + P^{-1} Bu \\ &= JZ + P^{-1} Bu \qquad .....(9.128)\end{aligned}$$

For a 7th order system with eigen values λ_1, λ_1, λ_2, λ_2, λ_2, λ_3, and λ_4, J is of the form,

$$J = \begin{bmatrix} \lambda_1 & 1 & 0 & 0 & 0 & 0 & 0 \\ 0 & \lambda_1 & 0 & 0 & 0 & 0 & 0 \\ 0 & 0 & \lambda_2 & 1 & 0 & 0 & 0 \\ 0 & 0 & 0 & \lambda_2 & 1 & 0 & 0 \\ 0 & 0 & 0 & 0 & \lambda_2 & 0 & 0 \\ 0 & 0 & 0 & 0 & 0 & \lambda_3 & 0 \\ 0 & 0 & 0 & 0 & 0 & 0 & \lambda_4 \end{bmatrix} \quad .....(9.129)$$

Each square block is called a Jordon block, as discussed earlier. J can thus be written as,

$$J = \begin{bmatrix} J_1 & & & \\ & J_2 & O & \\ & O & J_3 & \\ & & & J_4 \end{bmatrix}$$

where J_1 is of order 2

J_2 is of order 3

and J_3, J_4 are each of order 1

The condition for complete state controllability by this method can be stated as follows :

The system is completely state controllable if and only if,

1. No two Jordon blocks in eqn. (9.129), are associated with the same eigen value.
2. The elements of the row of P^{-1} B corresponding to the last row of any Jordon block of J are not all zero. If some of the Jordon blocks are of order 1, i.e., the corresponding eigen values are distinct, the rows in P^{-1} B corresponding to these eigen values must not contain all zeros.

   It is to be noted that application of Gilbert's method requires that the equations be put in the Jordon's form. Kalman's test can be applied to any representation of the given system.

Output Controllability

In practical systems, we may have to control the output rather than the states. In such cases, we can define output controllability. State controllability is neither necessary nor sufficient for controlling the output.

Consider the system

$$\dot{X} = AX + Bu$$
$$Y = CX + Du \quad .....(9.130)$$

X – is an n vector

u – is an m vector

Y – is a p vector

The above system is completely output controllable if it is possible to find an unconstrained control vector u(t) which will transfer any given initial output Y (t_0) to any final output Y (t_f) in finite time.

The test for complete controllability of the system in eqn. (9.130) is that the rank of the $[p \times (n + 1) m]$ matrix,

$$\left[CB \,\vdots\, CAB \,\vdots\, CA^2B \,\vdots\, ... \,\vdots\, CA^{n-1}B \,\vdots\, D\right] \quad .....(9.131)$$

be equal to p.

If in a particular system D = 0 i.e., the system output does not directly depend on the input, the rank of the matrix,

$$\left[CB \,\vdots\, CAB \,\vdots\, CA^2B \,\vdots\, ... \,\vdots\, CA^{n-1}B\right] \quad .....(9.132)$$

must be p.

Example 9.19

Test the following systems for controllability using (i) Kalman's test (ii) Gilbert's test.

(a) $A = \begin{bmatrix} 0 & 1 & 0 \\ 0 & 0 & 1 \\ -6 & -11 & -6 \end{bmatrix} \quad B = \begin{bmatrix} 0 \\ 0 \\ 1 \end{bmatrix}$ (b) $A = \begin{bmatrix} 1 & 0 & 2 \\ 0 & 2 & 1 \\ 0 & 0 & 3 \end{bmatrix} \quad B = \begin{bmatrix} 1 \\ -1 \\ 1 \end{bmatrix}$

(c) $A = \begin{bmatrix} 1 & 1 & 2 \\ 0 & 2 & 1 \\ 0 & 0 & 2 \end{bmatrix} \quad B = \begin{bmatrix} 1 \\ 1 \\ 0 \end{bmatrix}$ (d) $A = \begin{bmatrix} 1 & 1 & 2 \\ 0 & 2 & 0 \\ 0 & 0 & 2 \end{bmatrix} \quad B = \begin{bmatrix} 1 \\ -1 \\ 1 \end{bmatrix}$

Solution :

(a) $A = \begin{bmatrix} 0 & 1 & 0 \\ 0 & 0 & 1 \\ -6 & -11 & -6 \end{bmatrix} \quad B = \begin{bmatrix} 0 \\ 0 \\ 1 \end{bmatrix}$

(i) Kalman's test

The controllability matrix,

$$Q_C = \left[B \,\vdots\, AB \,\vdots\, A^2B\right]$$

$$= \begin{bmatrix} 0 & 0 & 1 \\ 0 & 1 & -6 \\ 1 & -6 & 25 \end{bmatrix}$$

The rank of Q_C is 3.

Hence the system is controllable.

(ii) Gilberts test

The eigen values are –1, –2, –3

The modal matrix P is given by,

$$P = \begin{bmatrix} 1 & 1 & 1 \\ \lambda_1 & \lambda_2 & \lambda_3 \\ \lambda_1^2 & \lambda_2^2 & \lambda_3^2 \end{bmatrix} = \begin{bmatrix} 1 & 1 & 1 \\ -1 & -2 & -3 \\ 1 & 4 & 9 \end{bmatrix}$$

$$P^{-1} = \begin{bmatrix} 3 & 2.5 & 0.5 \\ -3 & -4 & -1 \\ 1 & 1.5 & 0.5 \end{bmatrix}$$

$$P^{-1} B = \begin{bmatrix} 3 & 2.5 & 0.5 \\ -3 & -4 & -1 \\ 1 & 1.5 & 0.5 \end{bmatrix} \begin{bmatrix} 0 \\ 0 \\ 1 \end{bmatrix}$$

$$= \begin{bmatrix} 0.5 \\ -1 \\ 0.5 \end{bmatrix}$$

Since all the eigenvalues are distinct and none of the elements of $P^{-1} B$ is zero, the system is completely state controllable.

(b) $$A = \begin{bmatrix} 1 & 0 & 2 \\ 0 & 2 & 1 \\ 0 & 0 & 3 \end{bmatrix} \qquad B = \begin{bmatrix} 1 \\ -1 \\ 1 \end{bmatrix}$$

(i) Kalman's test

The controllability matrix,

$$Q_C = \begin{bmatrix} B \vdots AB \vdots A^2B \end{bmatrix}$$

$$= \begin{bmatrix} 1 & 3 & 9 \\ -1 & -1 & 1 \\ 1 & 3 & 9 \end{bmatrix}$$

The rank of this matrix is 2 and hence the system is not controllable.

(ii) Gilbert's method

From example 9.9, we have the eigen values to be 1, 2, and 3 and the modal matrix P to be,

$$P = \begin{bmatrix} 1 & 0 & 1 \\ 0 & 1 & 1 \\ 0 & 0 & 1 \end{bmatrix}$$

$$P^{-1} = \begin{bmatrix} 1 & 0 & -1 \\ 0 & 1 & -1 \\ 0 & 0 & 1 \end{bmatrix}$$

and $$P^{-1} B = \begin{bmatrix} 1 & 0 & -1 \\ 0 & 1 & -1 \\ 0 & 0 & 1 \end{bmatrix} \begin{bmatrix} 1 \\ -1 \\ 1 \end{bmatrix}$$

$$= \begin{bmatrix} 0 \\ -2 \\ 1 \end{bmatrix}$$

Since the first element in $P^{-1}B$ is zero, the system is not controllable.

(c) $$A = \begin{bmatrix} 1 & 1 & 2 \\ 0 & 2 & 1 \\ 0 & 0 & 2 \end{bmatrix} \qquad B = \begin{bmatrix} 1 \\ 1 \\ 0 \end{bmatrix}$$

(i) Kalman's test

The controllability matrix,

$$Q_C = \begin{bmatrix} 1 & 2 & 4 \\ 1 & 2 & 4 \\ 0 & 0 & 0 \end{bmatrix}$$

The rank of this matrix is 1 and hence the system is not controllable.

(ii) Gilbert's test

The eigen values are 1, 2 and 2.

From example 9.10, the modal matrix P is

$$P = \begin{bmatrix} 1 & 1 & 2 \\ 0 & 1 & 1 \\ 0 & 0 & 1 \end{bmatrix}$$

and the Jordon form of A is $J = \begin{bmatrix} 1 & 0 & 0 \\ 0 & 2 & 1 \\ 0 & 0 & 2 \end{bmatrix}$

$$P^{-1} = \begin{bmatrix} 1 & -1 & -1 \\ 0 & 1 & -1 \\ 0 & 0 & 1 \end{bmatrix}$$

and
$$P^{-1}B = \begin{bmatrix} 1 & -1 & -1 \\ 0 & 1 & -1 \\ 0 & 0 & 1 \end{bmatrix} \begin{bmatrix} 1 \\ 1 \\ 0 \end{bmatrix}$$

$$= \begin{bmatrix} 0 \\ 1 \\ 0 \end{bmatrix}$$

Since the first element of $P^{-1}B$ corresponding to the distinct eigen value, $\lambda_1 = 1$, is zero and the last row element of $P^{-1}B$ corresponding to the repeated eigen value, $\lambda_2 = 2$, is also zero the system is not controllable.

(d) $$A = \begin{bmatrix} 1 & 1 & 2 \\ 0 & 2 & 0 \\ 0 & 0 & 2 \end{bmatrix} \quad B = \begin{bmatrix} 1 \\ -1 \\ 1 \end{bmatrix}$$

(i) Kalman's test

The controllability matrix,

$$Q_C = \begin{bmatrix} 1 & 2 & 4 \\ -1 & -2 & -4 \\ 1 & 2 & 4 \end{bmatrix}$$

Since the rank of this is 1, the system is not controllable.

(ii) Gilbert's method

From example 9.11, the eigen values, the model matrix P, and its inverse are,

$$\lambda_1 = 1, \lambda_2 = \lambda_3 = 1\;;$$

$$P = \begin{bmatrix} 1 & 0 & 1 \\ 0 & 2 & 1 \\ 0 & -1 & 0 \end{bmatrix}; \quad P^{-1} = \begin{bmatrix} 1 & -1 & -2 \\ 0 & 0 & -1 \\ 0 & 1 & 2 \end{bmatrix}$$

The Jordon form of A is,

$$J = P^{-1}AP = \begin{bmatrix} 1 & 0 & 0 \\ 0 & 2 & 0 \\ 0 & 0 & 2 \end{bmatrix}$$

$$= \begin{bmatrix} J_1 & 0 & 0 \\ 0 & J_2 & 0 \\ 0 & 0 & J_3 \end{bmatrix}$$

Since the Jordon block J_2 and J_3 are associated with the same eigen value $\lambda_2 = \lambda_3 = 2$, the system is not controllable. In this case, it is not necessary to find the elements of $P^{-1}B$. For any B, the system is not controllable.

Example 9.20

For the following system, determine controllability.

$$\dot{X} = \begin{bmatrix} 2 & 0 & 0 \\ 0 & 2 & 0 \\ 0 & 3 & 1 \end{bmatrix} X + \begin{bmatrix} 0 & 1 \\ 1 & 0 \\ 0 & 1 \end{bmatrix} \begin{bmatrix} u_1 \\ u_2 \end{bmatrix}$$

Solution :

The controllability matrix,

$$Q_C = \left[B \ \vdots \ AB \ \vdots \ A^2B\right]$$

$$Q_C = \begin{bmatrix} 0 & 1 & \vdots & 0 & 2 & \vdots & 0 & 4 \\ 1 & 0 & \vdots & 2 & 0 & \vdots & 4 & 0 \\ 0 & 1 & \vdots & 3 & 1 & \vdots & 9 & 1 \end{bmatrix}$$

The rank of this matrix is 3 and hence the system is completely state controllable.

9.8.2 Observability

Now let us define the second aspect of the system, i.e., observability.

Definition

A system is said to be completely observable, if every state $X(t_0)$ can be determined from the measurement of the output Y(t) over a finite time interval, $t_0 \le t \le t_f$.

This property of the system is very useful, because the state vector can be constructed by observing the output variables over a finite interval of time. Therefore even if some states are not measurable, they can be estimated using the measurable output variables over a finite time. Estimation of all the state variables is required in some cases, when these state variables have to be fedback to obtain a desired control.

Kalman's test for observability

Consider the following system,

$$\dot{X} = AX + Bu$$
$$Y = CX \qquad .....(9.133)$$

The condition for complete state observability is that the observability matrix,

$$Q_0 = \begin{bmatrix} C \\ CA \\ CA^2 \\ \vdots \\ CA^{n-1} \end{bmatrix} \qquad .....(9.134)$$

has a rank n.

Gilbert's test for observability

The state equations represented by eqn. (9.133) can be transformed to its canonical form using the transformation,

$$X = PZ$$
$$\dot{Z} = P^{-1} APZ + P^{-1} Bu \qquad .....(9.135)$$
$$Y = CPZ$$

Since $P^{-1}AP$ is a diagonal matrix, the states are decoupled. Information about one state is not available in the other state equations. Since Y is a combination of these decoupled states, if any column of CP in eqn. (9.135) is zero, the corresponding state variable does not affect the output Y, i.e., if the jth column of CP contains all zeros, then the state variable Z_j is not observable from the output. This is the condition if all eigen values of A are distinct.

If some of the eigen values are repeated, the matrix P^{-1} AP is in Jordon form and the conditions for complete state observability are stated as follows :

The system is completely state observable if,

1. no two Jordon blocks in $J = P^{-1}$ AP are associated with the same eigen value.
2. no columns of CP that correspond to the first row of each Jordon block consist of zero elements. If some of the eigen values are distinct, the Jordon blocks corresponding to these eigen values are of order 1 and hence the columns of CP corresponding to these block must not be all zero.

The above conditions can be best illustrated by the following examples :

(a) $$\dot{X} = \begin{bmatrix} -2 & 0 \\ 0 & -4 \end{bmatrix} X \qquad Y = [2 \;\; -2]\, X$$

The system is observable since C has no zero elements.

(b)
$$\dot{X} = \begin{bmatrix} -1 & 0 \\ 0 & -1 \end{bmatrix} X \qquad Y = [1 \quad -1]\, X$$

Not observable since two Jordon blocks are associated with the same eigen value –1.

(c)
$$\dot{X} = \begin{bmatrix} -1 & 1 & 0 \\ 0 & -1 & 0 \\ 0 & 0 & -2 \end{bmatrix} X \qquad Y = [a \quad b \quad c]\, X$$

(i) If $a = 0$ and / or $c \neq 0$ the system is not observable.

(ii) If $a \neq 0$ and $c \neq 0$, the system is observable.

(d)
$$\dot{X} = \left[\begin{array}{c|ccc} -1 & 0 & 0 & 0 \\ \hline 0 & -2 & 1 & 0 \\ 0 & 0 & -2 & 1 \\ 0 & 0 & 0 & -2 \end{array}\right] X \qquad Y = \left[\begin{array}{c|ccc} 1 & 1 & -1 & 1 \\ 0 & 1 & -2 & 0 \end{array}\right] X$$

Since all the elements of first column of C of Jordon block corresponding to the eigenvalue $\lambda = -1$ are non zero and all the elements of first column of C of the Jordon block corresponding to the repeated eigenvalue $\lambda = -2$ are non zero, the system is completely observable.

(e)
$$\dot{X} = \left[\begin{array}{cc|cc} -3 & 1 & 0 & 0 \\ 0 & -3 & 0 & 0 \\ \hline 0 & 0 & -2 & 1 \\ 0 & 0 & 0 & -2 \end{array}\right] X \qquad Y = \left[\begin{array}{cc|cc} 0 & 1 & 2 & -1 \\ 0 & 0 & 0 & 1 \end{array}\right] X$$

Since all the elements of column 1 of C corresponding to the Jordon block of the repeated eigenvalue $\lambda = -3$ are zero, the system is not observable.

Problem 9.21

Comment on the complete state observability of the following systems using,

(i) Kalman's test

(ii) Gilbert's test

(a)
$$\dot{X} = \begin{bmatrix} 0 & 1 & 0 \\ 0 & 0 & 1 \\ -6 & -11 & -6 \end{bmatrix} X \qquad Y = [1 \quad -1 \quad 1]\, X$$

(b)
$$\dot{X} = \begin{bmatrix} 1 & 1 & 2 \\ 0 & 2 & 1 \\ 0 & 0 & 2 \end{bmatrix} X \qquad Y = [1 \quad -1 \quad 0]\, X$$

(c) $$\dot{X} = \begin{bmatrix} 1 & 1 & 2 \\ 0 & 2 & 0 \\ 0 & 0 & 2 \end{bmatrix} X \qquad Y = [1 \quad 0 \quad 0]\, X$$

(b) $$\dot{X} = \begin{bmatrix} 1 & 0 & 2 \\ 0 & 2 & 1 \\ 0 & 0 & 2 \end{bmatrix} X \qquad Y = \begin{bmatrix} 1 & 0 & 1 \\ 0 & 1 & 0 \end{bmatrix} X$$

Solution :

(a) $$\dot{X} = \begin{bmatrix} 0 & 1 & 0 \\ 0 & 0 & 1 \\ -6 & -11 & -6 \end{bmatrix} X \qquad Y = [1 \quad 0 \quad 0]\, X$$

(i) Kalman's test

The observability matrix,

$$Q_0 = \begin{bmatrix} C \\ CA \\ CA^2 \end{bmatrix} = \begin{bmatrix} 1 & 0 & 0 \\ 0 & 1 & 0 \\ 0 & 0 & 1 \end{bmatrix}$$

The rank of Q_0 is 3 and hence the system is observable.

(ii) Gilbert's test

The model matrix for this system is,

$$P = \begin{bmatrix} 1 & 1 & 1 \\ -1 & -2 & -3 \\ 1 & 4 & 9 \end{bmatrix}$$

$$\therefore \qquad CP = [1 \quad 0 \quad 0] \begin{bmatrix} 1 & 1 & 1 \\ -1 & -2 & -3 \\ 1 & 4 & 9 \end{bmatrix}$$

$$= [1 \quad 1 \quad 1]$$

Since none of the columns of CP are zero, the system is completely observable.

(b) $$\dot{X} = \begin{bmatrix} 1 & 1 & 2 \\ 0 & 2 & 1 \\ 0 & 0 & 2 \end{bmatrix} X \qquad Y = [1 \quad -1 \quad 0]\, X$$

(i) Kalman's test

The observability matrix,

$$Q_0 = \begin{bmatrix} C \\ CA \\ CA^2 \end{bmatrix} = \begin{bmatrix} 1 & -1 & 0 \\ 1 & -1 & 1 \\ 1 & -1 & 3 \end{bmatrix}$$

The rank of Q_0 is 2 and hence the system is not observable.

(ii) Gilbert's test

The modal matrix P is,

$$P = \begin{bmatrix} 1 & 1 & 2 \\ 0 & 1 & 1 \\ 0 & 0 & 1 \end{bmatrix}$$

The transformed equations are,

$$\dot{Z} = \left[\begin{array}{c|cc} 1 & 0 & 0 \\ \hline 0 & 2 & 1 \\ 0 & 0 & 2 \end{array}\right] Z$$

$$Y = CPZ$$

$$= [1 \quad -1 \quad 0] \begin{bmatrix} 1 & 1 & 2 \\ 0 & 1 & 1 \\ 0 & 0 & 1 \end{bmatrix} Z$$

$$= [1 \quad 0 \quad 1]\, Z$$

Since the first column of CP corresponding to the Jordon block of eigen value $\lambda = 2$ is zero, the system is not observable.

(c) $$\dot{X} = \begin{bmatrix} 1 & 1 & 2 \\ 0 & 2 & 0 \\ 0 & 0 & 2 \end{bmatrix} X\, C \qquad Y = [1 \quad 0 \quad 0]\, X$$

(i) Kalman's test

The observability matrix,

$$Q_0 = \begin{bmatrix} C \\ CA \\ CA^2 \end{bmatrix} = \begin{bmatrix} 1 & 0 & 0 \\ 1 & 1 & 2 \\ 1 & 3 & 6 \end{bmatrix}$$

Since the rank of Q_0 is 2, the system is not observable.

(ii) Gilbert's test

The modal matrix for the system is,

$$P = \begin{bmatrix} 1 & 0 & 1 \\ 0 & 2 & 1 \\ 0 & -1 & 0 \end{bmatrix}$$

The Jordon form of the system is

$$\dot{Z} = \begin{bmatrix} 1 & 0 & 0 \\ 0 & 2 & 0 \\ 0 & 0 & 2 \end{bmatrix} Z$$

$$Y = [1 \quad 0 \quad 0] \begin{bmatrix} 1 & 0 & 1 \\ 0 & 2 & 1 \\ 0 & -1 & 0 \end{bmatrix}$$

$$= [1 \quad 0 \quad 0]\, Z$$

Since two Jordon blocks correspond to the same eigen value $\lambda = 2$, the system is not observable.

(d) $$\dot{X} = \begin{bmatrix} 1 & 0 & 2 \\ 0 & 2 & 1 \\ 0 & 0 & 3 \end{bmatrix} X \qquad Y = \begin{bmatrix} 1 & 0 & 1 \\ 0 & 1 & 0 \end{bmatrix} X$$

(i) Kalman's test

The observability matrix,

$$Q_0 = \begin{bmatrix} 1 & 0 & 1 \\ 0 & 1 & 0 \\ 1 & 0 & 5 \\ 0 & 2 & 1 \\ 1 & 0 & 17 \\ 0 & 4 & 5 \end{bmatrix}$$

The rank of Q_0 is 3 and hence the system is observable.

(ii) Gilbert's test

The modal matrix P and the Jordon form of the equations are,

$$P = \begin{bmatrix} 1 & 0 & 1 \\ 0 & 1 & 1 \\ 0 & 0 & 1 \end{bmatrix}$$

and $$\dot{Z} = \begin{bmatrix} 1 & 0 & 0 \\ 0 & 2 & 0 \\ 0 & 0 & 3 \end{bmatrix} Z$$

$$Y = \begin{bmatrix} 1 & 0 & 1 \\ 0 & 1 & 0 \end{bmatrix} \begin{bmatrix} 1 & 0 & 1 \\ 0 & 1 & 1 \\ 0 & 0 & 1 \end{bmatrix} Z$$

$$= \begin{bmatrix} 1 & 0 & 2 \\ 0 & 1 & 1 \end{bmatrix} Z$$

Since none of the columns of Y contain all zero elements, and the eigenvalues of A are distinct, the system is observable.

9.8.3 Duality Principle

Kalman has introduced the duality principle to establish the relationship between controllability and observability.

Consider system 1,

$$\dot{X} = AX + Bu$$
$$Y = CX$$

and its dual defined by system 2,

$$\dot{Z} = A^* Z + C^* V$$
$$W = B^* Z$$

where A^*, B^* and C^* are conjugate transpose of A, B and C respectively.

The duality principle states that,

(i) if the pair (A, B) is controllable in system 1, the pair (A^*, B^*) is observable in its dual system 2.

and (ii) if the pair (A, C) is observable in system 1, the pair (A^* C^*) is controllable in its dual system 2.

Problems

9.1 Obtain the state space representation of the electrical system shown in Fig. P 9.1.

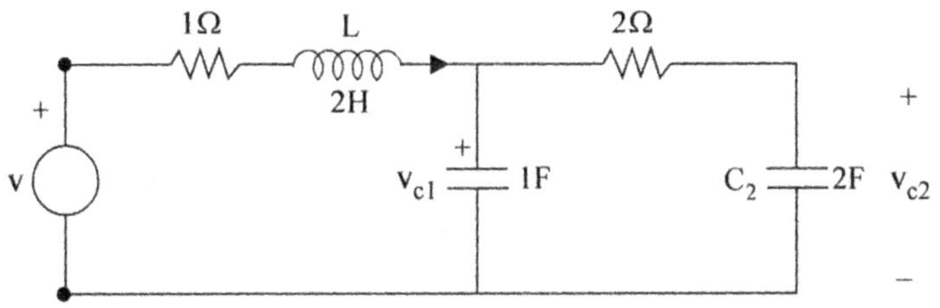

Fig. P. 1

Take $x_1 = i_L$ $x_2 = v_{c1}$ $x_3 = v_{c2}$

$v = u$ and $y = v_{c2}$

9.2 Obtain the state space representation for the mechanical system shown in Fig. P 9.2 taking the displacement and velocity of the mass as state variables

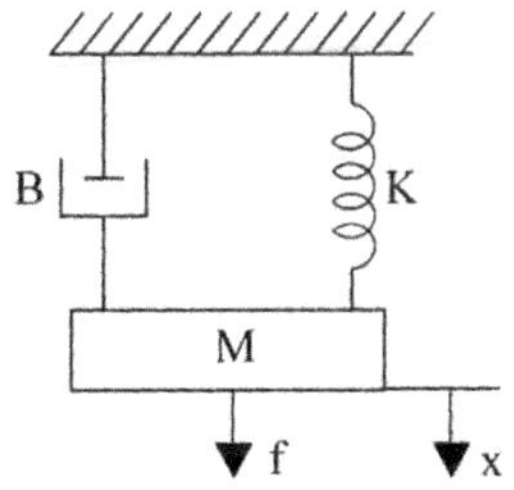

Fig. P. 2

9.3 The block diagram of a position control system is shown in Fig. P 9.3. Obtain the state space representation of the system.

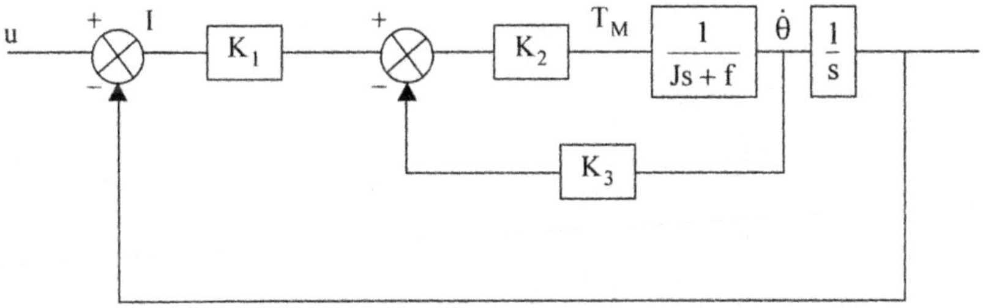

Fig. P. 3

Choose the state variables as

$$x_1 = \theta \qquad x_2 = \dot{\theta}$$

9.4 Obtain the state space representation of the system shown in Fig. P 9.4.
Take $c = x_1$, $\dot{c} = x_2$ and $m = x_3$

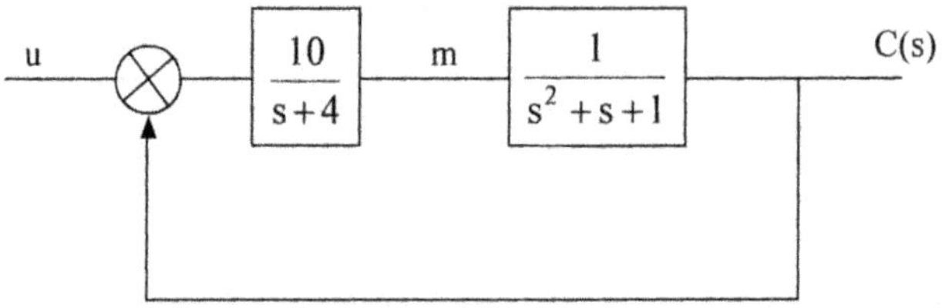

Fig. P. 4

9.5 Obtain the state space representation in,

(i) Companion form or Bush's form

(ii) Diagonal / Jordon form

for the following systems.

(a) $\dddot{y} + 9\ddot{x} + 23\dot{x} + 15 = u$

(b) $\dddot{y} + 9\ddot{x} + 23\dot{x} + 15 = \ddot{u} - 2\dot{u} + 4u$

9.6 Obtain the state space representation in,

(i) Phase variable form and

(ii) Jordon's form

for the systems whose transfer functions are given by,

(a) $\dfrac{s+1}{s(s+4)(s+5)}$ (b) $\dfrac{s+5}{(s+1)^2(s+4)}$ (c) $\dfrac{10}{(s+1)^3}$

9.7 Obtain the transfer functions of the following systems.

1. $\dot{X} = \begin{bmatrix} 0 & 1 & 0 \\ 0 & 0 & 1 \\ -2 & -3 & -4 \end{bmatrix} X + \begin{bmatrix} 0 \\ 0 \\ 1 \end{bmatrix} u$

$y = [1 \quad 0 \quad 0]\, X$

2. $\dot{X} = \begin{bmatrix} 0 & 1 & 0 \\ 0 & 0 & 1 \\ -4 & -6 & -8 \end{bmatrix} X + \begin{bmatrix} 0 \\ 0 \\ 1 \end{bmatrix} u$

$y = [-2 \quad 4 \quad 0]\, X + 2u$

3. $\dot{X} = \begin{bmatrix} -1 & 0 & 0 \\ 0 & -2 & 1 \\ 0 & 0 & -2 \end{bmatrix} X + \begin{bmatrix} 0 \\ 0 \\ 1 \end{bmatrix} u$

$y = [-1 \quad 2 \quad 4]\, X$

9.8 Find the transfer function of the systems given below.

1. $\dot{X} = \begin{bmatrix} 1 & -2 \\ -2 & 4 \end{bmatrix} X + \begin{bmatrix} 1 \\ 1 \end{bmatrix} u$ $\qquad$ 2. $\dot{X} = \begin{bmatrix} 0 & 1 & 1 \\ 1 & -1 & 0 \\ 0 & 1 & 0 \end{bmatrix} X + \begin{bmatrix} 0 \\ -1 \end{bmatrix} u$

$y = [1 \quad 0]\, X$ $\qquad$ $y = [1 \quad 1 \quad -1]\, X$

9.9 Transform the following matrices into Diagonal / Jordon form representations.

(i) $A = \begin{bmatrix} 0 & 1 & 0 \\ 0 & 0 & 1 \\ 6 & 5 & -2 \end{bmatrix}$ $\qquad$ (ii) $A = \begin{bmatrix} 0 & 1 & 0 \\ 0 & 0 & 1 \\ -1 & -3 & -3 \end{bmatrix}$

(iii) $A = \begin{bmatrix} -2 & 3 & 3 \\ -1 & 2 & 3 \\ 1 & -1 & -2 \end{bmatrix}$ $\qquad$ (iv) $A = \begin{bmatrix} -1 & 0 & 1 \\ 0 & -1 & 1 \\ 1 & -1 & -2 \end{bmatrix}$

(v) $A = \begin{bmatrix} -2 & 1 & 1 \\ -1 & 0 & 1 \\ 1 & -1 & -2 \end{bmatrix}$ $\qquad$ (vi) $A = \begin{bmatrix} 0 & -1 & 0 \\ 1 & -2 & 0 \\ 0 & 0 & -1 \end{bmatrix}$

9.10 Find the state transition matrix for the following systems.

(i) $\dot{X} = \begin{bmatrix} 1 & 0 \\ 1 & 1 \end{bmatrix} X$ $\qquad$ (ii) $\dot{X} = \begin{bmatrix} 1 & 0 \\ -1 & -2 \end{bmatrix} X$ $\qquad$ (iii) $\dot{X} = \begin{bmatrix} 0 & 1 & 0 \\ 0 & 0 & 1 \\ -2 & -5 & -4 \end{bmatrix} X$

9.11 Obtain the solution of the state equation

$$\dot{X} = \begin{bmatrix} 0 & 1 \\ 0 & -2 \end{bmatrix} X + \begin{bmatrix} 0 \\ 1 \end{bmatrix} u$$

$X_o = [1 \quad 1]^T$ and u(t) is a unit step input.

9.12 Obtain the Jordon's cononical form of the following system and obtain its solution for a step input

$$\dot{X} = \begin{bmatrix} 0 & 1 & 0 \\ 0 & 0 & 1 \\ 6 & 5 & -2 \end{bmatrix} X + \begin{bmatrix} 0 \\ 0 \\ 1 \end{bmatrix} u$$

$X(o) = [1 \quad 0 \quad 0]^T$

9.13 Determine whether the following systems are completely state controllable and observable using (i) Kalman's test and (ii) Gilbert's test.

(a) $\dot{X} = \begin{bmatrix} 0 & 1 & 0 \\ 0 & 0 & 1 \\ -6 & -11 & -6 \end{bmatrix} X + \begin{bmatrix} -1 \\ 1 \\ 1 \end{bmatrix} u \qquad Y = [1 \quad 1 \quad 0]\, X$

(b) $\dot{X} = \begin{bmatrix} 2 & 0 & 0 \\ 0 & 2 & 0 \\ 0 & 3 & 1 \end{bmatrix} X + \begin{bmatrix} 0 & 1 \\ 1 & 0 \\ 0 & 1 \end{bmatrix} u \qquad Y = \begin{bmatrix} 1 & 0 & 0 \\ 0 & 1 & 0 \end{bmatrix} X$

(c) $\dot{X} = \begin{bmatrix} 0 & 1 & 0 \\ 0 & 0 & 1 \\ 6 & 5 & -2 \end{bmatrix} X + \begin{bmatrix} 0 \\ 1 \\ -1 \end{bmatrix} u \qquad Y = [0 \quad 1 \quad -1]\, X$

(d) $\dot{X} = \begin{bmatrix} 0 & 1 & 0 \\ 0 & 0 & 1 \\ -1 & -3 & -3 \end{bmatrix} X + \begin{bmatrix} 0 \\ 0 \\ 1 \end{bmatrix} u \qquad Y = [0 \quad 1 \quad -1]\, X$

9.14 Determine output controllability of system in problem P. 9.13 (b).

Answers to Problems

Chapter 2

2.1 (a) $\left[\dfrac{s(L-R^2C)}{(LS+R)(RCS+1)}\right]$ (b) $\dfrac{s(2s^2+2s+1)}{(2s^4+14s^3+232s+20s+8)}$

2.2 (a) $\dfrac{K_1}{M_1M_2s^4+M_1B_2s^3+(K_1M_1+K_2M_1+K_1M_2)s^2+B_2K_1s+K_1K_2}$

(b) $\dfrac{Ms^2}{Ms^2+Bs+K}$ (c) $\dfrac{1}{Ms^2+(B_1+B_2)s+K_1}$

2.3 $T(s)=\dfrac{K}{s(J_{eq}s^2+B_{eq})}$

$$K=\frac{N_2N_4}{N_1N_3}$$

$$J_{eq}=J_L+J_2\left(\frac{N_4}{N_3}\right)^2+J_1\left(\frac{N_4}{N_3}\right)^2\left(\frac{N_2}{N_1}\right)^2$$

$$B_{eq}=B_L+B_2\left(\frac{N_4}{N_3}\right)^2+B_1\left(\frac{N_4}{N_3}\right)^2\left(\frac{N_2}{N_1}\right)^2$$

2.5 $\dfrac{K_f}{(R+LS)(Ms^2+Bs+K)+k_bK_fs}$

2.6 (i) $\dfrac{R}{RCS+1}$ (ii) $\dfrac{1}{RCS+1}$; $RC\dfrac{d\theta}{dt}+\theta=\theta_i+Rh_i$

2.7 $\theta=\theta_b\left(1-e^{-\frac{1}{RC}t}\right)$

2.8 $c/m=\dfrac{1}{\gamma\left[\dfrac{L}{g}s^2+\dfrac{B}{A\gamma}s+1\right]}$

2.9 $\dfrac{R_1}{[R_1R_2C_1C_2s^2+(R_2C_2+R_1C_1+R_1C_2)s+1]}$

2.10 $\dfrac{10.87K_A}{s^2+54.35K\,K_A+10.87K_A}$

2.11 $\dfrac{15}{s^2+6.5s+15}$

2.12 $\dfrac{G_1G_2G_3+G_4}{\Delta}$, $\Delta=1+G_1G_2H_1+G_2G_3H_2+G_1G_2G_3+G_4-G_4G_2H_2H_1$

2.13 1. $\dfrac{G_1G_2G_3}{1+G_2H_2+G_2G_3H_3+G_1G_2H_2+G_1G_2G_3H_1}$

2. $\dfrac{G_1G_2G_3(1+G_4)}{1+G_1G_2+G_4+G_1G_2G_4+G_1G_4G_5G_6H_1}$

3. $\dfrac{G_1(G_2+G_3)(1-G_4H_2)}{1+G_1(G_2+G_3)H_1H_2}$

2.14 (a) $P_1=G_1G_2G_3G_4G_5$ $P_2=G_1G_2G_6G_5$

$L_1=-G_1H_1$ $L_2=-G_1G_2$

$L_3=-G_3H_3$ $L_4=-G_4H_4$

$L_5=-G_5H_5$ $L_6=-G_1G_2G_3$

$L_7=G_6H_3H_4$

$\Delta=1-(L_1+L_2+L_3+L_4+L_5+L_6+L_7)+L_1L_3+L_1L_4+L_1L_5+L_2L_4+L_2L_5$
$+L_3L_5+L_1L_7-L_1L_3L_5$

$\Delta_1=\Delta_2=1$

$T=\dfrac{P_1\Delta_1+P_2\Delta_2}{\Delta}$

(b) $\dfrac{G_1G_2G_3G_4(1+G_7G_8H_2)+G_4G_6G_7G_8(1+G_2G_3H_1)+G_1G_2G_7G_8G_9}{1+G_2G_3H_1+G_7G_8H_2}$

2.15 $\frac{4s^3+2s^2+3s+2}{8s^4+17s^3+19s^2+15s+6}$

2.17 (a) 60 V, 190V (b) 0.1

(c) 197.14V (d) 81.14V

2.18 $S_G^T = 0.0125$ $S_H^T = 5.12$ $S_K^T = 0.128$

2.19 0.089

2.20 $\frac{1.367}{s(0.0118s+1)}$

Chapter 3

3.1 $T(s) = \frac{15K_A}{s^2+225.2s+15K_A}$ $\omega_n = 150$ rad/sec

$\delta = 0.75$ $t_p = 0.0317$ s

$M_p = 2.84\%$ $t_s = 0.0356$ sec

3.2 (i) $M_p = 10.07\%$ $t_p = 0.6$s $t_s = 1.043$ s

(ii) $K_A = 7$

(iii) $K_A = 23.64$

3.3 T = 1.44 sec K = 1.06

3.4 (i) $\omega_n = 3.16$ rad/sec $\delta = 0.316$

(ii) 0.1792

3.5 (i) $\delta = 0.316$ $\omega_n = 3.16$ rad/sec $\theta_{css} = 0.2$ rad

(ii) $K_f = 1.212$ $e_{ss} = 0.4424$ rad

(iii) $K_t = 2.6$ $K_A = 18$

3.6 K = 2.95 a = 0.47 sec

3.7 K = 16 a = 0.175 $c(t) = 1 - 1.25\, e^{-2.4t} \sin(3.2t + 53.13^{\circ})$

3.8 (i) $0, \frac{1}{3}, \infty$ (ii) $\frac{1}{11}, \infty, \infty,$

(iii) 0, 0, 0.01 (iv) $0, 0.1, \infty,$

3.9 $K = 23$ $a = 0.163$

3.10 $e_{ss}(t) = 0.54625 + 1.075t + 0.75t^2$

3.11 (a) $e_{ss}(t) = 0.01t + 0.0009$ (b) $e_{ss}(t) = 0.01t + 0.0209$

Chapter 4

4.1 (a) $a_1, a_2, > 0$

(b) $a_1, a_2, a_3, > 0$ $a_1, a_2, > a_3$

(c) $a_1, a_2, a_3, a_4, > 0$ $a_1\, a_2, > a_3,$ $a_4 < a_3 \dfrac{(a_1\, a_2 - a_3)}{a_1^{\,2}}$

4.2 (a) $-2.5 \pm j2.784$

(b) $-2, -2$

(c) $-1, -3, -6$

4.3 (a) $1 \pm j1, -1 \pm j1$

(b) $-4.495, 0.2475 \pm j1.471$

(c) $0.1789, -11.179$

4.4 (a) Stable

(b) Stable

(c) Two poles in RHS

(d) Four roots on $j\omega$-axis

(e) Two roots in RHS

4.5 (a) $k > 0.303$

(b) No real value of k

(c) > 1

4.6 $0 < k < 99$

4.7 (a) equal

(b) less

(c) greater

4.8 $-1 \pm j2;$ $-2 \pm j1$

Chapter 5

5.1 (i) (a) –5 (b) 60, 180, –60 (c) –1.79 (d) K = 660 (e) ω = 8.63 rad/sec

(ii) (a) –7 (b) 90, –90 (c) –7.37

(iii) (a) –1 (b) +60, 180, –60 (d) 48.6° (e) K = 8, ω = 2 rad/sec

(iv) (b) 180° (c) 0.414, –2.414 (e) K = 1, ω = 1.225 rad/sec

5.2 K = 19.05 δ = 0.189, ω_d = 2.6 rad/sec t_s = 7.986 sec

Closed loop poles for K = 19.05 : –5, –0.5 $\pm$ j 2.6

5.3 K = 2.96 $s_1 = s_2 = -3.19$, $s_3 = -0.58$

5.4 –5.52, –43.8°, –2.5

5.5 1. Stable for K < 30 ; root locus bends towards RHP

2. Stable for K < 70.68 ; bends towars RHP

5.6 (i) Root locus bends towards LHP, breakaway point moves to the left.

Original system is stable for K < 42, Modified system is stable for K > 0

(ii) System is stable for all positive values of K

5.7 K = 4 s = –1 $\pm$ j1, –3 $\pm$ j1

K = 64 ω_n = 2 rad/sec

5.8 (a) –385, 60°, 180°, –60°

(b) 60°, –60°

(c) $K_1 = 1.1 \times 10^6$, ω_1 = 16.6 rad/sec $K_2 = 1.75 \times 10^6$, ω_2 = 212 rad/sec

5.10 Root locus bends towards LHP

Stable for K < 22.75

Chapter 6

6.1 (i) M_r = 1.04 ω_r = 2.116 rad/sec ω_b = 4.592 rad/sec

(ii) M_r = 1.0 ω_r = 0 ω_b = 4 rad/sec

(iii) M_r = 1.0 ω_r = 0 ω_b = 8.7 rad/sec

(iv) M_r = 1.364 ω_r = 6.6 rad/sec ω_b = 11 rad/sec

6.2 (i) 4.1 rad/sec

(ii) 0.75 rad/sec

6.3 (i) K = 200

(ii) K = 398

6.4 (a) $\dfrac{10(0.5s+1)}{s(0.02s+1)}$

(b) $\dfrac{0.4s}{(0.2s+1)(0.05s+1)(0.005s+1)}$

6.5 $\dfrac{10(s+2)}{s(s+1)(s+10)}$

6.6 K = 14.91 $\quad \tau = 0.0616$ s

$M_p = 14.62 \quad t_s = 0.4924$ s

6.7 $M_r = 1.5 \quad \omega_r = 5.1$ rad/sec

Poles = $-2.11 \pm j5.53$

6.8 (a) Does not cross

(b) $\omega = 2.236$ rad/sec, $|G| = 0.077$

(c) $\omega = 4.795$ rad/sec, $|G| = 0.11$

(d) $\omega = 37.42$ rad/sec, $|G| = 2.38 \times 10^{-3}$

6.9 $\omega = 0.2$ rad/sec $\quad \angle G(j\omega) = 93.44^\circ$

Chapter 7

7.1 (a) 2 roots in RHP

(b) 2 roots on $j\omega$-axis

(c) 2 roots in RHP

7.2 At $\omega = 0$ and $\omega = \infty$ the curve is rotated by 90° in clockwise direction for addition of every pole at origin.

7.3 The behaviour at $\omega = 0$ is not altered but at $\omega = \infty$, the curve is rotated by 90° in clockwise direction for every non zero pole added.

7.4

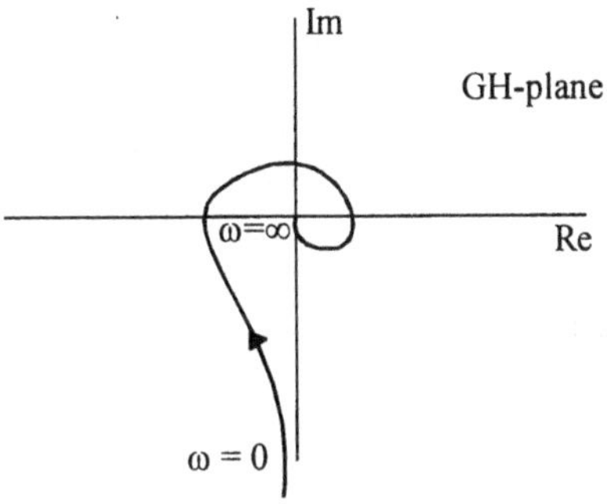

7.5 (a)

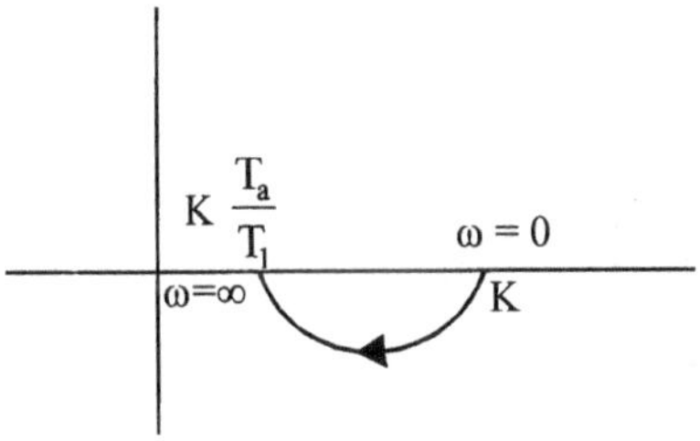

(b)

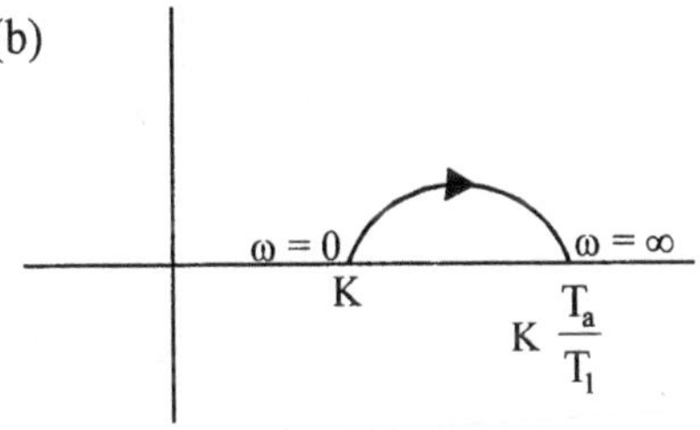

7.7 $K = 0.7$, The system becomes more stable, $K = 1.08$

7.8 (a) Unstable (b) Stable

7.9 GM = 9.54db $\phi_{pm} = 32.6^o$

7.10 (a) GM = 28 db $\phi_{pm} = 76^o$

(b) $K = 4.45$

(c) $K = 4.33$

7.12 $M_r = 1.78$ $\omega_r = 8$ rad/sec $\omega_b = 10$ rad/sec

Chapter 8

8.1 (i) $\phi_{pm} = 40^o$

(ii) $\omega_{gc} = 0.5$ rad/sec

(iii) $e_{ff} = 0.2$

(iv) Gain margin = 11 db

Chapter 9

9.1 $$\dot{X} = \begin{bmatrix} -\frac{1}{2} & -\frac{1}{2} & 0 \\ 1 & -\frac{1}{2} & \frac{1}{2} \\ 0 & \frac{1}{4} & -\frac{1}{4} \end{bmatrix} X + \begin{bmatrix} \frac{1}{2} \\ 0 \\ 0 \end{bmatrix} u \qquad y = [0 \quad 0 \quad 1]\, X$$

9.2 $$\dot{X} = \begin{bmatrix} 0 & 1 \\ \frac{-K}{M} & \frac{-B}{M} \end{bmatrix} X + \begin{bmatrix} 0 \\ \frac{1}{M} \end{bmatrix} u$$

9.3 $$\dot{X} = \begin{bmatrix} 0 & 1 \\ \frac{-K_1 K_2}{J} & -\left[\frac{f}{J} + \frac{K_2 K_3}{J}\right] \end{bmatrix} X + \begin{bmatrix} 0 \\ \frac{K_1 K_2}{J} \end{bmatrix} u$$

9.4 $$\dot{X} = \begin{bmatrix} 0 & 1 & 0 \\ -1 & -1 & 1 \\ -10 & 0 & -4 \end{bmatrix} X + \begin{bmatrix} 0 \\ 0 \\ 10 \end{bmatrix} u \qquad Y = [1 \quad 0 \quad 0]\, X$$

9.5 (a) (i) $$A = \begin{bmatrix} 0 & 1 & 0 \\ 0 & 0 & 1 \\ -15 & -23 & -9 \end{bmatrix} \qquad B = \begin{bmatrix} 0 \\ 0 \\ 1 \end{bmatrix} \qquad C = [1 \quad 1 \quad 0]$$

(ii) $$A = \begin{bmatrix} -1 & 0 & 0 \\ 0 & -3 & 0 \\ 0 & 0 & -5 \end{bmatrix} \qquad B = \begin{bmatrix} 1 \\ 1 \\ 1 \end{bmatrix} \qquad C = \begin{bmatrix} \frac{1}{8} & -\frac{1}{4} & \frac{1}{8} \end{bmatrix}$$

(b) (i) $$A = \begin{bmatrix} 0 & 1 & 0 \\ 0 & 0 & 1 \\ -15 & -23 & -9 \end{bmatrix} \qquad B = \begin{bmatrix} 0 \\ 0 \\ 1 \end{bmatrix} \qquad C = [4 \quad -2 \quad 1]$$

(ii) $$A = \begin{bmatrix} -1 & 0 & 0 \\ 0 & -3 & 0 \\ 0 & 0 & -5 \end{bmatrix} \qquad B = \begin{bmatrix} 0 \\ 0 \\ 1 \end{bmatrix} \qquad C = \begin{bmatrix} \frac{7}{8} & \frac{-19}{4} & \frac{39}{8} \end{bmatrix}$$

9.6 (a) (i) $A = \begin{bmatrix} 0 & 1 & 0 \\ 0 & 0 & 1 \\ 0 & -20 & -9 \end{bmatrix}$ $B = \begin{bmatrix} 0 \\ 0 \\ 1 \end{bmatrix}$ $C = [1 \quad 1 \quad 0]$

(ii) $A = \begin{bmatrix} 0 & 0 & 0 \\ 0 & -4 & 0 \\ 0 & 0 & -5 \end{bmatrix}$ $B = \begin{bmatrix} 1 \\ 1 \\ 1 \end{bmatrix}$ $C = \begin{bmatrix} \frac{1}{20} & \frac{3}{4} & \frac{-4}{5} \end{bmatrix}$

(b) (i) $A = \begin{bmatrix} 0 & 1 & 0 \\ 0 & 0 & 1 \\ -2 & -5 & -4 \end{bmatrix}$ $B = \begin{bmatrix} 0 \\ 0 \\ 1 \end{bmatrix}$ $C = [0 \quad 1 \quad 0]$

(ii) $A = \begin{bmatrix} -1 & 1 & 0 \\ 0 & -1 & 0 \\ 0 & 0 & -4 \end{bmatrix}$ $B = \begin{bmatrix} 0 \\ 1 \\ 1 \end{bmatrix}$ $C = \begin{bmatrix} \frac{4}{3} & \frac{-1}{9} & \frac{1}{9} \end{bmatrix}$

(c) (i) $A = \begin{bmatrix} 0 & 1 & 0 \\ 0 & 0 & 1 \\ -1 & -3 & -3 \end{bmatrix}$ $B = \begin{bmatrix} 0 \\ 0 \\ 1 \end{bmatrix}$ $C = [10 \quad 0 \quad 0]$

(ii) $A = \begin{bmatrix} -1 & 1 & 0 \\ 0 & -1 & 1 \\ 0 & 0 & -1 \end{bmatrix}$ $B = \begin{bmatrix} 0 \\ 0 \\ 1 \end{bmatrix}$ $C = [10 \quad 0 \quad 0]$

9.7 1. $\dfrac{1}{s^3 + 4s^2 + 3s + 2}$ 2. $\dfrac{2s^3 + 16s^2 + 16s + 6}{s^3 + 8s^2 + 6s + 4}$ 3. $\dfrac{3s^2 + 10s + 6}{(s+1)(s+2)^2}$

9.8 (i) $\dfrac{s-6}{s(s-5)}$ (ii) $\dfrac{2}{s+1}$

9.9 (i) $D = \begin{bmatrix} -1 & 0 & 0 \\ 0 & 2 & 0 \\ 0 & 0 & -3 \end{bmatrix}$ (ii) $J = \begin{bmatrix} -1 & 1 & 0 \\ 0 & -1 & 1 \\ 0 & 0 & -1 \end{bmatrix}$

(iii) $\begin{bmatrix} 1 & 0 & 0 \\ 0 & -1 & 0 \\ 0 & 0 & -2 \end{bmatrix}$ (iv) $\begin{bmatrix} -1 & 1 & 0 \\ 0 & -1 & 0 \\ 0 & 0 & -2 \end{bmatrix}$

(v) $A = \begin{bmatrix} -1 & 0 & 0 \\ 0 & -1 & 0 \\ 0 & 0 & -2 \end{bmatrix}$ (vi) $A = \begin{bmatrix} -1 & 1 & 0 \\ 0 & -1 & 0 \\ 0 & 0 & -1 \end{bmatrix}$

9.10 (i) $\begin{bmatrix} e^t & 0 \\ te^t & e^t \end{bmatrix}$

(ii) $\begin{bmatrix} (1+t)e^{-t} & te^{-t} \\ -te^{-t} & (1-t)e^{-t} \end{bmatrix}$

(iii) $\begin{bmatrix} e^{-2t}+2te^{-t} & 2e^{-2t}-2e^{-t}+3te^{-t} & e^{-2t}-e^{-t}+te^{-t} \\ -2e^{-2t}+2e^{-t}-2te^{-t} & -4e^{-2t}+5e^{-t}-3te^{-t} & -2e^{-2t}+2e^{-t}-te^{-t} \\ 4e^{-2t}-4e^{-t}+2te^{-t} & 8e^{-2t}-8e^{-t}+3te^{-t} & 4e^{-2t}-3e^{-t}+te^{-t} \end{bmatrix}$

9.11 $X(t) = \begin{bmatrix} \frac{5}{4}+\frac{1}{2}t-\frac{1}{4}e^{-2t} \\ \frac{1}{2}\left(1+e^{-2t}\right) \end{bmatrix}$

9.12 $X(t) = \begin{bmatrix} \frac{-1}{6}+\frac{7}{6}e^{-t}+\frac{7}{30}e^{2t}+\frac{1}{6}e^{-3t} \\ \frac{-7}{6}e^{-t}+\frac{7}{15}e^{2t}-\frac{1}{2}e^{-3t} \\ \frac{7}{6}e^{-t}+\frac{14}{15}e^{2t}+\frac{3}{2}e^{-3t} \end{bmatrix}$

9.13 (a) not controllable, not observable

(b) controllable, not observable

(c) not controllable, observable

(d) controllable, observable

9.14 Output controllable

Multiple Choice Questions (MCQs) from Competitive Examinations

Chapter 1

1. Which one of the following is open loop ?

(a) The respiratory system of man

(b) A system for controlling the movement of the slide of a copying milling machine

(c) A thermostatic control

(d) Traffic light control

2. Consider the following statements regarding a linear system y = f(x) :

1. $f(x_1 + x_2) = f(x_1) + f(x_2)$.
2. $f[x(t + T)] = f[x(t)] + f[x(T)]$.
3. $f(Kx) = Kf(x)$.

Of these statements

(a) 1, 2 and 3 are correct (b) 1 and 2 are correct

(c) 3 alone is correct (d) 1 and 3 are correct.

3. Which one of the following is an example of open-loop system ?

(a) A windscreen wiper

(b) Aqualung

(c) Respiratory system of an animal

(d) A system for controlling Anti-rocket-missiles

4. When a human being tries to approach an object, his brain acts as

(a) an error measuring device (b) a controller

(c) an actuator (d) an amplifier

5. In a continuous data system

(a) data may be a continuous function of time at all points in the system

(b) data is necessarily a continuous function of time at all points in the system

(c) data is continuous at the input and output parts of the system but not necessarily during intermediate processing of the data.

(d) Only the reference signal is a continuous function of time.

6. As compared to a closed-loop system, an open-loop system is

(a) more stable as well as more accurate

(b) less stable as well as less accurate

(c) more stable but less accurate

(d) less stable but more accurate

7. The principles of homogeneity and superposition are applied to

(a) linear time variant systems (b) non-linear time variant systems

(c) linear time invariant systems (d) non-linear time invariant systems

Chapter 2

1. For block diagram shown in Fig. C(s)/R(s) is given by

(a) $\dfrac{G_1 G_2 G_3}{1+H_2 G_2 G_3 + H_1 G_1 G_2}$

(b) $\dfrac{G_1 G_2 G_3}{1+G_1 G_2 G_3 H_1 H_2}$

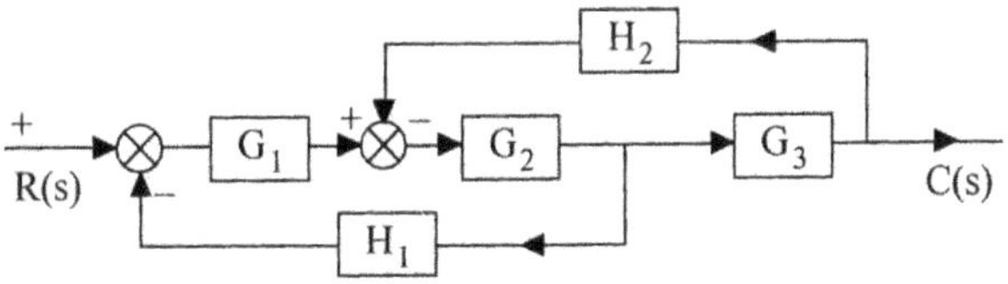

(c) $\dfrac{G_1 G_2 G_3}{1+G_1 G_2 G_3 H_1 + G_1 G_2 G_3 H_2}$

(d) $\dfrac{G_1 G_2 G_3}{1+G_1 G_2 G_3 H_1}$

2. The transfer function of the system described by $\dfrac{d^2y}{dt^2}+\dfrac{dy}{dt}=\dfrac{du}{dt}$ + 2u with 'u' as input and 'y' as output is

(a) $\dfrac{(s+2)}{(s^2+s)}$ (b) $\dfrac{(s+1)}{(s^2+s)}$ (c) $\dfrac{2}{(s^2+s)}$ (d) $\dfrac{2s}{(s^2+s)}$

3. Feedback control systems are
(a) Insensitive to both forward-and feedback-path parameter changes
(b) Less sensitive to feedback-path parameter changes than to forward-path parameter changes
(c) less sensitive to forward-path parameter changes than to feedback-path parameter changes.
(d) equally sensitive to forward and feedback-path parameter changes.

4. A system is represented by the block diagram given in the figure.
Which one of the following represents the input-output relationship of the above diagram

(a) R(s) → $\boxed{(G_1G_2)}$ → C(s)

(b) R(s) → $\boxed{(G_1+G_2)}$ → C(s)

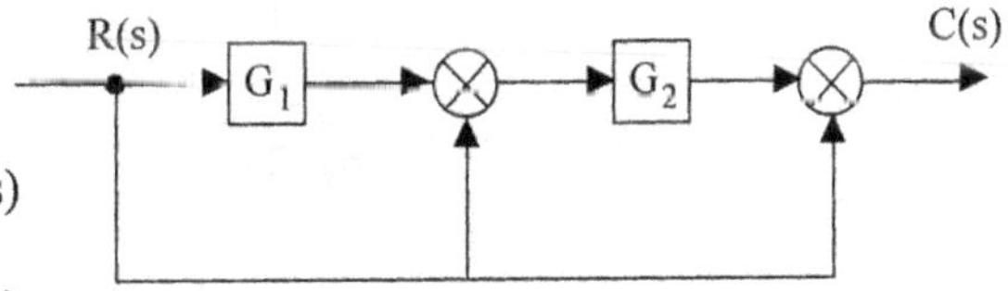

(c) R(s) → $\boxed{1+G_1+G_1G_2}$ → C(s)

(d) R(s) → $\boxed{1+G_2+G_1G_2}$ → C(s)

5. Consider a simple mass-spring-friction system as given in the figure.

K_1, K_2 - Spring Constants, f – Friction

M – Mass, F – Force, x – Displacement

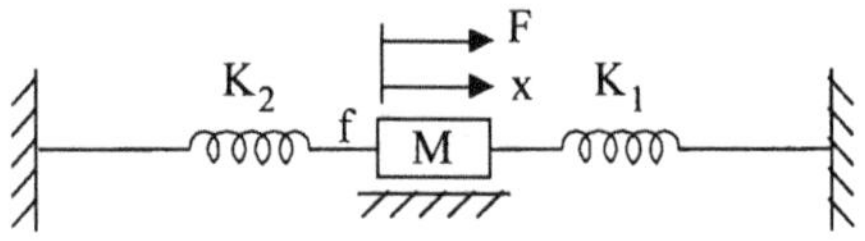

The transfer function $\frac{X(s)}{F(s)}$ of the given system will be

(a) $\frac{1}{Ms^2 + fs + K_1 . K_2}$

(b) $\frac{1}{Ms^2 + fs + K_1 + K_2}$

(c) $\frac{1}{Ms^2 + fs + \frac{K_1 . K_2}{K_1 + K_2}}$

(d) $\frac{K_2}{Ms^2 + fs + K_1}$

6. In the following block diagram, $G_1 = 10/s$, $G_2 = \frac{10}{(s+1)}$, $H_1 = s + 3$ and $H_2 = 1$. The overall transfer function C/R is given by

(a) $\frac{10}{11s^2 + 31s + 10}$

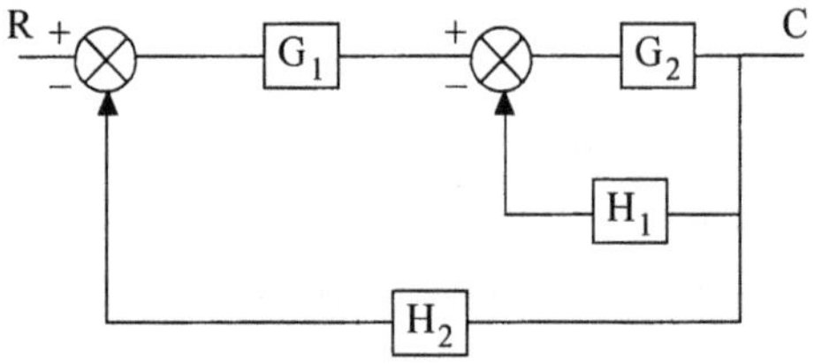

(b) $\frac{100}{11s^2 + 31s + 100}$

(c) $\frac{100}{11s^2 + 31s + 10}$

(d) $\frac{100}{11s^2 + 31s}$

7. The signal flow of a system is shown in the given figure. In this graph, the number of three non-touching loops is

(a) zero

(b) 1

(c) 2

(d) 3

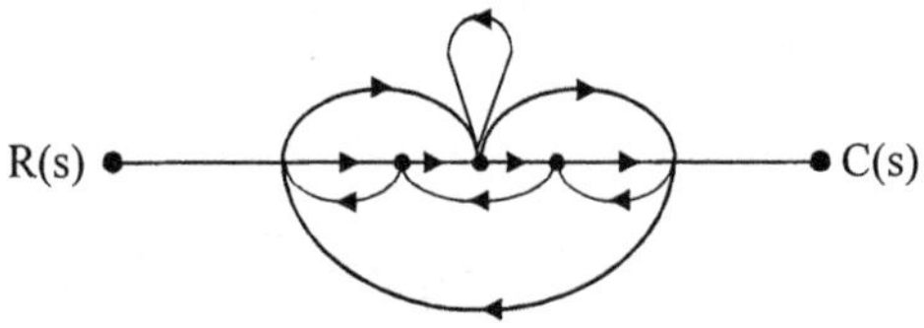

8. The sum of products of two non-touching loops in the following signal flow graph is

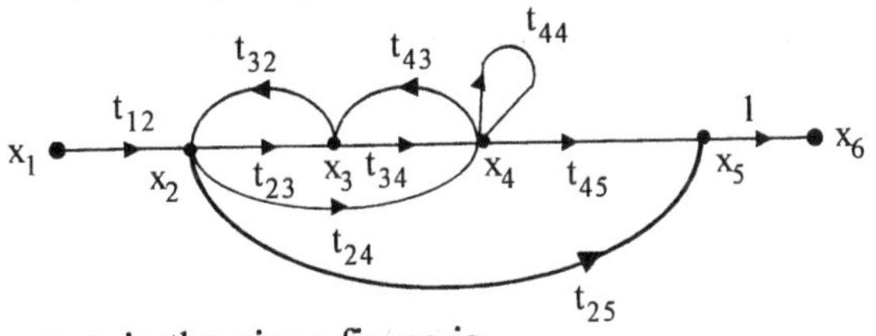

(a) $t_{23}\ t_{32}\ t_{44}$

(b) $t_{23}\ t_{32} + t_{34}\ t_{43}$

(c) $t_{23}\ t_{32} + t_{34}\ t_{43} + t_{44}$

(d) $t_{24}\ t_{43}\ t_{32} + t_{44}$

9. The closed-loop gain of the system in the given figure is

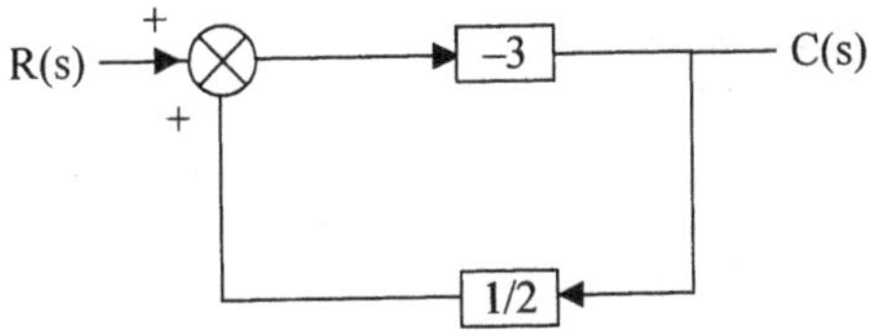

(a) $-\frac{9}{5}$ (b) $-\frac{6}{5}$

(c) $\frac{6}{5}$ (d) $\frac{9}{5}$

10. The response C(t) of a system to an input r(t) is given by the following differential equation : $\frac{d^2C(t)}{dt^2} + 3\frac{dC(t)}{dt} + 5C(t) = 5\ r(t)$

The transfer function of the system is given by

(a) $G(s) = \frac{5}{s^2 + 3s + 5}$ (b) $G(s) = \frac{1}{s^2 + 3s + 5}$

(c) $G(s) = \frac{3s}{s^2 + 3s + 5}$ (d) $\frac{s + 3}{s^2 + 3s + 5}$

11. For the RC circuit shown in the given figure, V_i and V_o are the input and output of the system respectively. The block diagram of the system is represented by

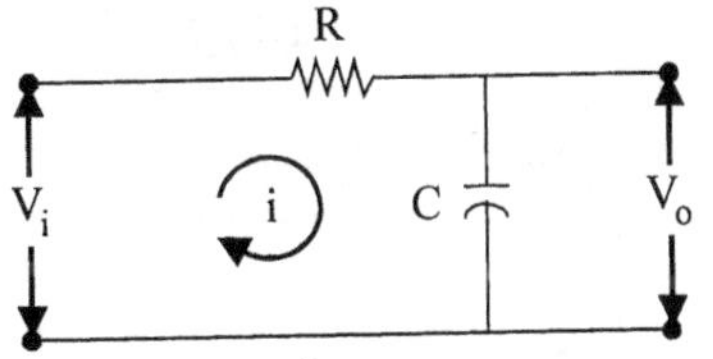

System

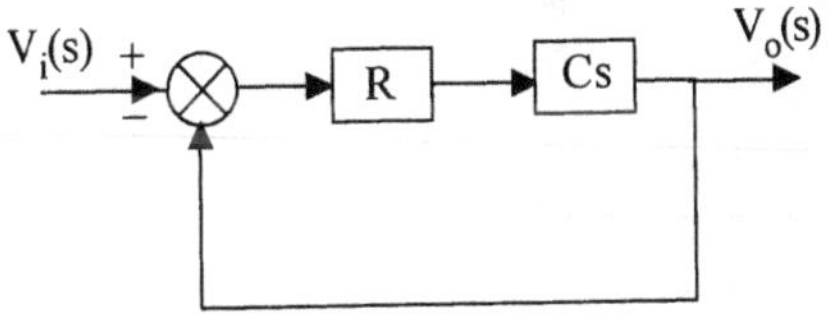

(a)

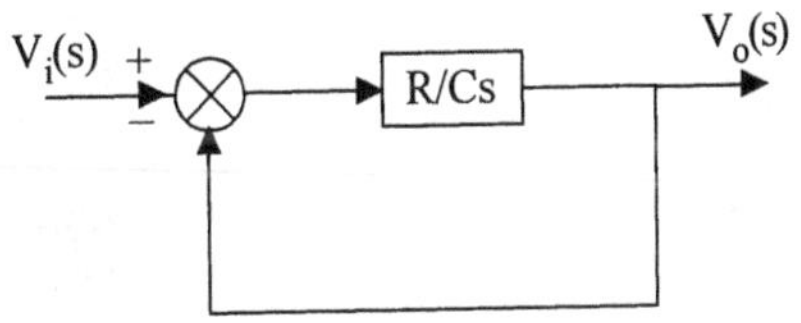

(b)

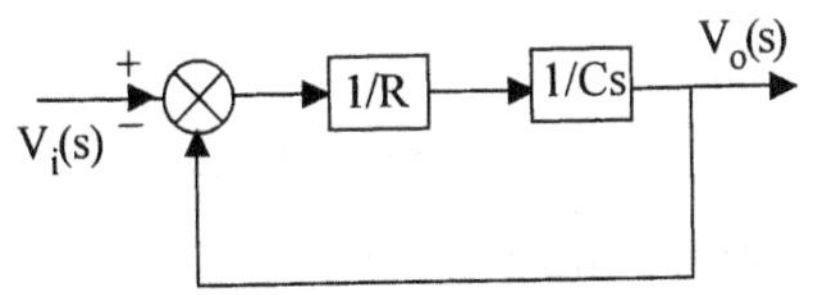

(c)

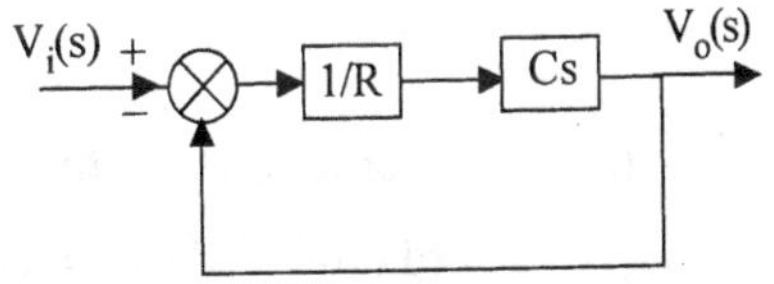

(d)

12. Given : $KK_t = 99$; $s = j1$ rad/s, the sensitivity of the closed-loop system (shown in the given figure) to variation in parameter K is approximately

(a) 0.01 (b) 0.1

(c) 1.0 (d) 10

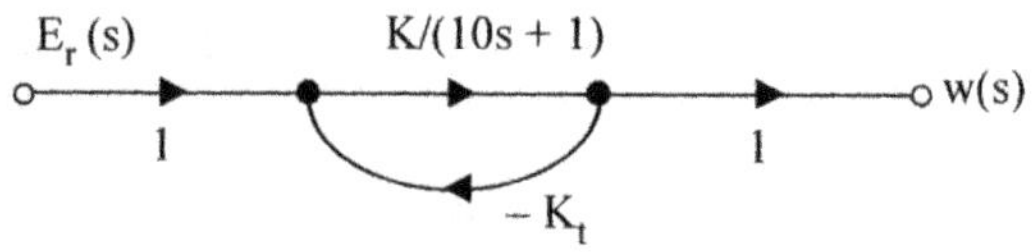

13. The transfer function C/R of the system shown in the figure is :

(a) $\dfrac{G_1 G_2 H_2}{H_1(1+G_1 G_2 H_2)}$

(b) $\dfrac{G_1 H_2}{H_1(1+G_1 G_2 H_2)}$

(c) $\dfrac{G_2 H_2}{H_1(1+G_1 G_2 H_1)}$

(d) $\dfrac{G_2 H_1}{H_2(1+G_1 G_2 H_2)}$

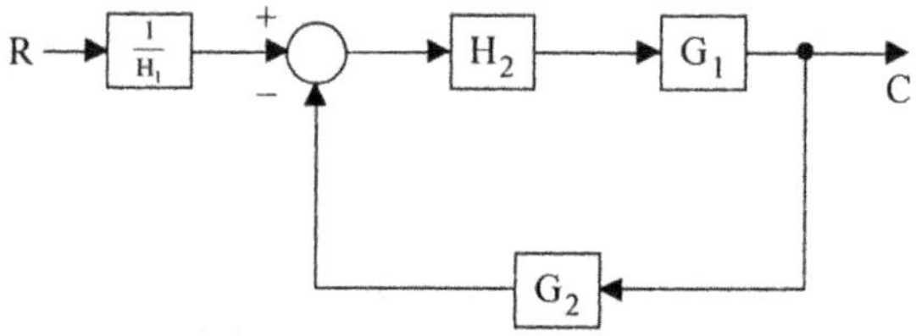

14. Laplace transform of the output response of a linear system is the system transfer function when the input is

(a) a step signal (b) a ramp signal

(c) an inpulse signal (d) a sinusoidal signal

15. A simple electric water heater is shown in the given figure. The system can be modelled by

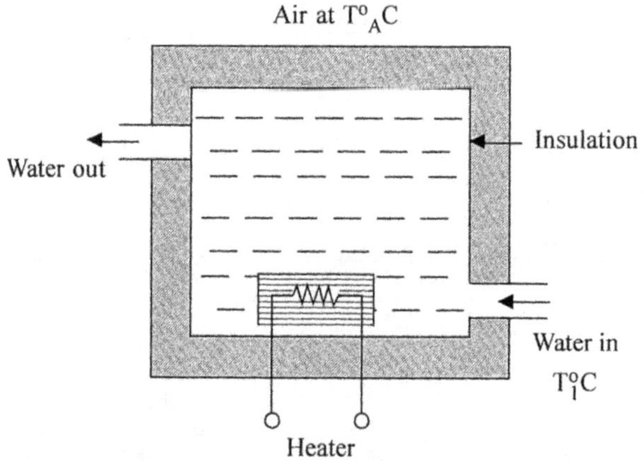

(a) a first order differential equation (b) a second order differential equation

(c) a third order differential equation (d) an algebraic equation

16. Which of the following is used in digital position control systems

(a) Stepper motor (b) AC servo motor

(c) Synchro (d) DC servo motor

17. Four speed-torque curves (labelled I, II, III and IV) are shown in the given figure. That of an ac servomotor will be as in the curve tabelled

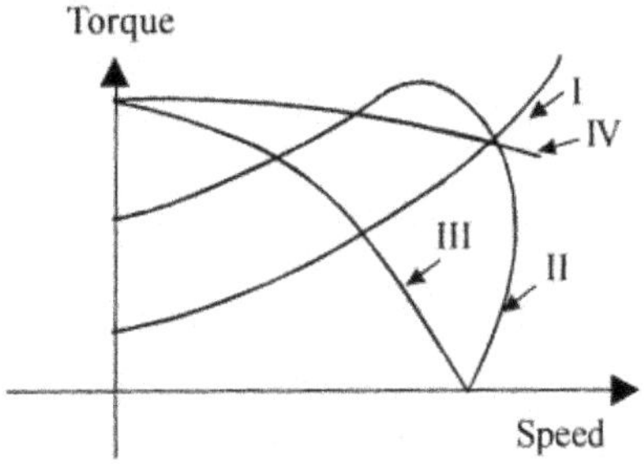

(a) I

(b) II

(c) III

(d) IV

18. Match List I with List II and select the correct answer using the codes given below the lists.

List I	List II
A. Hydraulic actuator	1. Linear device
B. Flapper valve	2. AC servo systems
C. Potentiometer error detector	3. Large power to weight ratio
D. Dumb bell rotor	4. Pneumatic systems

Codes :

	A	B	C	D
(a)	4	3	2	1
(b)	3	4	2	1
(c)	3	4	1	2
(d)	4	3	1	2

19. When the signal flow graph is as shown in the figure, the overall transfer function of the systems, will be

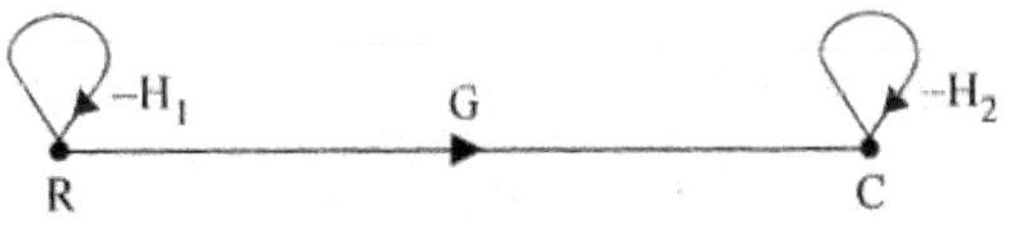

(a) $\frac{C}{R} = G$

(b) $\frac{C}{R} = \frac{G}{1+H_2}$

(c) $\frac{C}{R} = \frac{G}{(1+H_1)(1+H_2)}$ (d) $\frac{C}{R} = \frac{G}{1+H_1+H_2}$

20. The block diagram shown in Fig. 1 is equivalent to

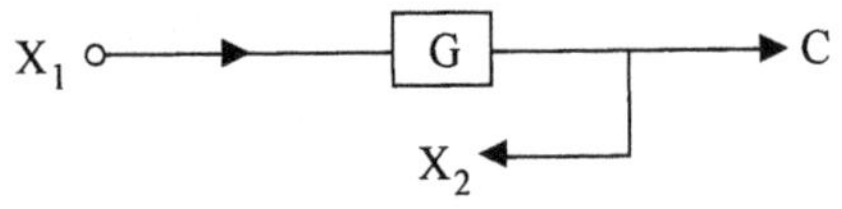

(a) 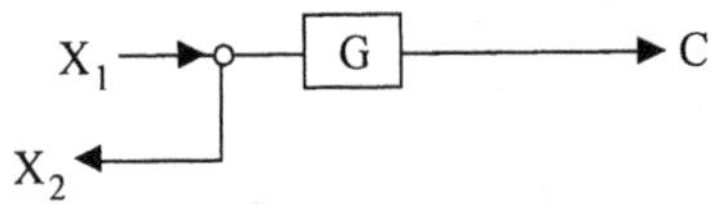

(b) 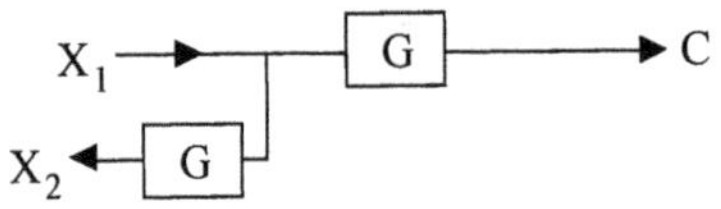

(c) 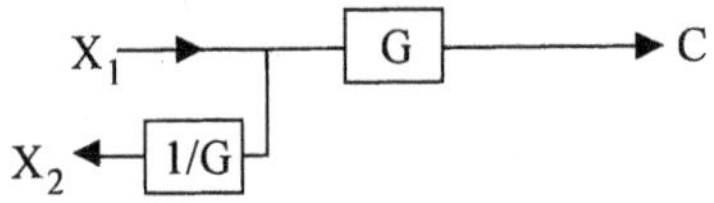

(d) 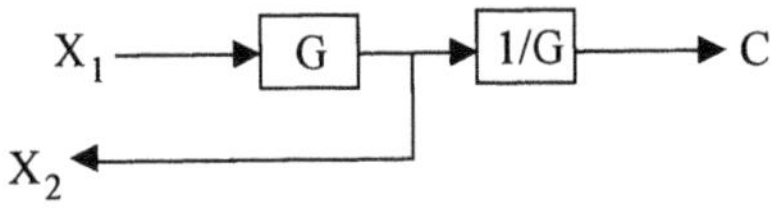

21. Consider the system shown in figure-I and figure-II. If the forward path gain is reduced by 10% in each system, then the variation in C_1 and C_2 will be respectively

(a) 10% and 10%

(b) 2% and 10%

(c) 5% and 1%

(d) 10% and 1%

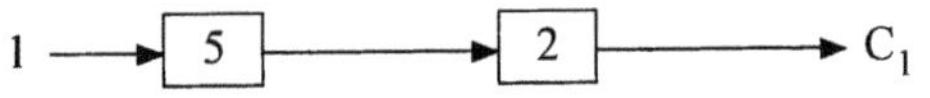

Figure I

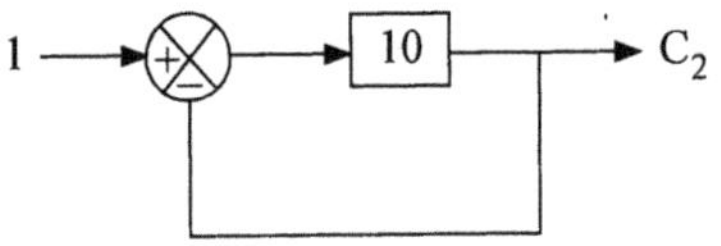

Figure II

22. The block diagrams shown in figure-I and figure-II are equivalent if 'X' (in figure-II) is equal to

(a) 1

(b) 2

(c) s + 1

(d) s + 2

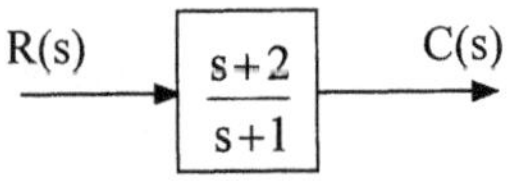

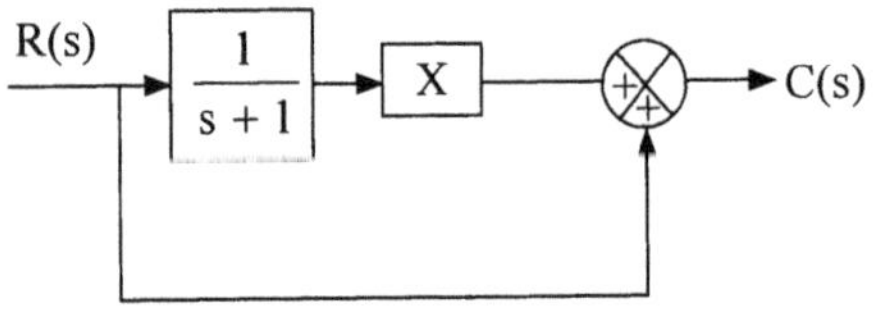

23. The signal flow graph shown in the given figure has

(a) three forward paths and two non-touching loops

(b) three forward paths and two loops

(c) two forward paths and two non-touching loops

(b) two forward paths and three loops

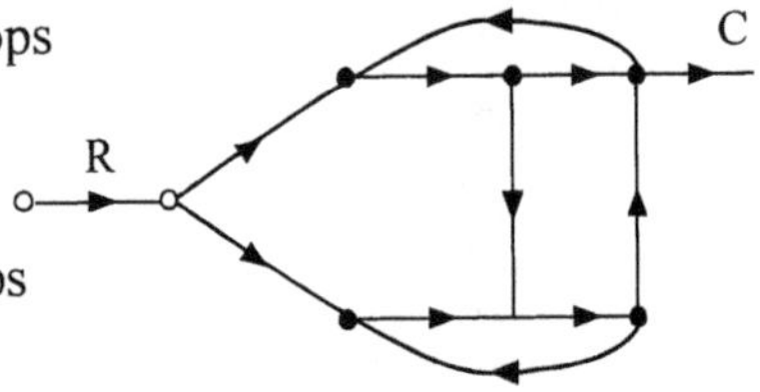

24. In the system shown in the given figure, to eliminate the effect of disturbance D(s) on C(s), the transfer function G_d (s) should be

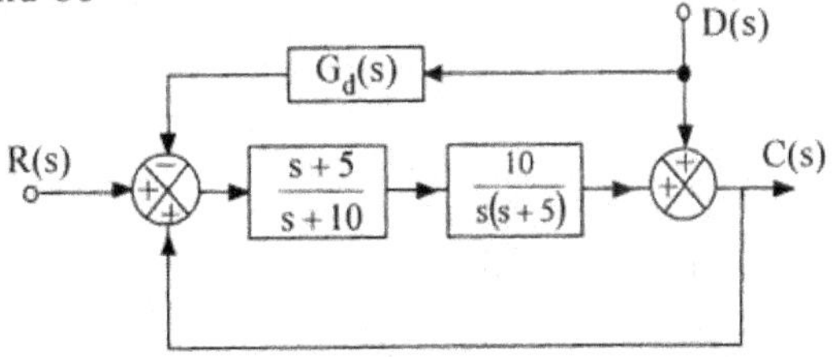

(a) $\frac{(s+10)}{10}$ (b) $\frac{s(s+10)}{10}$

(c) $\frac{10}{s+10}$ (d) $\frac{10}{s(s+10)}$

25. Match List-I (Component) with List-II (Transfer functions) and select the correct answer using the codes given below the lists :

List-I		List-II	
A.	ac servo motor	1.	$\frac{K}{s(1+s\tau_m)}$
B.	Field controlled dc servo motor	2.	$\frac{K}{s(1+s\tau_e)(1+s\tau_m)}$
C.	Tacho generator	3.	Ks
D.	Integrating gyro	4.	$\frac{K}{1+s\tau}$

Codes :

	A	B	C	D
(a)	1	2	3	4
(b)	1	2	4	3
(c)	2	1	3	4
(d)	2	1	4	3

26. A closed-loop system is shown in the given figure. The noise transfer function $\frac{C_n(s)}{N(s)}$

[C_n (s) = output corresponding to noise input N(s)] is approximately

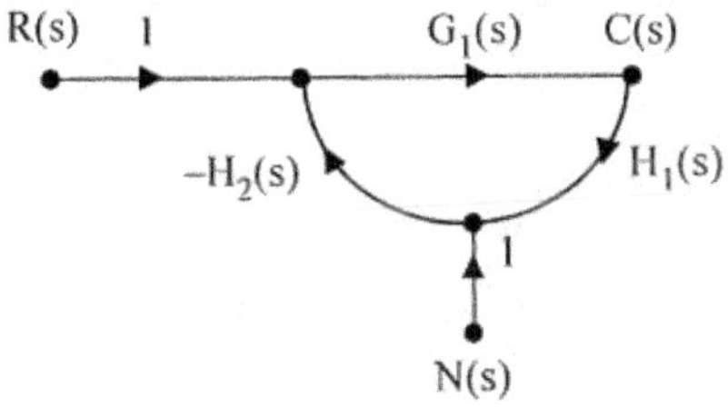

(a) $\frac{1}{G(s)H_1(s)}$ for $|G_1(s)H_1(s)H_2(s)| << 1$

(b) $-\frac{1}{H_1(s)}$ for $|G_1(s)H_1(s)H_2(s)| >> 1$

(c) $-\frac{1}{H_1(s)H_2(s)}$ for $|G_1(s)H_1(s)H_2(s)| >> 1$

(d) $\frac{1}{G(s)H_1(s)H_2(s)}$ for $|G_1(s)H_1(s)H_2(s)| << 1$

27. A signal flow graph is shown in the given figure. The number of forward paths M and the number of individual loops P for this signal flow graph would be

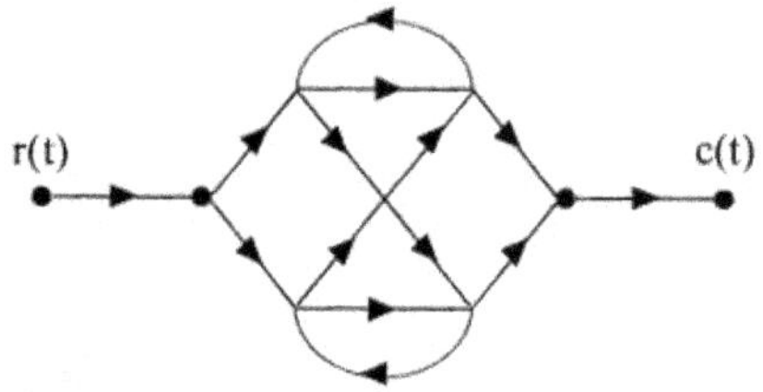

(a) M = 4 and P = 3

(b) M = 6 and P = 3

(c) M = 4 and P = 6

(d) M = 6 and P = 6

28. The mechanical system is shown in the given figure

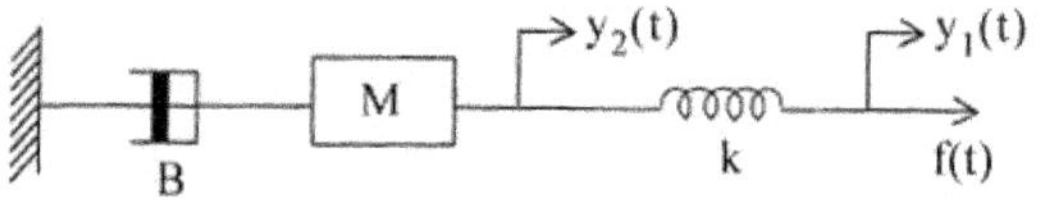

The system is described as

(a) $M\dfrac{d^2y_1(t)}{dt^2} + B\dfrac{dy_1(t)}{dt} = k[y_2(t) - y_1(t)]$

(b) $M\dfrac{d^2y_2(t)}{dt^2} + B\dfrac{dy_2(t)}{dt} = k[y_2(t) - y_1(t)]$

(c) $M\dfrac{d^2y_1(t)}{dt^2} + B = k[y_1(t) - y_2(t)]$

(d) $M\dfrac{d^2y_2(t)}{dt^2} + B\dfrac{dy_2(t)}{dt} = k[y_1(t) - y_2(t)]$

29. A synchro transmitter consists of a

(a) salient pole rotor winding excited by an ac supply and a three-phase balanced stator winding

(b) three-phase balanced stator winding excited by a three-phase balanced ac signal and rotor connected to a dc voltage source

(c) salient pole rotor winding excited by a dc signal

(d) cylindrical rotor winding and a stepped stator excited by pulses

30. The torque-speed characteristic of two-phase induction motor is largely affected by

(a) voltage (b) $\dfrac{R}{X}$ and speed (c) $\dfrac{X}{R}$ (d) supply voltage frequency

31. Consider the following statements regarding A.C. servometer :

1. The torque-speed curve has negative slope.
2. It is sensitive to noise.
3. The rotor has high resistance and low inertia
4. It has slow acceleration.

Which of the following are the characteristic of A.C. servomotor as control component?

(a) 1 and 2 (b) 2 and 3 (c) 1 and 3 (d) 2 and 4

32. Which of the following are the characteristics of closed-loop systems ?

1. It does not compensate for disturbances.
2. It reduces the sensitivity of plant-parameter variations.
3. It does not involve output measurements.
4. It has the ability to control the system transient response.

Select the correct answer using the codes given below:

(a) 1 and 4 (b) 2 and 4 (c) 1 and 3 (d) 2 and 3

33. The number of forward paths and the number of non-touching loop pairs for the signal flow graph given in the figure are, respectively,

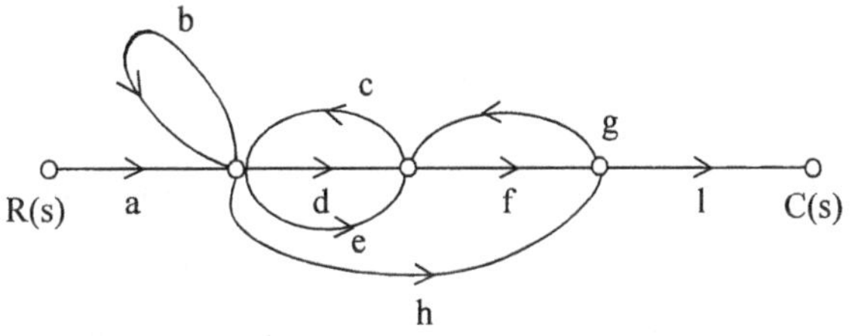

(a) 1, 3

(b) 3, 2

(c) 3, 1

(d) 2, 4

34. From the signal flow graph shown in the figure, the value of x_6 is :

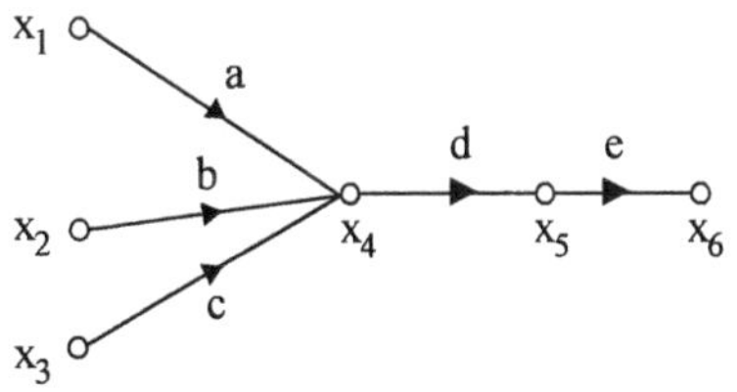

(a) de $(ax_1 + bx_2 + cx_3)$

(b) $(a + b + c)(x_1 + x_2 + x_3)(d + e)$

(c) $(ax_1 + bx_2 + cx_3)(d + e)$

(d) abcde $(x_1 + x_2 + x_3)$

35. From the figure shown, the transfer function of the signal flow graph is

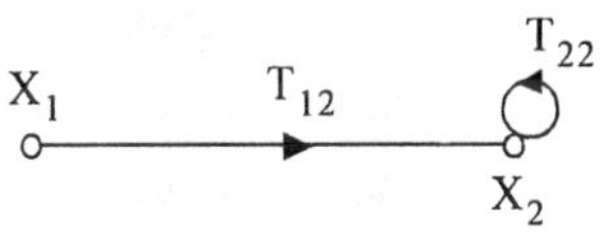

(a) $\dfrac{T_{12}}{1-T_{22}}$ (b) $\dfrac{T_{22}}{1-T_{12}}$

(c) $\dfrac{T_{12}}{1+T_{22}}$ (d) $\dfrac{T_{22}}{1+T_{12}}$

36. A stepper motor

(a) is a two phase induction motor

(b) is a kind of rotating amplifier

(c) is an electromagnetic transducer commonly used to convert an angular position of a shaft into an electrical system

(d) is an electromechanical device which-actuates a train of step angular (or linear) movements in response to a train of input pulses on one to one basis.

37. Match List-I with List-II and select the correct answer by using the codes given below the lists :

List-I	List-II
A. Synchro	1. Amplifier
B. Amplidyne	2. Actuator
C. Servo	3. Compensator
D. RC Network	4. Transducer

Codes :

	A	B	C	D		A	B	C	D
(a)	1	2	3	4	(b)	4	3	2	1
(c)	3	2	4	1	(d)	4	1	2	3

38. The open-loop transfer function of a unity feedback control system is $G(s) = \frac{1}{(s+2)^2}$

The closed-loop transfer function will have poles at

(a) −2, −2 (b) −2, −1 (c) $-2 \pm j1$ (d) −2, 2

39. The signal flow diagram of a system is shown in the given figure. The number of forward paths and the number of pairs of non-touching loops are respectively

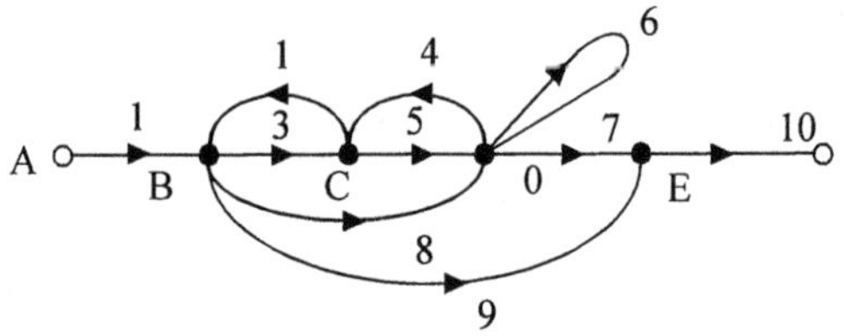

(a) 3, 1

(b) 3, 2

(c) 4, 2

(d) 2, 4

40. The ac motor used in servo applications is a

(a) single-phase induction motor (b) two-phase induction motor

(c) three-phase induction motor (d) synchronous motor

41. A synchro transmitter-receiver unit is a

(a) two-phase ac device (b) 3-phase ac device

(c) dc device (d) single-phase ac voltage device

42. Match the control system components in List-I with their functions in List-II and select the correct answer using the codes given below the lists :

	List-I		**List-II**
A.	Servo motor	1.	Error detector
B.	Amplidyne	2.	Transducer
C.	Potentiometer	3.	Actuator
D.	Flapper valve	4.	Power amplifier

Codes :

	A	B	C	D		A	B	C	D
(a)	3	1	2	4	(b)	3	4	1	2
(c)	4	3	2	1	(d)	3	4	2	1

43. An electromechanical device which actuates a train of step angular movements in response to a train of input pulses on one to one basis is

(a) synchro control transformer (b) LVDT
(c) stepper motor (d) ac tachogenerator

44. For a two-phase servo motor which one of the following statements is not true ?

(a) The rotor diameter is small
(b) The rotor resistance is low
(c) The applied voltages are seldom balanced
(d) The torque speed characteristics are linear

45. Which one of the following transducers is used to obtain the output position in a position control system ?

(a) Strain Gauge (b) Load cell (c) Synchro (d) Thermistor

46. For the system shown in the given figure the transfer function $\frac{X(s)}{F(s)}$ is

(a) $\frac{1}{Ms^2 + K}$ (b) $\frac{Ms^2 + 1}{K}$

(c) $\frac{K}{Ms^2 + 1}$ (d) $\frac{1}{Ks^2 + M}$

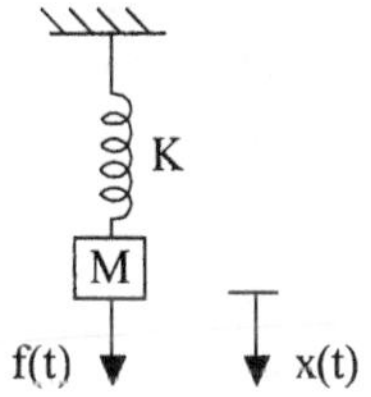

47. In case of synchro error detector, the electrical zero position of control-transformer is obtained when angular displacement between rotors is

(a) zero (b) 45° (c) 90° (d) 180°

48. The transfer function T(s) of the system shown in the following figure is given by

(a) $T(s) = \dfrac{G_1(s)G_2(s)}{1-G_2(s)}$

(b) $T(s) = \dfrac{G_1(s)}{1-G_1(s)G_2(s)}$

(c) $T(s) = \dfrac{G_2(s)}{1-G_1(s)G_2(s)}$

(d) $T(s) = \dfrac{G_2(s)}{1+G_1(s)G_2(s)}$

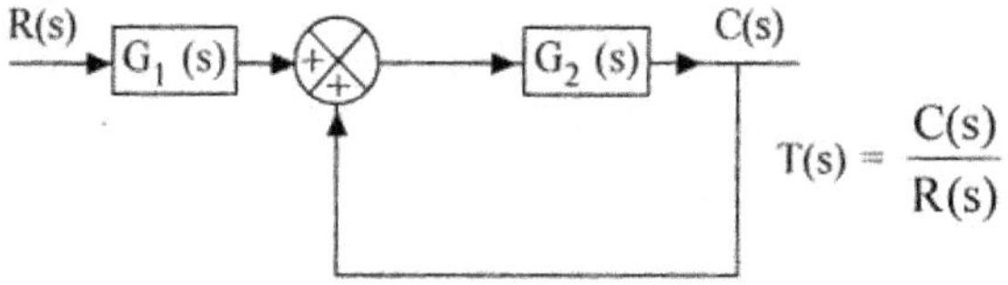

49. In a two-phase ac servometer, the rotor has a resistance R and a reactance X. The torque-speed characteristic of the servomotor will be linear provided that

(a) $\dfrac{X}{R} << 1$ (b) $\dfrac{X}{R} >> 1$ (c) $\dfrac{X}{R} = 1$ (d) $X^2 = R$

50. The $\dfrac{C(s)}{R(s)}$ for the system shown in the following block diagram is

(a) $\dfrac{G_1(s)G_2(s)}{1+G_1(s)[G_2(s)+H_1(s)]}$

(b) $\dfrac{G_1(s)G_2(s)}{1+G_2(s)[G_1(s)+H_1(s)]}$

(c) $\dfrac{G_1(s)+G_2(s)}{1+G_2(s)[G_2(s)+H_1(s)]}$

(d) none of the above

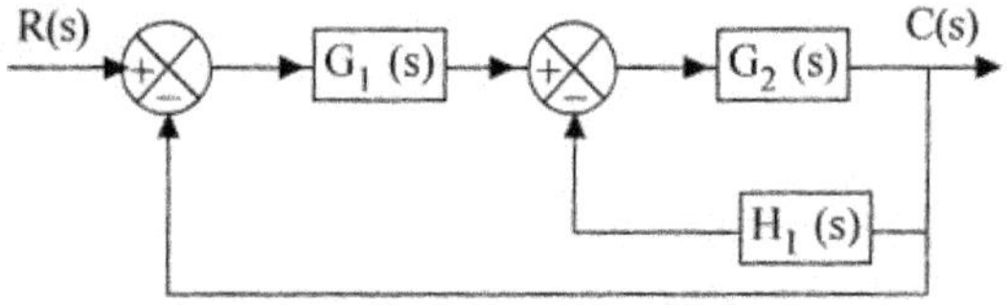

51. Which of the following can work as error detecting devices ?

1. A pair of potentiometers
2. A pair of synchros
3. A metadyne
4. A control transformer

Select the correct answer using the codes given below :

Codes :

(a) 1, 2 (b) 2, 3, 4

(c) 1, 3, 4 (d) 1, 2, 4

52. The sensitivity S_G^M of a system with the transfer function $M = \dfrac{G}{1+GH}$ is given by

(a) $\dfrac{1}{1+GH}$ (b) $\dfrac{1+GH}{H}$ (c) $\dfrac{1+G}{H}$ (d) H

53. The signal graph of a closed-loop system is shown in the figure, wherein T_D represents the distrubance in the forward path :

The effect of the disturbance can be reduced by

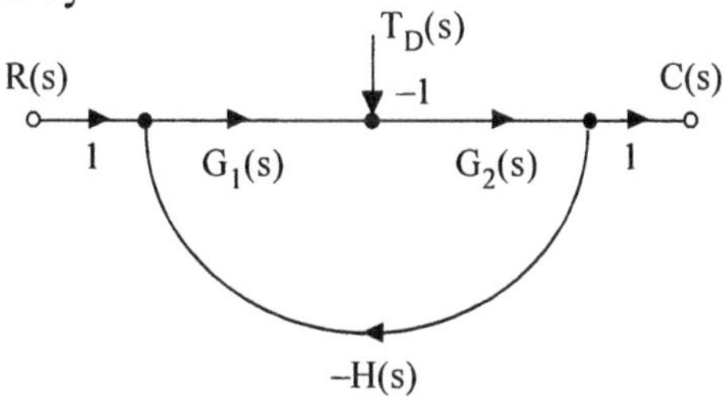

(a) increasing G_2 (s)

(b) decreasing G_2 (s)

(c) increasing G_1 (s)

(d) decreasing G_1 (s)

54. Which is the following relate to rational transfer function of a system ?

1. Ratio of Fourier transform of output to input with zero initial conditions
2. Ratio of Laplace transform of output to input with zero initial conditions
3. Laplace transform of system impulse response
4. Laplace transform of system unit step response

Select the correct answer using the codes given below codes :

(a) 1 and 4 (b) 2 and 3 (c) 1 and 3 (d) 2 and 4

55. The closed-loop system shown in the figure is subjected to a disturbance N(s). The transfer function $\dfrac{C(s)}{N(s)}$ is given by

(a) $\dfrac{G_1(s)G_2(s)}{1+G_1(s)G_2(s)H(s)}$

(b) $\dfrac{G_1(s)}{1+G_1(s)H(s)}$

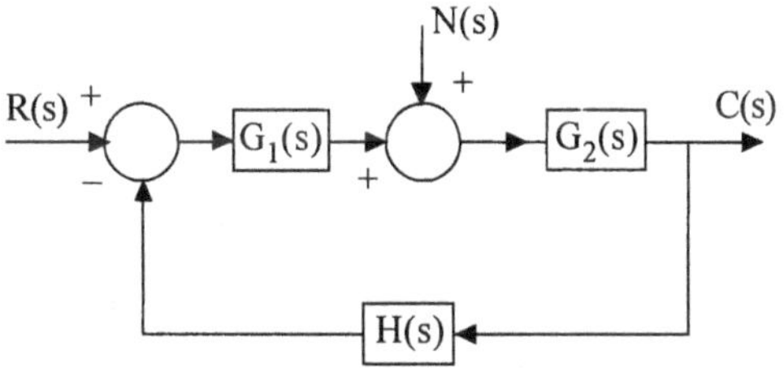

(c) $\dfrac{G_2(s)}{1-G_2(s)H(s)G_1(s)}$

(d) $\dfrac{G_2(s)}{1+G_1(s)G_2(s)H(s)}$

56. Which of the following components can be used as a rotating amplifier in a control system ?

1. An amplidyne
2. A separatively excited dc generator
3. A self-excited dc generator
4. A synchro.

Select the correct answer using the codes given below

Codes :

(a) 3 and 4 (b) 1 and 2 (c) 1, 2 and 3 (d) 1, 2, 3 and 4

57. The transfer function of the system shown in the given figure is :

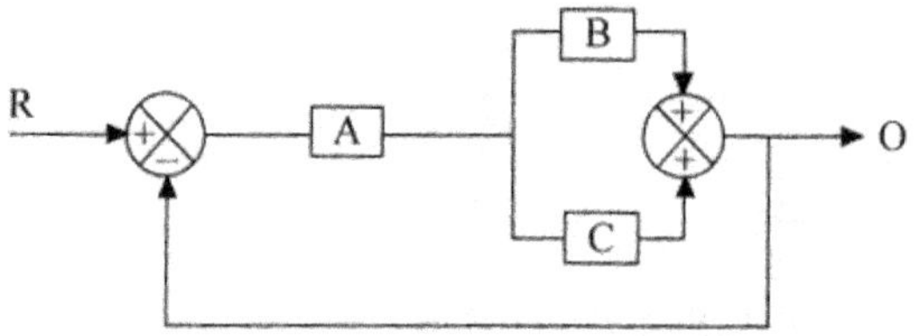

(a) $O/R = \dfrac{ABC}{1+ABC}$

(b) $O/R = \dfrac{A+B+C}{1+AB+AC}$

(c) $O/R = \dfrac{AB+AC}{ABC}$

(d) $O/R = \dfrac{AB+AC}{1+AB+AC}$

58. A signal flow graph is shown in the following figure :

Consider the following statements regarding the signal flow graph :

1. There are three forward paths.
2. There are three individual loops.
3. There are three sets of two non-touching loops.

Of these statements

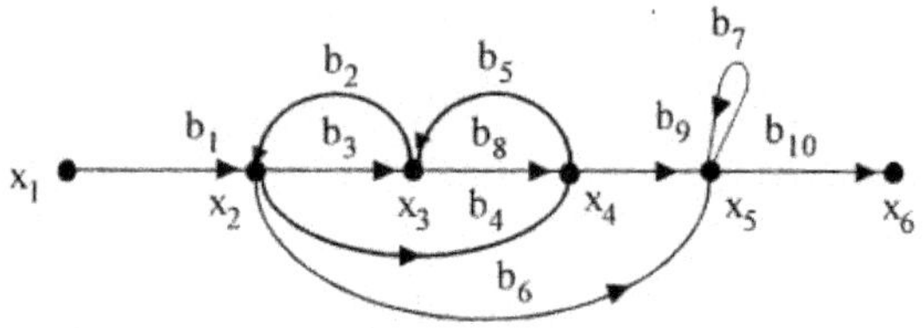

(a) 1, 2, and 3 are correct

(b) 1 and 2 are correct

(c) 2 and 3 are correct

(d) 1 and 3 are correct

59. In the field-controlled motor, the entire damping comes from

(a) the armature resistance (b) the back emf

(c) the motor friction and load (d) field resistance

60. Which of the following rotors are used in a two-phase ac servomotor ?

1. Solid iron motor 2. Squirrel cage rotor 3. Drag cup rotor

Select the correct answer using the codes given below.

Codes :

(a) 1, 2 and 3 (b) 1 and 2 (c) 2 and 3 (d) 1 and 3

61. For two phase a.c. servomotor, if the rotor's resistance and reactance are respectively R and X, its length and diameter are respectively L and D, then

(a) $\dfrac{X}{R}$ and $\dfrac{L}{D}$ are both small (b) $\dfrac{X}{R}$ is large but $\dfrac{L}{D}$ is small

(c) $\dfrac{X}{R}$ is small but $\dfrac{L}{D}$ is large (d) $\dfrac{X}{R}$ and $\dfrac{L}{D}$ are both large

62. Consider the following statements relating to synchros :

1. The rotor of the control transformer is either disc shaped or umbrella shaped.
2. The rotor of the transmitter is so constructed as to have a low magnetic reluctance.
3. Transmitter and control transformer pair is used as an error detector.

Which of these statemens are correct ?

(a) 1, 2 and 3 (b) 1 and 2 (c) 2 and 3 (d) 1 and 3

63. Consider the following servomotors :

1. a.c. two-phase servomotor
2. d.c. servomotor
3. Hydraulic servomotor
4. Pneumatic servomotor

The motor which has highest power handling capacity is

(a) 2 (b) 1 (c) 3 (d) 4

64. Match List I (Functional components) with List II (Devices) and select the correct answer using the codes given below the Lists :

List I	List II
A. Error detector	1. Three-phase FHP induction motor
B. Servomotor	2. A pair of synchronous transmitter and control transformer
C. Amplifier	3. Tachogenerator
D. Feedback	4. Armature controlled FHP d.c. motor
	5. Amplidyne

Codes :

(a)	A	B	C	D
	2	4	1	5
(b)	A	B	C	D
	4	2	5	3
(c)	A	B	C	D
	2	4	5	3
(d)	A	B	C	D
	1	2	3	5

65. For the signal flow diagram shown in the given figure, the transmittance between x_2 and x_1 is

(a) $\frac{rsu}{1-st}+\frac{efh}{1-fg}$

(b) $\frac{rsu}{1-fg}+\frac{efh}{1-st}$

(c) $\frac{efh}{1-ru}+\frac{rsu}{1-eh}$

(d) $\frac{rst}{1-eh}+\frac{rsu}{1-st}$

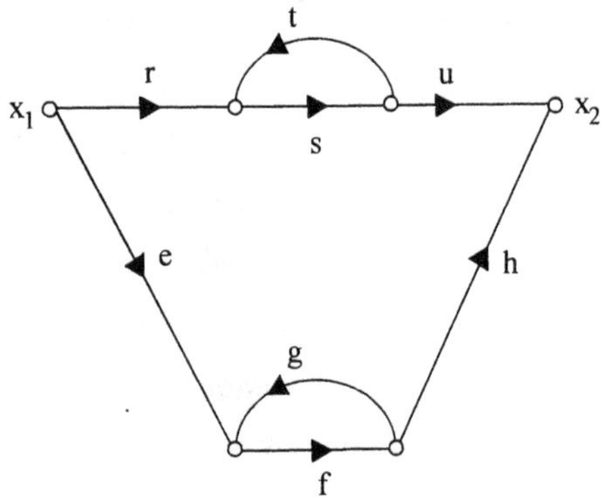

66. Consider the mechanical system shown in the given figure. If the system is set into motion by unit impulse force, the equation of the resulting oscillation will be

(a) $x(t) = \sin t$

(b) $x(t) = \sqrt{2} \sin t$

(c) $x(t) = \frac{1}{2} \sin 2t$

(d) $x(t) = \sqrt{2}\, t$

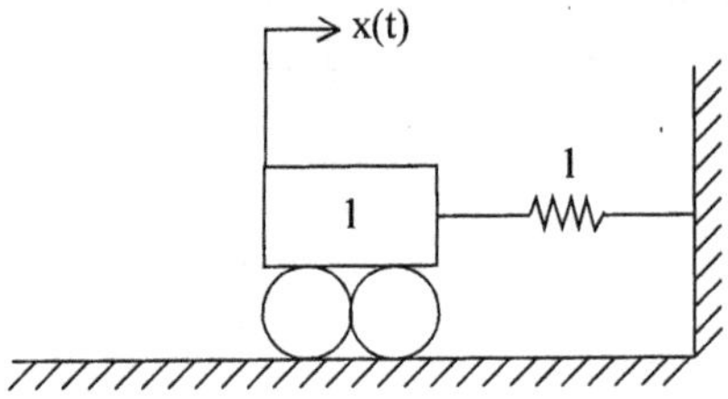

67. Which one of the following relations holds goods for the tachometer shown in the given figure ?

(a) $V_2(s) = s\, k_t\, \omega(s)$

(b) $V_2(s) = k_t\, s^2\, \theta(s)$

(c) $V_2(s) = k_t\, s^2\, \omega(s)$

(d) $V_2(s) = k_t\, s\, \theta(s)$

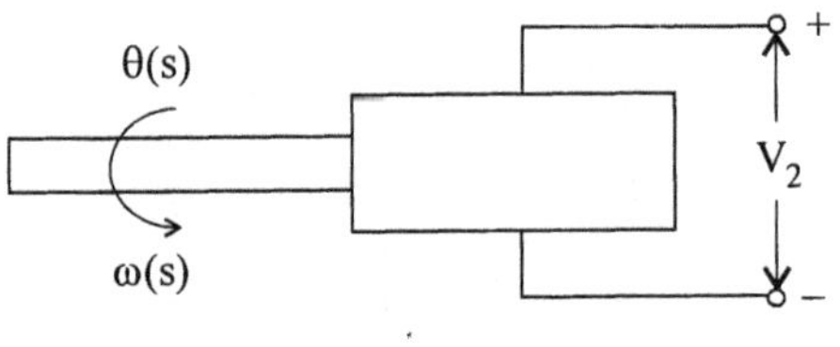

68. The unit step response of a particular control system is given by $c(t) = 1 - 10e^{-t}$. Then its transfer function is

(a) $\frac{10}{s+1}$ (b) $\frac{s-9}{s+1}$ (c) $\frac{1-9s}{s+1}$ (d) $\frac{1-9s}{s(s+1)}$

69. 1. Transfer function can be obtained from the signal flow graph of the system

2. Transfer function typically characterizes linear time variant system

3. Block diagram of the system can be obtained from its transfer function given the ratio of output to input in frequency domain of the system.

4. Transfer function gives the ratio of output to input in frequency domain of the system.

Which of the following is the correct combination about the four statements stated above.

(a) only (1) and (2) are correct
(b) only (2), (3) and (4) are correct
(c) only (3) and (4) are correct
(d) only (1), (2) and (4) are correct

70. Which of the following is not valid in case of signal flow graph ?

(a) in signal flow graph signals travel along branches only in the marked direction
(b) nodes are arranged from right to left in a sequence
(c) signa flow graph is applicable to linear systems only
(d) for signal flow graph, the algebraic equations must be in the form of cause and effect relationship.

71. The sum of the gains of the feedback paths in the signal flow graph below is

(a) af + be + cd + cbef + abcdef
(b) af + be + cd + abef + bcde
(c) af + be + cd + abef + abcdef
(d) af + be + cd

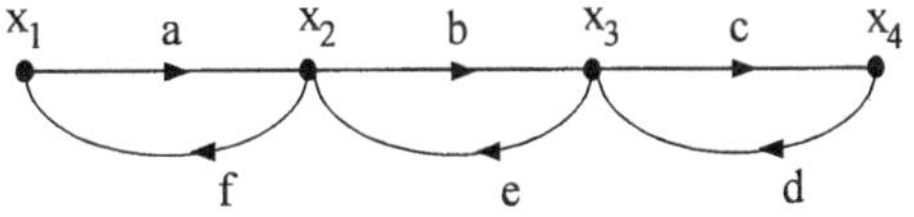

72. The signal flow graph shown in the figure has

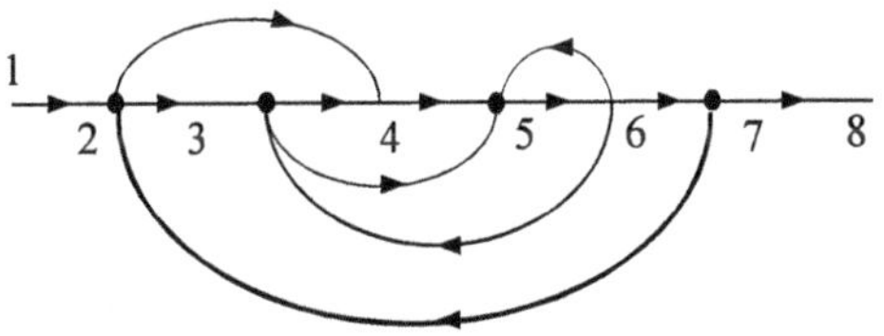

(a) two forward paths, four loops and no non-touching loops
(b) three forward paths, four loops and no non-touching loops
(c) three forward paths, three loops and no non-touching loops
(d) two forward paths, four loops and two non-touching loops

73. Consider the system shown in the block diagram in the figure. The signal flow diagram of the system is best represented as

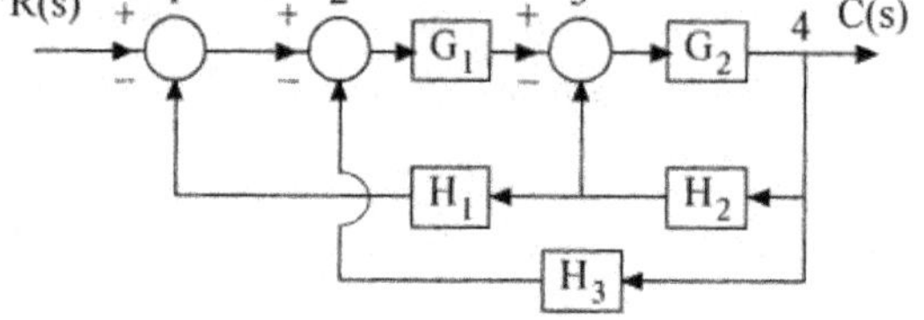

(a)

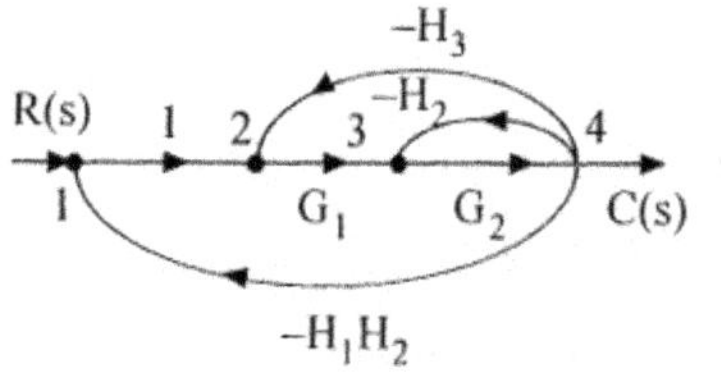

(b)

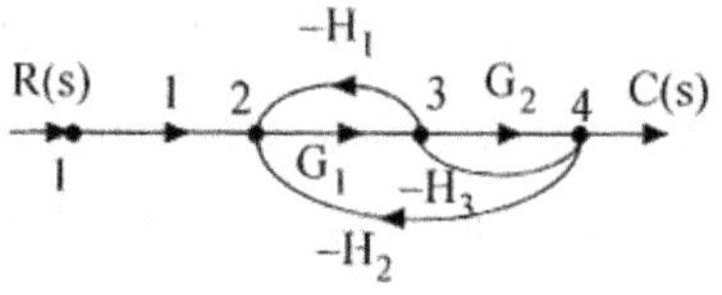

(c)

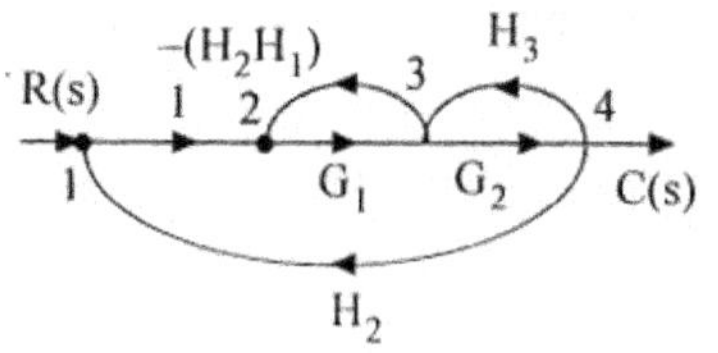

(d)

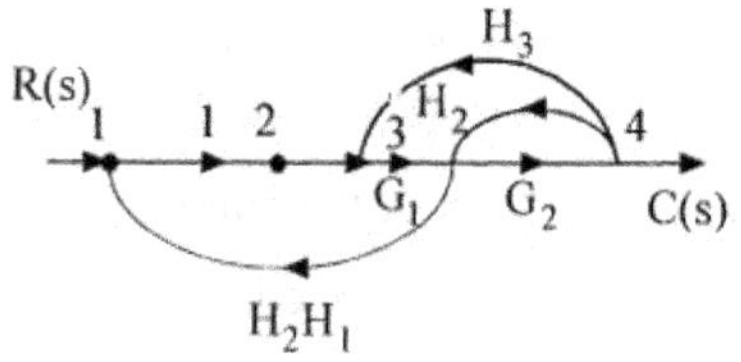

74. The transfer function C(s)/R(s) of the system, whose block diagram is given below is

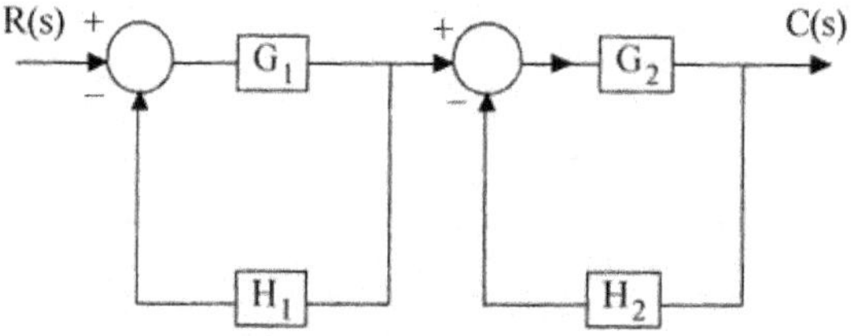

(a) $\dfrac{G_1G_2}{1+G_1H_1+G_2H_2-G_1G_2H_1H_2}$

(b) $\dfrac{G_1G_2}{1+G_1H_1+G_2H_2+G_1G_2H_1H_2}$

(c) $\dfrac{G_1G_2}{1+G_1H_1+G_2H_2}$

(d) $\dfrac{G_1(1+G_2H_2)+G_2(1+G_1H_1)}{1+G_1H_1+G_2H_2+G_1G_2H_1H_2}$

75. Which one of the four signal flow graphs shown in (a), (b), (c) and (d) represents the bock diagram shown in the given figure ?

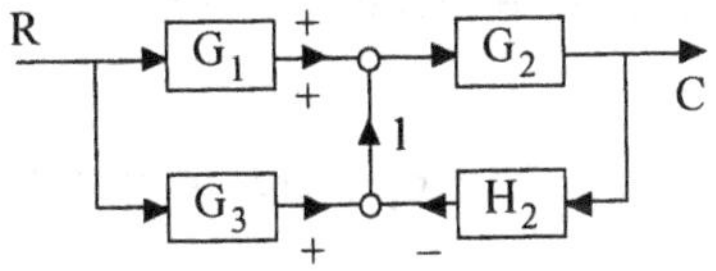

(a)
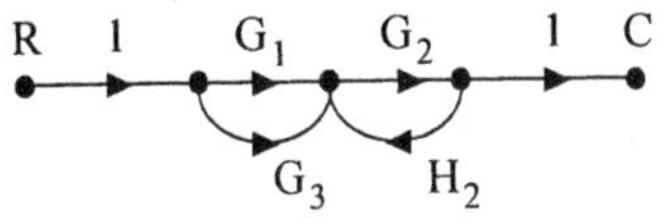

(b)
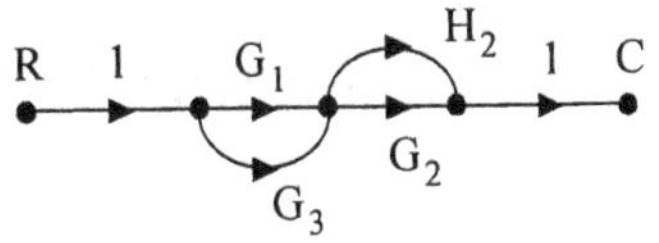

(c)
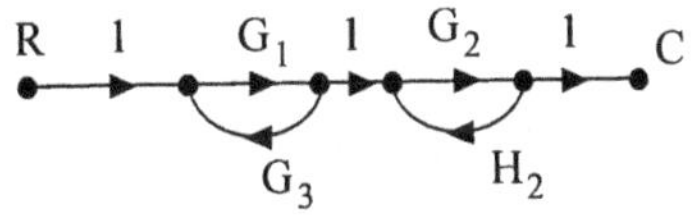

(d)
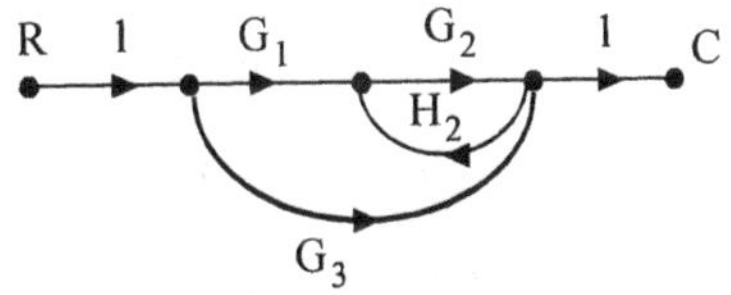

76. In the figure alongside, spring constant is K, viscous friction coefficient is B, mass is M and the system output motion is x(t) corresponding to input force F(t). Which of the following parameters relate to the above system ?

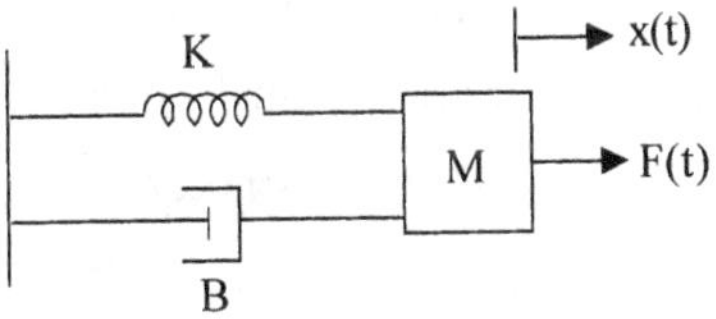

1. Time constant $= \dfrac{1}{M}$

2. Damping coefficient $= \dfrac{B}{2\sqrt{KM}}$

3. Natural frequency of oscillation $= \sqrt{\dfrac{K}{M}}$

Select the correct answer using the codes given below :

Codes :

(a) 1, 2 and 3 (b) 1 and 2 (c) 2 and 3 (d) 1 and 2

77. In the feedback system shown in the given figure, the noise component of output is given by (assume high loop gain at frequencies of interest)

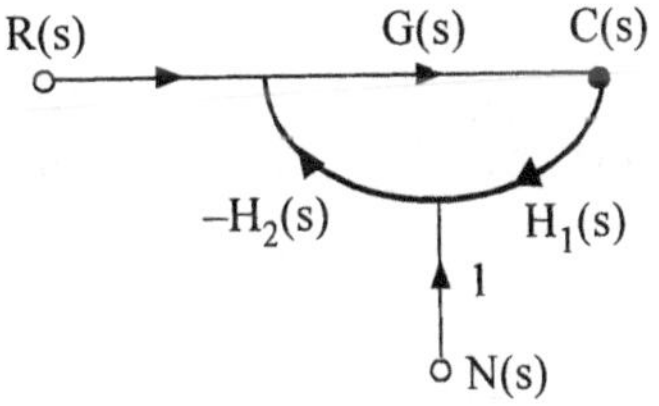

(a) $\dfrac{-N(s)}{H_1(s)}$ (b) $\dfrac{N(s)}{H_1(s)}$

(c) $\dfrac{N(s)}{H_1(s)H_2(s)}$ (d) $\dfrac{-N(s)}{H_1(s)H_2(s)}$

78. Consider the following statements regarding the advantages of closed-loop negative feedback control-systems over open-loop systems :

1. The overall reliability of the closed-loop systems is more than that of open-loop system.
2. The transient response in the closed-loop system decays more quickly than in open-loop system.
3. In an open-loop system, closing of the loop increases the overall gain of the system.
4. In the closed-loop system, the effect of variation of component parameters on its performance is reduced.

Of these statements :

(a) 1 and 3 are correct (b) 1 and 2 are correct

(c) 2 and 4 are correct (d) 3 and 4 are correct

79. Match List I with List II and select the correct answer using the codes given below the lists :

List I (Unit)	List II (Type of rotor)
A. Synchro transmitter	1. Dumb-bell rotor
B. Control Transformer	2. Drag-cup rotor
C. A.C. Servo-motor	3. Cylindrical rotor
D. Stepper motor	4. Toothed rotor
	5. Phase wound rotor

Codes :

	A	B	C	D
(a)	1	3	2	4
(b)	1	5	3	2
(c)	2	4	3	1
(d)	3	2	1	5

80. In the signal flow graph shown in the figure, the value of the C/R ratio is

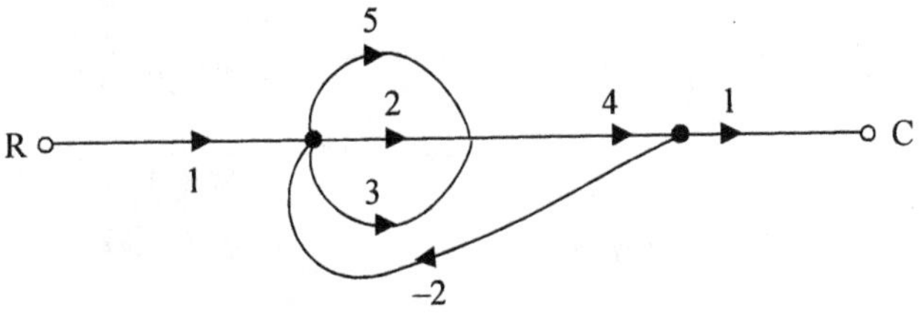

(a) $\frac{28}{57}$ (b) $\frac{40}{57}$

(c) $\frac{40}{81}$ (d) $\frac{28}{81}$

81. Consider the following statements :

1. The effect of feedback is to reduce the system error.
2. Feedback increases the gain of the system in one frequency range but decreases in another.
3. Feedback can cause a system that is originally stable to become unstable.

(a) 1, 2 and 3 (b) 1 and 2 (c) 2 and 3 (d) 1 and 3

82. Select the correct transfer function $\frac{V_0(s)}{V_1(s)}$ from the following, for the given system,

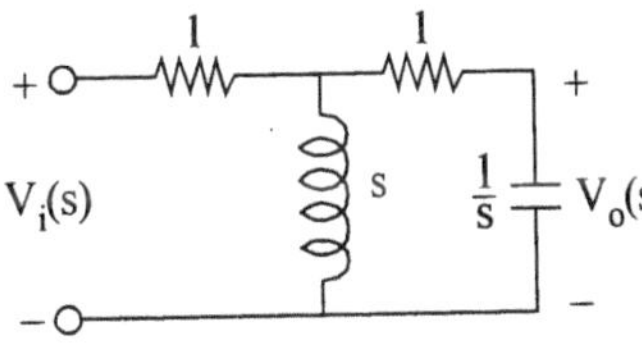

(a) $\frac{1}{2(s^2+s+1)}$ (b) $\frac{s}{2(s+1)^2}$

(c) $\frac{s}{2s^2+2s+1}$ (d) $\frac{1}{2s^2+2s+1}$

83. Consider a control system shown in the given figure. For a slight variation in G, the ratio of open-loop sensitivity to closed-pool-sensitivity will be given by

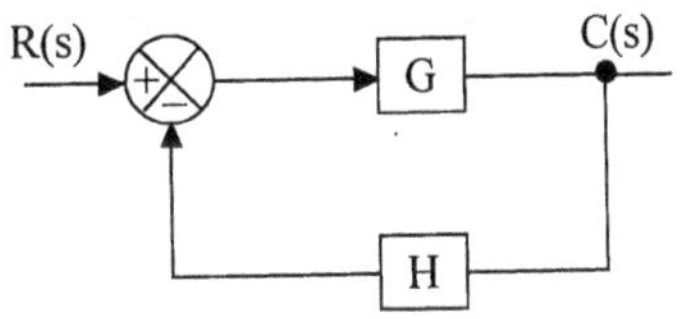

(a) $1 : (1 + GH)$

(b) $1 : (1 + GH)^{-1}$

(c) $1 : (1 - GH)$

(d) $1 : (1 - GH)^{-1}$

84. Consider the following statements relating to synchros :

1. The rotor of the control transformer is cylindrical.
2. The rotor of the transmitter is so constructed as to have a low magnetic reluctance.
3. Transmitter and control transformer pair is used as an error detector.

Which of these statemens are correct ?

(a) 1, 2 and 3 (b) 1 and 2 (c) 2 and 3 (d) 1 and 3

Chapter 3

1. Match List (transfer functions) with List II (impulse response) and select the correct answer using the codes given below the lists :

List I

A. $\frac{1}{s(s+1)}$

B. $\frac{1}{(s+1)^2}$

C. $\frac{1}{s(s+1)+1}$

D. $\frac{1}{s^2+1}$

List II

1.
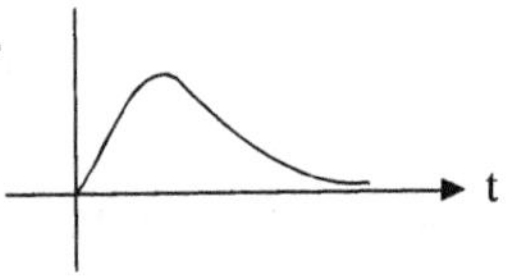

2.
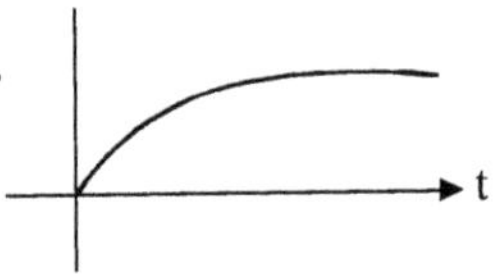

3.
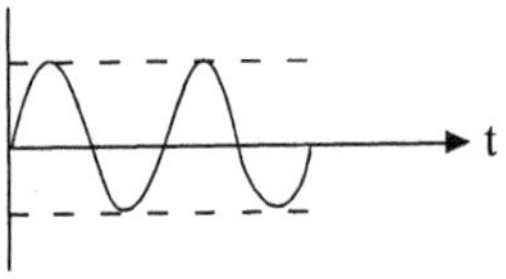

4.
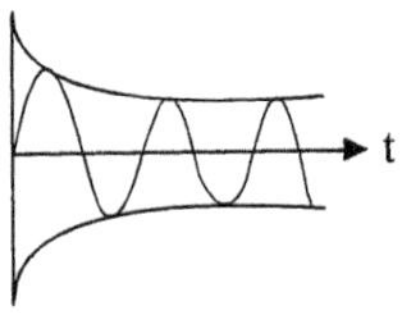

Codes :

	A	B	C	D
(a)	2	1	4	3
(b)	1	2	4	3
(c)	2	1	3	4
(d)	1	2	3	4

2. The unit-impulse response of a unity-feedback system is given by $c(t) = -te^{-t} + 2e^{-t}$, $(t \geq 0)$ The open-loop transfer function is equal to

(a) $\frac{s+1}{(s+2)^2}$ (b) $\frac{2s+1}{s^2}$ (c) $\frac{s+2}{(s+1)^2}$ (d) $\frac{s+1}{s^2}$

3. Consider the unit-step response of a unity-feedback control system whose open-loop transfer function is $G(s) = \frac{1}{s(s+1)}$. The maximum overshoot is equal to

(a) 1.143 (b) 0.153 (c) 0.163 (d) 0.173

4. For a feedback control system of type 2, the steady state error for a ramp input is

(a) infinite (b) constant (c) zero (d) indeterminate

5. The closed-loop transfer function of a control system is given by $\frac{C(s)}{R(s)} = \frac{1}{1+s}$. For the input $r(t) = \sin t$, the steady state value of c (t) is equal to

(a) $\frac{1}{\sqrt{2}} \cos t$ (b) 1 (c) $\frac{1}{\sqrt{2}} \sin t$ (d) $\frac{1}{\sqrt{2}} \sin\left(t - \frac{\pi}{4}\right)$

6. For the system shown in figure with a damping ratio ξ of 0.7 and an undamped natural frequency ω_n of 4 rad/sec, the values of K and a are

(a) K = 4, a = 0.35

(b) K = 8, a = 0.455

(c) K = 16, a = 0.225

(d) K = 64, a = 0.9

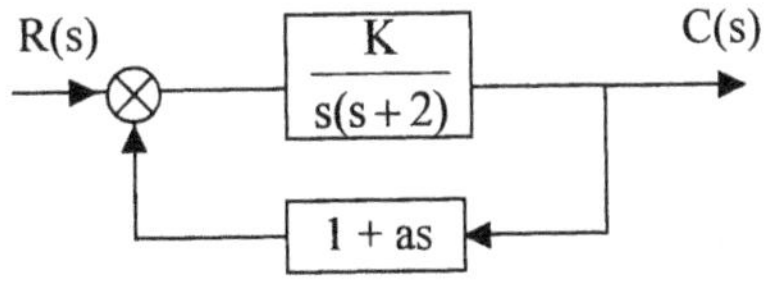

7. A unity feedback system has open-loop transfer function G(s). The steady-state error is zero for

(a) step input and type-1 G(s)
(b) ramp input and type-1 G(s)
(c) step input and type-0 G(s)
(d) ramp input and type-0 G(s)

8. A linear time-invariant system initially at rest, when subjected to a unit-step input, gives a response $y(t) = te^{-t}$, $t > 0$. The transfer function of the system is,

(a) $\dfrac{1}{(s+1)^2}$ (b) $\dfrac{1}{s(s+1)^2}$ (c) $\dfrac{s}{(s+1)^2}$ (d) $\dfrac{1}{s(s+1)}$

9. A unity feedback system has open-loop transfer function G(s) = {25/[s (s+6)]}. The peak overshoot in the step-input response of the system is approximately equal to

(a) 5% (b) 10% (c) 15% (d) 20%

10. Introduction of integral action in the forward path of a unity feedback system results in a

(a) Marginally stable system
(b) System with no steadystate error
(c) System with increased stability margin
(d) System with better speed of response

11. For a unit step input, a system with forward path transfer function $G(s) = \dfrac{20}{s^2}$ and feedback path transfer function
H(s) = (s + 5), has a steady state output of

(a) 20 (b) 5 (c) 0.2 (d) zero

12. Consider a system shown in the given figure :

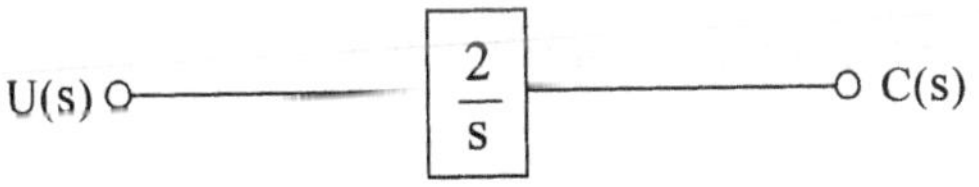

If the system is disturbed so that c(0) = I, then c(t) for a unit step input will be

(a) 1 + t (b) 1 − t (c) 1 + 2t (d) 1 − 2t

13. What will be the closed-loop transfer function of a unity feedback control system whose step response is given by $c(t) = k[1 - 1.66\, e^{-8t} \sin (6t + 37^o)]$?

(a) $\frac{100k}{s^2 + 16s + 100}$ (b) $\frac{10}{s^2 + 16s + 100}$

(c) $\frac{k}{s^2 + 16s + 100}$ (d) $\frac{10\, k}{s^2 + 8s + 100}$

14. The transfer function of a control system is given as $T(s) = \frac{K}{s^2 + 4s + K}$

where K is the gain of the system in radians/Amp.

For this system to be critically damped, the value of K should be

(a) 1 (b) 2 (c) 3 (d) 4

15. A linear system, initially at rest, is subject to an input signia $r(t) = 1 - e^{-t}\ (t \geq 0)$

The response of the system for $t > 0$ is given by $c(t) = 1 - e^{-2t}$

The transfer function of the system is

(a) $\frac{(s+2)}{(s+1)}$ (b) $\frac{(s+1)}{(s+2)}$ (c) $\frac{2(s+1)}{(s+2)}$ (d) $\frac{1(s+1)}{2(s+2)}$

16. If the time response of a system is given by the following equation

$$y(t) = 5 + 3 \sin (\omega t + \delta_1) + e^{-3t} \sin (\omega t + \delta_2) + e^{-5t}$$

then the steady-state part of the above response is given by

(a) $5 + 3 \sin (\omega t + \delta_1)$ (b) $5 + 3 \sin (\omega t + \delta_1) + e^{-3t} \sin (\omega t + \delta_2)$

(c) $5 + e^{-5t}$ (d) 5

17. The impulse response of a system is $5\, e^{-10t}$. Its step response is equal to

(a) $0.5\, e^{-10t}$ (b) $5(1 - e^{-10t})$ (c) $0.5\, (1 - e^{-10t})$ (d) $10(1 - e^{-10t})$

18. The transfer function of a system is $\frac{10}{1+s}$. When operated as a unity feedback system, the steady-state error to a unit step input will be

(a) zero (b) $\frac{1}{11}$ (c) 10 (d) infinity

19. A unity feedback second order control system is characterised by $G(s) = \dfrac{K}{s(Js + B)}$

Where J = moment of inertia, K = system gain, B = viscous damping coefficient. The transient response specification which is NOT affected by variation of system gain is the

(a) peak overshoot (b) rise time
(c) settling time (d) damped frequency of oscillations

20. In the control system shown in the given figure, the controller which can give zero steady-state error to a ramp input, with K = 9 is

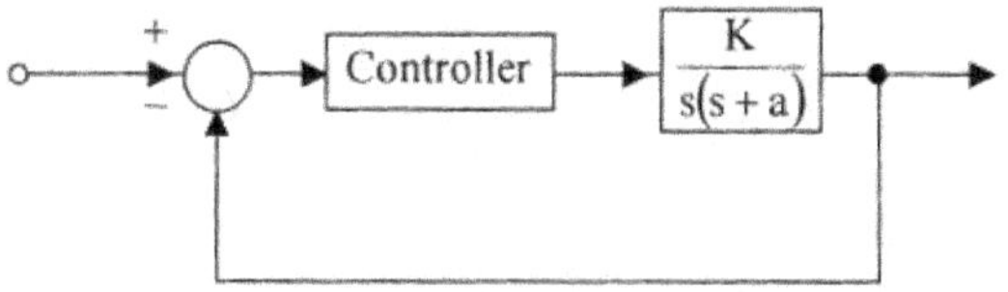

(a) proportional type
(b) integral type
(c) derivative type
(d) proportional plus derivative type.

21. A linear second-order system with the transfer function $G(s) = \dfrac{49}{s^2 + 16s + 49}$ is initially at rest and is subject to a step input signal. The response of the system will exhibit a peak overshoot of

(a) 16% (b) 9% (c) 2% (d) zero

22. A system has the following transfer function : $G(s) = \dfrac{100(s+5)(s+50)}{s^4(s+10)(s^2+3s+10)}$

The type and order of the systems are respectively

(a) 4 and 9 (b) 4 and 7 (c) 5 and 7 (d) 7 and 5

23. When the input to a system was withdrawn at t = 0, its output was found to decrease exponentially from 1000 units to 500 units in 1.386 seconds. The time constant of the system is

(a) 0.500 (b) 0.693 (c) 1.386 (d) 2.000

24. For the system shown in the given figure, the steady-state value of the output c(t) is

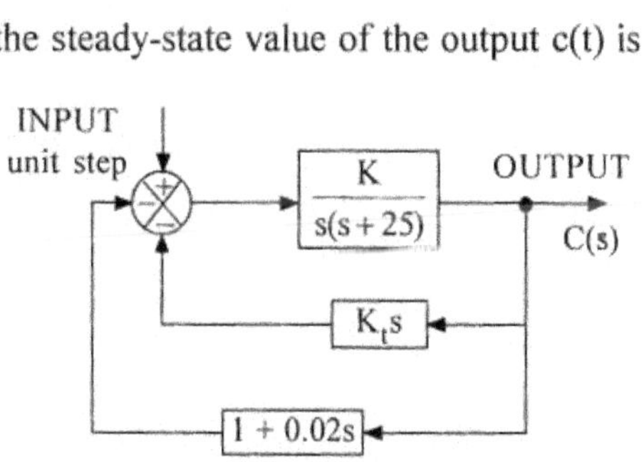

(a) 0

(b) 1

(c) ∞

(d) dependent of the values of K and K_1

25. The unit impulse response of a linear time-invariant second-order system is :

$$g(t) = 10\, e^{-8t} \sin 6t \ (t \geq 0).$$

The natural frequency and the damping factor of the system are respectively

(a) 10 rad/s and 0.6 (b) 10 rad/s and 0.8

(c) 5 rad/s and 0.6 (d) 6 rad/s and 0.8

26. $[-a \pm jb]$ are the complex conjugate roots of the characteristic equation of a second order system. Its damping coefficient and natural frequency will be respectively

(a) $\frac{b}{\sqrt{a^2+b^2}}$ and $\sqrt{a^2+b^2}$ (b) $\frac{b}{\sqrt{a^2+b^2}}$ and $a^2 + b^2$

(c) $\frac{a}{\sqrt{a^2+b^2}}$ and $\sqrt{a^2+b^2}$ (d) $\frac{a}{\sqrt{a^2+b^2}}$ and $a^2 + b^2$

27. A unity feedback control system has a forward path transfer function $G(s) = \frac{10(1+4s)}{s^2(1+s)}$.

If the system is subjected to an input $r(t) = 1 + t + \frac{t^2}{2}$ $(t \geq 0)$, the steady-state error of the system will be

(a) zero (b) 0.1 (c) 10 (d) infinity

28. In the system shown in the given figure, $r(t) = 1 + 2t$ $(t > 0)$. The steady-state value of the error e(t) is equal to

(a) zero

(b) 2/10

(c) 10/2

(d) infinity

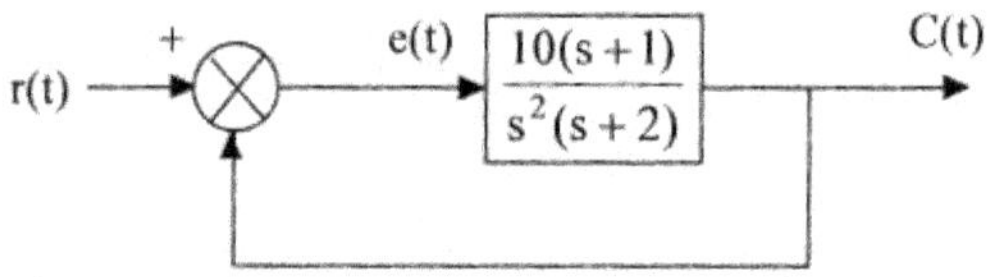

29. The steady state error due to a ramp input for a type two system is equal to

(a) zero (b) infinite

(c) non-zero number (d) constant

30. A second order control system is defined by the following differential equation :

$$4\frac{d^2 c(t)}{dt^2} + 8\frac{d\,c(t)}{dt^2} + 16\, c(t) = 16\, u(t)$$

The damping ratio and natural frequency for this system are respectively

(a) 0.25 and 2 rad/s (b) 0.50 and 2 rad/s

(c) 0.25 and 4 rad/s (d) 0.50 and 4 rad/s

31. The open loop transfer function of a unity feedback system is given by $\frac{K}{s(s+1)}$. If the value of gain K is such that the system is critically damped, the closed loop poles of the system will lie at

(a) −0.5 and −0.5 (b) $\pm j0.5$ (c) 0 and −1 (d) $0.5 \pm j0.5$

32. A linear time invariant system, initially at rest when subjected to a unit step input gave a response $c(t) = te^{-t}$ $(t \geq 0)$. The transfer function of the syste is

(a) $\frac{s}{(s+1)^2}$ (b) $\frac{1}{s(s+1)^2}$ (c) $\frac{1}{(s+1)^2}$ (d) $\frac{1}{s(s+1)}$

33. The steady-state error resulting from input $r(t) = 2 + 3t + 4t^2$ for given system is

(a) 2.4

(b) 4.0

(c) zero

(d) 3.2

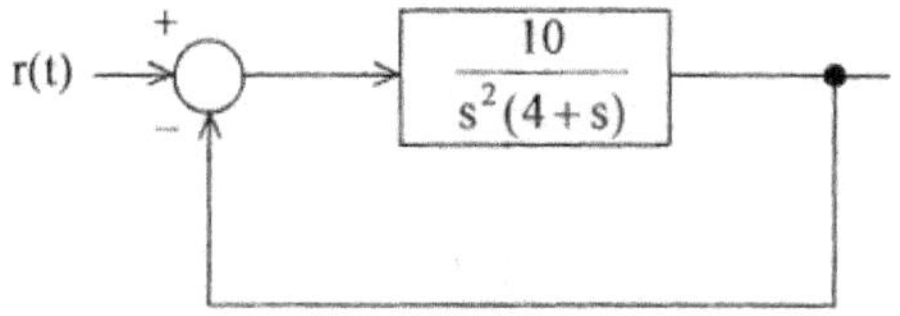

34. In the derivative error compensation

(a) damping decreases and settling time decreases

(b) damping increases and settling time increases

(c) damping decreases and settling time increases

(d) damping increases and settling time decreases

35. A second order system exhibits 100% overshoot. Its damping coefficient is

(a) equal to 0 (b) equal to 1 (c) less than 1 (d) grater than 1

36. For a second order system $2\frac{d^2y}{dt^2} + 4\frac{dy}{dt} + 8y = 8x$

The damping ratio is

(a) 0.1 (b) 0.25 (c) 0.333 (d) 0.5

37. The feedback control system shown in the given figure represents a

(a) Type 0 system

(b) Type 1 system

(c) Type 2 system

(d) Type 3 system

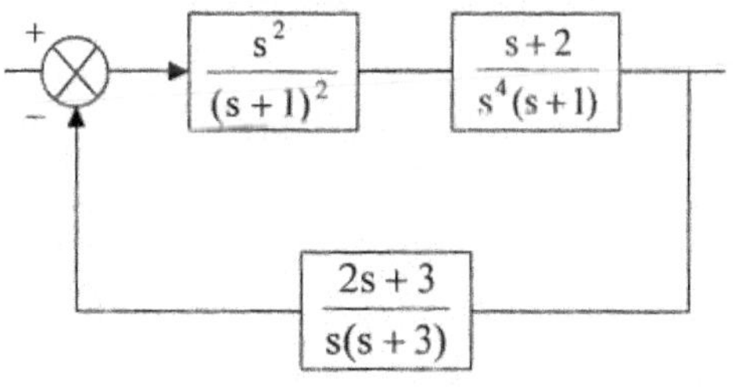

38. In the type 1 system, the velocity error is

(a) Inversely proportional to bandwidth

(b) Directly proportional to error constant

(c) Inversely proportional to error constant

(d) Independent of error constant

39. In position control systems, the device used for providing rate-feedback voltage is called

(a) potentiometer (b) synchro transmitter

(c) synchro transformer (d) technogenerator

40. A unity feedback control system has a forward path transfer function equal to $\frac{42.25}{s(s+6.5)}$

The unit step response of this system starting from rest, will have its maximum value at a time equal to

(a) 0 sec (b) 0.56 sec (c) 5.6 sec (d) infinity

41. Match the system open-loop transfer functions given in List-I with the steady-state errors produced for a unit ramp input. Select the correct answer using the code given below the lists :

List-I	List-II
A. $\frac{30}{s^2+6s+9}$	1. Zero
B. $\frac{30}{s^2+6s}$	2. 0.2
C. $\frac{30}{s^2+9s}$	3. 0.3
D. $\frac{s+1}{s^2}$	4. Infinity

Codes :

	A	B	C	D		A	B	C	D
(a)	1	2	3	4	(b)	4	3	1	2
(c)	1	3	2	4	(d)	4	2	3	1

42. A transfer function G(s) has the pole-zero plot as shown in the given figure. Given that the DC gain is 2, the transfer function G(s) will be given by

(a) $\frac{2(s+1)}{s^2+4s+5}$ (b) $\frac{5(s+1)}{s^2+4s+4}$

(c) $\frac{10(s+1)}{s^2+4s+5}$ (d) $\frac{10(s+1)}{(s+2)^2}$

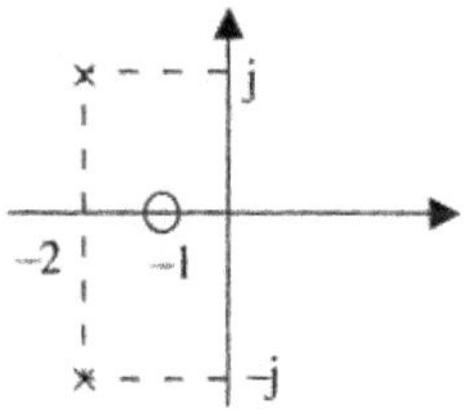

43. A plant has the following transfer function $G(s) = \frac{1}{(s^2+0.2s+1)}$

For a step input it is required that the response settles to within 2% of its final value. The plant settling time is

(a) 20 sec (b) 40 sec (c) 35 sec (d) 45 sec

44. Match List-I with List-II and select the correct answer using the codes given below the lists :

List-I (Transfer function)	List-II (Controller)
A. $\frac{K_1 s + K_2}{K_3}$	1. P-controller
B. $\frac{K_1 s^2 + K_2 s + K_3}{K_4 s}$	2. PI-controller
C. $\frac{K_1 s + K_2}{K_3 s}$	3. PD-controller
D. $\frac{K_1 s}{K_2 s}$	4. PID-controller

Codes :

	A	B	C	D
(a)	3	4	2	1
(b)	4	3	2	1
(c)	3	2	4	1
(d)	4	1	2	3

45. Consider the following expressions which indicate the step or impulse response of an initially relaxed control system :

1. $(5 - 4e^{-2t})\ u(t)$
2. $(e^{-2t} + 5)\ u\ (t)$
3. $\delta(t) + 8e^{-2t}\ u(t)$
4. $\delta(t) + 4e^{-2t}\ u(t)$

Those which correspond to the step and impulse response of the same system include

(a) 1 and 3 (b) 1 and 4 (c) 2 and 4 (d) 2 and 3

46. Assuming the transient response of a second-order system to be given by

$c(t) = 1 - \dfrac{e^{-4t}}{\sqrt{1-\delta^2}} \sin(\omega_n \sqrt{1-\delta^2} + \theta)$ the settling time for the 5% criterion will be

(a) $\frac{1}{4}$ sec (b) $\frac{3}{4}$ sec (c) $\frac{5}{4}$ sec (d) 4 sec

47. Consider the systems with the following open-loop transfer functions :

1. $\dfrac{36}{s(s+3.6)}$
2. $\dfrac{100}{s(s+5)}$
3. $\dfrac{6.25}{s(s+4)}$

The correct sequence of these systems in increasing order of the time taken for the unit-step response to settle is

(a) 1, 2, 3 (b) 3, 1, 2 (c) 2, 3, 1 (d) 3, 2, 1

48. Match List I with List II and select the correct answer using the codes given below the lists :

List I (Characteristic equations)	**List II (Nature of damping)**
A. $s^2 + 15s + 56.25$	1. Undamped
B. $s^2 + 5s + 6$	2. Underdamped
C. $s^2 + 20.25$	3. Critically damped
D. $s^2 + 4.5s + 42.25$	4. Overdamped

Codes :

	A	B	C	D
(a)	3	4	1	2
(b)	2	3	1	4
(c)	4	3	1	2
(d)	3	4	2	1

49. For the control system in the given figure to be critically damped the value of the gain 'K' required is :

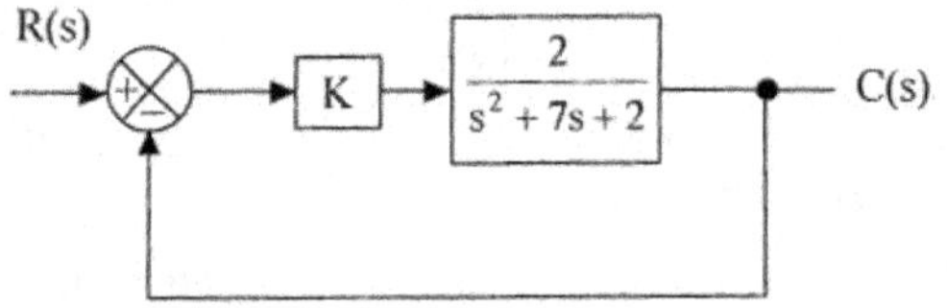

(a) 1 (b) 5.125

(c) 6.831 (d) 10

50. Match List I with List II and select the correct answer using the codes given below the lists :

List I	**List II**
A. Derivative control	1. Improved overshoot response
B. Integral control	2. Less steady-state errors
C. Rate feedback control	3. Less stable
D. Proportional control	4. More damping

Codes :

	A	B	C	D
(a)	1	2	3	4
(b)	2	3	1	4
(c)	4	3	1	2
(d)	1	2	4	3

51. A typical control system is shown in the given figure.

Assuming $R(s) = \frac{1}{s}$, the steady-state error is given by

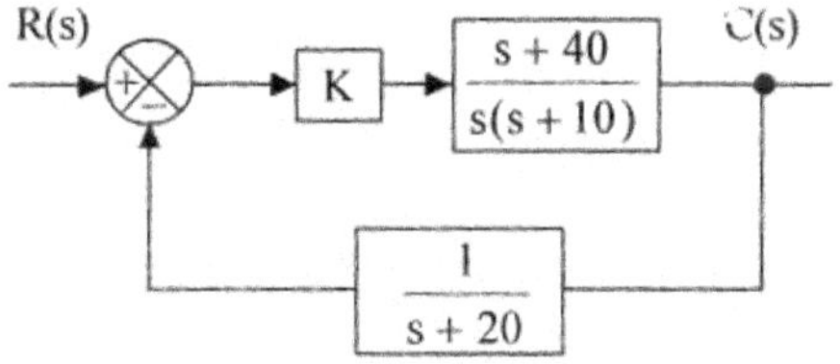

(a) $\frac{1}{1+K}$ (b) K

(c) zero (d) 1

52. A system has open-loop transfer function $G(s) = \frac{10}{s(s+1)(s+2)}$

What is the steady state error when it is subjected to the input $r(t) = 1 + 2t + \frac{3}{2}t^2$?

(a) zero (b) 0.4 (c) 4 (d) infinity

53. Consider a unit feedback control system shown in the given figure. The ratio of time constants of open-loop response to closed loop response will be

(a) 1 : 1

(b) 2 : 1

(c) 3 : 2

(d) 2 : 3

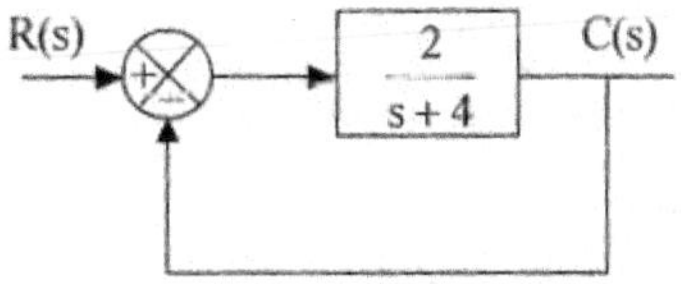

54. Consider the following overall transfer function for a unity feedback system :

$$\frac{4}{s^2 + 4s + 4}$$

Which of the following statements regarding this system are correct ?

1. Position error constant K_p for the system is 4.
2. The system type one.
3. The velocity error constant K_v for the system is finite.

Select the correct answer using the codes given below :

Codes :

(a) 1, 2 and 3 (b) 1 and 2 (c) 2 and 3 (d) 1 and 3

55. A first order system is shown in the given figure. Its time response to a unit step input is given by

(a) $c(t) = [1/T]\,[e^{-t/T}]$

(b) $c(t) = T\,(1 - e^{-t/T})$

(c) $c(t) = (1 - e^{-t/T})$

(d) $c(t) = Te^{-t/T}$

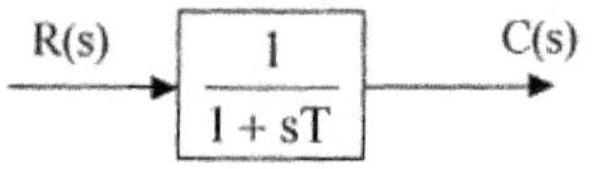

56. For a unity feedback system, the open-loop transfer function is $G(s) = \dfrac{16(s+2)}{s^2(s+1)(s+4)}$

What is the steady-state error if the input is, $r(t) = (2 + 3t + 4t^2)\,u(t)$?

(a) 0 (b) 1 (c) 2 (d) 3

57. A system has a transfer function $\dfrac{C(s)}{R(s)} = \dfrac{4}{s^2 + 1.6s + 4}$

For the unit step response, the setling time (in seconds) for 2% tolerance band is

(a) 1.6 (b) 2.5 (c) 4 (d) 5

58. Consider the following statements :

1. The derivative control improves the overshoot of a given system.
2. The derivative control reduces steady-state error.
3. Integral control reduces steady-state error.
4. Integral control does not affect stability of the system.
5. Integral control improves the overshoot of the system.

Of these statements

(a) 1 and 3 are correct (b) 1, 2 and 5 are correct

(c) 2, 4 and 5 are correct (d) 1, 3 and 5 are correct

59. In the case of a second order system described by the differential equation,

$$J\frac{d^2\theta_o}{dt^2} + F\frac{d\theta_o}{dt} + K\theta_o = K\theta_i$$

(where θ_i and θ_o are the input and output shaft angles), the natural frequency is given by

(a) $\sqrt{\frac{K}{J}}$ (b) $\sqrt{\frac{J}{K}}$ (c) $\sqrt{KJ}$ (d) $\sqrt{K-J}$

60. A transducer has two poles as shown in the figure. The zeros are at infinity and the DC gain is 1. The steady-state output of the transducer for a unit step input will be

(a) $\frac{1}{4}$

(b) $\frac{1}{2}$

(c) $\frac{1}{\sqrt{2}}$

(d) 1

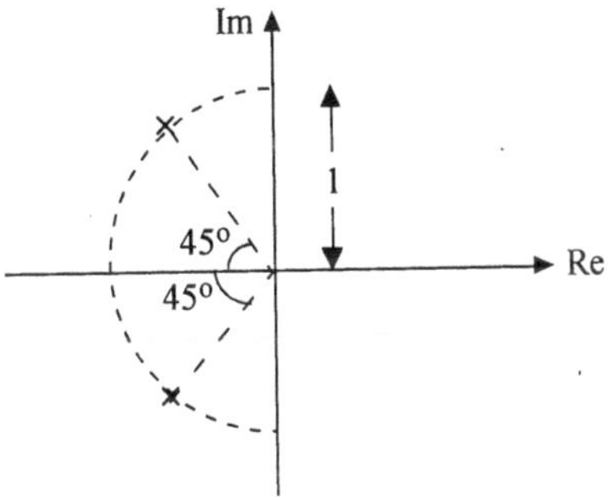

61. The velocity error constant K_v of a feedback system with closed-loop transfer function

$$\frac{C(s)}{R(s)} = \frac{G(s)}{1+G(s)H(s)}$$ is

(a) $K_v = \lim_{s\to 0} s\,G(s)\,H(s)$ (b) $K_v = \lim_{s\to 0} s\frac{G(s)}{1+G(s)H(s)}$

(c) $K_v = \lim_{s\to 0} s\,G(s)$ (d) $K_v = \lim_{s\to 0} s\,[1+G(s)H(s)]$

62. The settling time of a feedback system with the closed-loop transfer function

$$\frac{C(s)}{R(s)} = \frac{\omega_n^2}{s^2 + 2\xi\omega_n s + \omega_n^2}$$

(a) $t_s = 2/(\xi\,\omega_n)$ (b) $t_s = \frac{\xi\omega_n}{2}$

(c) $t_s = 4/(\xi\,\omega_n)$ (d) $t_s = 4\,\xi\,\omega_n$

63. A second order under-damped system exhibited a 15% maximum overshoot on being excited by a step input r (t) = 2u(t), and then attained a steady-state value of 2 (see figures given). If, at t = t_0, the input were changed to a unit step r (t) = u (t), then its time response C (t) would be similar to

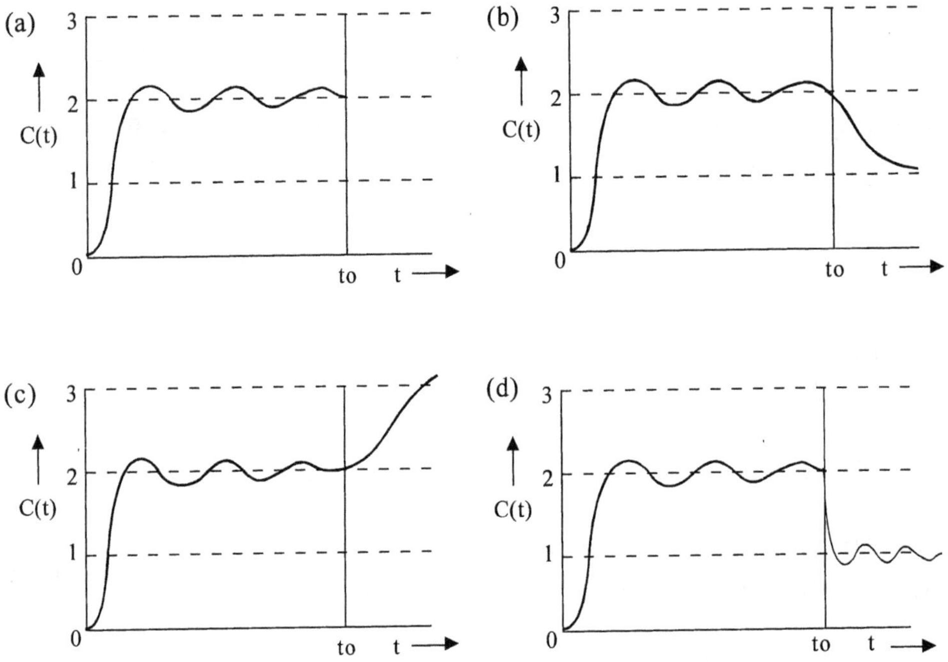

64. The response c(t) of a system is described by the differential equation

$$\frac{d^2 c(t)}{dt^2} + 4\frac{dc(t)}{dt} + 5c(t) = 0$$

The system response is

(a) undamped (b) underdamped

(c) critically damped (d) oscillatory

65. The system with the open-loop transfer function $G(s)\,H(s) = \frac{1}{s(1+s)}$ is

(a) type 2 and order 1 (b) type 1 and order 1

(c) type 0 and order 0 (d) type 1 and order 2

66. The transfer function G (s) of a PID controller is

(a) $K\left(1+\frac{1}{T_i s}+T_d s\right)$ (b) $K(1 + T_i s + T_d s)$

(c) $K\left(1+\frac{1}{T_i s}+\frac{1}{T_d s}\right)$ (d) $K\left(1+T_i s+\frac{1}{T_d s}\right)$

67. The industrial controller having the best steady-state accuracy is

(a) a derivative controller (b) an integral controller

(c) a rate feedback controller (d) a proportional controller

68. A step input is applied to a system with the transfer function $G(s) = \frac{e^{-s}}{1+0.5s}$. The output response will be

(a)

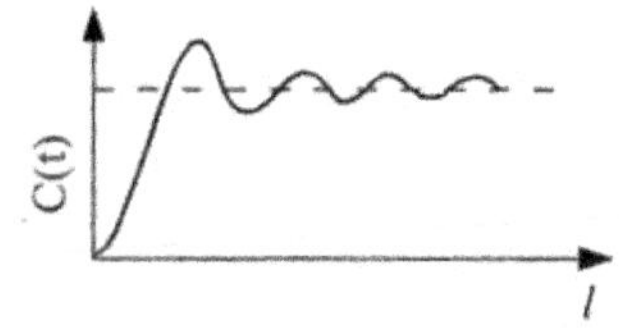

(b)

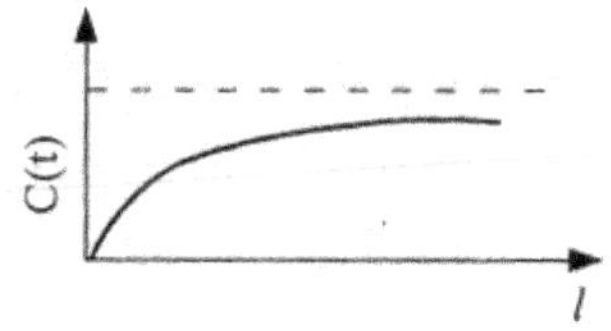

(c)

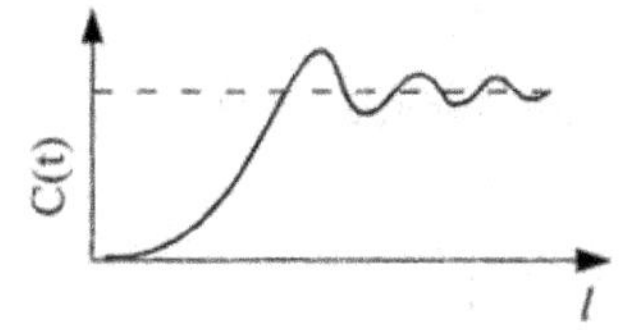

(d)

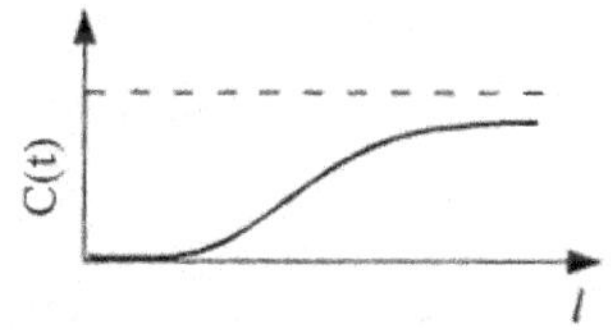

69. The open-loop transfer function G(s) of a unity feedback control system is $\frac{1}{s(s+1)}$

The system is subjected to an input r(t) = sin t. The steady-state error will be

(a) zero (b) 1

(c) $\sqrt{2}\sin\left(t-\frac{\pi}{4}\right)$ (d) $\sqrt{2}\sin\left(t+\frac{\pi}{4}\right)$

70. A second order system has the damping ratio ξ and undamped natural frequency of oscillation ω_n. The settling time at 2% tolerance band of the system is

(a) $\frac{2}{\xi\omega_n}$ (b) $\frac{3}{\xi\omega_n}$ (c) $\frac{4}{\xi\omega_n}$ (d) $\xi\omega_n$

71. Consider the following statements :

A proportional plus derivative controller

1. has high sensitivity
2. increases the stability of the system
3. improves the steady-state accuracy

Which of these statements are correct ?

(a) 1, 2 and 3 (b) 1 and 2 (c) 2 and 3 (d) 1 and 3

72. Which one of the following is the steady-state error for a step input applied to a unity feedback system with the open loop transfer function $G(s) = \frac{10}{s^2 + 14s + 50}$?

(a) $e_{ss} = 0$ (b) $e_{ss} = 0.83$ (c) $e_{ss} = 0.2$ (d) $e_{ss} = \infty$

73. With derivative output compensation, for a specified velocity error constant,

(a) Overshoot for a step input increases

(b) The settling time is decreased

(c) The natural frequency decreases

(d) Peak time increases

74. Which of the following is not a desirable feature of a modern control system

(a) Quick response (b) accuracy

(c) correct power level (d) no oscillation

75. Damping factor and undamped natural frequency for the position control system is given by

(a) $2\sqrt{KJ}$, $\sqrt{KJ}$ respectively. (b) $\frac{K}{2fJ}$, $\sqrt{K/J}$ respectively

(c) $\frac{f}{2\sqrt{KJ}}$, $\sqrt{K/J}$ respectively. (d) $\frac{J}{2\sqrt{Kf}}$, $\sqrt{KJ}$ respectively.

Where, K = torsional stiffness

J = Moment of Inertia

and f = Coefficient viscous friction

76. In the derivative error compensation

(a) damping decreases and settling time decreases

(b) damping increases and settling time increases

(c) damping decreases and settling time increases

(d) damping increases and settling time decreases.

77. A control system having unit damping factor will give

(a) critically damped response (b) oscillatory response

(c) undamped response (d) no response.

78. Which of the following will not decrease as a result of introduction of negative feedback?

(a) Instability (b) Band width

(c) Overall gain (d) Distortion

79. Consider the following statements regarding time domain analysis of control systems :

1. Derivative control improves systems transient performance.
2. integral control does not improve system steady state performance
3. integral control can convert a second order system into a third order system

Of these statements

(a) 1 and 2 are correct (b) 1 and 3 are correct

(c) 2 and 3 are correct (d) 1, 2 and 3 are correct

80. The open-loop transfer transfer function of a unity feedback control system is given by

$$G(s) = \frac{K}{s(s+1)}$$

If the gain K is increased to infinity, then the damping ratio will tend to become

(a) $\frac{1}{\sqrt{2}}$ (b) 1 (c) 0 (d) ∞

81. The system shown in the given figure has a unit step input. In order that the steady-state error is 0.1, the value of K required is

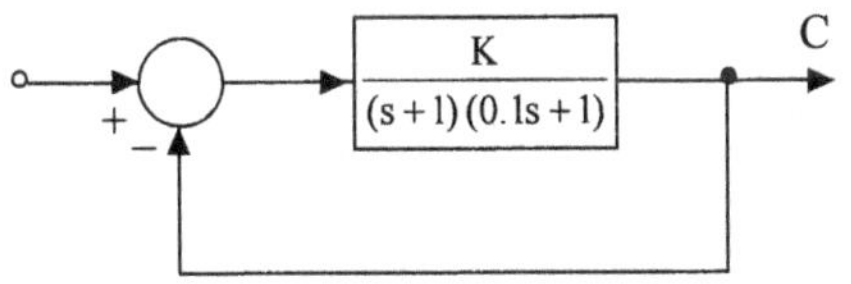

(a) 0.1 (b) 0.9

(c) 1.0 (d) 9.0

82. The system shown in the given figure has second order response with a damping ratio of 0.6 and a frequency of damped oscillations of 10 rad/sec. The values of K_1 and K_2 are respectively

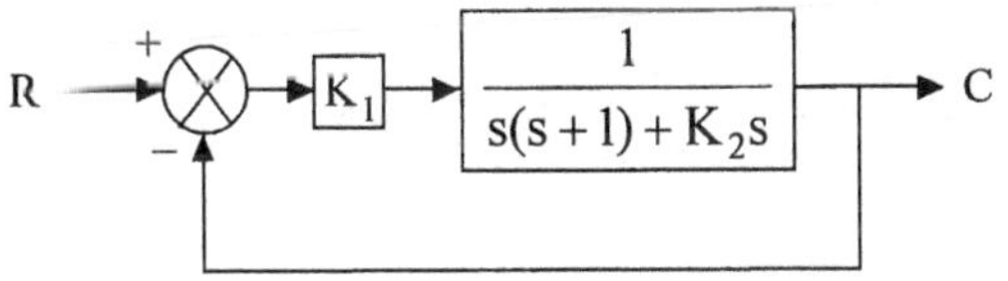

(a) 12.5 and 15

(b) 156.25 and 15

(c) 156.25 and 14

(d) 12.5 and 14

Chapter 4

1. None of the poles of a linear control system lies in the right half of s plane. For a bounded input the output of this system
   (a) is always bounded (b) could be unbounded
   (c) always tends to zero (d) none of the above
2. The number of roots of the equation $2s^4 + s^3 + 3s^2 + 5s + 7 = 0$ that lie in the right half of s plane is
   (a) zero (b) one (c) two (d) three
3. The characteristic equation of a feedback control system is $2s^4 + s^3 + 3s^2 + 5s + 10 = 0$. The number of roots in the right half of s-plane are
   (a) zero (b) 1 (c) 2 (d) 3
4. Consider a system shown in the given figure with

$$G(s) = \frac{K(s+1)}{s^3 + as^2 + 2s + 1}$$

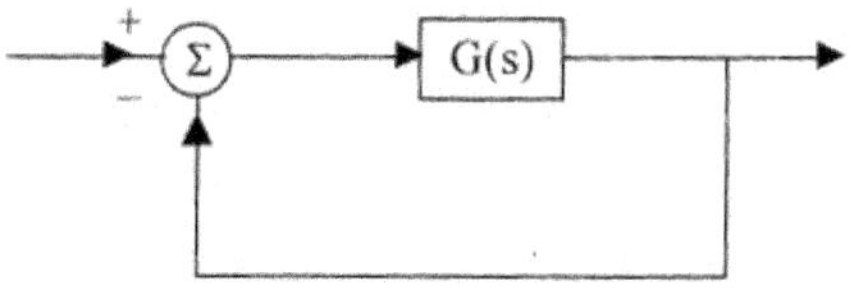

   What values of 'K' and 'a' should be chosen so that the system oscillates ?
   (a) K = 2, a = 1 (b) K = 2, a = 0.75 (c) K = 4, a = 1 (d) K = 4, a = 0.75
5. The open loop transfer function of a system is given by $G(s) = \frac{k}{s(s+2)(s+4)}$. The maximum value of k for which the unity feedback system will be stable, is
   (a) 16 (b) 32 (c) 48 (d) 64
6. The characteristic equation 1 + G(s) H(s) = 0 of a system is given by

$$s^4 + 6s^3 + 11s^2 + 6s + K = 0$$

   For the system to remain stable, the value of gain K should be
   (a) zero (b) greater than zero but less than 10
   (c) greater than 10 but less than 20 (d) greater than 20 but less than 30
7. The open-loop transfer function of a unity feedback control system is

   $G(s)\,H(s) = \frac{30}{s(s+1)(s+T)}$ where T is a variable parameter. The closed loop system will be stable for all values of
   (a) $T > 0$ (b) $0 < T < 3$ (c) $T > 5$ (d) $3 < T < 5$

8. The open-loop transfer function of a unity-feedback control system is :

$$G(s) = \frac{K(s+10)(s+20)}{s^2(s+2)}$$

The closed-loop system will be stable if the value of K is

(a) 2 (b) 3 (c) 4 (d) 5

9. The characteristic equation of a feedback control system is $s^3 + Ks^2 + 5s + 10 = 0$. For the system to be critically stable, the value of K should be

(a) 1 (b) 2 (c) 3 (d) 4

10. The control system shown in the given figure has an internal rate feedback loop. The closed-loop system for open and close conditions of switch will be respectively

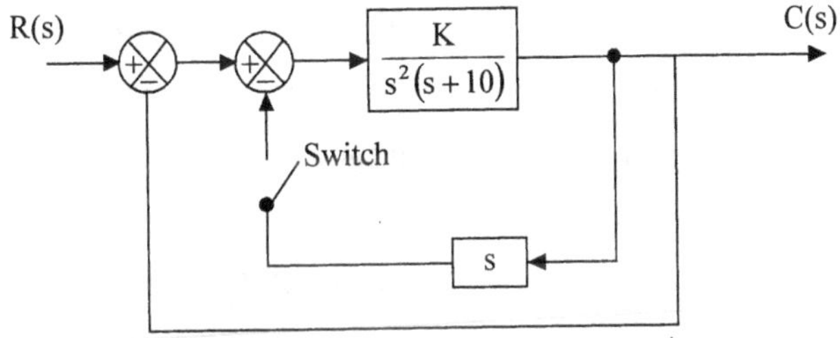

(a) unstable and stable

(b) unstable and unstable

(c) stable and unstable

(d) stable and stable

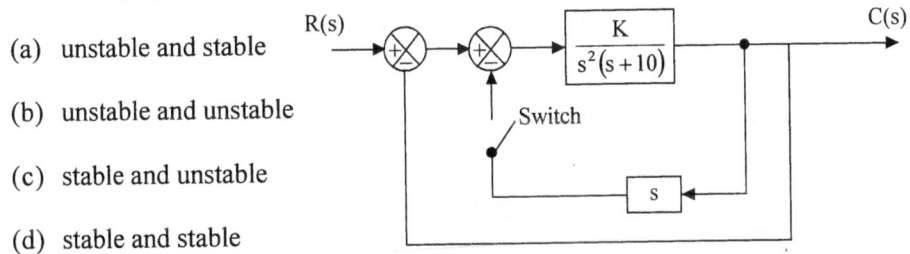

11. For the block diagram shown in the given figure, the limiting values of K for stability of inner loop is found to be X < K < Y. The overall system will be stable if and only if

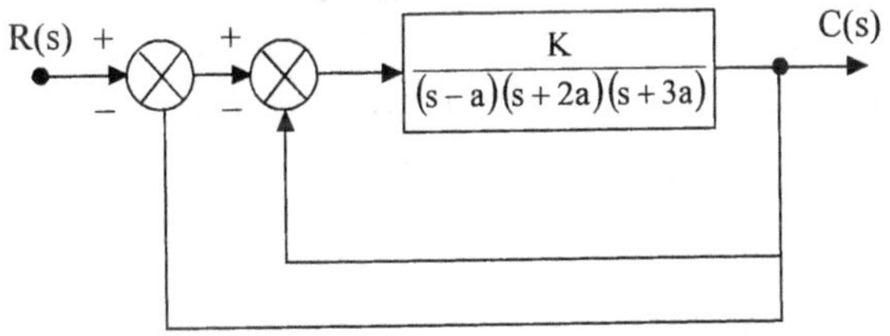

(a) 4X < K < 4Y

(b) 2X < K < 2Y

(c) X < K < Y

(d) $\frac{X}{2} < K < \frac{Y}{2}$

12. A feedback control system is shown in the given figure. The system is stable for all positive values of K, if

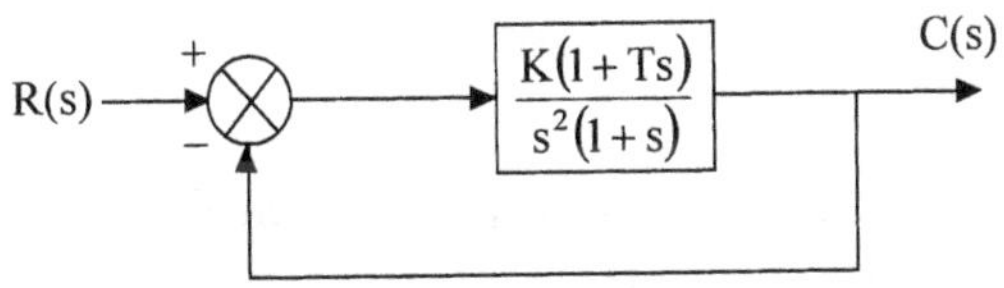

(a) T = 0

(b) T < 0

(c) T > 1

(d) 0 < T < 1

13. The characteristic equation of a system is given by $3s^4 + 10s^3 + 5s^2 + 2 = 0$. This system is

(a) stable (b) marginaly stable (c) unstable (d) neither (a), (b) nor (c)

14. The characteristic equation $s^3 + 3s^2 + 3s + k = 0$ is stable for which value of k ?

(a) – 6 (b) 15 (c) 5 (d) 12

15. Which of the following statement about the equation below, for Routh Hurwitz criterion is true ?

$2s^4 + s^3 + 3s^2 + 5s + 10 = 0$

(a) It has only one root on the imaginary axis

(b) It has one root in the right half of the s-plane

(c) The system is unstable

(d) The system is stable.

16. When all the roots of the characteristic equation are found in the left half of s-plane, the system response due to initial condition will

(a) increase to infinity as time approaches infinity

(b) decreases to zero as time approaches infinity

(c) remain constant for all time

(d) be oscillating

17. Match List-I with List-II and select the correct answer by using the codes given below the lists :

List-I (Characteristic Root Location)	List-II (System characteristic)
A. $(-1 + j), (-1 - j)$	1. Marginally stable
B. $(-2 + j), (-2 - j), (2j), (-2j)$	2. Unstable
C. $-j, j, -1, 1$	3. Stable

Codes :

	A	B	C		A	B	C
(a)	1	2	3	(b)	2	3	1
(c)	3	1	2	(d)	1	3	2

18. A control system is as shown in the given figure. The maximum value of gain K for which the system is stable is

(a) $\sqrt{3}$

(b) 3

(c) 4

(d) 5

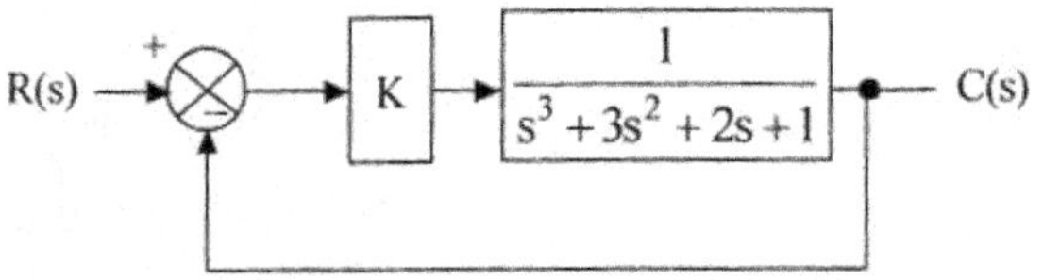

19. If the open-loop transfer function of the system is $G(s)\,H(s) = \dfrac{K(s+10)}{s\,(s+8)\,(s+16)\,(s+72)}$,

then a closed loop pole will be located at $s = -12$ when the value of K is

(a) 4355 (b) 5760 (c) 9600 (d) 9862

20. Consider the following statements regarding the number of sign change in the first column of Routh array in respect of the characteristic equation $s^2 + 2as + 4$:

1. If $a = +\varepsilon$, where ε = near zero, number of sign changes will be equal to zero.
2. If $a = 0$, the number of sign change will be equal to one.
3. If $a = -\varepsilon$, where ε = near zero, the number of sign changes will be equal to two

Of these statements

(a) 1, 2 and 3 are correct (b) 1 and 2 are correct

(c) 2 and 3 are correct (d) 1 and 3 are correct

21. How many roots of the characteristic equation $s^5 + s^4 + 2s^3 + 2s^2 + 3s + 15 = 0$ line in the left half of the s-plane ?

(a) 1 (b) 2 (c) 3 (d) 5

22. The first column of a Routh array is

s^5	1
s^4	2
s^3	$\dfrac{3}{2}$
s^2	$-\dfrac{1}{3}$
s^1	10
s^0	2

How many roots of the corresponding characteristic equation are there in the left-half of the s-plane ?

(a) 2 (b) 3 (c) 4 (d) 5

23. Consider a negative feedback system where, $G(s) = \dfrac{1}{(s+1)}$, $H(s) = \dfrac{K}{s(s+2)}$

The closed-loop system is stable for

(a) $K > 20$ (b) $15 < K < 10$

(c) $8 \leq K \leq 14$ (d) $K < 6$

24. The value of K for which the unity feedback system $G(s) = \dfrac{K}{s\,(s+2)\,(s+4)}$ crosses the

imaginary axis is

(a) 2 (b) 4 (c) 6 (d) 48

25. While forming Routh's array, the situation of a row of zeros indicates that the system
    (a) has symmetrically located roots (b) is not sensitive to variations in gain
    (c) is stable (d) unstable

26. The closed loop system shown above becomes marginally stable if the constant K is chosen to be

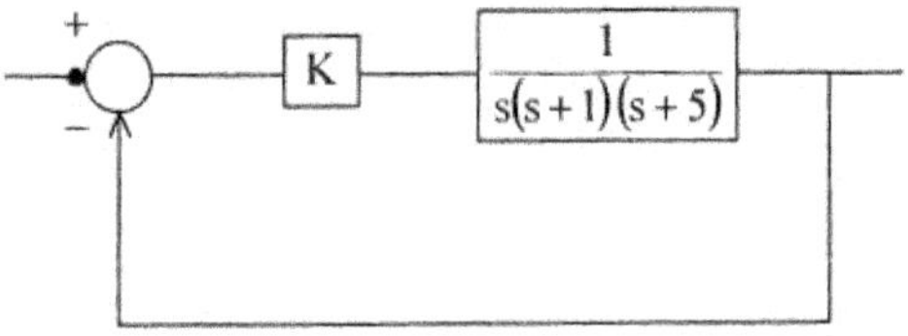

    (a) 10
    (b) 20
    (c) 30
    (d) 40

27. A closed-loop system is shown in the following figure :
    The largest possible value of β for which this system would be stable is :

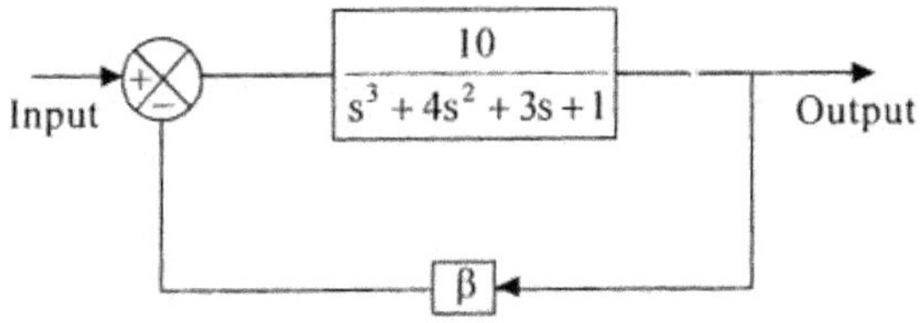

    (a) 1
    (b) 1.1
    (c) 1.2
    (d) 2.3

28. First column elements of the Routh's tabulation are 3, 5 $\frac{-3}{4}$, $\frac{1}{2}$, 2. It means that there
    (a) is one root in the left half of s-plane
    (b) are two roots in the left half of s-plane
    (c) are two roots in the right half of s-plane
    (d) is one root in the right half of s-plane

29. The open-loop transfer fucntion of unity feedback control system is

$$G(s) = \frac{K}{s(s+a)(s+b)}, \; 0 < a \leq b$$

    The system is stable if
    (a) $0 < K < \frac{(a+b)}{ab}$ (b) $0 < K < \frac{ab}{(a+b)}$
    (c) $0 < K < ab\,(a+b)$ (d) $0 < K < \frac{a}{b}\,(a+b)$

30. Which one of the following characteristic equations can result in the stable operation of the feedback system ?

(a) $s^3 + 4s^2 + s - 6 = 0$ (b) $s^3 - s^2 + 5s + 6 = 0$

(c) $s^3 + 4s^2 + 10s + 11 = 0$ (b) $s^4 + s^3 + 2s^2 + 4s + 6 = 0$

31. Consider the following statements :

Routh-Hurwitz criterion gives

1. absolute stability
2. the number of roots lying on the right half of the s-plane
3. the gain margin and phase margin

Which of these statements are correct ?

(a) 1, 2 and 3 (b) 1 and 2 (c) 2 and 3 (d) 1 and 3

32. The given characteristic polynomial $s^4 + s^3 + 2s^2 + 2s + 3 = 0$ has

(a) zero roots in RHS of s-plane (b) one root in RHS of s-plane

(c) two roots in RHS of s-plane (d) three roots in RHS of s-plane

33. For making an unstable system stable

(a) gain of the system should be increased

(b) gain of the system should be decreased

(c) the number of zeros in the loop transfer function should be increased

(d) the number of poles in the loop transfer function should be increased.

34. Which of the following system is unstable ?

(a) $G(s)\,H(s) = \dfrac{K}{(T_1 s + 1)(T_2 s + 1)}$ $T_1, T_2 > 0$

(b) $G(s)\,H(s) = \dfrac{K(s+1)}{s^2(s+4)(s+5)}$ $K > 99$

(c) $G(s)\,H(s) = \dfrac{K(s+2)}{(s+1)(s-3)}$ $K > 2$

(d) $G(s)\,H(s) = \dfrac{K}{(Ts+1)^3}$ $-1 < K < 8$ $T > 0$

35. The characteristic equation of a closed-loop system is given by :

$s^4 + 6s^3 + 11s^2 + 6s + K = 0.$

Stable closed-loop behaviour can be ensured when gain K is such that

(a) $0 < K < 10$ (b) $K > 10$ (c) $-\infty \le K < \infty$ (d) $0 < K \le 20$

36. By a suitable choice of the scalar paramter K, the system shown in the given figure can be made to oscillate continuously at a frequency of

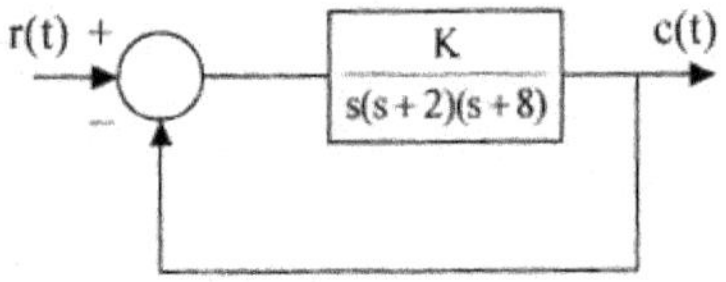

(a) 1 rad/s

(b) 2 rad/s

(c) 4 rad/s

(d) 8 rad/s

37. The open-loop transfer functions with unity feedback are given below for different systems :

1. $G(s) = \dfrac{2}{s+2}$ 2. $G(s) = \dfrac{2}{s(s+2)}$

3. $G(s) = \dfrac{2}{s^2(s+2)}$ 4. $G(s) = \dfrac{2(s+1)}{s(s+2)}$

Among these systems the unstable system is

(a) 1 (b) 2 (c) 3 (d) 4

38. The open-loop transfer function of a control system is given by $\dfrac{K(s+10)}{s(s+2)(s+a)}$

The smallest possible value of 'a' for which this system is stable under unity feedback closed-loop condition for all positive values of K is

(a) 0 (b) 8 (c) 10 (d) 12

39. The open-loop transfer function of a unity negative feedback control system is given by

$$G(s) = \frac{K(s+2)}{(s+1)(s-7)}$$

For $K > 6$, the stability characteristic of the open-loop and closed-loop configurations of the system are respectively

(a) stable and stable (b) unstable and stable

(c) stable and unstable (d) unstable and unstable

Chapter 5

1. A unity feedback has on open loop transfer function, $G(s) = \frac{K}{s^2}$. The root locus plot is:

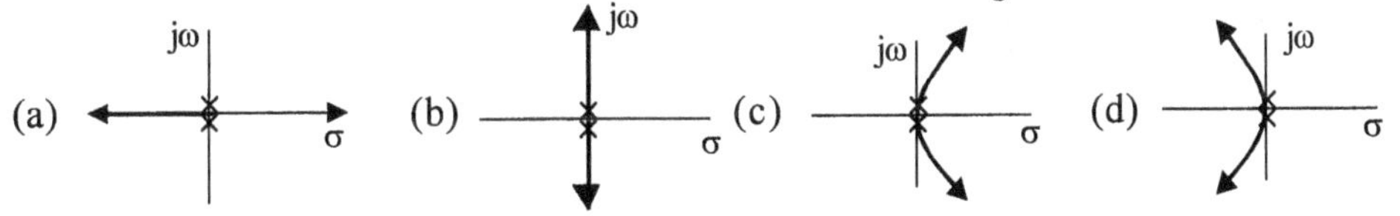

2. Given a unity feedback system with open-loop transfer function $G(s) = \frac{K(s+2)}{(s+1)^2}$

The correct root-locus plot of the system is

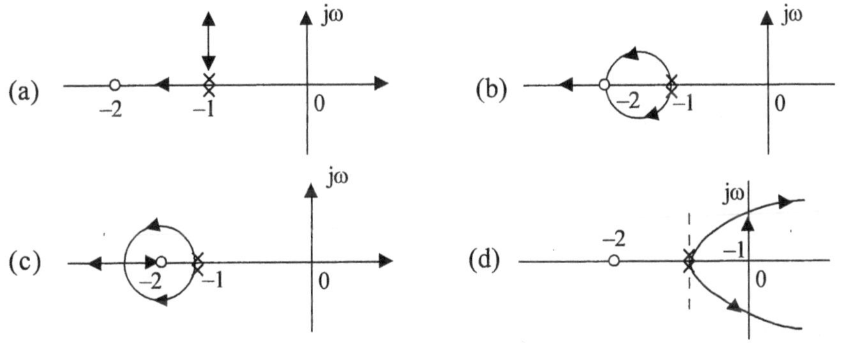

3. The closed-loop transfer function of a feedback control system is given by

$$\frac{C(s)}{R(s)} = \frac{K}{s^2 + (3+K)s + 2}$$

Which one of the following diagrams represents the root-locus diagram of the system for K > 0?

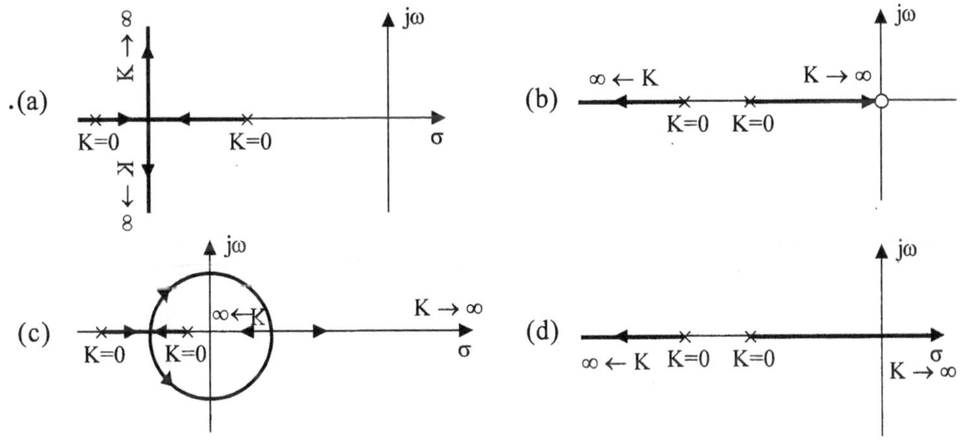

4. The open-loop transfer function of a feedback control system is given by

$$G(s)\,H(s) = \frac{K(s+2)}{s(s+4)(s^2+4s+8)}$$

In the root-locus diagram of the system, the asymptotes of the root loci for large values of K meet a point in the s-plane. Which one of the following is the set of co-ordinates of that point ?

(a) (–1.0) (b) (–2.0) (c) $\left(-\frac{10}{3}, 0\right)$ (d) (2, 0)

5. If the characteristic equation of a closed-loop system is $1 + \frac{K}{s(s+1)(s+2)} = 0$ the centroid of the asymptotes in root-locus will be

(a) zero (b) 2 (c) – 1 (d) – 2

6. The root-locus of a unity feedback system is shown in the given figure. The open-loop transfer function of the system is

(a) $\frac{K}{s(s+1)(s+3)}$

(b) $\frac{K(s+1)}{s(s+3)}$

(c) $\frac{K(s+3)}{s(s+1)}$

(d) $\frac{Ks}{(s+1)(s+3)}$

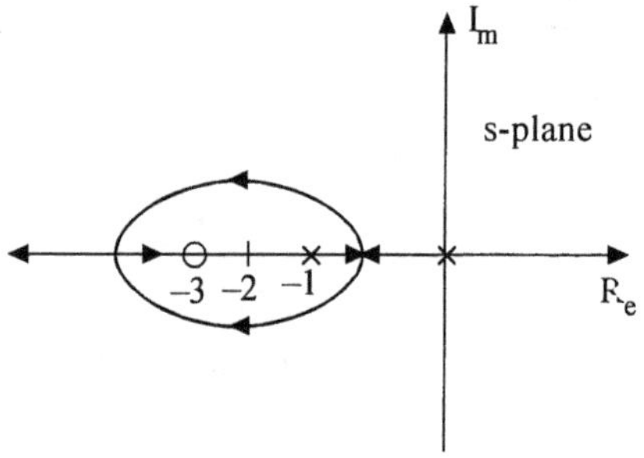

7. The open-loop transfer function of a feedback control system is $\frac{K}{s(s^2+3s+6)}$.

The break-away point(s) of its root-locus plot

(a) exist at $(-1 \pm j\,1)$ (b) exist at $\left(-\frac{3}{2} \pm \sqrt{\frac{15}{16}}\right)$

(c) exists at origin (d) do not exist

8. Match List-I with List-II in respect of the open-loop transfer function

$$G(s)\,H(s) = \frac{K(s+10)(s^2+20s+500)}{s(s+20)(s+50)(s^2+4s+5)}$$ and select the correct answer using the codes given below the lists :

List-I (Types of loci)	**List-II (Numbers)**
(A) Separate loci	1. One
(B) Loci on the real axis	2. Two
(C) Asymptotes	3. Three
(D) Breakaway points	4. Five

Codes :

	A	B	C	D
(a)	3	4	2	1
(b)	3	4	1	2
(c)	4	3	1	2
(d)	4	3	2	1

9. The characteristic equation of a linear control system is $s^2 + 5Ks + 10 = 0$

The root-loci of the system for $0 < K < \infty$ is

(a)

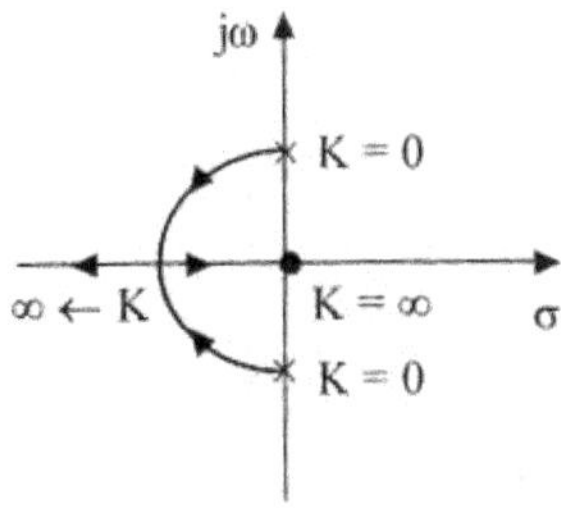

(b)

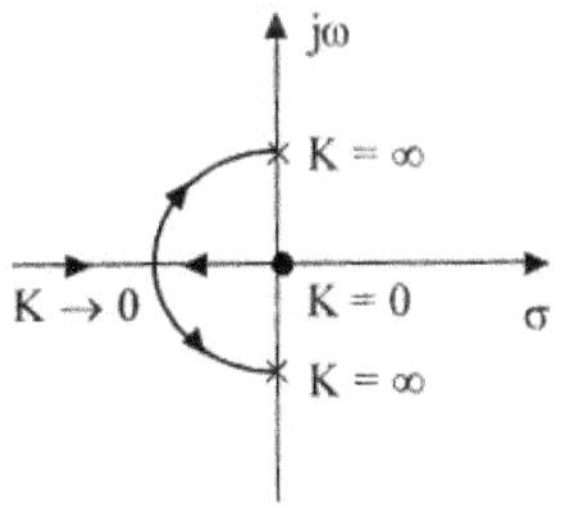

(c)

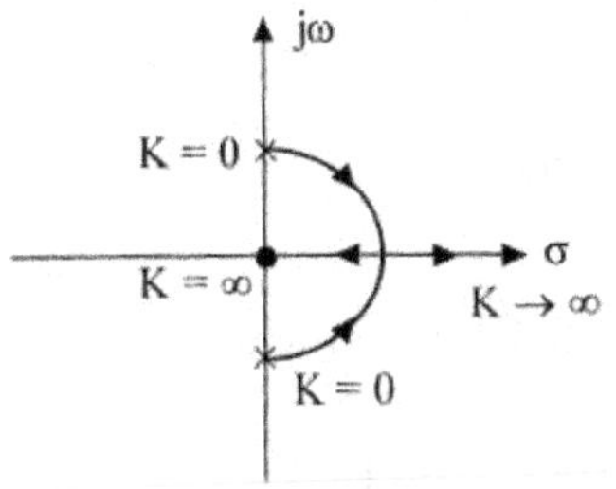

(d)

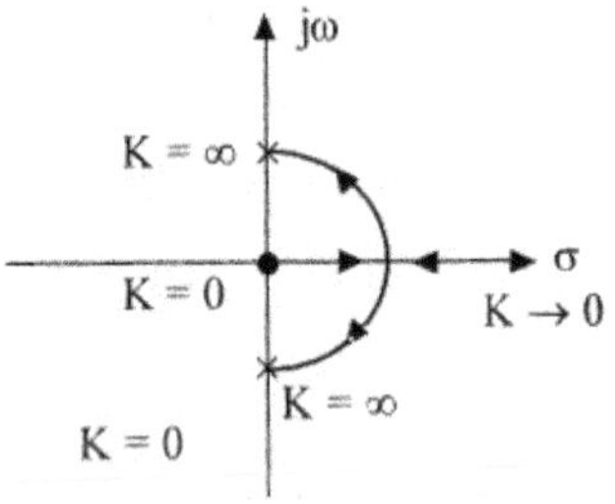

10. The characteristic equation of a feedback control system is given by $s^3 + 5s^2 + (K + 6)s + K = 0$. In the root loci diagram, the asymptotes of the root loci for large 'K' meet at a point in the s-plane whose coordinates are

(a) (2, 0) (b) (–1, 0) (c) (–2, 0) (d) (–3, 0)

11. Which of the following are the characteristics of the root locus of

$$G(s)\,H(s) = \frac{K(s+5)}{(s+1)(s+3)}$$

1. It has one asymptote
2. It has intersection with jω-axis
3. It has two real axis intersections
4. It has two zeros at infinity

Select the correct answer using the codes given below :

Codes :

(a) 1 and 2 (b) 2 and 3 (c) 3 and 4 (d) 1 and 3

12. Identify the correct root locus from the figures given below referring to poles and zeros at $\pm j\,8$ and $\pm j\,10$ respectively of G(s) H(s) of a single-loop control system.

(a)
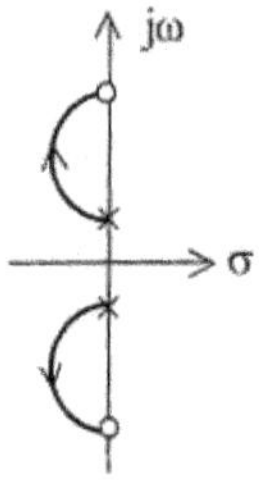

(b)
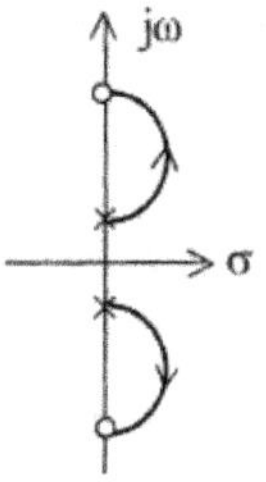

(c)
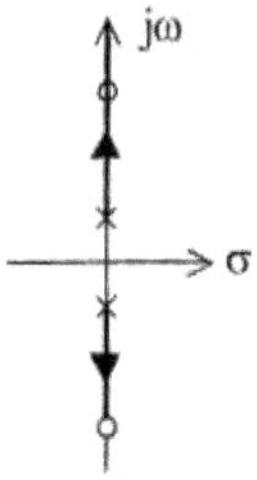

(d)
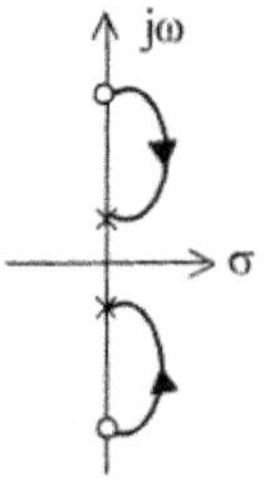

13. Consider the Root Locus Diagram of a system and the following statements:

1. The open loop system is a second order system
2. The system is overdamped for K > 1
3. The system is absolutely stable for all values of K

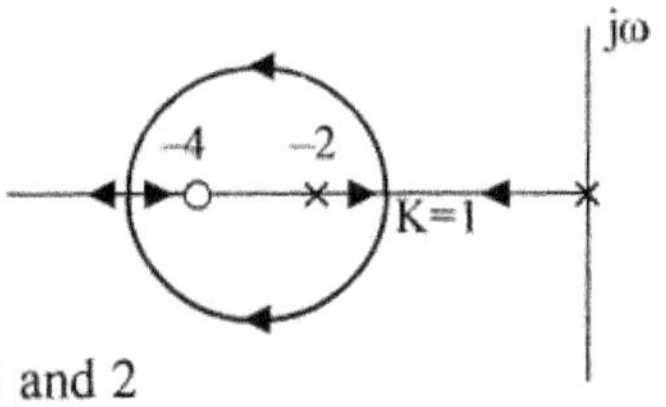

Which of these statements are correct ?

(a) 1, 2, and 3 (b) 1 and 3 (c) 2 and 3 (d) 1 and 2

14. In the root-locus for open-loop transfer function $G(s)\,H(s) = \dfrac{K(s+6)}{(s+3)(s+5)}$, the break away and break in points are located respectively at

(a) – 2 and – 1
(b) – 2.47 and – 3.77
(c) – 4.27 and – 7.73
(d) – 3 and – 4

15. A unity feedback system has $G(s) = \dfrac{K}{s(s+1)(s+2)}$

In the root-locus, the break-away point occurs between

(a) s = 0 and –1 (b) s = – 1 and – ∞

(c) s = – 1 and – 2 (d) s = – 2 and – ∞

16. The loop transfer function of a feedback control systme is given by

$$G(s)\,H(s) = \frac{k}{s(s+2)(s^2+2s+2)}$$

Number of asymptotes of its root loci is

(a) 1 (b) 2 (c) 3 (d) 4

17. Which of the following are the features of the break away point in the root-locus of a closed-loop control system with the characteristic equation $I + KG_1(s)\,H_1(s) = 0$?

1. It need not always occur only on the real axis.
2. At this point $G_1(s)\,H_1(s) = 0$.
3. At this point $\dfrac{dK}{ds} = 0$

Select the correct answer using the codes given below :

(a) 1, 2 and 3 (b) 1 and 2

(c) 2 and 3 (d) 1 and 3

18. An open-loop transfer function is given by $\dfrac{K(s+3)}{s(s+5)}$

Its root-loci will be as in

(a)

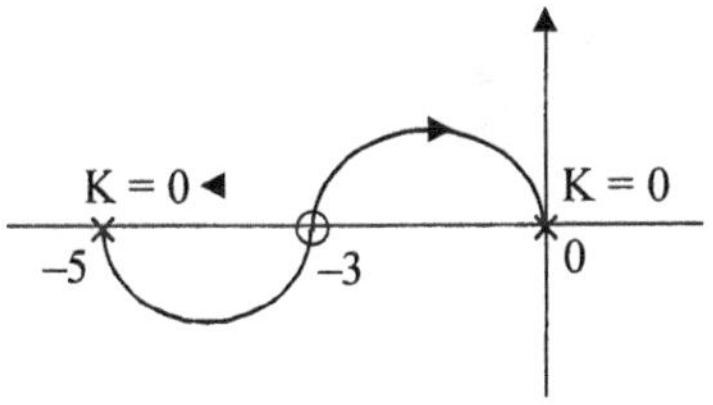

(b)

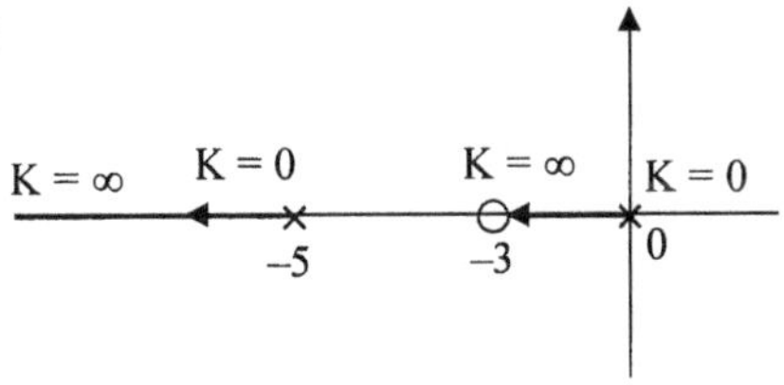

(c)

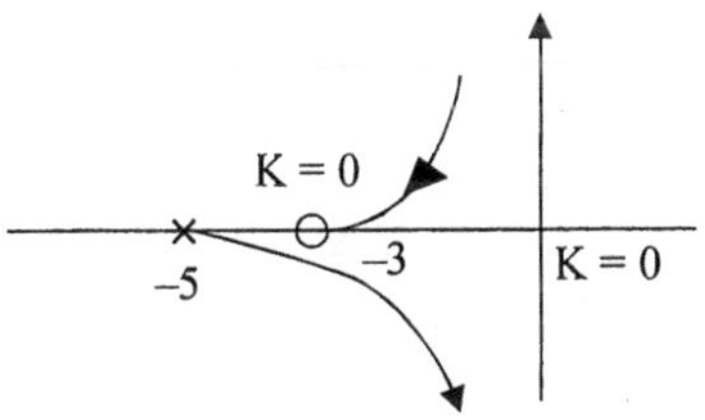

(d)

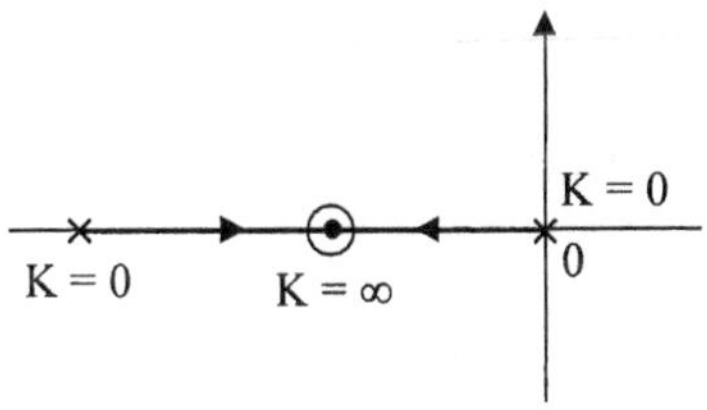

19. The loop transfer function of a closed-loop system is given by

$$G(s)\,H(s) = \frac{K}{s^2\,(s^2 + 2s + 2)}$$

The angle of departure of the root locus at s = – 1 + j is

(a) zero (b) 90° (c) – 90° (d) – 180°

20. Consider the loop transfer function $G(s)\,H(s) = \dfrac{K(s+6)}{(s+3)(s+5)}$

In the root-locus diagram, the centroid will be located at

(a) – 4 (b) – 1 (c) – 2 (d) – 3

21. For a unity negative feedback control system, the open-loop transfer function is

$$G\,(s) = \frac{K}{s(s+1)(s+2)}$$

The root-locus plot of the system is

(a)

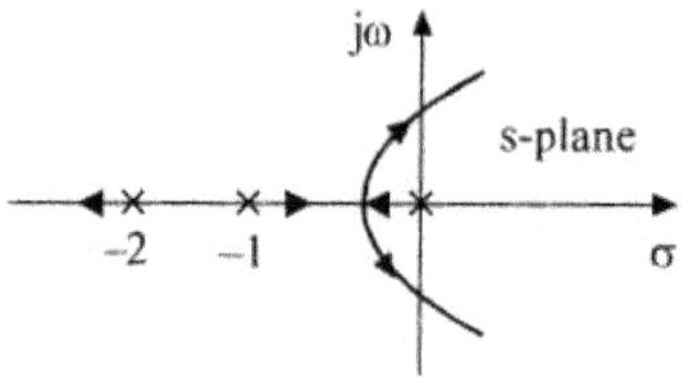

(b)

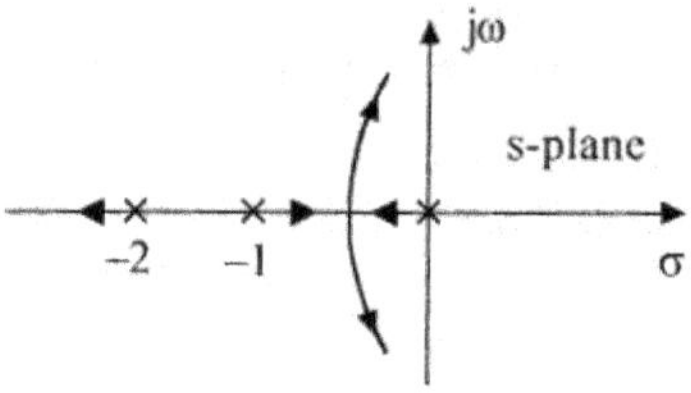

(c)

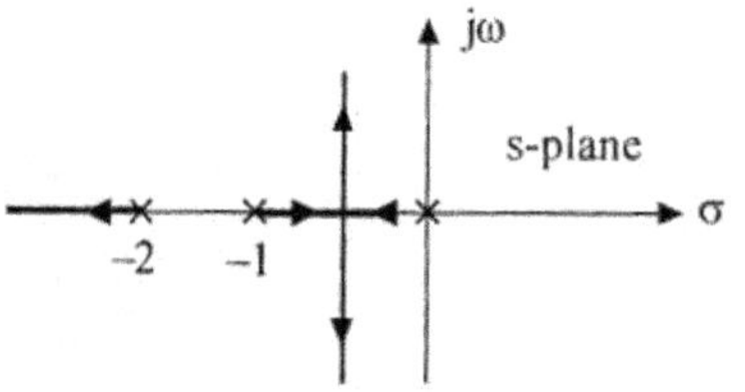

(d)

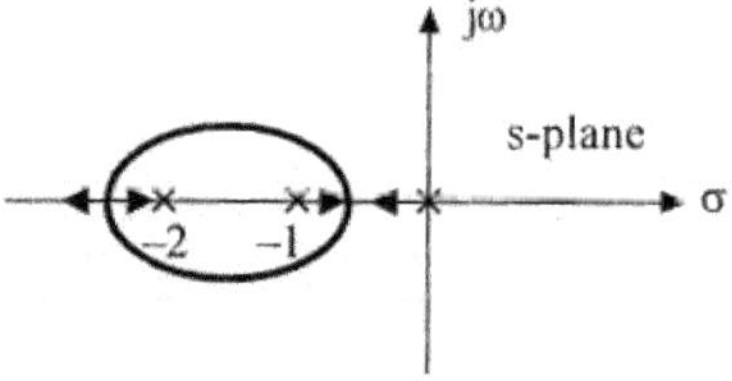

22. The intersection of asymptotes of root-loci of a system with open-loop transfer function

$$G(s)\,H(s) = \frac{K}{s(s+1)(s+3)} \text{ is}$$

(a) 1.44 (b) 1.33 (c) – 1.44 (d) – 1.33

23. The root locus plot of the system having the loop transfer function

$$G(s)\,H(s) = \frac{K}{s\,(s+4)\,(s^2+4s+5)} \text{ has}$$

(a) no breakaway point (b) three real breakaway points

(c) only one breakaway point (d) one real and two complex breakaway points

24. An open loop transfer function is given by $G(s)\,H(s) = \dfrac{k(s+1)}{s\,(s+2)\,(s^2+2s+2)}$. It has

(a) one zero at infinity (b) two zeros at infinity

(c) three zeros at infinity (d) four zeros at infinity

25. The intersection of root locus branches with the imaginary axis can be determined by the use of

(a) Nyquist criterion (b) Routh's criterion

(c) Polar plot (d) None of the above.

26. The root-locus of a unity feedback system is shown in the figure. The open loop transfer function is given by

(a) $\dfrac{K}{s(s+1)(s+2)}$ (b) $\dfrac{K(s+1)}{s(s+2)}$

(c) $\dfrac{K(s+2)}{s(s+1)}$ (d) $\dfrac{Ks}{(s+1)\,(s+2)}$

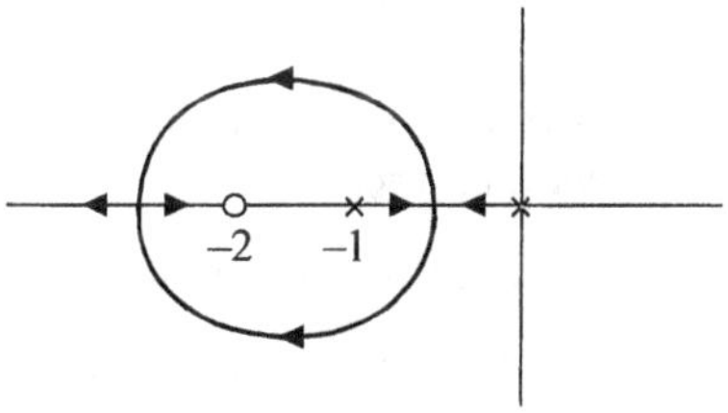

27. The characteristic equation of a unity feedback control system is given by

$s^3 + K_1 s^2 + s + K_2 = 0$

Consider the following statements in this regard :

1. For a given value of K_1, all the root-locus branches will terminate at infinity for variable K_2 in the positive direction.
2. For a given value of K_2, all the root-locus branches will terminate at infinity for variable K_1 in the positive direction.
3. For a given value of K_2, only one root-locus branch will terminate at infinity for variable K_1 in the positive direction.

Of these statments :

(a) 1 and 2 are correct (b) 3 alone is correct

(c) 2 alone is correct (d) 1 and 3 are correct

28. The breakaway point of the root locus for the system $G(s)\,H(s) = \dfrac{k}{s(s+1)(s+4)}$ is

(a) –0.465 (b) –2.87 (c) –1.0 (d) –2.0

29. The given figure shows the root-locus of open-loop transfer function of a control system.

Index :

PQ = 2.6 = PQ'

RP = 1.4

OR = 2.0

OQ = 1.4 = OQ'

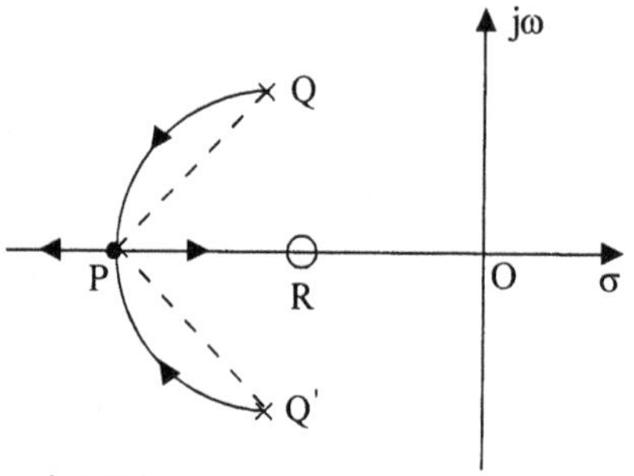

The value of the forward path gain K at the point P is

(a) 0.2 (b) 1.4 (c) 3.1 (d) 4.8

30. The loop transfer function GH of a control system is given by

$$GH = \frac{K}{s(s+1)(s+2)(s+3)}$$

Which of the following statements regarding the conditions of the system root loci diagram is/are correct ?

1. There will be four asymptotes,
2. There will be three separate root loci.
3. Asymptotes will intersect at real axis at $\sigma_A = -2/3$

Select the correct answer using the codes given below :

Codes :

(a) 1 alone (b) 2 alone (c) 3 alone (d) 1, 2 and 3

Chapter 6

1. The magnitude plot for a transfer function is shown in figure.
What is the steady-state error corresponding to a unit step input ?

(a) $\dfrac{1}{101}$ (b) $\dfrac{1}{100}$

(c) $\dfrac{1}{41}$ (d) $\dfrac{1}{40}$

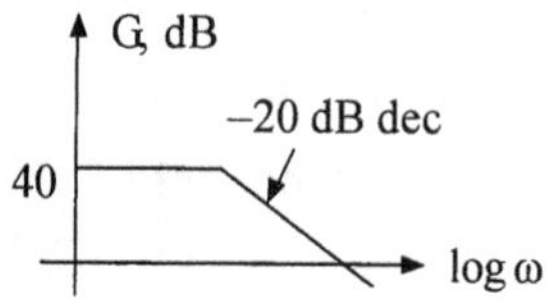

2. The function corresponding to the Bode plot of figure is

(a) $A = j\,{}^{f}\!/_{f_1}$

(b) $A = {}^{1}\!/\!\left(1 - jf_1/f\right)$

(c) $A = {}^{1}\!/\!\left(1 + jf_1/f\right)$

(d) $A = 1 + {}^{jf}\!/_{f_1}$

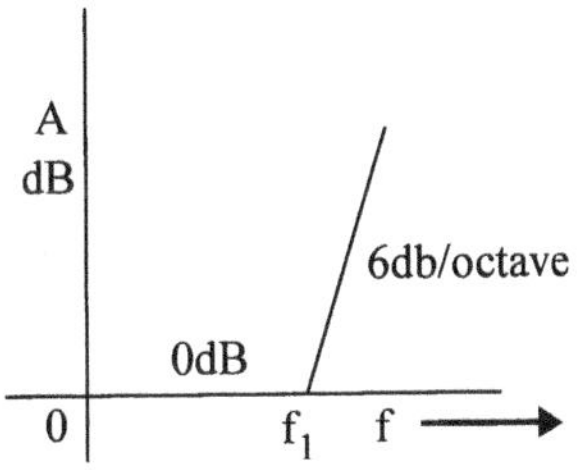

3. The asymptotic approximation of the log-magnitude versus frequency plot of minimum phase system with real poles and one zero is shown in Fig. Its transfer function is

(a) $\dfrac{20(s+5)}{s(s+2)(s+25)}$

(b) $\dfrac{10(s+5)}{(s+2)^2(s+25)}$

(c) $\dfrac{20(s+5)}{s^2(s+2)(s+25)}$

(d) $\dfrac{50(s+5)}{s^2(s+2)(s+25)}$

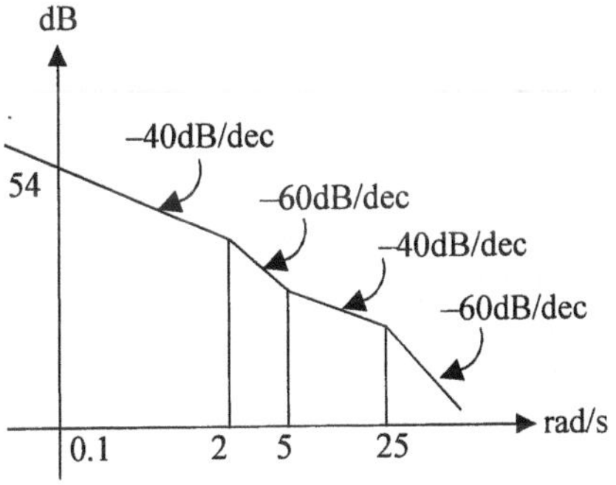

4. A differentiator has a transfer function whose

(a) phase increases with frequency (b) Amptitude remains constant

(c) Amptitude increases linearly (d) Amptitude decreases linearly with 'f'

5. The magnitude-frequency response of a control system is shown in the figure. The value of ω_1 and ω_2 are respectively

(a) 10 and 200

(b) 20 and 200

(c) 20 and 400

(d) 100 and 400

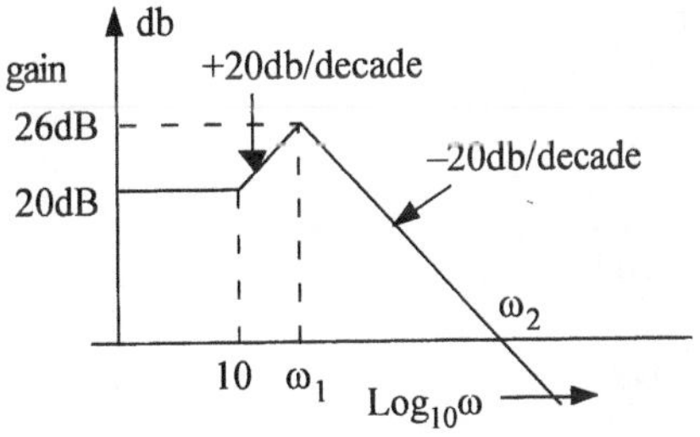

6. The transfer function of a system is given by $G(j\omega) = \dfrac{K}{(j\omega)(j\omega T + 1)}$, $K < \dfrac{1}{T}$

Which one of the following is the Bode plot of this function ?

(a)

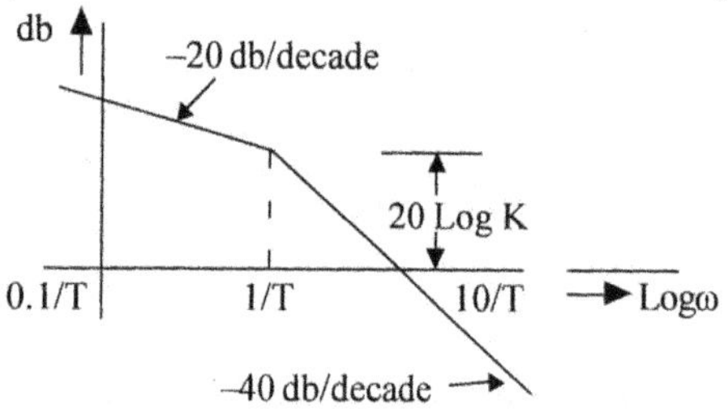

(b)

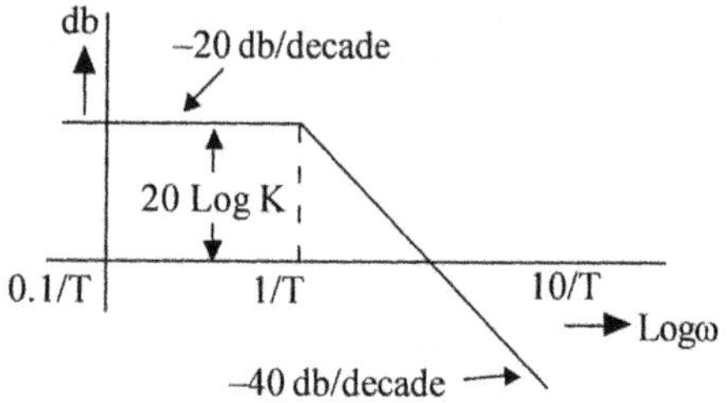

(c)

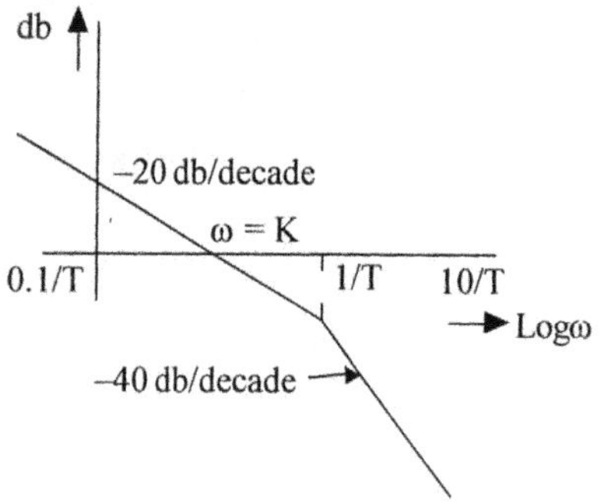

(d)

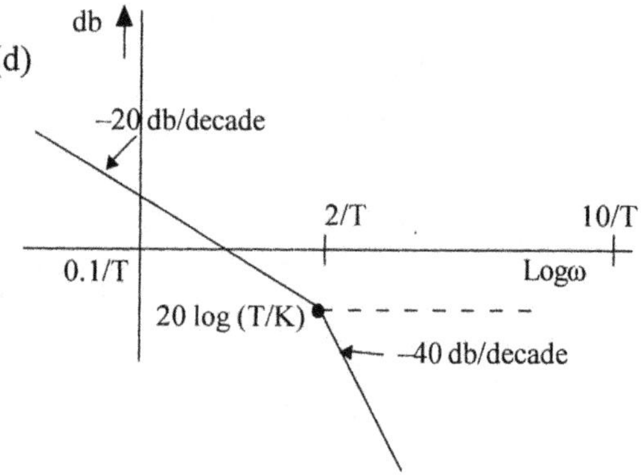

7. The Bode plot shown in the given figure has $G(j\omega)$ as

(a) $\dfrac{100}{j\omega(1 + j0.5\,\omega)(1 + j0.1\,\omega)}$

(b) $\dfrac{100}{j\omega\,(2 + j\,\omega)(10 + j\omega)}$

(c) $\dfrac{10}{j\omega\,(1 + 2j\,\omega)(1 + 10j\omega)}$

(d) $\dfrac{10}{j\omega\,(1 + 0.5j\,\omega)(1 + 0.1j\omega)}$

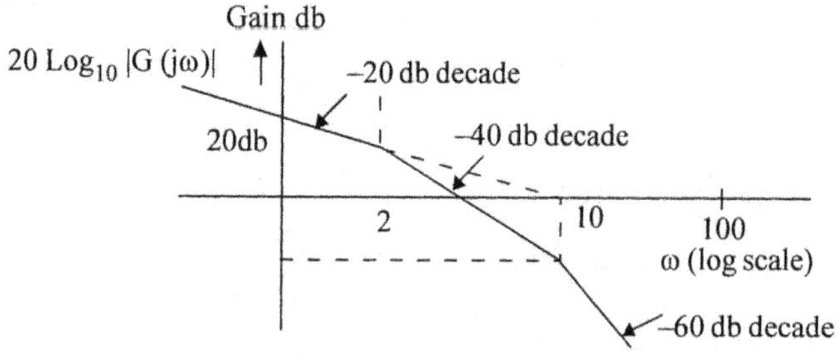

8. The polar plot of $G(s) = \dfrac{1+s}{1+4s}$ for $0 \le \omega \le \infty$ in G-plane is

(a)
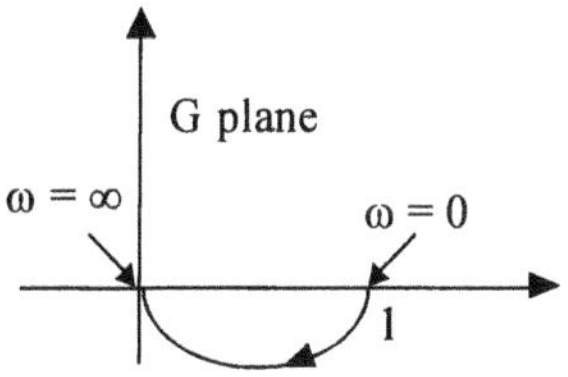

(b)
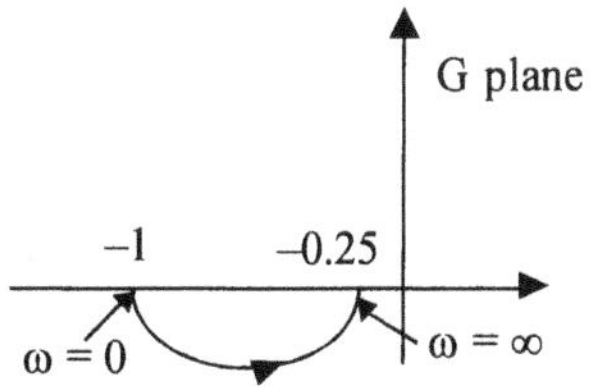

(c)
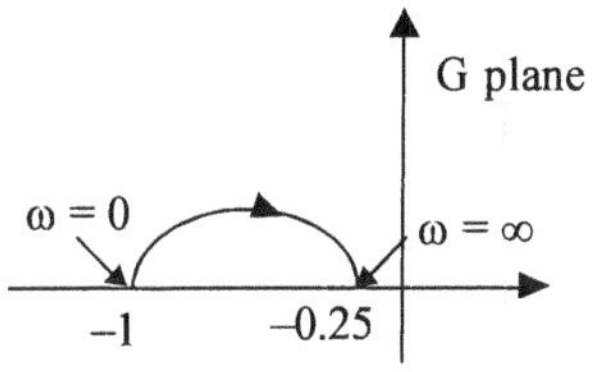

(d)
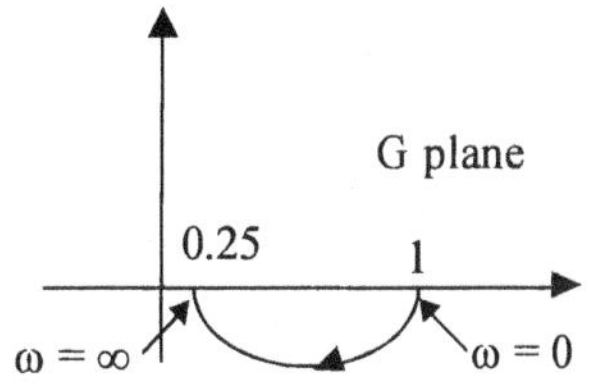

9. A system with transfer function $G(s) = \dfrac{s}{(1+s)}$ is subjected to a sinusoidal input $r(t) = \sin \omega t$. In steady-state, the phase angle of the output relative to the input at $\omega = 0$ and $\omega = \infty$ will be respectively

(a) 0° and -90° (b) 0° and 0° (c) 90° and 0° (d) 90° and -90°

10. A system has fourteen poles and two zeros. The slope of its highest frequency asymptote in its magnitude plot is

(a) – 40 dB/decade (b) – 240 dB/decade

(c) – 280 dB/decade (d) – 320 dB/decade

11. The phase angle of the system $G(s) = \dfrac{s+5}{s^2+4s+9}$ varies between

(a) 0° and 90° (b) 0° and -90° (c) 0° and -180° (d) -90° and -180°

12. The following Bode plot represents :

(a) $\dfrac{100\,s^2}{(.1+0.1s)^3}$

(b) $\dfrac{1000\,s^2}{(1+0.1s)^3}$

(c) $\dfrac{100\,s^2}{(1+0.1s)^5}$

(d) $\dfrac{1000\,s^2}{(1+0.1s)^5}$

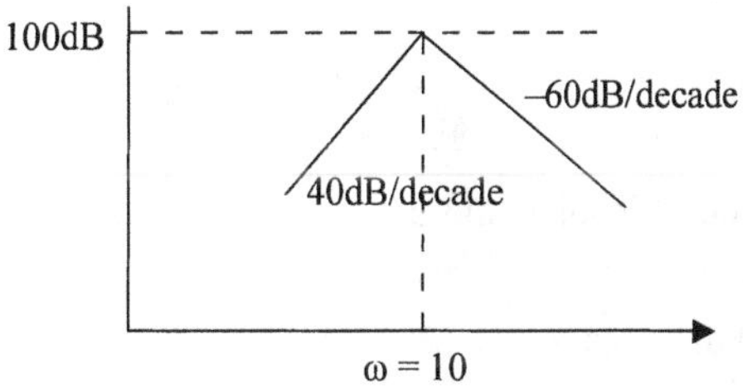

13. List I and List II show the transfer function and polar plots respectively. Match List-I with List-II and select the correct answer using the codes given below the lists :

List-I **List-II**

A. $\frac{1}{s(1+sT)}$ 1.

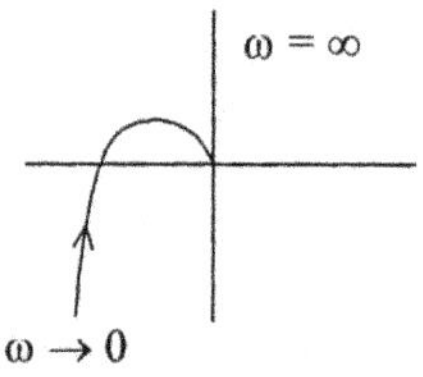

B. $\frac{1}{(1+sT_1)(1+sT_2)}$ 2.

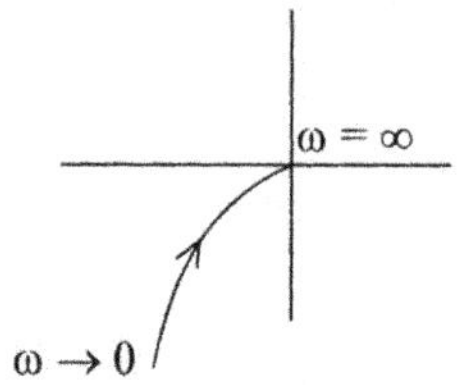

C. $\frac{1}{s(1+sT_1)(1+sT_2)}$ 3.

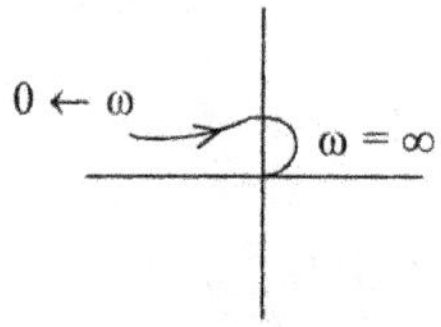

D. $\frac{1}{s^2(1+sT_1)(1+sT_2)}$ 4.

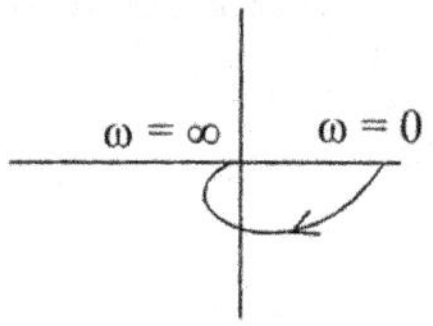

Codes :

	A	B	C	D		A	B	C	D
(a)	2	1	4	3	(b)	3	4	1	2
(c)	2	4	1	3	(d)	3	1	4	2

14. Octave frequency range is given by :

(a) $\frac{\omega_1}{\omega_2} = 2$ (b) $\frac{\omega_1}{\omega_2} = 4$ (c) $\frac{\omega_1}{\omega_2} = 8$ (d) $\frac{\omega_1}{\omega_2} = 10$

15. A system has 12 poles and 2 zeros. Its high frequency asymptote in its magnitude plot has a slope of

(a) – 200 dB/decade (b) – 240 dB/decade

(c) – 280 dB/decade (d) – 320 dB/decade

16. A linear stable time-invariant system is forced with an input $x(t) = A \sin \omega t$

Under steady-state conditions, the output $y(t)$ of the system will be

(a) $A \sin (\omega t + \phi)$, where $\phi = \tan^{-1} | G (j\omega) |$

(b) $| G (j\omega) | A \sin [\omega t + \angle G (j\omega)]$

(c) $| G (j\omega) | A \sin [2 \omega t + \angle G (j\omega)]$

(d) $AG (j\omega) \sin [\omega t + \angle G (j\omega)]$

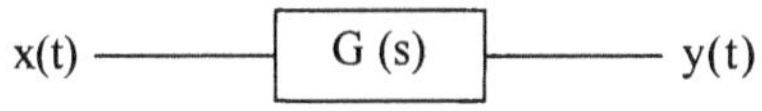

17. For the second-order transfer function $T(s) = \dfrac{4}{s^2 + 2s + 4}$

The maximum resonance peak will be

(a) 4 (b) $\dfrac{4}{3}$ (c) 2 (d) $\dfrac{2}{\sqrt{3}}$

18. The magnitude plot for a minimum phase function is shown in the figure. The phase plot for this function.

(a) cannot be uniquely determined

(b) will be monotonically increasing from 0° to 180°

(c) will be monotonically decreasing from 180° to 0°

(d) will be monotonically decreasing from 180° to -90°

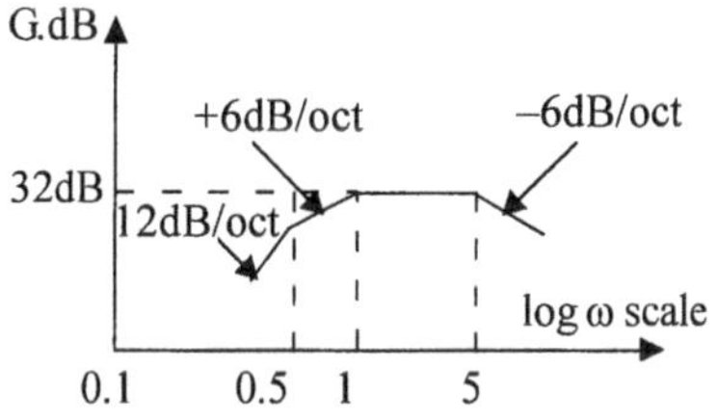

19. What is the slope change at $\omega = 10$ of the magnitude v/s frequency characteristic of a unity feedback system with the following open-loop transfer function ?

$$G(s) = \frac{5(1 + j0.1\omega)}{j\omega(1 + j0.5\omega)[1 + j0.6(\omega/50) + (j\omega/50)^2]}$$

(a) – 40 dB/dec to – 20 dB/dec (b) 40 dB/dec to 20 dB/dec

(c) – 20 dB/dec to – 40 dB/dec (d) 40 dB/dec to – 20 dB/dec

20. Match List I with List II and select the correct answer using the codes given below the lists :

List I (Transfer Functions)		**List II (Description)**
A. $\dfrac{1-s}{1+s}$	1.	Non-minimum phase-system
B. $\dfrac{1-s}{(1+s)(1+2s)(1+3s)}$	2.	Minimum phase system
C. $\dfrac{1+3s}{(1+4s)(1+2s)(1+s)}$	3.	All pass system

Codes :

	A	B	C
(a)	1	3	2
(b)	3	2	1
(c)	3	1	2
(d)	2	1	3

21. Which one of the following is the polar plot of a typical type zero system with open-loop transfer function $G(j\omega) = \dfrac{k}{(1+j\omega T_1)(1+j\omega T_1)}$

(a)

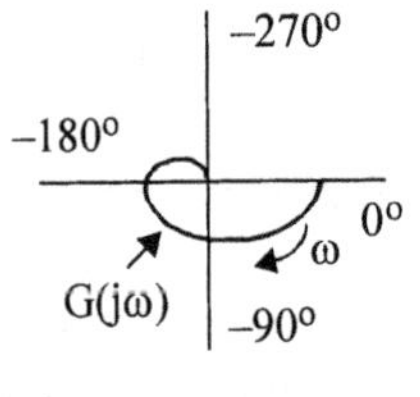

(b)

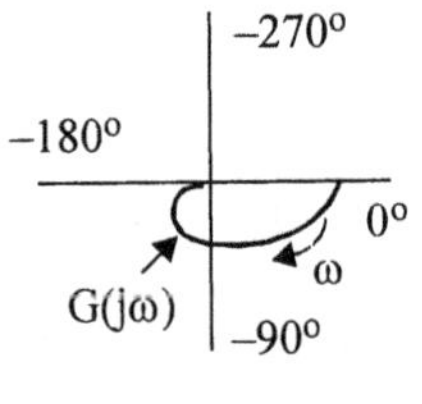

(c)

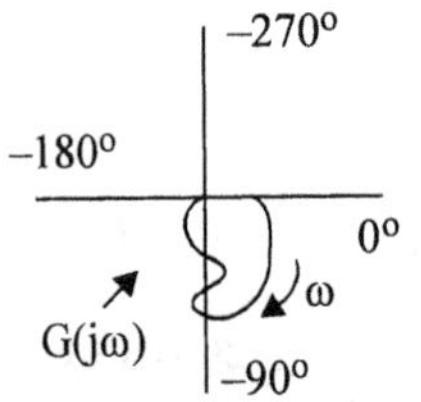

(d)

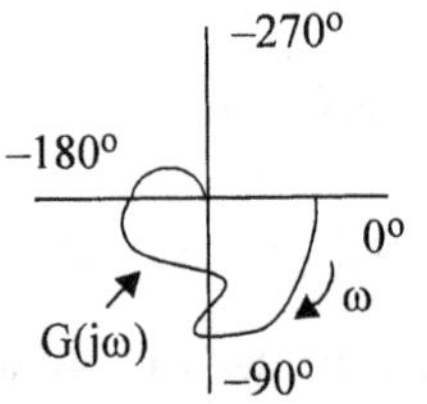

22. A second-order overall transfer function is given by $\frac{4}{s^2 + 2s + 4}$

Its resonant frequency is

(a) 2 (b) $\sqrt{2}$ (c) $\sqrt{3}$ (d) 3

23. The phase angle for the transfer function $G(s) = \frac{1}{(1 + sT)^3}$ at corner frequency is

(a) – 45° (b) – 90° (c) – 135° (d) – 270°

24. The magnitude plot of a transfer function is shown in the figure. The transfer function in question is

(a) $\frac{4\left(1+\frac{s}{2}\right)}{s\left(1+\frac{s}{10}\right)}$ (b) $\frac{4s\left(1+\frac{s}{2}\right)}{\left(1+\frac{s}{10}\right)}$

(c) $\frac{4(1+2s)}{s(1+10s)}$ (d) $\frac{4s(1+2s)}{(1+10s)}$

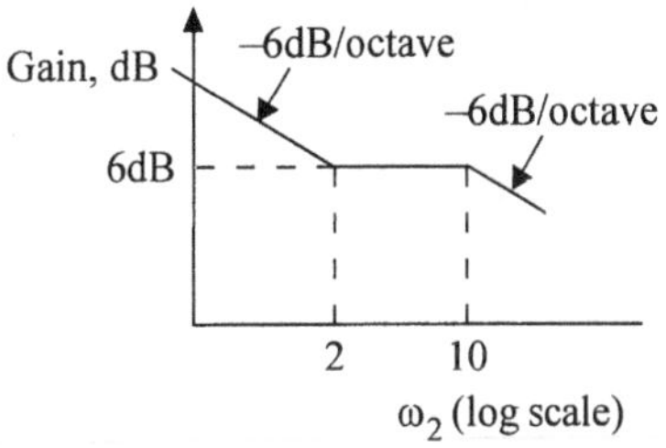

25. The open-loop transfer function of a unity negative feedback system is

$$G(s) = \frac{K(s+10)(s+20)}{s^3(s+100)(s+200)}$$

The polar plot of the system will be

(a)

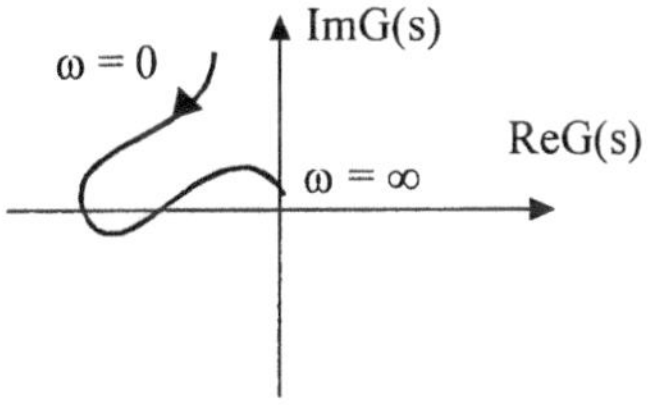

(b)

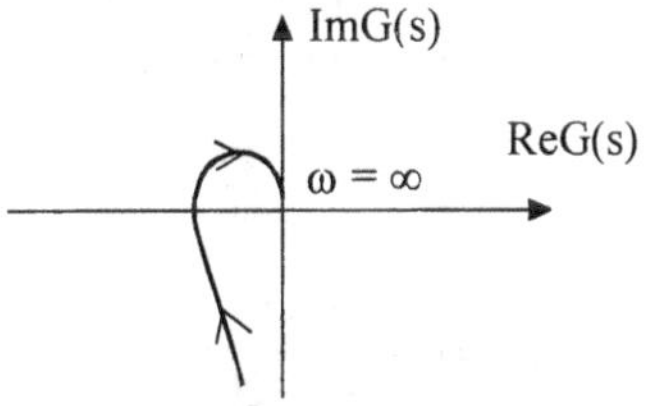

(c)

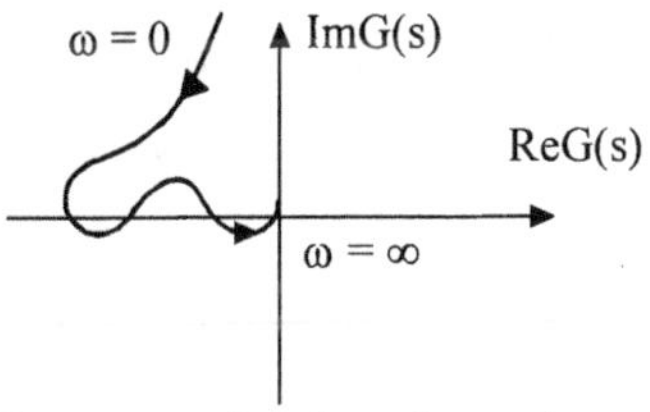

(d)

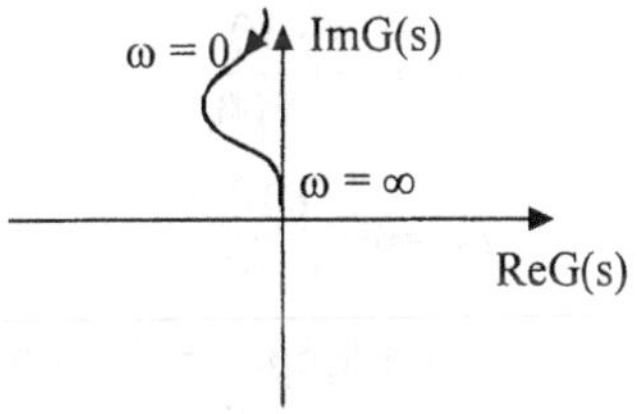

26. Open loop transfer function of a system having one zero with a positive real value is called

(a) zero phase function (b) negative phase function

(c) positive phase function (d) non-minimum phase function

27. An open loop transfer function of a unity feedback control system has two finite zeros, two poles at origin and two pairs of complex conjugate poles. The slope of high frequency asymptote in Bode magnitude plot will be

(a) + 40 dB/decade (b) 0 dB/decade

(c) – 40 dB/decade (b) –80 dB/decade

28. A decade frequency range is specified by

(a) $\frac{\omega_2}{\omega_1} = 2$ (b) $\frac{\omega_2}{\omega_1} = 10$ (c) $\frac{\omega_2}{\omega_1} = 8$ (d) none of the above

29. Which one of the following transfer functions represents the Bode plot shown in the figure ?

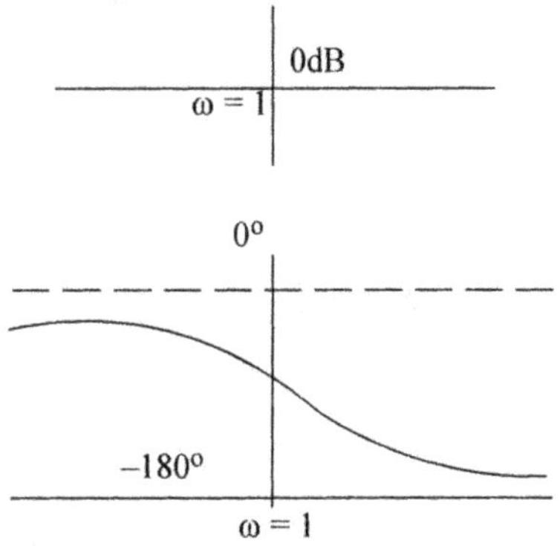

(a) $G = \frac{1-s}{1+s}$

(b) $G = \frac{1}{(1+s)^2}$

(c) $G = \frac{1}{s^2}$

(d) $G = \frac{1}{s(1+s)}$

30. The characteristic equation of a closed-loop control system is given by $s^2 + 4s + 16 = 0$ The resonant frequency in radians/sec of the system is

(a) 2 (b) $2\sqrt{3}$ (c) 4 (d) $2\sqrt{2}$

31. The log-magnitude Bode plot of a minimum phase system is shown in the figure. Its transfer function is given by

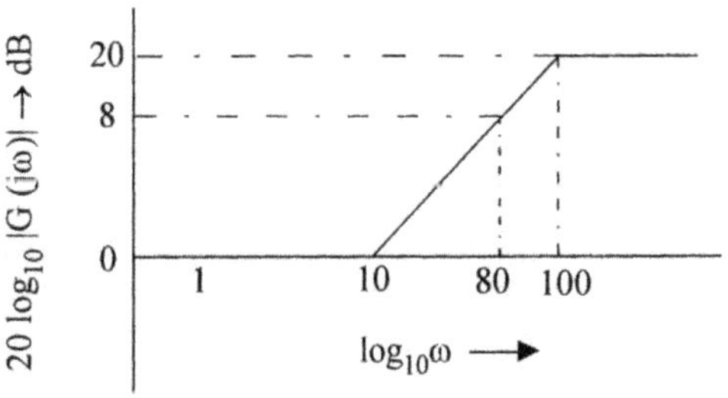

(a) $G(s) = \frac{s-10}{s+100}$

(b) $G(s) = \frac{s+10}{s-100}$

(c) $G(s) = \frac{s-10}{s-100}$

(d) $G(s) = \frac{s+10}{s+100}$

32. For a closed-loop transfer function given below

$\frac{C(s)}{R(s)} = \frac{2600K(s+25)}{s^4 + 125s^3 + 5100s^2 + 65000s + 65000K}$ the imaginary axis intercepts of the root loci will be

(a) $\pm j22.8$ (b) $\pm j2.28$ (c) $\pm j1.14$ (d) $\pm j11.4$

33. Match List I with List II and select the correct answer using the codes given below the lists

List I [G(s) H(s)]	List II (Nyquist plot)
A. $\frac{K}{1+sT_1}$	1. 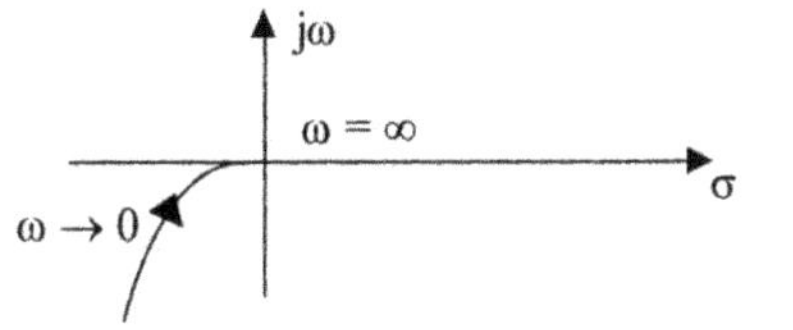
B. $\frac{K}{s(1+sT_1)}$	2. 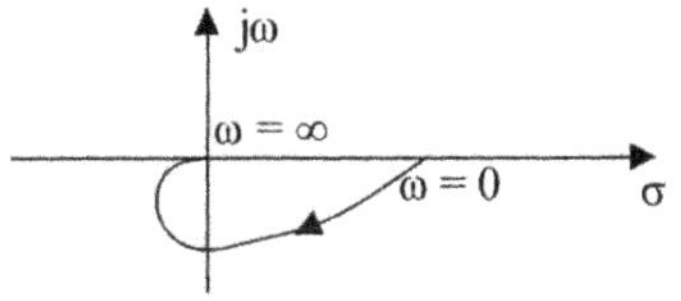
C. $\frac{K}{(1+T_1s)(1+T_2s)}$	3. 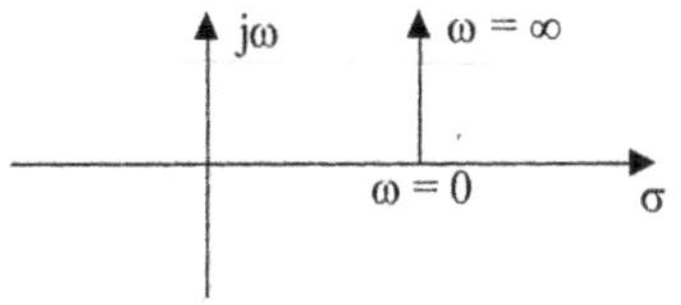
D. $\frac{K}{s^2(1+sT_1)}$	4. 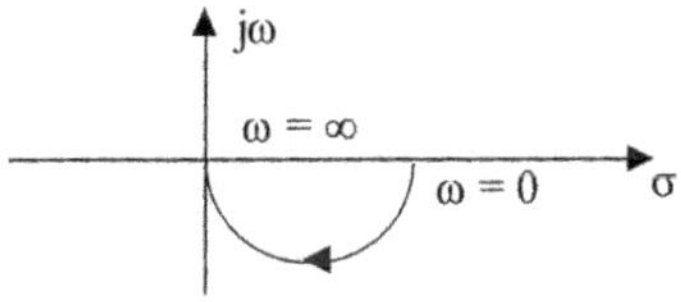
	5. 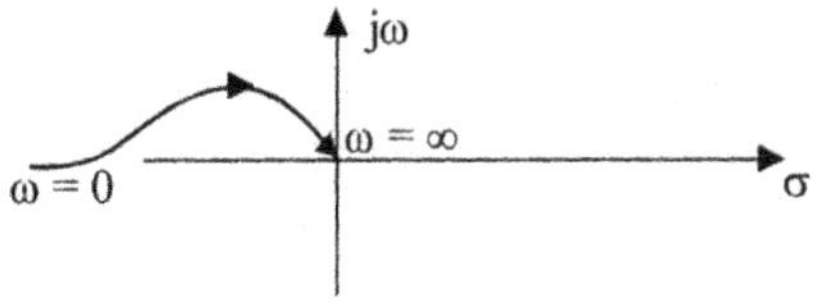

Codes :

(a)	A	B	C	D	(b)	A	B	C	D
	1	4	5	3		1	4	3	5
(c)	A	B	C	D	(d)	A	B	C	D
	4	1	2	3		4	1	2	5

34. The Nyquist plot for a control system is shown in Figure. I. The Bode plot for the same system will be as in

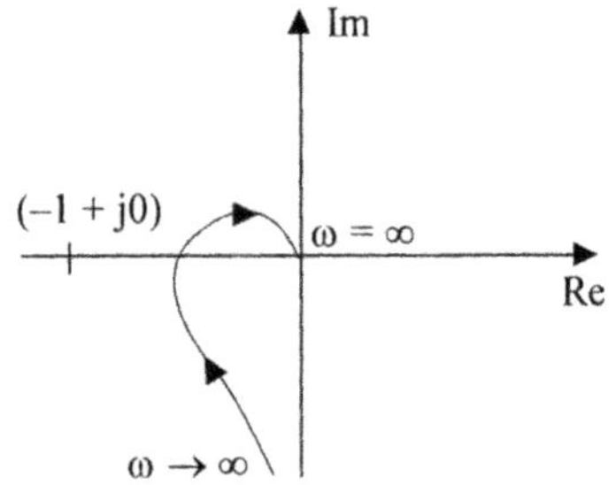

Nyquist plot (Figure I)

(a)

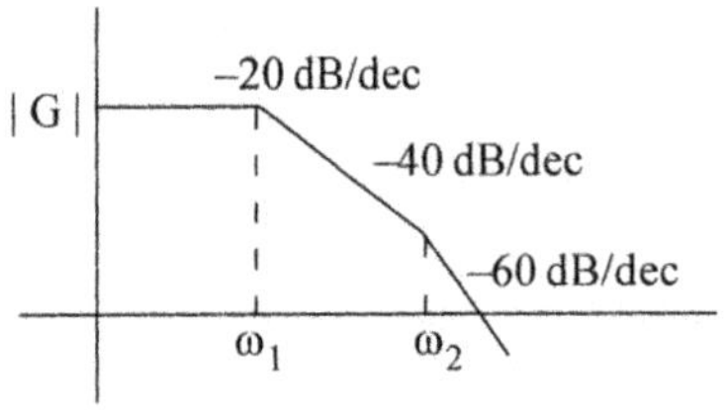

(b)

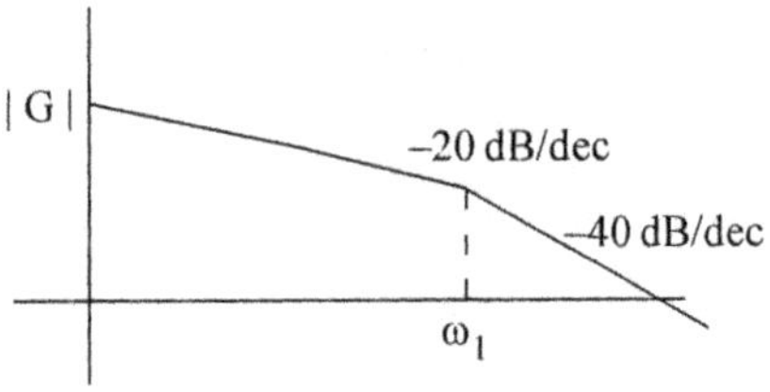

(c)

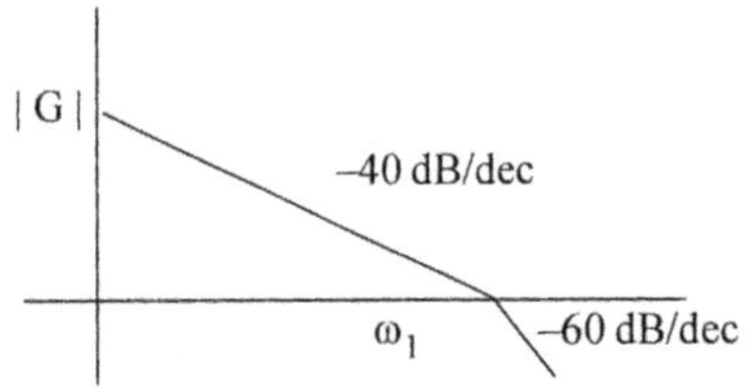

(d)

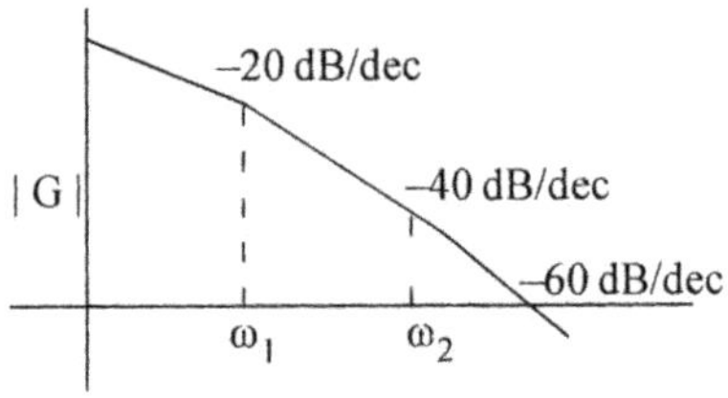

Chapter 7

1. The polar plot of a type-1, 3-pole, open-loop system is shown in Fig. The closed-loop system is

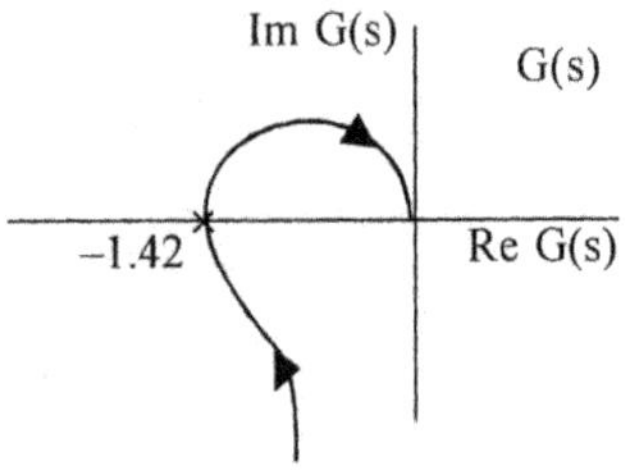

(a) always stable

(b) marginally stable

(c) unstable with one pole on the right half of s-plane

(d) unstable with two poles on the right half of s-plane

2. Which one of the following statements is true for gain margin and phase margin of two closed-loop systems having loop transfer functions G(s) H(s)and exp(–s) G(s) H(s) ?
   (a) Both gain and phase margins of the two systems will be identical
   (b) Both gain and phae margins of G(s) H(s) will be more
   (c) Gain margins of the two system are the same but phase margin of G(s) H(s) will be more
   (d) Phase margins of the two system are the same but gain margin of G(s) H(s) will be less

3. An effect of phase-lag compensation on servo system performance is that
   (a) for a given relative stability, the velocity constant is increased
   (b) for a given relative stability, the velocity constant is decreased
   (c) the bandwidth of the system is increased
   (d) the time response is made faster

4. If the compensated system shown has a phase margin of 60° at the crossover frequency of 1 rad/sec, the value of the gain k is
   (a) 0.366
   (b) 0.732
   (c) 2.738
   (d) 1.366

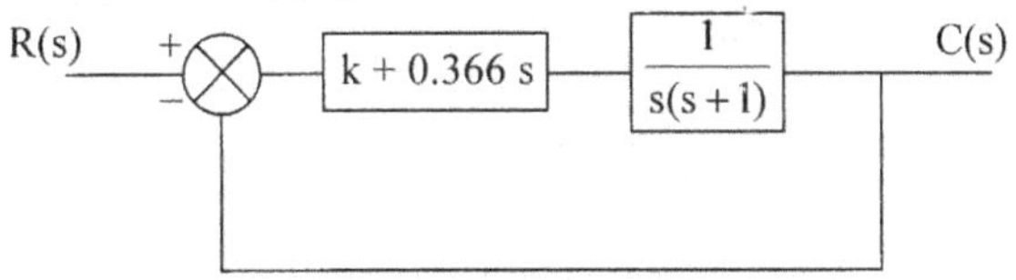

5. The open-loop transfer function of a unity feedback control system is $\frac{10}{(s+5)^3}$. The gain margin of the system will be
   (a) 20 dB (b) 40 dB (c) 60 dB (d) 80 dB

6. The Nyquist plot of the open-loop transfer function of a feedback control system is shown in the given figure. If the open-loop poles and zeros are all located in the left half of s-plane, then the number of closed-loop poles in the right half of s-plane will be
   (a) zero
   (b) 1
   (c) 2
   (d) 3

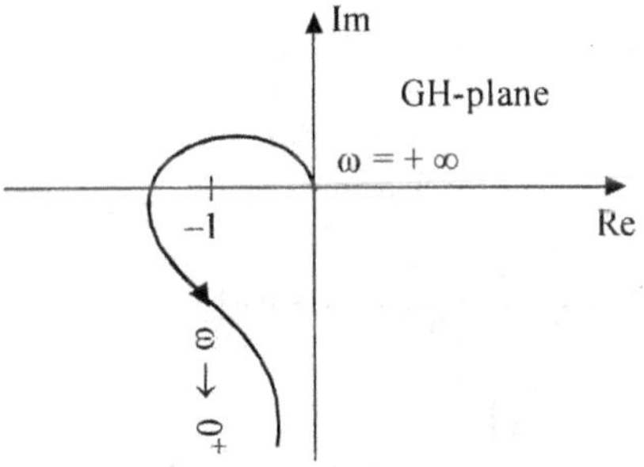

7. Match List-I (Plot/diagram/chart) with List-II (Characteristics) and select the correct answer using the codes given below the Lists :

List-I	**List-II**
A. Constant M loci	1. Constant gain and phase shift loci of the closed-loop system
B. Constant N loci	2. Plot of loop gain with variation of ω
C. Nichol's chart	3. Circles of constant gain for closed-loop transfer function
D. Nyquist plot	4. Circles of constant phase shaft of closed-loop transfer function

Codes :

	A	B	C	D		A	B	C	D
(a)	3	4	2	1	(b)	3	4	1	2
(c)	4	3	2	1	(d)	4	3	1	2

8. Consider the Nyquist diagram for given KG(s) H(s). The transfer function KG(s) H(s) has no poles and zeros in the right half of s-plane. If the (–1, j0) point is located first in region I and then in region II, the change in stability of the system will be from

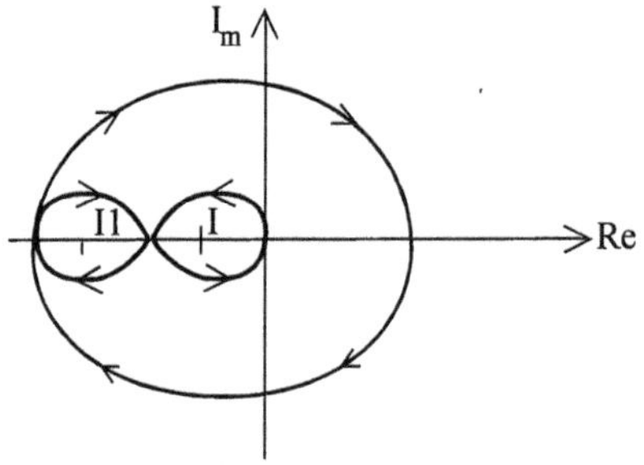

(a) unstable to stable

(b) stable to stable

(c) unstable to unstable

(d) stable to unstable

9. Consider the following Nyquist plots of different control systems :

Which of these plot(s) represent(s) a stable system ?

1. 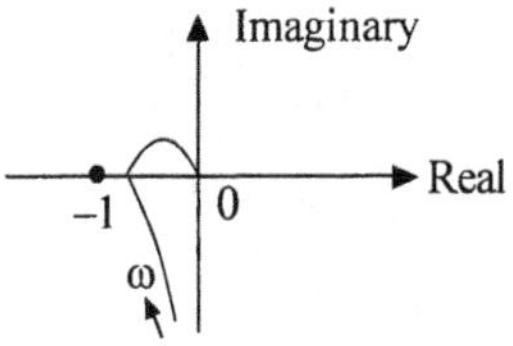

2. 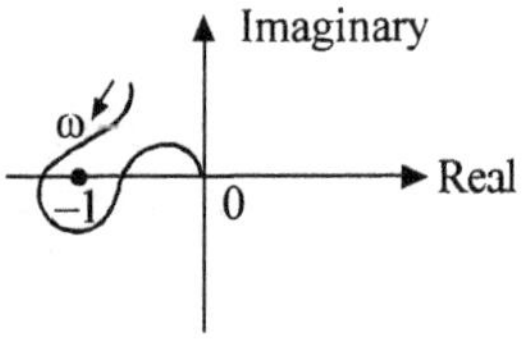

3. 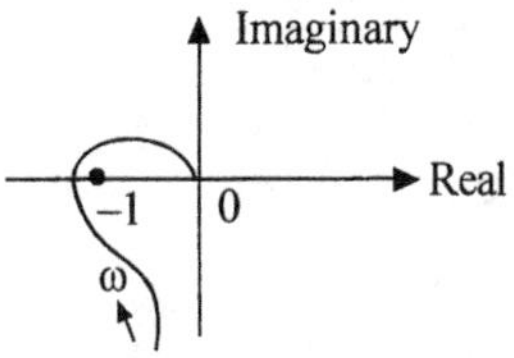

4. 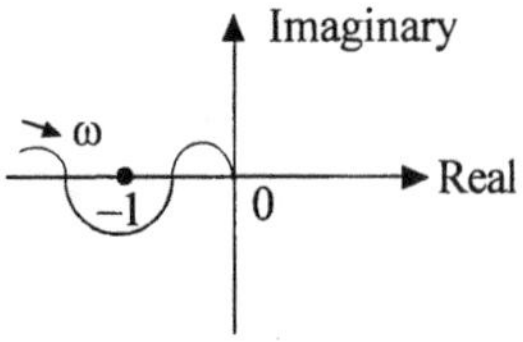

(a) 1 alone (b) 2, 3 and 4 (c) 1, 3 and 4 (d) 1, 2 and 4

10. The transfer function of a certain system is given by $G(s) = \frac{s}{(1+s)}$

The Nyquist plot of the system is

(a)

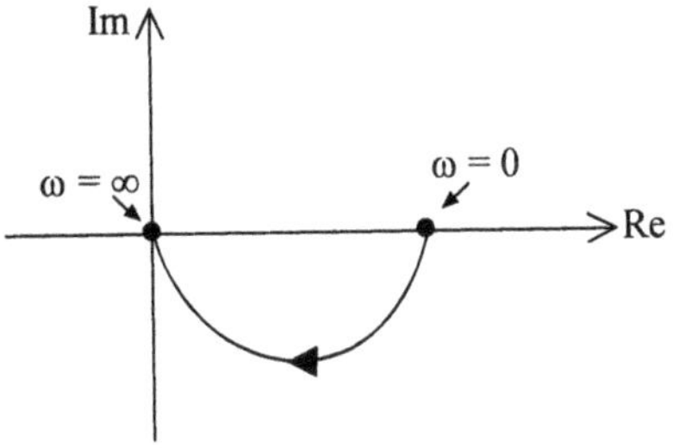

(b)

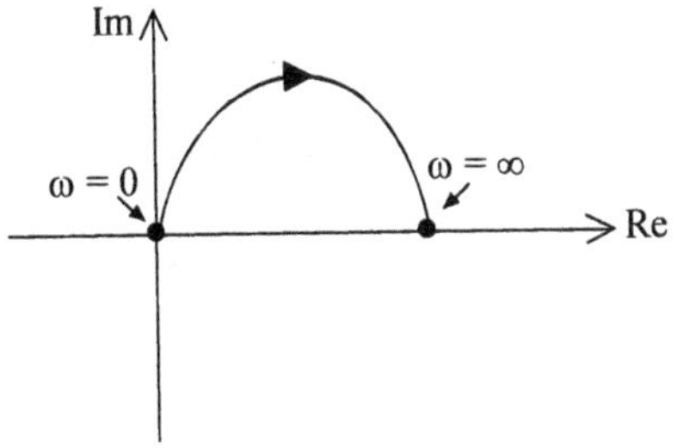

(c)

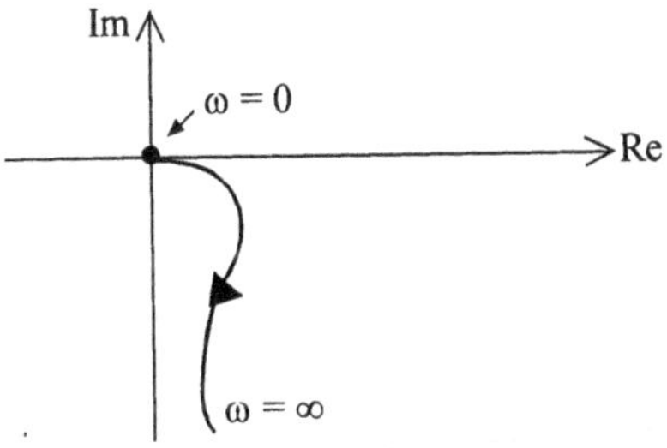

(d)

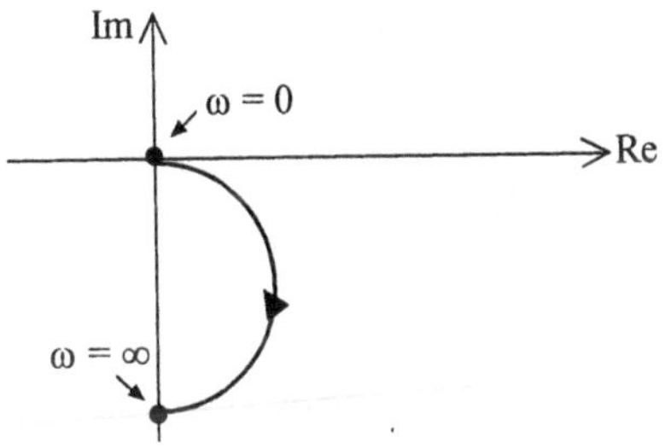

11. The polar plot of $G(s) = \frac{10}{s(s+1)^2}$ intersects real axis at $\omega = \omega_0$, Then, the real part and ω_0 are respectively given by :

(a) –5, 1 (b) –2.5, 1 (c) –5, 0.5 (d) –5, 2

12. For the transfer function $G(s)\,H(s) = \frac{1}{s(s+1)(s+0.5)}$ the phase cross-over frequency is

(a) 0.5 rad/sec (b) 0.707 rad/sec

(c) 1.732 rad/sec (d) 2 rad/sec

13. The gain phase plot of open loop transfer function of four different systems labelled A, B, C and D are shown in the figure. The correct sequence of the increasing order of stability of the four systems will be

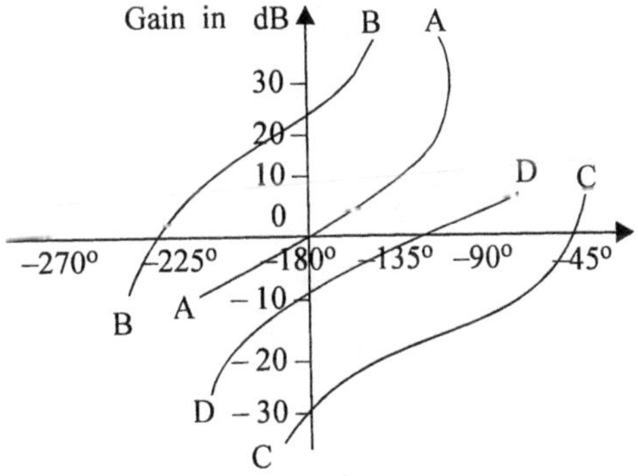

(a) A, B, C, D

(b) D, C, B, A

(c) B, A, D, C

(d) B, C, D, A

14. For the gain-phase plot shown in the given figure, for the open-loop transfer function G (s), gain margin, gain crossover frequency, phase margin and phase crossover frequency are respectively

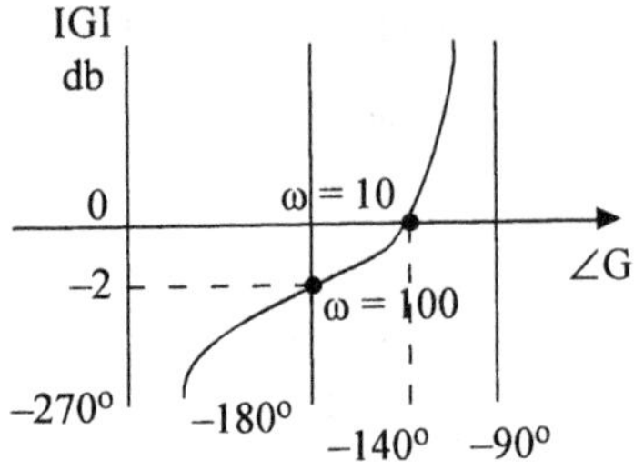

(a) 2 db, 100 rad/sec, 40°, 10 rad/sec

(b) 0 db, 10 rad/sec, – 40°, 100 rad/sec

(c) 2 db, 10 rad/sec, 40°, 100 rad/sec

(d) – 2 db, 10 rad/sec, – 40°, 100 rad/sec

15. The type and order of the system whose Nyquist-plot is shown in the given figure are respectively

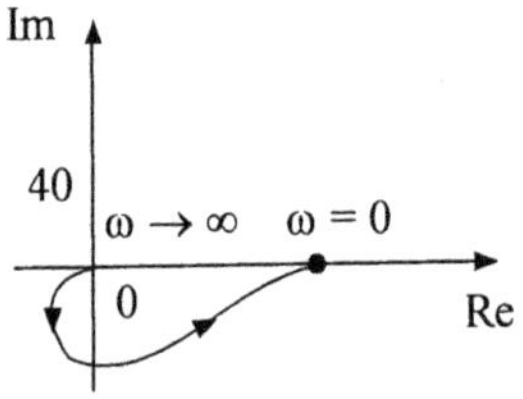

(a) 0, 1
(b) 1, 2
(c) 0, 2
(d) 2, 1

16. A unity feedback system has the following open-loop frequency response :

ω	| G (jω) |	∠| G (jω) |
2	7.5	– 118°
3	4.8	– 130°
4	3.15	– 140°
5	2.25	– 150°
6	1.70	– 157°
8	1.00	– 170°
10	0.64	– 180°

The gain margin and phase margin of the system are

(a) 0 dB, – 170°
(b) 3.86 dB, – 180°
(c) 0 dB, 10°
(d) 3.86 dB, 10°

17. The radius and the centre of M circles are given respectively by

(a) $\frac{M}{M^2-1}, \left(\frac{-M^2}{M^2-1}, 0\right)$

(b) $\frac{M^2}{M^2-1}, \left(\frac{-M}{M^2-1}, 0\right)$

(c) $\frac{M^2}{M-1}, \left(\frac{-M^2}{M-1}, 0\right)$

(d) $\frac{M^2}{M^2-1}, \left(\frac{-M^2}{M^2-1}, 0\right)$

18. For a given G(s) H(s) the complete polar plot is shown in the given figure. If all T's are positive, then how many poles of the closed-loop system will be there in the RHS ?

$$G(s)\,H(s) = \frac{K_0(1+T_1s)^2}{(1+T_2s)(1+T_3s)(1+T_4s)(1+T_5s)^2}$$

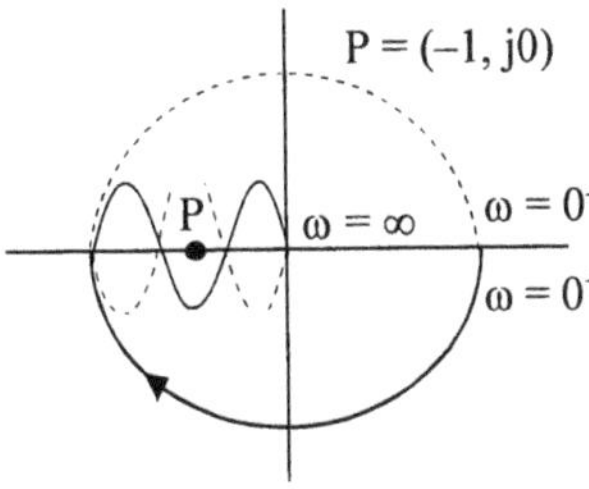

(a) zero

(b) 1

(c) 2

(d) 3

19. The polar plot of the open-loop transfer function of a feedback control system intersects the real exis at –2. The gain margin of the system is

(a) –5 dB (b) 0 dB (c) –6 dB (d) 40 dB

20. Which one of the following statements regarding the stability of a feedback control system is correct ?

(a) Gain margin (GM) gives complete information about the relative stability of the system

(b) Phase margin (PM) gives complete information about the relative stability of the system

(c) GM and PM together gives information about the relative stability of the system

(d) Gain cross-over and phase cross-over frequencies give the required information about the relative stability of the system.

21. The constant M loci plot is symmetrical with respect to

(a) real axis and imaginary axis

(b) M = 1 straight line and the real axis

(c) M = 1 straight line and the imaginary axis

(d) M = 1 straight line

22. The radius of constant-N circle for N = 1 is

(a) 2 (b) $\sqrt{2}$ (c) 1 (d) $\frac{1}{\sqrt{2}}$

23. The constant M circle for M = 1 is the

(a) straight line $x = -\frac{1}{2}$ (b) critical point (– 1, j0)

(c) circle with r = 0.33 (d) circle with r = 0.67

24. The polar plot of a transfer function passes through the critical point (–1, 0). Gain margin is

(a) zero (b) – 1 dB (c) 1 dB (d) infinity

25. The polar plot (for positive frequencies) for the open-loop transfer function of a unity feedback control system is shown in the given figure

The phase margin and the gain margin of the system are respectively

(a) 150° and 4

(b) 150° and $\frac{3}{4}$

(c) 30° and 4

(d) 30° and $\frac{3}{4}$

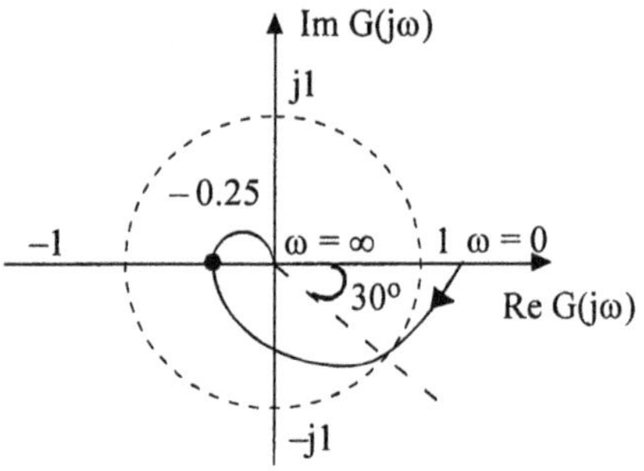

26. Which one of the following features is NOT associated with Nichols chart ?

(a) (0 dB, – 180°) point on Nichols chart represents the critical point (–1 + jO)

(b) It is symmetric about –180°

(c) The M loci are centred about (0 dB, – 180°) point

(d) The frequency at the intersection of the G(jω) locus and M = +3 dB locus gives bandwidth of the closed-loop system.

27. Which one of the following equations represents the constant magnitude locus in G-plane for M = 1 ? {x-axis is Re G (jω) and y-axis is Im G (jω)}

(a) $x = -0.5$ (b) $x = 0$

(c) $x^2 + y^2 = 1$ (d) $(x + 1)^2 + y^2 = 1$

28. The open loop transfer function of a system is $G(s)\,H(s) = \dfrac{K}{(1+s)(1+2s)(1+3s)}$

The phase crossover frequency ω_c is

(a) $\sqrt{2}$ (b) 1 (c) zero (d) $\sqrt{3}$

29. Nyquist plot shown in the given figure is a type

(a) zero system

(b) one system

(c) two system

(d) three system

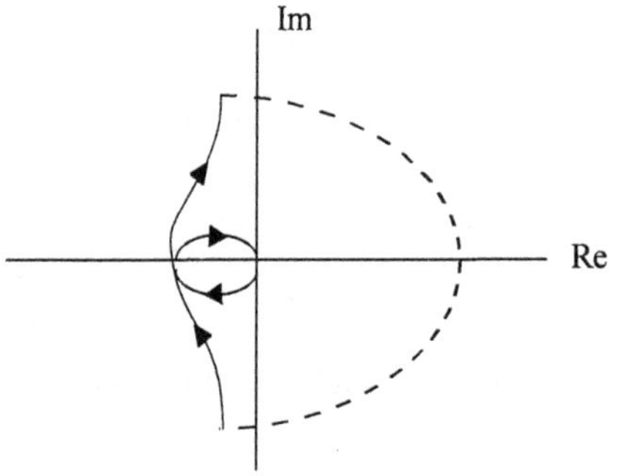

30. The open loop transfer function of a unity feedback control system is given as

$$G(s) = \frac{1}{s(1+sT_1)(1+sT_2)}$$

The phase crossover frequency and the gain margin are, respectively,

(a) $\frac{1}{\sqrt{T_1 T_2}}$ and $\frac{T_1+T_2}{T_1T_2}$ (b) $\sqrt{T_1T_2}$ and $\frac{T_1+T_2}{T_1T_2}$

(c) $\frac{1}{\sqrt{T_1 T_2}}$ and $\frac{T_1T_2}{T_1+T_2}$ (d) $\sqrt{T_1T_2}$ and $\frac{T_1T_2}{T_1+T_2}$

31. A constant N-circle having centre at $(-\frac{1}{2} + j0)$ in the G-plane, represents the phase angle equal to

(a) 180° (b) 90° (c) 45° (d) 0°

32. The constant M-circle represented by the equation $x^2 + 2.25x + y^2 = -1.125$ where $x = \text{Re}[G(j\omega)]$ and $y = \text{Im}[G(j\omega)]$ has the value of M equal to

(a) 1 (b) 2 (c) 3 (d) 4

33. The advantages of Nyquist stability test are

(a) it guides in stabilising an unstable system

(b) it enables to predict closed loop stability from open loop results

(c) is is applicable to experimental results of frequency response of open loop system

(d) all of the above

34. The Nyquist locus of a transfer function G(s), $H(s) = \frac{K}{1+sT_1}$ is given in Fig. I.

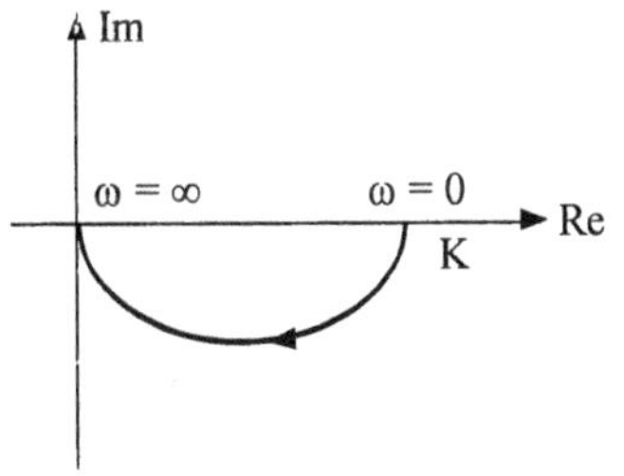

Fig. I.

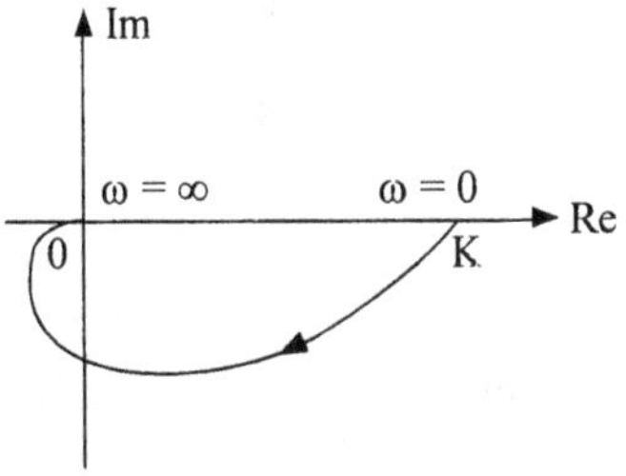

Fig. II.

The locus is modified as shown is Fig. II on addition of pole or poles to the original G(s) H(s) Then, the modified transfer function of the modified locus is

(a) $G(s)\,H(s) = \frac{K}{s(1+sT_1)}$ (b) $G(s)\,H(s) = \frac{K}{(1+sT_1)(1+sT_2)}$

(c) $G(s)\,H(s) = \frac{K}{(1+sT_1)(1+sT_2)}$ (d) $G(s)\,H(s) = \frac{K}{(1+sT_1)(1+sT_2)(1+sT_3)}$

35. The constant M-circles corresponding to the magnitude (M) of the closed-loop transfer of linear system for values of M greater than one lie in the G-plane and to the

(a) right of the M = 1 line (b) left of the M = 1 line

(c) upper side of the M = + j 1 line (d) lower side of the m = –j 1 line

36. The Nyquist plot of servo system is shown in the Figure-I. The root loci for the system would be

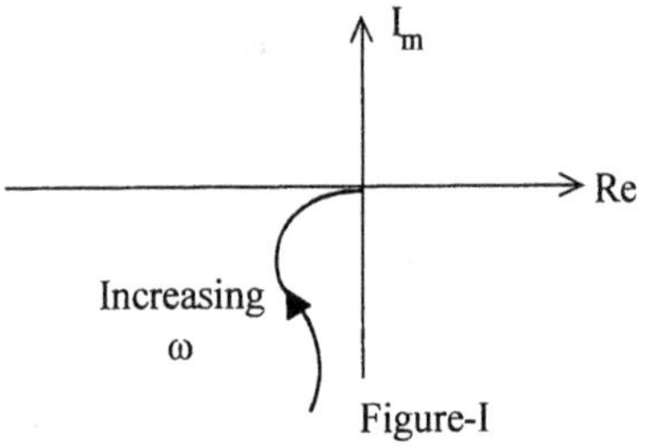

Figure-I

(a)

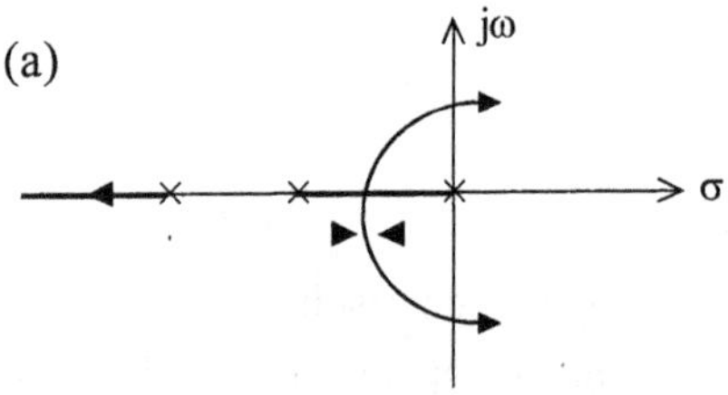

(b)

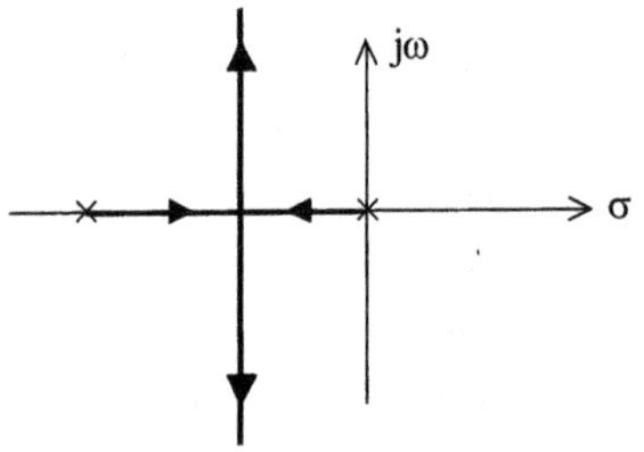

(c)

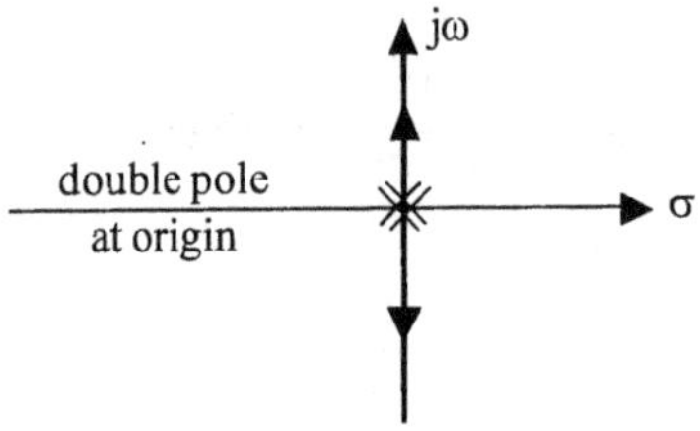

(d) None of the drawn plot of (a), (b), (c) of the question

37. Consider the following Nyquist plot of a open loop stable system :
The feedback system will be stable if and only if the critical point lies in the region

(a) I (OP)

(b) II (PQ)

(c) III (QR)

(d) None of these

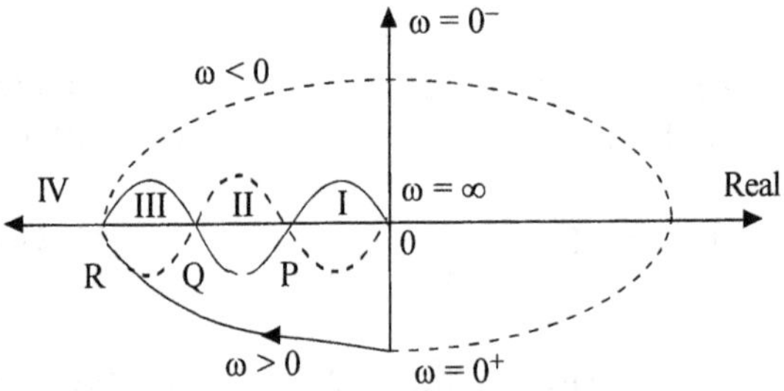

Chapter 8

1. The gain cross-over frequency and bandwidth of a control system are ω_{cu} and ω_{bu} respectively. A phase-lag network is employed for compensating the system. If the gain cross-over frequency and band width of the compensated system are ω_{cc} and ω_{bc} respectively, then

   (a) $\omega_{cc} < \omega_{cu}$; $\omega_{bc} < \omega_{bu}$
   (b) $\omega_{cc} > \omega_{cu}$; $\omega_{bc} < \omega_{bu}$
   (c) $\omega_{cc} < \omega_{cu}$; $\omega_{bc} > \omega_{bu}$
   (d) $\omega_{cc} > \omega_{cu}$; $\omega_{bc} > \omega_{bu}$

2. A portion of the polar plot of an open-loop transfer function is shown in the given figure The phase margin and gain margin will be respectively

   (a) 30° and 0.75
   (b) 60° and 0.375
   (c) 60° and 0.375
   (d) 60° and $\frac{1}{0.75}$

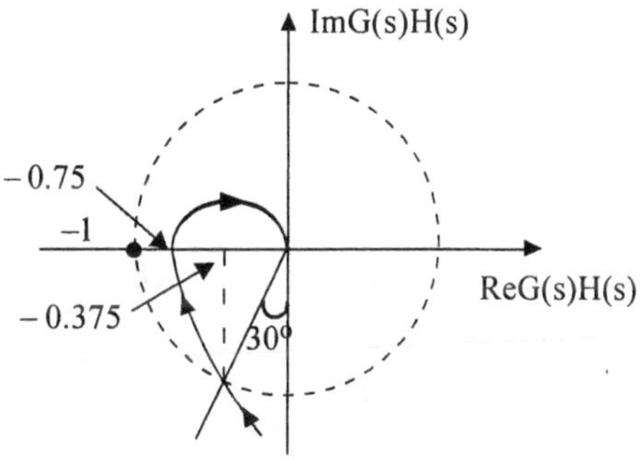

3. If the transfer function of a phase lead compensator is $\frac{s+a}{s+b}$ and that of a lag compensator is $\frac{s+p}{s+q}$ then which one of the following sets of conditions must be satisfied ?

   (a) a > b and p > q
   (b) a > b and p < q
   (c) a < b and p < q
   (d) a < b and p > q

4. The phase lead compensation is used to

   (a) increase rise time and decrease overshoot
   (b) decrease both rise time and overshoot
   (c) increase both rise time and overshoot
   (d) decrease rise time and increase overshoot

5. Maximum phase-lead of the compensator $D(s) = \frac{(0.5s+1)}{(0.05s+1)}$, is

   (a) 52 deg at 4 rad/sec
   (b) 52 deg at 10 rad/sec
   (c) 55 deg at 12 rad/sec
   (d) None of the answer is correct

6. Match List I with List II and select the correct answer using the codes given below the lists :

List I	List II
A. Transfer function of a lag network	1. $L^{-1}(sI - A)^{-1}$
B. State-transition matrix for $\dot{X} = AX$	2. $\frac{(s+Z)}{(s+P)}, \frac{Z}{P} > 1$
C. DC amplifier transfer function	3. k
D. Steady state error of type-1 system for step input	4. $L^{-1}(sA - 1)^{-1}$
	5. $\frac{s+z}{s+p}, \frac{z}{p} < 1$
	6. K(1 + sT)
	7. zero
	8. ∞

Codes :

	A	B	C	D
(a)	1	3	4	5
(b)	2	3	1	8
(c)	5	4	6	8
(d)	2	1	3	7

7. Match List-I with List-II and select the correct answer using the codes given below the lists :

List-I	List-II
A. Phase lag controller	1. Improvement in transient response
B. Addition of zero at origin	2. Reduction in steady-state error
C. Derivative output compensation	3. Reduction in settling time
D. Derivative error compensation	4. Increase in damping constant.

Codes :

	A	B	C	D
(a)	4	3	1	2
(b)	2	1	3	4
(c)	4	1	3	2
(d)	2	3	1	4

8. A phase-lead compensator has the transfer function $G_c(s) = \dfrac{10(1+.04s)}{(1+.01s)}$

The maximum phase-angle lead provided by this compensator will occur at a frequency ω_m equal to

(a) 50 rad/sec (b) 25 rad/sec (c) 10 rad/sec (d) 4 rad/sec

9. Match List I with List II and select the correct answer using codes given below the lists:

List I (Circuit diagram) **List II (Name)**

A. 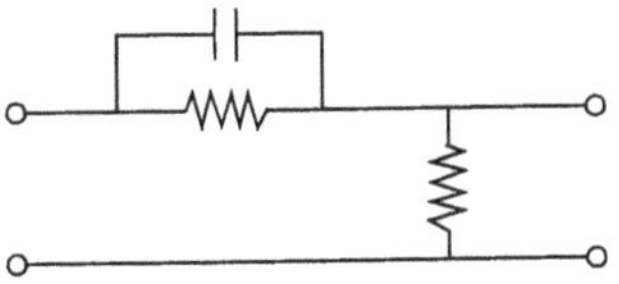

1. Lag network

B. 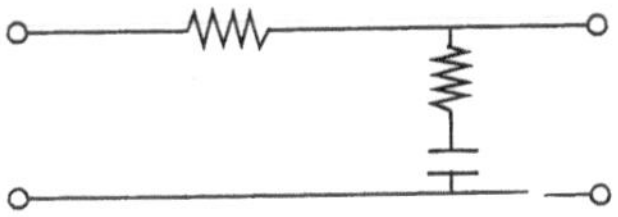

2. Lead network

C. 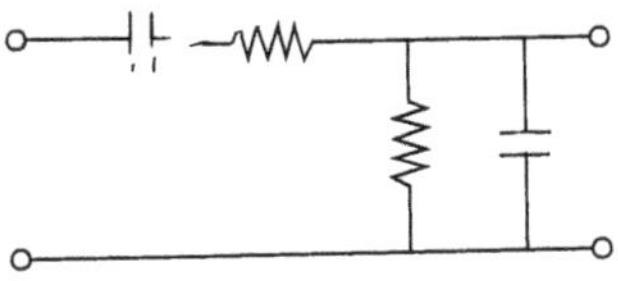

3. Lag-lead network

D. 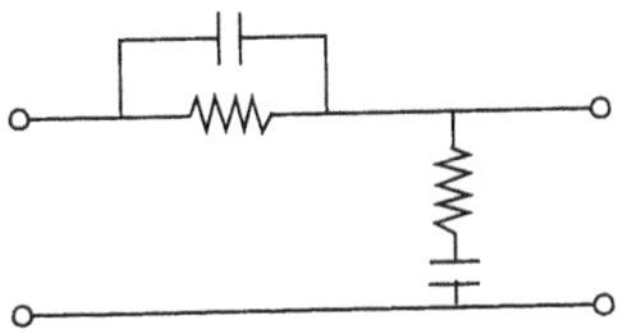

4. Lead-lag network

Codes :

	A	B	C	D
(a)	1	2	3	4
(b)	2	1	3	4
(c)	1	2	4	3
(d)	2	1	4	3

10. The root-locus plot for an uncompensated unstable system is shown in the given figure. The system is to be compensated by a compensating zero. The most desirable location of the compensating zero would be the point marked,

(a) A

(b) B

(c) C

(d) D

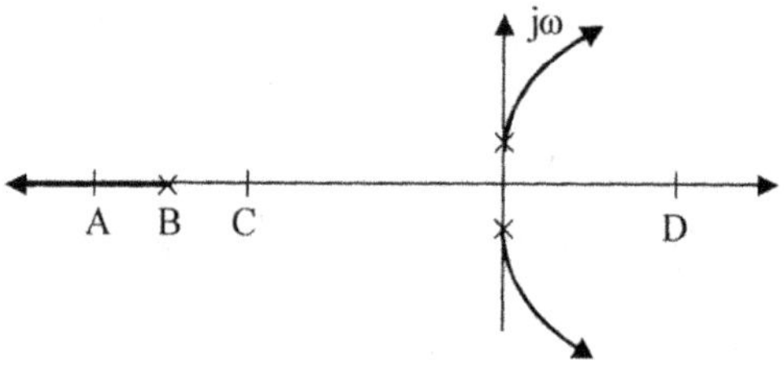

11. The transfer function of a lead conpensator is $G_C(s) = \frac{1+0.12s}{1+0.04s}$

The maximum phase shift that can be obtained from this compensator is

(a) 60° (b) 45° (c) 30° (d) 15°

12. The transfer function of a phase lead compensator is given by $\frac{1+aTs}{1+Ts}$ where $a > 1$ and $T > 0$. The maximum phase shift provided by such a compensator is

(a) $\tan^{-1}\left(\frac{a+1}{a-1}\right)$ (b) $\tan^{-1}\left(\frac{a-1}{a+1}\right)$ (c) $\sin^{-1}\left(\frac{a+1}{a-1}\right)$ (d) $\sin^{-1}\left(\frac{a-1}{a+1}\right)$

13. Indicate which one of the following transfer functions represents phase lead compensator?

(a) $\frac{s+1}{s+2}$ (b) $\frac{6s+3}{6s+2}$ (c) $\frac{s+5}{3s+2}$ (d) $\frac{s+8}{s^2+5s+6}$

14. The maximum phase shift that can be provided by a lead compensator with transfer function, $G(s) = \frac{1+6s}{1+2s}$

(a) 15° (b) 30° (c) 45° (d) 60°

15. The transfer function of a compensating network is of the form $\frac{1+\alpha Ts}{(1+Ts)}$.

If this is a phase-lag network the value of α should be

(a) exactly equal to 0 (b) between 0 and 1

(c) exactly equtal to 1 (d) greater than 1

16. For the given phase-lead network, the maximum possible phase lead is

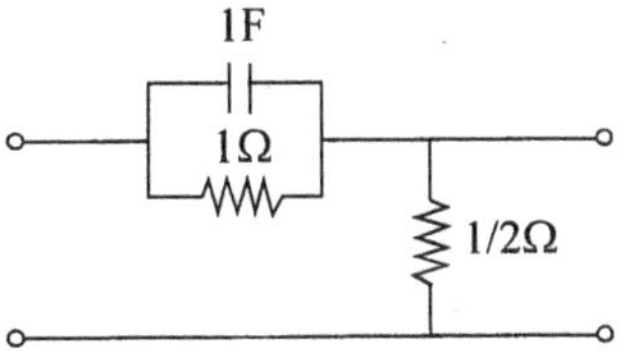

(a) $\sin^{-1}(1/3)$

(b) 30°

(c) 45°

(d) 60°

17. When phase-lag compensation is used in a system, gain crossover frequency, band width and undamped frequency are respectively

(a) increased, increased, increased (b) increased, increased, decreased

(c) increased, decreased, decreased (d) decreased, decreased, decreased

18. A lag compensator is basically a

(a) high pass filter (b) band pass filter

(c) low pass filter (d) band elimination filter

19. Which one of the following is a phase-lead compensation network ?

(a) 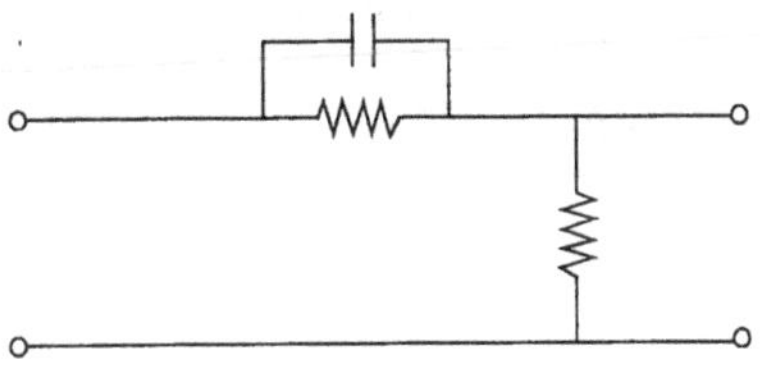

(b) 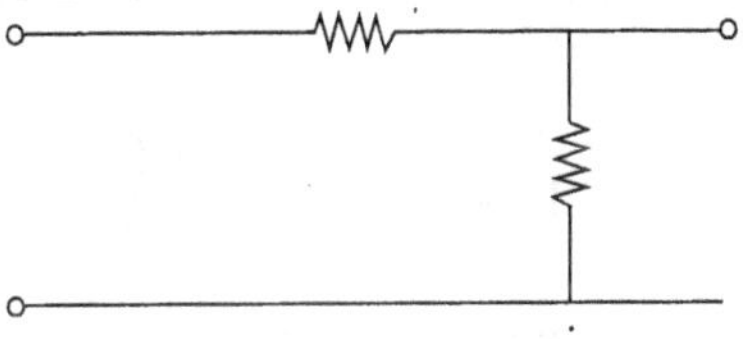

(c) 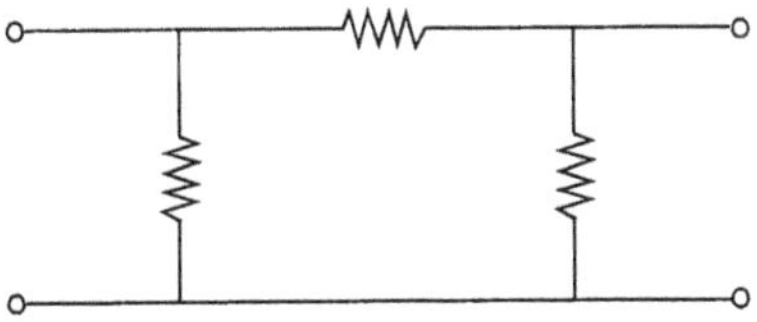

(d) 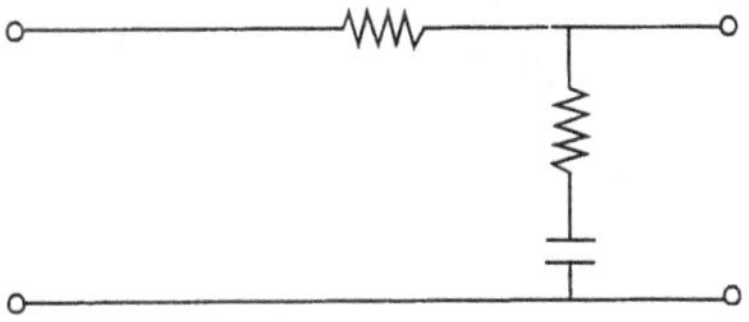

20. The phase-lead network function $G_C(s) = \dfrac{s+\frac{1}{T}}{s+\frac{1}{aT}}$, where a < 1 would provide maximum phase-lead at a frequency of

(a) $\dfrac{1}{T}$ (b) $\dfrac{1}{aT}$ (c) $\dfrac{1}{T\sqrt{a}}$ (d) $\dfrac{1}{a\sqrt{T}}$

21. The compensator $G_c(s) = \frac{5(1+0.3s)}{1+0.1s}$ would provide a maximum phase shift of

(a) 20° (b) 30° (c) 45° (d) 60°

22. For the given network, the maximum phase lead ϕ_m of V_o with respect to V_i is

(a) $\sin^{-1}\left(\frac{R_1}{2R_2}\right)$

(b) $\sin^{-1}\left(\frac{R_1}{R_1+2R_2}\right)$

(c) $\sin^{-1}\left(\frac{R_1}{R_1+3R_2}\right)$

(d) $\sin^{-1}\left(\frac{R_1}{2R_2C_1}\right)$

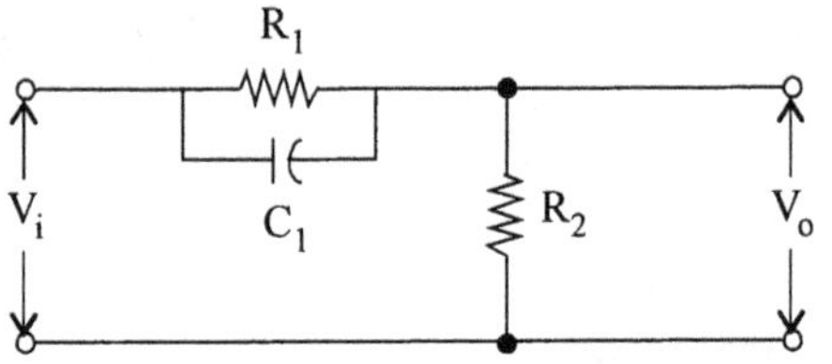

23. The transfer function of a phase lead network can be written as

(a) $\frac{1+sT}{1+s\beta T}$; $\beta > 1$ (b) $\frac{\alpha(1+sT)}{1+s\alpha T}$; $\alpha < 1$

(c) $\frac{\beta(1+sT)}{1+s\beta+T}$; $\beta < 1$ (d) $\frac{(1+sT)}{\alpha(1+s\alpha T)}$; $\alpha > 1$

24. Which one of the following compensations is adopted for improving transient response of a negative unity feedback system ?

(a) Phase lead compensation

(b) Phase lag compensation

(c) Gain compensation

(d) Both phase lag compensation and gain compensation

25. For a given gain constant K, the phase-lead compensator

(a) reduces the slope of the magnitude curve in the entire range of frequency

(b) decreases the gain cross-over frequency

(c) reduces the phase margin

(d) reduces the resonance peak M_P

26. Consider the following statements :

In a feedback control system, lead compensator

1. increases the margin of stability
2. speeds up transient response
3. does not affect the system error constant

Of these statements

(a) 2 and 3 are correct (b) 1 and 2 are correct

(c) 1 and 3 are correct (d) 1, 2 and 3 are correct

27. A phase-lag compenstion will

(a) improve relative stability (b) increase the speed of response

(c) increase bandwidth (d) increase overshoot

28. Maximum phase lead of $\dfrac{4(1+0.15s)}{(1+0.05s)}$ is equal to

(a) 15° (b) 30° (c) 45° (d) 60°

Chapter 9

1. The state transition matrix for the system $\dot{X} = AX$ with initial state X (0) is

(a) $(sI - A)^{-1}$

(b) e^{At} X (0)

(c) Laplace inverse of $[(sI - A)^{-1}]$

(d) Laplace inverse of $[(sI - A)^{-1} X(0)]$

2. For the system $\dot{X} = \begin{bmatrix} 2 & 0 \\ 0 & 4 \end{bmatrix} X + \begin{bmatrix} 1 \\ 1 \end{bmatrix} u$; $y = [4 \quad 0]$ X, with u as unit impulse and with zero initial state the output, y becomes :

(a) $2e^{2t}$ (b) $4e^{2t}$ (c) $2e^{4t}$ (d) $4e^{4t}$

3. The eigenvalues of the system represented by $\dot{X} = \begin{bmatrix} 0 & 1 & 0 & 0 \\ 0 & 0 & 1 & 0 \\ 0 & 0 & 0 & 1 \\ 0 & 0 & 0 & 1 \end{bmatrix} X$ are

(a) 0,0,0,0 (b) 1,1,1,1 (c) 0,0,0,–1 (d) 1,0,0,0

4. Given the homogeneous state-space equation $\dot{X} = \begin{bmatrix} -3 & 1 \\ 0 & -2 \end{bmatrix} X$ the steady state value of $x_{ss} = \lim_{t \to \infty} x(t)$, given the initial state value of $x(0) = [10 \quad -10]^T$, is

(a) $x_{ss} = \begin{bmatrix} 0 \\ 0 \end{bmatrix}$ (b) $\begin{bmatrix} -3 \\ -2 \end{bmatrix}$ (c) $\begin{bmatrix} -10 \\ -10 \end{bmatrix}$ (d) $\begin{bmatrix} \infty \\ \infty \end{bmatrix}$

5. Consider the system $\dot{X}(t) = \begin{bmatrix} 1 & 1 \\ 0 & 1 \end{bmatrix} X(t) + \begin{bmatrix} b_1 \\ b_2 \end{bmatrix} U(t)$; $y(t) = [d_1 \; d_2]\, X(t)$

The conditions for complete state controllability and complete observability is

(a) $d_1 \neq 0$, $b_2 \neq 0$, b_1 and d_2 can be anything

(b) $d_1 > 0$, $d_2 > 0$, b_1 and b_2 can be anything

(c) $b_1 \neq 0$, $b_2 \neq 0$, d_1 and d_2 can be anything

(d) $b_1 > 0$, $b_2 > 0$, b_2 and d_1 can be anything

6. The transfer function of a multi-input multi-output system, with the state-space representation of

$\dot{X} = AX + BU$ $\qquad Y = CX + DU$

where X represents the state, Y the output and U the input vector, will be given by

(a) $C(sI - A)^{-1} B$ (b) $C(sI - A)^{-1} B + D$

(c) $(sI - A)^{-1} B + D$ (d) $(sI - A)^{-1} B + D$

7. The state variable description of a single-input single-output linear system is given by

$\underline{\dot{X}} = A\,\underline{X}(t) + \underline{b}\,u(t)$

$y(t) - \underline{c}\,\underline{X}(t)$

where $A = \begin{bmatrix} 1 & 1 \\ 2 & 0 \end{bmatrix}$, $b = \begin{bmatrix} 0 \\ 1 \end{bmatrix}$ and $c = [1, -1]$

The system is

(a) controllable and observable

(b) controllable but unobservable

(c) uncontrollable but observable

(d) uncontrollble and unobservable

8. Which of the following properties are associated with the state transition matrix $\phi(t)$?

1. $\phi(-t) = \phi^{-1}(t)$ 2. $\phi(t_1/t_2) = \phi(t_1).\ \phi^{-1}(t_2)$

3. $\phi(t_1 - t_2) = \phi(-t_2).\ \phi(t_1)$

Select the correct answer using the codes given below :

Codes :

(a) 1, 2 and 3 (b) 1 and 2 (c) 2 and 3 (d) 1 and 3

9. The second order system $\dot{X} = AX$ has $A = \begin{bmatrix} -1 & -1 \\ 1 & 0 \end{bmatrix}$

The values of its damping factor ζ and natural frequency ω_n are respectively

(a) 1 and 1 (b) 0.5 and 1 (c) 0.707 and 2 (d) 1 and 2

10. A liner system is described by the state equations, $\begin{bmatrix} \dot{x}_1 \\ \dot{x}_2 \end{bmatrix} = \begin{bmatrix} 1 & 0 \\ 1 & 1 \end{bmatrix} \begin{bmatrix} x_1 \\ x_2 \end{bmatrix} + \begin{bmatrix} 0 \\ 1 \end{bmatrix} r$

$$c = x_2$$

Where r and c are the input and output respectively. The transfer function is :

(a) $1/(s+1)$ (b) $1/(s+1)^2$ (c) $1/(s-1)$ (d) $1/(s-1)^2$

11. The transfer function of a certain system is $\dfrac{Y(s)}{U(s)} = \dfrac{1}{s^4 + 5s^3 + 7s^2 + 6s + 3}$.

The A, B matrix pair of the equivalent state-space model will be

(a) $\begin{bmatrix} 0 & 1 & 0 & 0 \\ 0 & 0 & 1 & 0 \\ 0 & 0 & 0 & 1 \\ -3 & -6 & -7 & -5 \end{bmatrix};\ \begin{bmatrix} 0 \\ 0 \\ 0 \\ 1 \end{bmatrix}$ (b) $\begin{bmatrix} 0 & 1 & 0 & 0 \\ 0 & 0 & 1 & 0 \\ 0 & 0 & 0 & 1 \\ -3 & -5 & -6 & -7 \end{bmatrix};\ \begin{bmatrix} 0 \\ 0 \\ 0 \\ 1 \end{bmatrix}$

(c) $\begin{bmatrix} 0 & 1 & 0 & 0 \\ 0 & 0 & 1 & 0 \\ 0 & 0 & 0 & 1 \\ -5 & -7 & -6 & -3 \end{bmatrix};\ \begin{bmatrix} 1 \\ 0 \\ 0 \\ 0 \end{bmatrix}$ (d) $\begin{bmatrix} 1 & 0 & 0 & 0 \\ 0 & 1 & 0 & 0 \\ 0 & 0 & 1 & 0 \\ -3 & -6 & -7 & -5 \end{bmatrix};\ \begin{bmatrix} 0 \\ 0 \\ 0 \\ 1 \end{bmatrix}$

12. The state equation of a dynamic system is given by $\dot{X}(t) = A\ X(t)$

$$A = \begin{bmatrix} -1 & 1 & 0 & 0 & 0 \\ 0 & -1 & 1 & 0 & 0 \\ 0 & 0 & -1 & 0 & 0 \\ 0 & 0 & 0 & -3 & 4 \\ 0 & 0 & 0 & -4 & -3 \end{bmatrix}$$

The eigen values of the system would be

(a) real non-repeated only
(b) real non-repeated and complex
(c) real repeated
(d) real repeated and complex

13. The value of A matrix in $\dot{X} = AX$ for the system described by the differential equation $\ddot{y} + 2\dot{y} + 3y = 0$ form is

(a) $\begin{bmatrix} 1 & 0 \\ -2 & -1 \end{bmatrix}$ (b) $\begin{bmatrix} 1 & 0 \\ -1 & -2 \end{bmatrix}$ (c) $\begin{bmatrix} 0 & 1 \\ -2 & -3 \end{bmatrix}$ (d) $\begin{bmatrix} 0 & 1 \\ -3 & -2 \end{bmatrix}$

14. The state and output equations of a system are as under :

State equation : $\begin{bmatrix} \dot{x}_1(t) \\ \dot{x}_2(t) \end{bmatrix} = \begin{bmatrix} 0 & 1 \\ -1 & -2 \end{bmatrix} \begin{bmatrix} x_1(t) \\ x_2(t) \end{bmatrix} + \begin{bmatrix} 0 \\ 1 \end{bmatrix} u(t)$

Output equation : $C(t) = [1\quad 1] \begin{bmatrix} x_1(t) \\ x_2(t) \end{bmatrix}$

The system is

(a) neither state controllable nor output controllable
(b) state controllable but not output controllable
(c) output controllable but not state controllable
(d) both state controllable and output controllable

15. The minimum number of states necessary to describe the network shown in the figure in a state variable form is

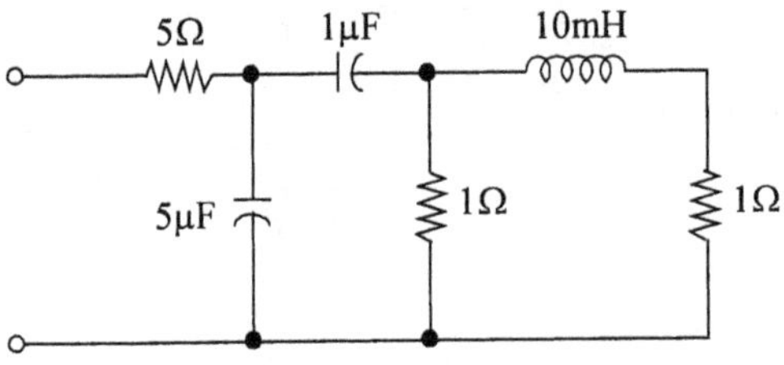

(a) 2
(b) 3
(c) 4
(d) 6

16. A system is represented by $\dddot{y} + 2\ddot{y} + 5\dot{y} + 6y = 5x$

If state variables are $x_1 = y$, $x_2 = \dot{y}$ and $x_3 = \ddot{y}$, then the coefficient matrix 'A' will be

(a) $\begin{bmatrix} 0 & 1 & 0 \\ 0 & 0 & 1 \\ -6 & -5 & -2 \end{bmatrix}$ (b) $\begin{bmatrix} 0 & 1 & 0 \\ 0 & 0 & 1 \\ -2 & -5 & -6 \end{bmatrix}$ (c) $\begin{bmatrix} 0 & 0 & 1 \\ 0 & 1 & 0 \\ -6 & -5 & -2 \end{bmatrix}$ (d) $\begin{bmatrix} 0 & 0 & 1 \\ 0 & 1 & 0 \\ -2 & -5 & -6 \end{bmatrix}$

17. The state equation of a linear system is given by $\dot{X} = AX + BU$, where

$A = \begin{bmatrix} 0 & 2 \\ -2 & 0 \end{bmatrix}$ and $B = \begin{bmatrix} 0 \\ 1 \end{bmatrix}$

The state transition matrix of the system is

(a) $\begin{bmatrix} e^{2t} & 0 \\ 0 & e^{2t} \end{bmatrix}$ (b) $\begin{bmatrix} e^{-2t} & 0 \\ 0 & e^{2t} \end{bmatrix}$ (c) $\begin{bmatrix} \sin 2t & \cos 2t \\ -\cos 2t & \sin 2t \end{bmatrix}$ (d) $\begin{bmatrix} \cos 2t & \sin 2t \\ -\sin 2t & \cos 2t \end{bmatrix}$

18. The control system shown in the given figure is represented by the equation

$$\begin{bmatrix} y_1(s) \\ y_2(s) \end{bmatrix} = [\text{Matrix 'G'}] \begin{bmatrix} u_1(s) \\ u_2(s) \end{bmatrix}$$

The matrix G is

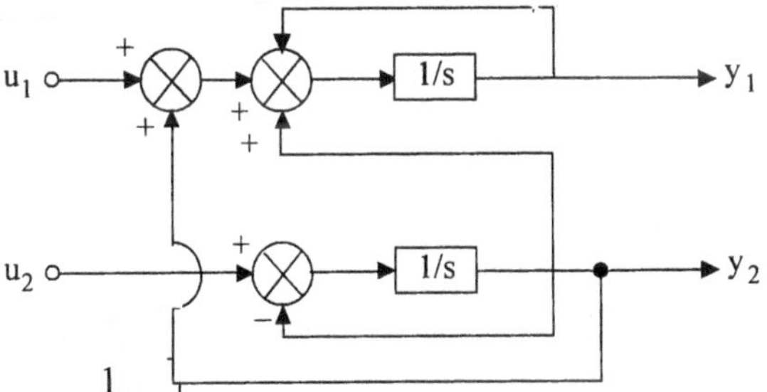

(a) $\begin{bmatrix} \frac{1}{s} & -\frac{1}{s^2} \\ 0 & \frac{1}{s} \end{bmatrix}$ (b) $\begin{bmatrix} \frac{1}{s} & \frac{1}{s^2} \\ 0 & -\frac{1}{s} \end{bmatrix}$

(c) $\begin{bmatrix} \frac{1}{(s+1)} & \frac{2}{(s+1)^2} \\ 0 & \frac{1}{(s+1)} \end{bmatrix}$ (d) $\begin{bmatrix} \frac{1}{(s+1)} & -\frac{1}{(s+1)^2} \\ 0 & -\frac{1}{(s+1)} \end{bmatrix}$

19. The state diagram of a system is shown in the given figure :
The system is

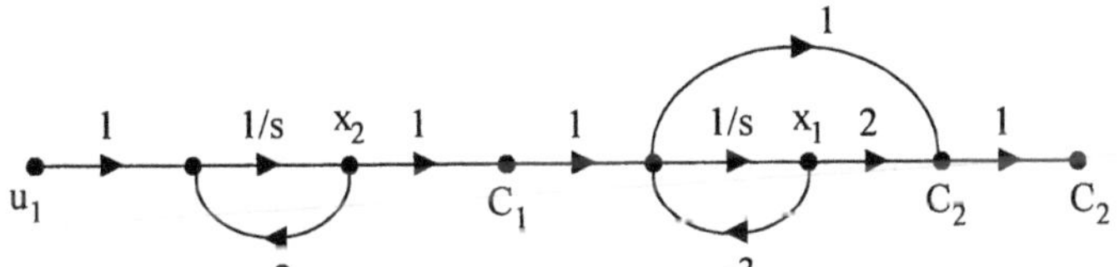

(a) controllable and observable (b) controllable but not observable

(c) observable but not controllable (d) neither controllable nor observable

20. Match List-I (Matrix) with List-II (Dimensions) for the state equations : $\dot{X}(t) = PX(t) + QU(t)$ and $Y(t) = RX(t) + SU(t)$ and select the correct answer using the codes given below the lists :

List-I		List-II	
A.	P	1.	$(n \times p)$
B.	Q	2.	$(q \times n)$
C.	R	3.	$(n \times n)$
D.	S	4.	$(q \times p)$

Codes :

	A	B	C	D		A	B	C	D
(a)	1	3	4	2	(b)	1	3	2	4
(c)	3	1	4	2	(d)	3	1	2	4

21. The state-variable description of a linear autonomous system is $\dot{X} = AX$ where X is a state vector and $A = \begin{bmatrix} 0 & 2 \\ 2 & 0 \end{bmatrix}$.

The poles of the system are located at

(a) –2 and + 2 (b) –2j and + 2j (c) –2 and –2 (d) + 2 and +2.

22. Consider the state transition matrix : $\phi(s) = \begin{bmatrix} \dfrac{s+6}{s^2+6s+5} & \dfrac{1}{s^2+6s+5} \\ \dfrac{-5}{s^2+6s+5} & \dfrac{s}{s^2+6s+5} \end{bmatrix}$

The eigenvalues of the system are

(a) 0 and –6 (b) 0 and +6 (c) 1 and -5 (d) –1 and –5

23. A particular control system is described by the following state equations :

$X = \begin{bmatrix} 0 & 1 \\ -2 & -3 \end{bmatrix} X + \begin{bmatrix} 0 \\ 1 \end{bmatrix} U$ and $Y = [2 \quad 0]X$. The transfer function of this system is :

(a) $\dfrac{Y(s)}{U(s)} = \dfrac{1}{2s^2+3s+1}$ (b) $\dfrac{Y(s)}{U(s)} = \dfrac{2}{2s^2+3s+1}$

(c) $\dfrac{Y(s)}{U(s)} = \dfrac{1}{s^2+3s+2}$ (d) $\dfrac{Y(s)}{U(s)} = \dfrac{2}{s^2+3s+2}$

24. A transfer function of a control system does not have pole-zero cancellation. Which one of the following statement is true ?
(a) System is neither controllable nor observable
(b) System is completely controllable and observable
(c) System is observable but uncontrollable
(d) System is controllable but unobservable.

25. The state-space representation in phase-variable form for the transfer function

$$G(s) = \frac{2s+1}{s^2+7s+9} \text{ is}$$

(a) $x = \begin{bmatrix} 0 & 1 \\ -9 & -7 \end{bmatrix} x + \begin{bmatrix} 0 \\ 1 \end{bmatrix} u;\ y = [1\ \ 2]x$

(b) $x = \begin{bmatrix} 1 & 0 \\ -9 & -7 \end{bmatrix} x + \begin{bmatrix} 0 \\ 1 \end{bmatrix} u;\ y = [0\ \ 1]x$

(c) $x = \begin{bmatrix} -9 & 0 \\ 0 & -7 \end{bmatrix} x + \begin{bmatrix} 0 \\ 1 \end{bmatrix} u;\ y = [2\ \ 0]x$

(d) $x = \begin{bmatrix} 9 & -7 \\ 1 & 0 \end{bmatrix} x + \begin{bmatrix} 0 \\ 1 \end{bmatrix} u;\ y = [1\ \ 2]x$

26. Let $\dot{X} = \begin{bmatrix} 1 & 2 \\ 0 & 1 \end{bmatrix} X + \begin{bmatrix} 0 \\ 1 \end{bmatrix} U$

$Y = [b\ \ 0]\ X$

where b is an unknown constant. This system is
(a) observable for all values of b
(b) unobservable for all values of b
(c) observable for all non-zero valuesof b
(d) unobservable for all non-zero values of b

27. The state representation of a second order system is $\dot{x}_1 = -x_1 + u,\ \dot{x}_2 = x_1 - 2x_2 + u$
Consider the following statements regarding the above system :
1. The system is completely state controllable.
2. If x_1 is the output, then the system is completely output controllable.
3. If x_2 is the output, then the system is completely output controllable.

Of these statements
(a) 1, 2 and 3 are correct
(b) 1 and 2 are correct
(c) 2 and 3 are correct
(d) 1 and 3 are correct

28. A state variable system

$$\dot{X}(t) = \begin{bmatrix} 0 & 1 \\ 0 & -3 \end{bmatrix} X(t) + \begin{bmatrix} 1 \\ 0 \end{bmatrix} u$$

with initial condition $X(0) = [-1 \quad 3]^T$ and a unit step input has the state transition matrix.

(a) $\begin{bmatrix} 1 & \frac{1}{3}(e^{-t} - e^{-3t}) \\ 0 & e^{-t} \end{bmatrix}$ (b) $\begin{bmatrix} 1 & \frac{1}{3}(e^{-t} - e^{-3t}) \\ 0 & e^{-3t} \end{bmatrix}$

(c) $\begin{bmatrix} 1 & \frac{1}{3}(1 - e^{-3t}) \\ 0 & e^{-3t} \end{bmatrix}$ (d) $\begin{bmatrix} 1 & 1 - e^{-t} \\ 0 & e^{-3t} \end{bmatrix}$

29. Consider the following properties attributed to state model of a system :

1. State model is unique.
2. State model can be derived from the system transfer function.
3. State model can be derived for time variant systems.

Of these statements :

(a) 1, 2 and 3 are correct (b) 1 and 2 are correct

(c) 2 and 3 are correct (d) 1 and 3 are correct

30. A system is described by the state equation $\begin{bmatrix} \dot{x}_1 \\ \dot{x}_2 \end{bmatrix} = \begin{bmatrix} 2 & 0 \\ 0 & 2 \end{bmatrix} \begin{bmatrix} x_1 \\ x_2 \end{bmatrix} + \begin{bmatrix} 1 \\ 1 \end{bmatrix} u$

The state transition matrix of the system is

(a) $\begin{bmatrix} e^{2t} & 0 \\ 0 & e^{2t} \end{bmatrix}$ (b) $\begin{bmatrix} e^{-2t} & 0 \\ 0 & e^{-t} \end{bmatrix}$

(c) $\begin{bmatrix} e^{2t} & 1 \\ 0 & e^{2t} \end{bmatrix}$ (d) $\begin{bmatrix} e^{-2t} & 0 \\ 0 & e^{-2t} \end{bmatrix}$

31. For the system $\dot{X} = \begin{bmatrix} 2 & 3 \\ 0 & 5 \end{bmatrix} X + \begin{bmatrix} 1 \\ 0 \end{bmatrix} u$, which of the following statements is true ?

(a) the system is controllable but unstable

(b) the system is uncontrollable and unstable

(c) the system is controllable and stable

(d) the system is uncontrollable and stable.

32. A linear time invariant system is described by the following dynamic equation

$$\dot{X} = AX + Bu \qquad y = cX$$

$$A = \begin{bmatrix} 0 & 1 \\ -2 & -3 \end{bmatrix}, \quad B = \begin{bmatrix} 0 \\ 1 \end{bmatrix}, \quad C = [1 \quad 1]$$

The system is

(a) Both controllable and observable (b) Controllable but unobservable

(c) Observable but not controllable (d) Both unobservable and uncontrollable

33. Consider the single input, single output system with its state variable representation :

$$X = \begin{bmatrix} -1 & 0 & 0 \\ 0 & -2 & 0 \\ 0 & 0 & -3 \end{bmatrix} X + \begin{bmatrix} 1 \\ 1 \\ 0 \end{bmatrix} U; Y = [1 \quad 0 \quad 2]X$$

The system is

(a) neither controllable nor observable (b) controllable but not observable

(c) uncontrollable but observable (d) both controllable and observable

34. Consider the closed-loop system shown in the given figure. The state model of the system is

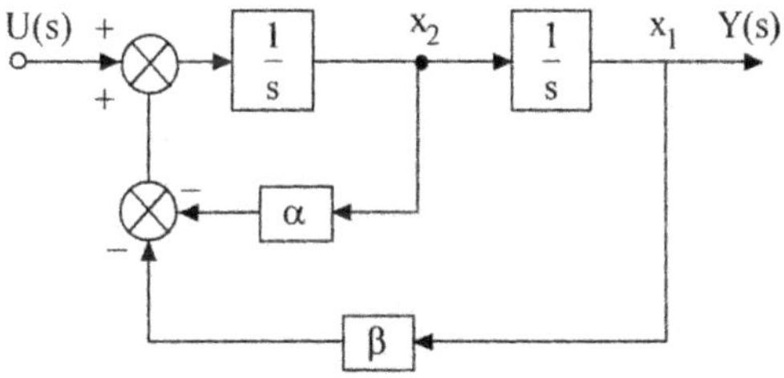

(a) $$\begin{bmatrix} \dot{x}_1 \\ \dot{x}_2 \end{bmatrix} = \begin{bmatrix} 1 & 0 \\ -\beta & -\alpha \end{bmatrix} \begin{bmatrix} x_1 \\ x_2 \end{bmatrix} + \begin{bmatrix} 0 \\ 1 \end{bmatrix} u$$

$$y = [1 \; 0] \begin{bmatrix} x_1 \\ x_2 \end{bmatrix}$$

(b) $$\begin{bmatrix} \dot{x}_1 \\ \dot{x}_2 \end{bmatrix} = \begin{bmatrix} 0 & 1 \\ -\alpha & -\beta \end{bmatrix} \begin{bmatrix} x_1 \\ x_2 \end{bmatrix} + \begin{bmatrix} 0 \\ 1 \end{bmatrix} u$$

$$y = [1 \; 0] \begin{bmatrix} x_1 \\ x_2 \end{bmatrix}$$

(c) $$\begin{bmatrix} \dot{x}_1 \\ \dot{x}_2 \end{bmatrix} = \begin{bmatrix} 0 & 1 \\ -\beta & -\alpha \end{bmatrix} \begin{bmatrix} x_1 \\ x_2 \end{bmatrix} + \begin{bmatrix} 0 \\ 1 \end{bmatrix} u$$

$$y = [1 \; 0] \begin{bmatrix} x_1 \\ x_2 \end{bmatrix}$$

(d) $$\begin{bmatrix} \dot{x}_1 \\ \dot{x}_2 \end{bmatrix} = \begin{bmatrix} 0 & 1 \\ -\beta & -\alpha \end{bmatrix} \begin{bmatrix} x_1 \\ x_2 \end{bmatrix} + \begin{bmatrix} 0 \\ 1 \end{bmatrix} u$$

$$y = [0 \quad 1] \begin{bmatrix} x_1 \\ x_2 \end{bmatrix}$$

35. The zero input response of a system,

$$\dot{X} = \begin{bmatrix} 1 & 0 \\ 1 & 1 \end{bmatrix} X \text{ with } X(0) = [1 \quad 0]^T \text{ is}$$

(a) $\begin{bmatrix} te^t \\ t \end{bmatrix}$ (b) $\begin{bmatrix} e^t \\ te^t \end{bmatrix}$ (c) $\begin{bmatrix} e^t \\ t \end{bmatrix}$ (d) $\begin{bmatrix} t \\ te^t \end{bmatrix}$

36. A second order system starts with an initial condition of $\begin{bmatrix} 2 \\ 3 \end{bmatrix}$ without any external input. The state transition matrix for the system is given by

$$\begin{bmatrix} e^{-2t} & 0 \\ 0 & e^{-t} \end{bmatrix}$$

The state of the system at t = 1 sec is given by

(a) $\begin{bmatrix} 0.135 \\ 0.368 \end{bmatrix}$ (b) $\begin{bmatrix} 0.271 \\ 1.1 \end{bmatrix}$ (c) $\begin{bmatrix} 0.271 \\ 0.736 \end{bmatrix}$ (d) $\begin{bmatrix} 0.135 \\ 1.1 \end{bmatrix}$

Miscellaneous

1. For the function $X(s) = \dfrac{1}{s(s+1)^3(s+2)}$, the residues associated with the simple poles at s = 0 and s = –2 are respectively

(a) $\frac{1}{2}$ and $\frac{1}{2}$ (b) 1 and 1 (c) –1 and –1 (d) $-\frac{1}{2}$ and $\frac{1}{2}$

2. The Laplace transform of a transportation lag of 5 seconds is

(a) exp (–5s) (b) exp (5s) (c) $\dfrac{1}{s+5}$ (d) exp $\left(\dfrac{-s}{5}\right)$

3. A unit impulse function on differentiation results in

(a) unit doublet (b) unit triplet
(c) unit parabolic function (d) unit ramp function

4. Laplace transform of $\dfrac{e^{-at} - e^{-bt}}{b-a}$ is

(a) $\dfrac{1}{(s+a)(s-b)}$ (b) $\dfrac{1}{(s-a)(s+b)}$

(c) $\dfrac{1}{(s-a)(s-b)}$ (d) $\dfrac{1}{(s+a)(s+b)}$

5. Match List-I with List-II and select the correct answer by using the Codes given below the lists :

List-I (Function)	**List-II** (Laplace Transform)
A. sin ωt	1. $\omega/(s^2 + \omega^2)$
B. cos ωt	2. $s/(s^2 + \omega^2)$
C. sinh ωt	3. $\omega/(s^2 - \omega^2)$
D. cosh ωt	4. $s/(s^2 - \omega^2)$

Codes :

	A	B	C	D
(a)	1	2	3	4
(b)	2	1	4	3
(c)	3	1	4	2
(d)	4	1	2	3

6. Match List-I (System) with List-II (Transfer function) and select the correct answer using the codes given below the Lists :

List-I	List-II
A. AC servomotor	1. $\dfrac{s+z}{s+p}$ $(z < p)$
B. DC amplifier	2. $\dfrac{1+T_1 s}{1+T_2 s}$ $(T_1 < T_2)$
C. Lead network	3. K
D. Lag network	4. $\dfrac{K}{s(1+Ts)}$

Codes :

	A	B	C	D
(a)	3	4	1	2
(b)	4	3	1	2
(c)	3	4	2	1
(d)	4	3	2	1

7. The output of a linear, time invariant control system is c (t) for a certain input r (t). If r (t) is modified by passing it through a block whose transfer function is e^{-s} and then applied to the system, the modified output of the system would bc

(a) $\dfrac{c(t)}{1+e^{t}}$ (b) $\dfrac{c(t)}{1-e^{-t}}$ (c) c (t – 1) u (t – 1) (d) c (t) u (t – 1)

8. Match List-I (Mathematical expression) with List-II (Nomenclature) and select the correct answer using the codes given below the lists :

List-I	List-II
A. $\int_0^{\infty} h(t-\tau)\, x(\tau)\, d\tau$	1. Step function
B. $\int_0^{\infty} x(t)\, e^{-st}\, dt$	2. Convolution integral
C. $\int_0^{\infty} x(t)\, e^{-j\omega t}\, dt$	3. Fourier transform
D. $\int_0^{\infty} \delta(t)\, dt$	4. Laplace transform

Codes :

	A	B	C	D
(a)	1	3	4	2
(b)	1	4	3	2
(c)	2	3	4	1
(d)	2	4	3	1

9. Match List I with List II and select the correct answer using the codes given below the lists :

List I (Controller)	List II (Suitable application)
A. On-off	1. Steam kettle with different time settings
B. Proportional	2. Elevator
C. Cascade	3. Robot positioning
D. Digital	4. Domestic refrigerator

Codes :

	A	B	C	D
(a)	4	2	3	1
(b)	2	4	1	3
(c)	3	1	4	2
(d)	4	2	1	3

10. Match List-I (Roots in the 's' plane) with List-II (Impulse response) and select the correct answer using the codes given below the lists :

List-I	List-II
(A) A single root at the origin	1. h(t) 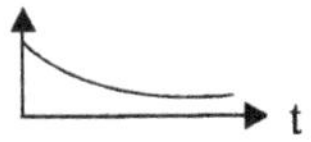
(B) A single root on the negative real axis	2. h(t) 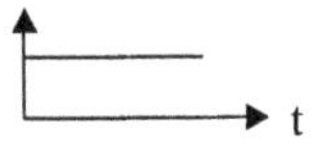
(C) Two imaginary roots	3. h(t) 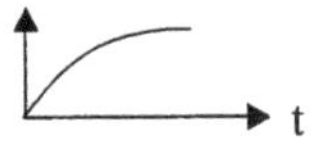
(D) Two complex roots in the right half plane	4. h(t) 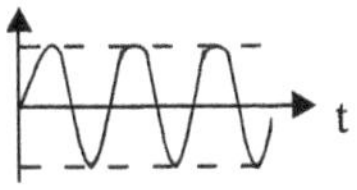
	5. h(t) 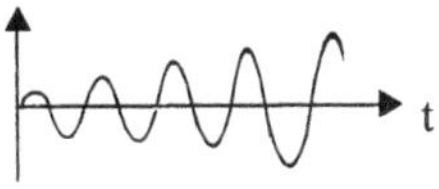

Codes :

	A	B	C	D		A	B	C	D
(a)	2	1	5	4	(b)	3	2	4	5
(c)	3	2	5	4	(d)	2	1	4	5

11. The correct sequence of steps needed to improve system stability is

(a) insert derivative action, use negative feedback, reduce gain

(b) reduce gain, use negative feedback, insert derivative action

(c) reduce gain, insert derivative action, use negative feedback

(d) use negative feedback, reduce gain, insert derivative action

12. Consider the vectors drawn from the poles and zero at $j\omega = j1$ on the imaginary axis as shown in the given figure. The transfer function $G(j1)$ is given by

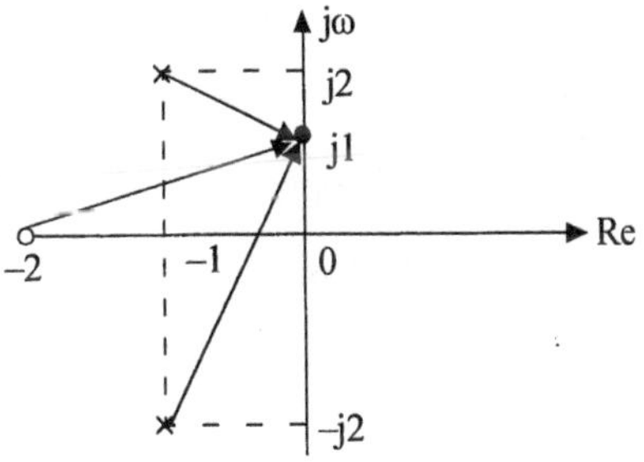

(a) $\frac{1}{2} \angle 0^{\circ}$

(b) $2.7 \angle 31^{\circ}$

(c) $2 \angle 45^{\circ}$

(d) $2 \angle -67.4^{\circ}$

13. The effect of adding poles and zeros can be determined quickly by

(a) Nicholas chart (b) Nyquist plot

(c) Bode plot (d) Root locus

14. Match List-I with List-II and select the correct answer by using the codes given below the lists :

List-I (Roots in the 's' plane)	**List-II (Impulse Response)**
A. Two imaginary roots	1. 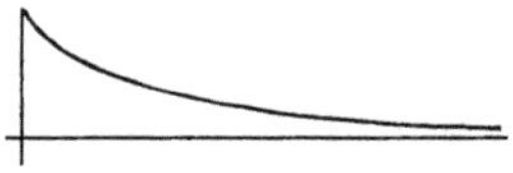
B. Two complex roots in the right half plane	2. 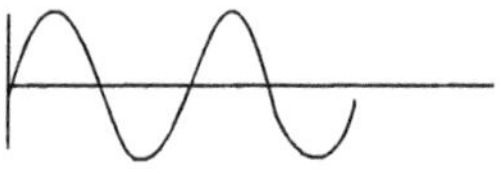
C. A single root on the negative real axis	3. 
D. A single root at the origin	4.  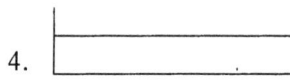

Codes :

	A	B	C	D
(a)	2	3	1	4
(b)	2	1	3	4
(c)	4	3	2	1
(d)	3	2	4	1

15. In control systems, we have

I. Nyquist criterion II. Bode plot

III. Root Locus plot IV. Routh Hurwitz criterion

Which of the above are in time domain ?

(a) I and II only (b) II and IV only

(c) I and III only (d) III and IV only

16. A system with transfer function, $\frac{600}{s(s+1)(s+15)(s+20)}$ can be approximated by the system

(a) $\frac{2}{s(s+1)}$ (b) $\frac{40}{s(s+20)}$ (c) $\frac{600}{s(s+15)(s+20)}$ (d) $\frac{40}{(s+1)(s+20)}$

17. 1. Nyquist criterion is in frequency domain

2. Bode Plot is in frequency domain

3. Root locus plot is in time domain

4. Routh Huwitz's criterion is in time domain.

(a) 1, 2, and 3 are correct (b) 2, 3 and 4 are correct

(c) 1 and 2 are correct (d) all four are correct

18. Match List I (Scientist) with List II (Contribution in the area of) and select the correct answer using the codes given below the Lists :

List I	**List II**
A. Bode	1. Asymptotic plots
B. Evans	2. Polar plots
C. Nyquist	3. Root-locus technique
	4. Constant M and N plots

Codes :

	A	B	C
(a)	1	4	2
(b)	2	3	4
(c)	3	1	4
(d)	1	3	2

19. The system shown in the given figure relates to temperature control of air flow.

Equation of heat exchanger is

$$10 \frac{dT_A}{dt} + T_A = u$$

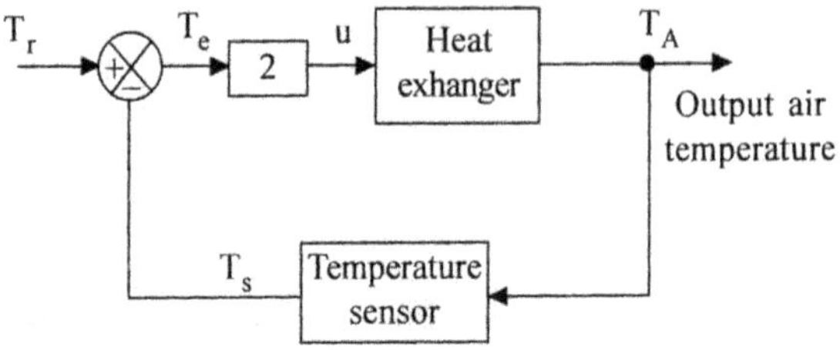

Temperature sensor equation is

$$\frac{dT_s}{dt} + T_s = T_A$$

The closed-loop transfer function $\frac{T_A(s)}{T_r(s)}$ of the system is

(a) $\frac{4s+2}{20s^2+12s+3}$ (b) $\frac{4s+2}{20s^2+12s+1}$ (c) $\frac{2}{20s^2+12s+1}$ (d) $\frac{2}{20s^2+12s+3}$

20. The state equation of a system is $X = \begin{bmatrix} 0 & 1 \\ -20 & -9 \end{bmatrix} X + \begin{bmatrix} 0 \\ 1 \end{bmatrix} u$

The poles of this system are located at

(a) $-1, -9$ (b) $-1, -20$ (c) $-4, -5$ (d) $-9, -20$

21. Given the relationship between the input u(t) and the output y(t) to be

$y(t) = \int_0^t (2 + t - \tau)\, e^{-3(t-\tau)}\, u(\tau)\, d\tau$ the transfer function Y(s)/U(s) is

(a) $\frac{2e^{-2s}}{s+3}$ (b) $\frac{s+2}{(s+3)^2}$ (c) $\frac{2s+5}{s+3}$ (d) $\frac{2s+7}{(s+3)^2}$

Answers to MCQs from Competitive Examinations

Chapter 1

1. (d)	2. (d)	3. (a)	4. (b)	5. (b)	6. (c)
7. (c)					

Chapter 2

1. (a)	2. (a)	3. (c)	4. (d)	5. (b)	6. (b)
7. (b)	8. (a)	9. (b)	10. (a)	11. (c)	12. (b)
13. (b)	14. (c)	15. (a)	16. (a)	17. (c)	18. (c)
19. (b)	20. (b)	21. (d)	22. (a)	23. (a)	24. (b)
25. (a)	26. (b)	27. (b)	28. (d)	29. (a)	30. (c)
31. (c)	32. (b)	33. (c)	34. (a)	35. (a)	36. (d)
37. (d)	38. (c)	39. (a)	40. (b)	41. (d)	42. (b)
43. (c)	44. (b)	45. (c)	46. (a)	47. (c)	48. (a)
49. (a)	50. (b)	51. (a)	52. (a)	53. (c)	54. (b)
55. (d)	56. (b)	57. (d)	58. (b)	59. (c)	60. (c)
61. (c)	62. (d)	63. (c)	64. (c)	65. (a)	66. (a)
67. (d)	68. (c)	69. (d)	70. (b)	71. (d)	72. (b)
73. (a)	74. (b)	75. (a)	76. (c)	77. (a)	78. (c)
79. (a)	80. (c)	81. (d)	82. (c)	83. (b)	84. (d)

Chapter 3

1. (a)	2. (b)	3. (c)	4. (c)	5. (d)	6. (c)
7. (a)	8. (c)	9. (b)	10. (b)	11. (d)	12. (c)
13. (a)	14. (d)	15. (c)	16. (a)	17. (c)	18. (b)
19. (c)	20. (b)	21. (d)	22. (b)	23. (d)	24. (a)
25. (b)	26. (c)	27. (b)	28. (a)	29. (a)	30. (b)
31. (a)	32. (a)	33. (d)	34. (d)	35. (a)	36. (d)
37. (d)	38. (c)	39. (d)	40. (b)	41. (d)	42. (c)
43. (b)	44. (a)	45. (a)	46. (b)	47. (c)	48. (a)
49. (b)	50. (c)	51. (c)	52. (d)	53. (c)	54. (c)
55. (c)	56. (b)	57. (d)	58. (a)	59. (a)	60. (d)
61. (a)	62. (c)	63. (d)	64. (b)	65. (d)	66. (a)
67. (b)	68. (d)	69. (d)	70. (c)	71. (b)	72. (b)
73. (b)	74. (d)	75. (c)	76. (d)	77. (a)	78. (b)
79. (b)	80. (c)	81. (d)	82. (c)		

Chapter 4

1. (b)	2. (c)	3. (c)	4. (b)	5. (c)	6. (b)
7. (c)	8. (d)	9. (b)	10. (a)	11. (d)	12. (c)
13. (c)	14. (c)	15. (c)	16. (b)	17. (c)	18. (d)
19. (b)	20. (d)	21. (b)	22. (b)	23. (d)	24. (d)
25. (a)	26. (c)	27. (b)	28. (c)	29. (c)	30. (c)
31. (b)	32. (c)	33. (b)	34. (c)	35. (a)	36. (c)
37. (c)	38. (b)	39. (b)			

Chapter 5

1. (b)	2. (c)	3. (b)	4. (b)	5. (c)	6. (c)
7. (d)	8. (d)	9. (a)	10. (c)	11. (d)	12. (c)
13. (b)	14. (c)	15. (a)	16. (d)	17. (d)	18. (b)
19. (d)	20. (c)	21. (a)	22. (d)	23. (b)	24. (c)
25. (b)	26. (c)	27. (d)	28. (a)	29. (d)	30. (a)

Chapter 6

1. (a)	2. (d)	3. (d)	4. (c)	5. (c)	6. (c)
7. (d)	8. (d)	9. (c)	10. (b)	11. (b)	12. (d)
13. (c)	14. (a)	15. (a)	16. (b)	17. (d)	18. (d)
19. (a)	20. (c)	21. (b)	22. (b)	23. (c)	24. (a)
25. (a)	26. (d)	27. (d)	28. (b)	29. (a)	30. (d)
31. (d)	32. (a)	33. (d)	34. (d)		

Chapter 7

1. (d)	2. (b)	3. (a)	4. (d)	5. (b)	6. (c)
7. (b)	8. (d)	9. (d)	10. (b)	11. (a)	12. (b)
13. (c)	14. (c)	15. (c)	16. (d)	17. (a)	18. (a)
19. (c)	20. (c)	21. (b)	22. (d)	23. (a)	24. (a)
25. (a)	26. (d)	27. (a)	28. (b)	29. (b)	30. (a)
31. (b)	32. (c)	33. (d)	34. (b)	35. (b)	36. (b)
37. (b)					

Chapter 8

1. (a)	2. (d)	3. (d)	4. (b)	5. (d)	6. (d)
7. (b)	8. (a)	9. (d)	10. (c)	11. (c)	12. (d)
13. (a)	14. (b)	15. (b)	16. (b)	17. (d)	18. (c)
19. (a)	20. (c)	21. (b)	22. (b)	23. (b)	24. (a)
25. (d)	26. (b)	27. (a)	28. (b)		

Chapter 9

1. (c)	2. (b)	3. (d)	4. (a)	5. (a)	6. (b)
7. (b)	8. (d)	9. (b)	10. (c)	11. (a)	12. (d)
13. (d)	14. (d)	15. (b)	16. (a)	17. (d)	18. (c)
19. (a)	20. (d)	21. (a)	22. (d)	23. (d)	24. (b)
25. (a)	26. (c)	27. (c)	28. (c)	29. (c)	30. (a)
31. (b)	32. (b)	33. (c)	34. (b)	35. (a)	36. (b)

Miscellaneous

1. (a)	2. (a)	3. (a)	4. (d)	5. (a)	6. (b)
7. (c)	8. (d)	9. (d)	10. (d)	11. (d)	12. (a)
13. (b)	14. (a)	15. (d)	16. (a)	17. (d)	18. (d)
19. (a)	20. (c)	21. (d)			

MCQs from Gate Examination, Yearwise from 2003 to 2014

2003

1. The figure shows the Nyquist plot of the open-loop transfer function G(s)H(s) of a system. If G(s)H(s) has one right-hand pole, the closed-loop system is

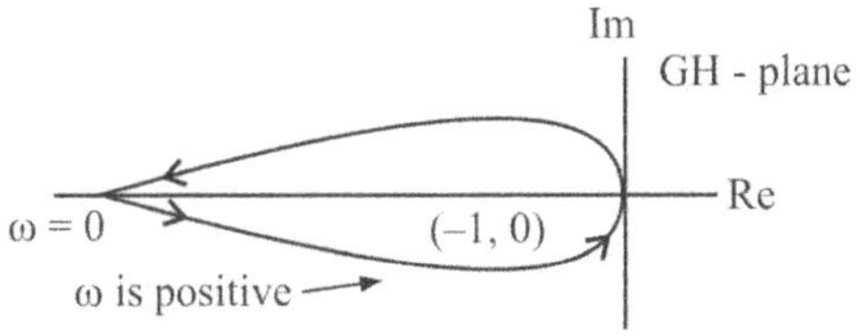

(a) Always stable

(b) Unstable with one closed-loop right hand pole

(c) Unstable with two closed-loop right hand poles

(d) Unstable with three closed-loop right hand poles

2. The signal flow graph of a system is shown in the figure. the transfer function $\frac{C(s)}{R(s)}$ of the system is

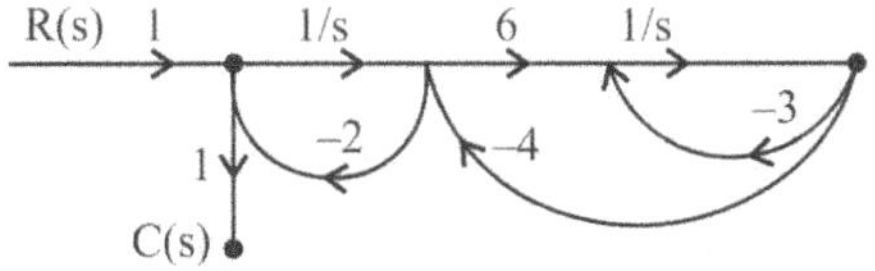

(a) $\frac{6}{s^2+29s+6}$ (b) $\frac{6s}{s^2+29s+6}$

(c) $\frac{s(s+2)}{s^2+29s+6}$ (d) $\frac{s(s+27)}{s^2+29s+6}$

3. The root locus of the system

$$G(s)H(s)=\frac{K}{s(s+2)(s+3)}$$ has the break-away point located at

(a) (– 0.5, 0) (b) (– 2.548, 0)

(c) (– 4, 0) (d) (– 0.784, 0)

4. The approximate Bode magnitude plot of a minimum-phase system is shown in the figure. The transfer function of the system is

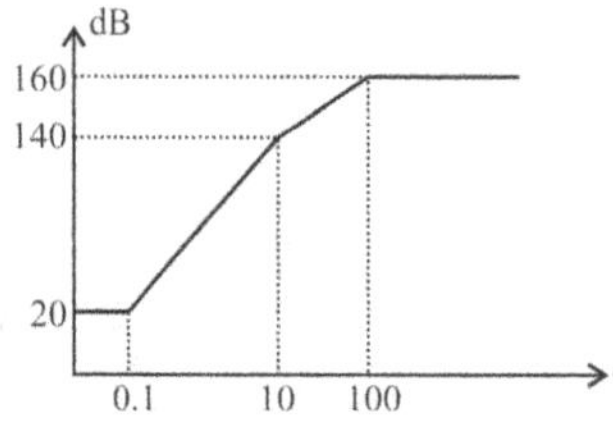

(a) $10^8 \dfrac{(s+0.1)^3}{(s+10)^2(s+100)}$ (b) $10^7 \dfrac{(s+0.1)^3}{(s+10)(s+100)}$

(c) $10^8 \dfrac{(s+0.1)^2}{(s+10)^2(s+100)}$ (d) $10^9 \dfrac{(s+0.1)^3}{(s+10)(s+100)^2}$

5. A second-order system has the transfer function $\dfrac{C(s)}{R(s)} = \dfrac{4}{s^2+4s+4}$. With r(t) as the unit-step function, the response c(t) of the system is represented by

(a)

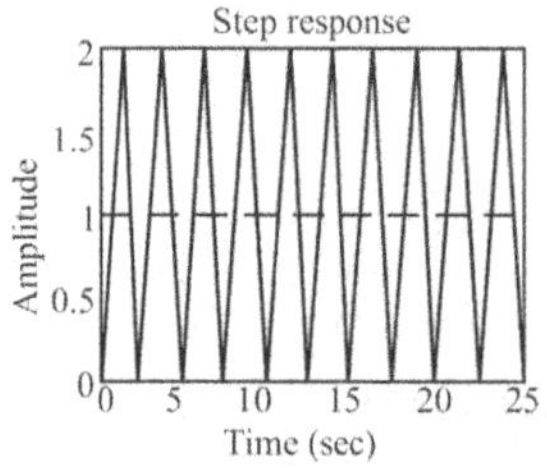

(b)

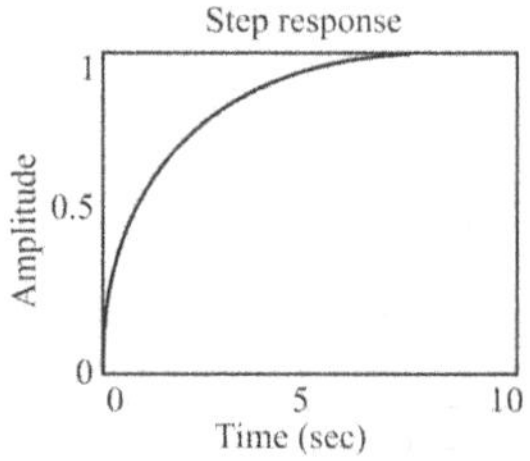

(c)

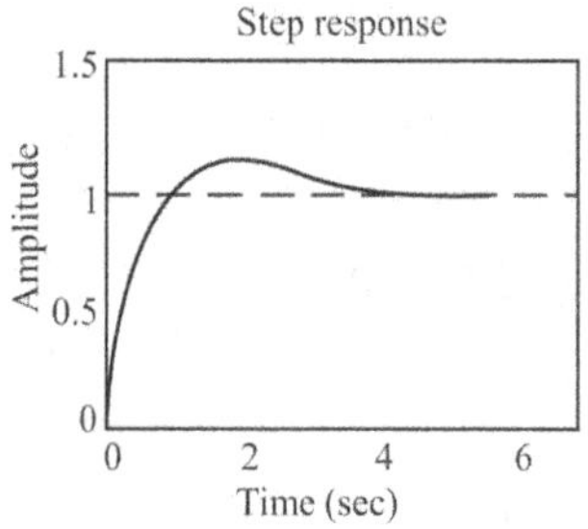

(d)

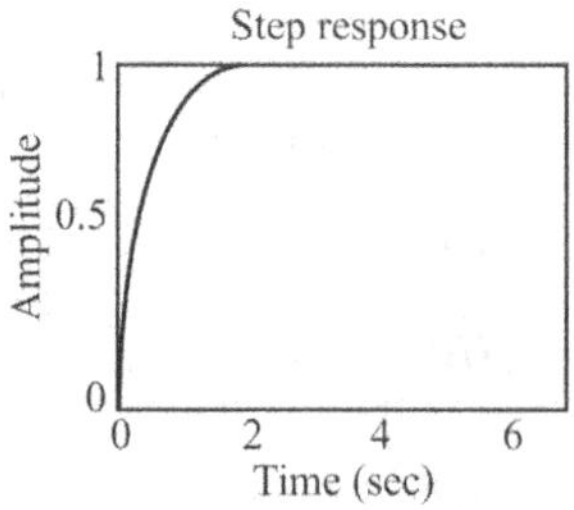

6. A control system is defined by the following mathematical relationship

$$\frac{d^2x}{dt^2}+6\frac{dx}{dt}+5x=12\left(1-e^{-2t}\right)$$

The response of the system as $t \to \infty$ is

(a) $x = 6$ (b) $x = 2$ (c) $x = 2.4$ (d) $x = -2$

7. A lead compensator used for a closed loop controller has the following transfer function $\dfrac{K\left(1+\dfrac{s}{a}\right)}{\left(1+\dfrac{s}{b}\right)}$

For such a lead compensator, the condition will be

(a) $a < b$ (b) $b < a$

(c) $a > Kb$ (d) $a < Kb$

8. A second order system starts with an initial condition of $\begin{bmatrix} 2 \\ 3 \end{bmatrix}$ without any external input. The state transition matrix for the system is given by $\begin{bmatrix} e^{-2t} & 0 \\ 0 & e^{-t} \end{bmatrix}$. The state of the system at the end of 1 second is given by

(a) $\begin{bmatrix} 0.271 \\ 1.100 \end{bmatrix}$ (b) $\begin{bmatrix} 0.135 \\ 0.368 \end{bmatrix}$

(c) $\begin{bmatrix} 0.271 \\ 0.736 \end{bmatrix}$ (d) $\begin{bmatrix} 0.135 \\ 1.100 \end{bmatrix}$

9. The block diagram shown in the figure below gives a unity feedback closed loop control system. The steady state error in the response of the above system to unit step input is

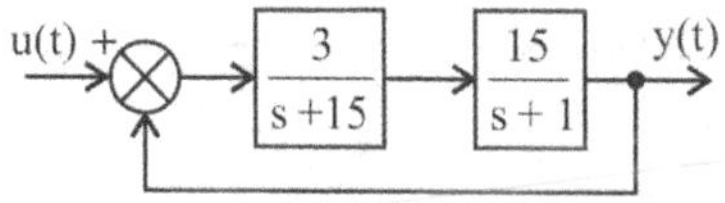

(a) 25% (b) 0.75 %

(c) 6 % (d) 33%

10. The roots of the closed loop characteristic equation of the system shown in above figure

(a) – 1 and – 15 (b) 6 and 10 (c) – 4 and – 15 (d) – 6 and – 10

11. The following equation defines a separately excited dc motor in the form of a differential equation $\frac{d^2\omega}{dt^2}+\frac{B}{J}\frac{d\omega}{dt}+\frac{K^2}{LJ}\omega=\frac{K}{LJ}V_\alpha$

The above equation may be organized in the state space form as follows

$$\begin{bmatrix} \frac{d^2\omega}{dt^2} \\ \frac{d\omega}{dt} \end{bmatrix} = P\begin{bmatrix} \frac{d\omega}{dt} \\ \omega \end{bmatrix} + QV_\alpha$$

Where the P matrix is given by

(a) $\begin{bmatrix} -\frac{B}{J} & -\frac{K^2}{LJ} \\ 1 & 0 \end{bmatrix}$ (b) $\begin{bmatrix} -\frac{K^2}{LJ} & -\frac{B}{J} \\ 0 & 1 \end{bmatrix}$ (c) $\begin{bmatrix} 0 & 1 \\ -\frac{K^2}{LJ} & -\frac{B}{J} \end{bmatrix}$ (d) $\begin{bmatrix} 1 & 0 \\ -\frac{B}{J} & -\frac{K^2}{LJ} \end{bmatrix}$

12. The loop gain GH of the closed loop system is given by the following expression $\frac{K}{s(s+2)(s+4)}$

The value of K for which the system just becomes unstable is

(a) $K=6$ (b) $K=8$ (c) $K=48$ (d) $K=96$

13. The asymptotic Bode plot of the transfer function $\frac{K}{1+\frac{s}{a}}$ is given in Fig. The error in phase angle and dB gain at a frequency of $\omega = 0.5$ a are respectively

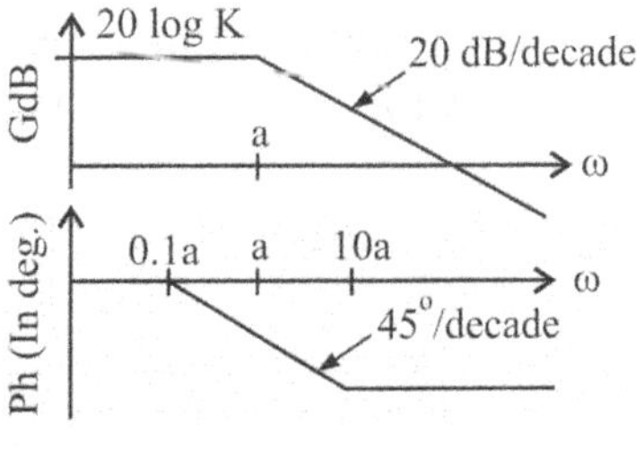

(a) 4.9°, 0.97 dB (b) 5.7°, 3 dB

(c) 4.9°, 3 dB (d) 5.7°, 0.97 dB

14. The block diagram of a control system is shown in the figure below. The transfer function G(s) = Y(s)/U(s) of the system is

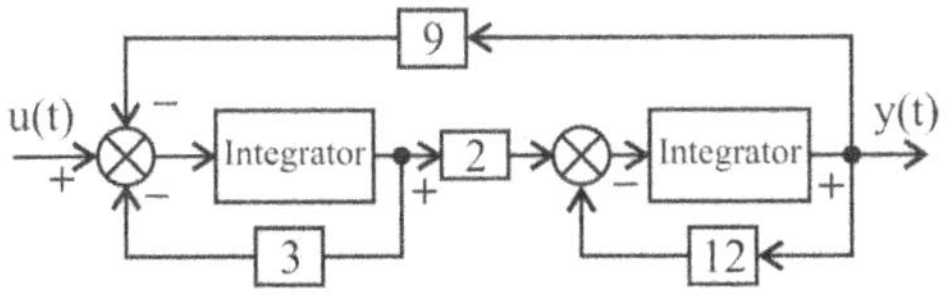

(a) $\frac{1}{18\left(1+\frac{s}{12}\right)\left(1+\frac{s}{3}\right)}$ (b) $\frac{1}{27\left(1+\frac{s}{6}\right)\left(1+\frac{s}{9}\right)}$

(c) $\frac{1}{27\left(1+\frac{s}{12}\right)\left(1+\frac{s}{9}\right)}$ (d) $\frac{1}{27\left(1+\frac{s}{9}\right)\left(1+\frac{s}{3}\right)}$

15. A control system with certain excitation is governed by the following mathematical equation

$$\frac{d^2x}{dt^2}+\frac{1}{2}\frac{dx}{dt}+\frac{1}{18}x=10+5e^{-4t}+2e^{-5t}$$

The natural time constants of the response of the system are

(a) 2s and 5s (b) 3s and 6s

(c) 4s and 5s (d) 1/3s and 1/6s

2004

1. The gain margin for the system with open-loop transfer function $G(s)H(s)=\frac{2(1+s)}{s^2}$, is

(a) ∞ (b) 0 (c) 1 (d) – ∞

2. Given $G(s)H(s)=\frac{K}{s(s+1)(s+3)}$, the point of intersection of the asymptotes of the root loci with the real axis is

(a) – 4 (b) 1.33 (c) – 1.33 (d) 4

3. Consider the Bode magnitude plot shown in the given figure. The transfer function H(s) is

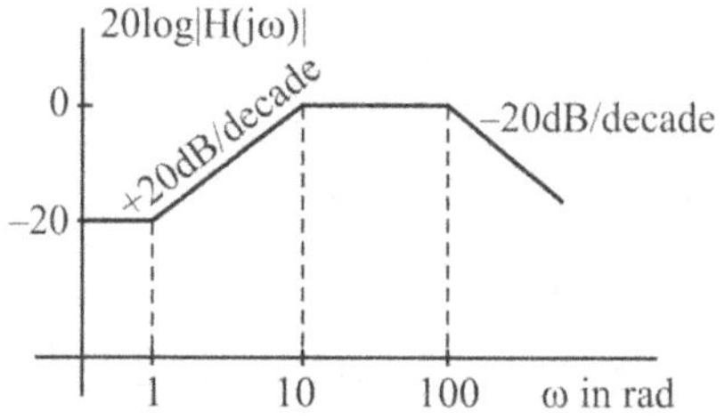

(a) $\dfrac{(s+10)}{(s+1)(s+100)}$ (b) $\dfrac{10(s+1)}{(s+1)(s+100)}$

(c) $\dfrac{10^2(s+1)}{(s+10)(s+100)}$ (d) $\dfrac{10^3(s+100)}{(s+1)(s+10)}$

4. A causal system having the transfer function $H(s)=\dfrac{1}{s+2}$ is excited with 10 u(t). The time at which the output reaches 99% of its steady state value is

(a) 2.7 sec (b) 2.5 sec (c) 2.3 sec (d) 2.1 sec

5. A system has poles at 0.01 Hz, 1 Hz and 80 Hz; zeros at 5 Hz, 100 Hz and 200 Hz. The approximate phase of the system response at 20 Hz is

(a) -90° (b) 0° (c) 90° (d) -180°

6. Consider the signal flow graph shown in the figure below. The gain $\dfrac{X_5}{X_1}$ is

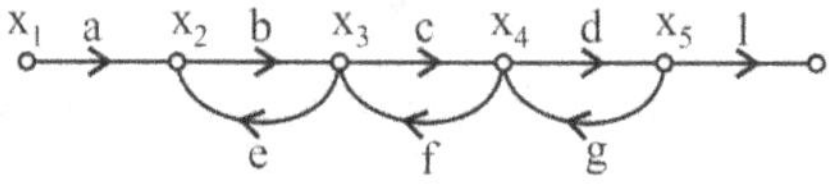

(a) $\dfrac{1-(be+cf+dg)}{abc}$ (b) $\dfrac{bedg}{1-(be+cf+dg)}$

(c) $\dfrac{abcd}{1-(be+cf+dg)+bedg}$ (d) $\dfrac{1-(be+cf+dg)+bedg}{abcd}$

7. The open-loop transfer function of a unity feedback system is $G(s)=\dfrac{K}{s(s^2+s+2)(s+3)}$. The range of K for which the system is stable is

(a) $\dfrac{21}{4}>K>0$ (b) $13>K>0$ (c) $\dfrac{21}{4}<K<\infty$ (d) $-6<K<\infty$

8. For the polynomial $P(s) = s^5 + s^4 + 2s^3 + 2s^2 + 3s + 15$, the number of roots which lie in the right half of the s-plane is

(a) 4 (b) 2 (c) 3 (d) 1

9. The state variable equations of a system are:

1. $\dot{x}_1 = -3x_1 - x_2 + u$

2. $\dot{x}_2 = 2x_1$

$y = x_1 + u$

The system is

(a) Controllable but not observable (b) Observable but not controllable

(c) Neither controllable nor observable (d) Controllable and observable

10. Given $A = \begin{bmatrix} 1 & 0 \\ 0 & 1 \end{bmatrix}$, the state transition matrix e^{At} is given by

(a) $\begin{bmatrix} 0 & e^{-t} \\ e^{-t} & 0 \end{bmatrix}$ (b) $\begin{bmatrix} e^{t} & 0 \\ 0 & e^{t} \end{bmatrix}$ (c) $\begin{bmatrix} e^{-t} & 0 \\ 0 & e^{-t} \end{bmatrix}$ (d) $\begin{bmatrix} 0 & e^{t} \\ e^{t} & 0 \end{bmatrix}$

11. The Nyquist plot of loop transfer function G(s) H(s) of a closed loop control sytem passes through the point (– 1, j0) in the G(s) H(s) plane. The phase margin of the system is

(a) 0° (b) 45° (c) 90° (d) 180°

12. For a tachometer, if θ(t) is the rotor displacement is radians, e(t) is the output voltage and K_t is the tachometer constant in V/rad/sec, then the transfer function, $\frac{E(s)}{\theta(s)}$ will be

(a) $K_t s^2$ (b) $\frac{K_t}{s}$ (c) $K_t s$ (d) K_t

13. For the block diagram shown below the transfer function $\frac{C(s)}{R(s)}$ is equal to

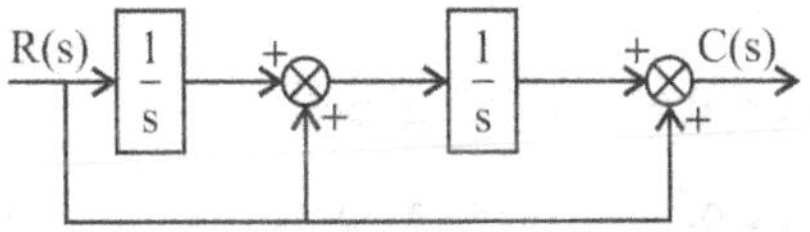

(a) $\frac{s^2+1}{s^2}$ (b) $\frac{s^2+s+1}{s^2}$ (c) $\frac{s^2+s+1}{s}$ (d) $\frac{1}{s^2+s+1}$

14. The state variable description of a linear autonomous system is, $\dot{X} = AX$, Where X is the two dimensional state vector and A is the system matrix given by $A = \begin{bmatrix} 0 & 2 \\ 2 & 0 \end{bmatrix}$. The roots of the characteristic equation are

(a) -2 and $+2$ (b) $-j2$ and $+j2$

(c) -2 and -2 (d) $+2$ and $+2$

15. The block diagram of a closed loop control system is given in figure. The values of K and P such that the system has a damping ratio of 0.7 and an undamped natural frequency ω_n of 5 rad/sec, are respectively equal to

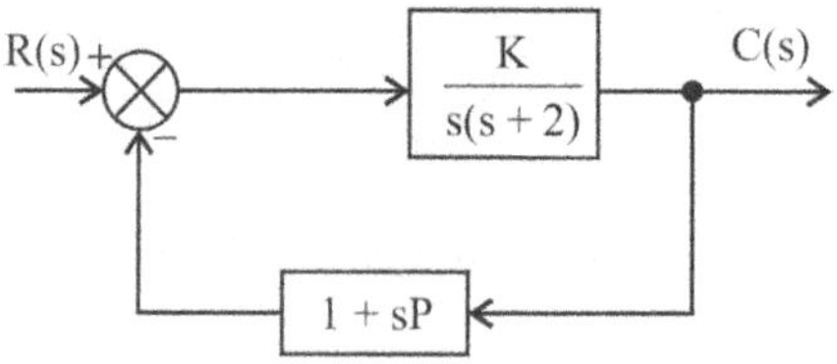

(a) 20 and 0.3 (b) 20 and 0.2

(c) 25 and 0.3 (d) 25 and 0.2

16. The unit impulse response of a second order under-damped system starting from rest is given by $c(t) = 12.5\, e^{-6t} \sin 8\,t,\ t \geq 0$

The steady-state value of the unit step response of the system is equal to

(a) 0 (b) 0.25 (c) 0.5 (d) 1.0

17. In the system shown in the figure, the input x(t) = sin t. In the steady-state, the response y(t) will be

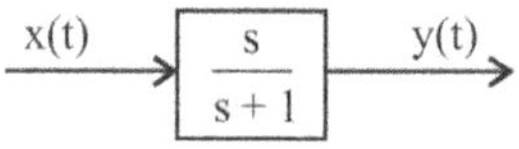

(a) $\frac{1}{\sqrt{2}} \sin\left(t - 45^\circ\right)$ (b) $\frac{1}{\sqrt{2}} \sin\left(t + 45^\circ\right)$

(c) $\sin\left(t - 45^\circ\right)$ (d) $\sin\left(t + 45^\circ\right)$

18. The open loop transfer function of a unity feedback control system is given as $G(s) = \frac{as + 1}{s^2}$. The value of 'a' to give a phase margin of 45° is equal to

(a) 0.141 (b) 0.441 (c) 0.841 (d) 1.141

2005

1. A linear system is equivalently represented by two sets of state equations.

$$\dot{X} = AX + BU$$

and $\dot{W} = CW + DU$.

The eigen values of the representations are also computed as $[\lambda]$ and $\{\mu\}$. Which one of the following statements is true?

(a) $[\lambda] = [\mu]$ and $X = W$

(b) $[\lambda] = [\mu]$ and $X \neq W$

(c) $[\lambda] \neq [\mu]$ and $X = W$

(d) $[\lambda] \neq [\mu]$ and $X \neq W$

2. Which one of the following polar diagrams corresponds to a lag network?

(a)

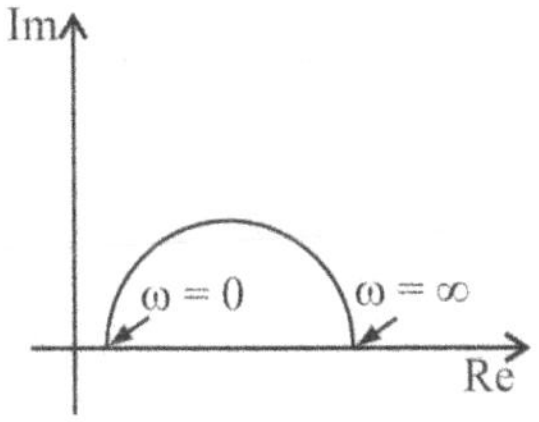

(b)

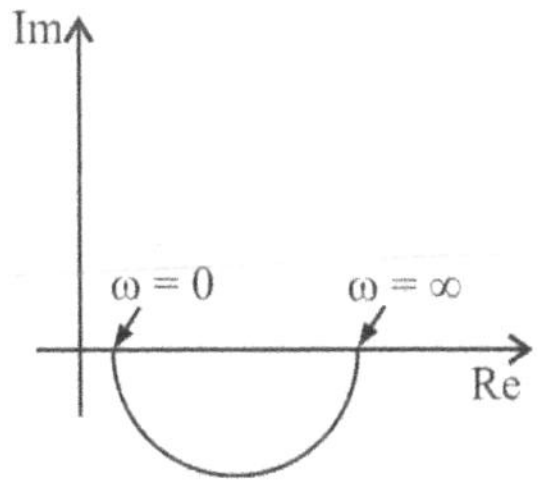

(c)

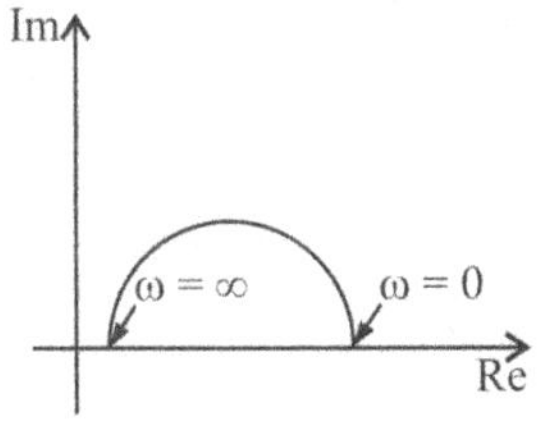

(d)

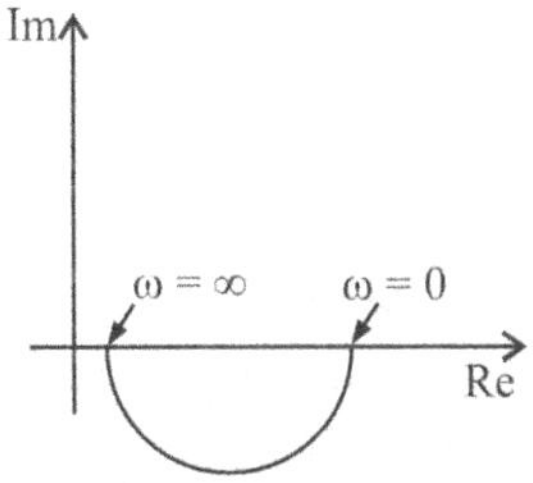

3. Despite the presence of negative feedback, control systems still have problems of instability because the

(a) Components used have non-linearities

(b) Dynamic equations of the subsystems are not known exactly

(c) Mathematical analysis involves approximations

(d) System has large negative phase angle at high frequencies.

4. The polar diagram of a conditionally stable system for open loop gain K = 1 is shown in the figure. The open loop transfer function of the system is known to be stable. The closed loop system is stable for

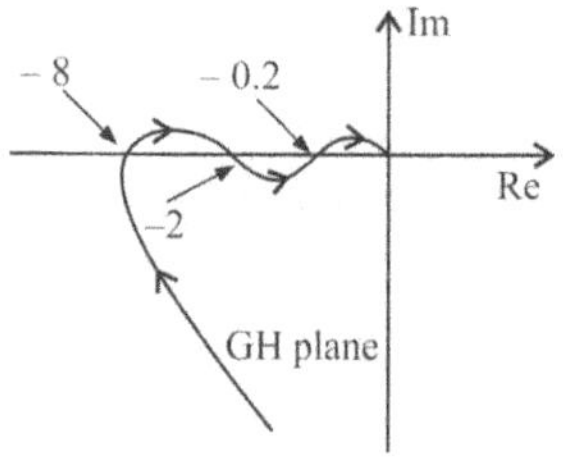

(a) $K < 5$ and $\frac{1}{2} < K < \frac{1}{8}$.   (b) $K < \frac{1}{8}$ and $\frac{1}{2} < K < 5$.

(c) $K < \frac{1}{8}$ and $5 < K$.   (d) $K > \frac{1}{8}$ and $K < 5$.

5. In the derivation of expression for peak percent overshoot. $M_p = \exp\left(\frac{-\pi\zeta}{\sqrt{1-\zeta^2}}\right) \times 100\%$, which one of the following conditions is NOT required?

(a) System is linear and time invariant.

(b) The system transfer function has a pair of complex conjugate poles and no zeroes.

(c) There is no transportation delay in the system

(d) The system has zero initial conditions.

6. A ramp input applied to an unity feedback system results in 5% steady state error. The type number and zero frequency gain of the system are respectively

(a) 1 and 20   (b) 0 and 20

(c) 0 and $\frac{1}{20}$ (d) 1 and $\frac{1}{20}$

7. A double integrator plant, $G(s) = \frac{K}{s^2}$, $H(s) = 1$ is to be compensated to achieve the damping ratio $\zeta = 0.5$, and an undamped natural frequency, $\omega_n = 5$ rad/s. Which one of the following compensator $G_c(s)$ will be suitable

(a) $\frac{s+3}{s+9.9}$ (b) $\frac{s+9.9}{s+3}$ (c) $\frac{s-6}{s+8.33}$ (d) $\frac{s-6}{s}$

8. An unity feedback system is given as, $G(s)=\dfrac{K(1-s)}{s(s+3)}$. Indicate the correct root locus diagram.

(a)

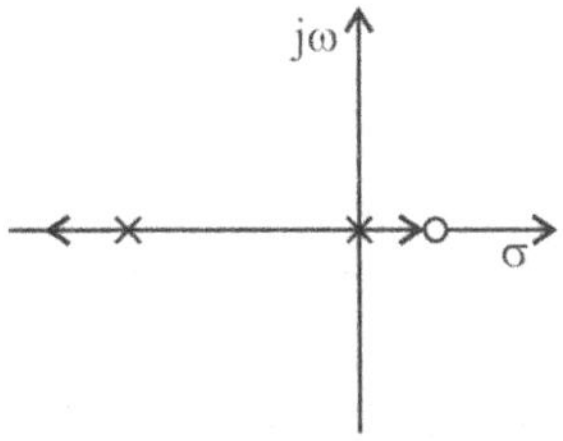

(b)

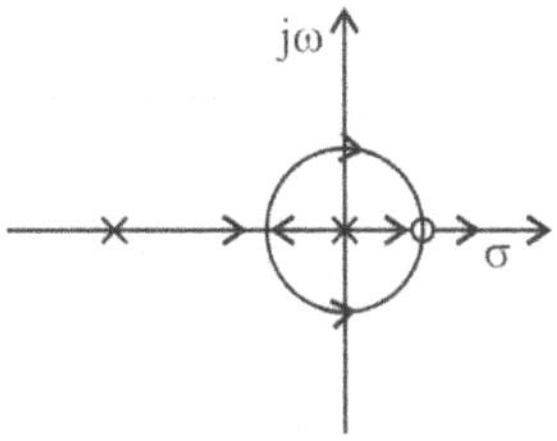

(c)

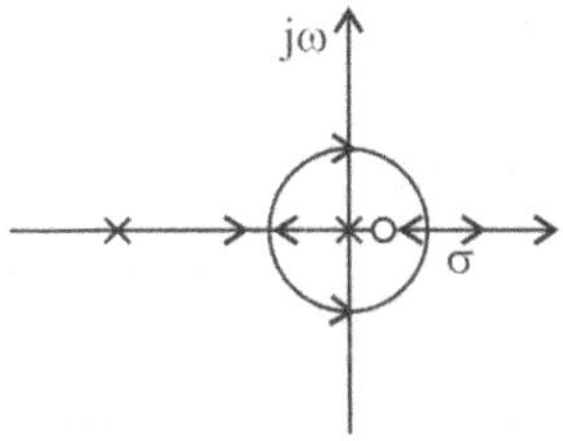

(d)

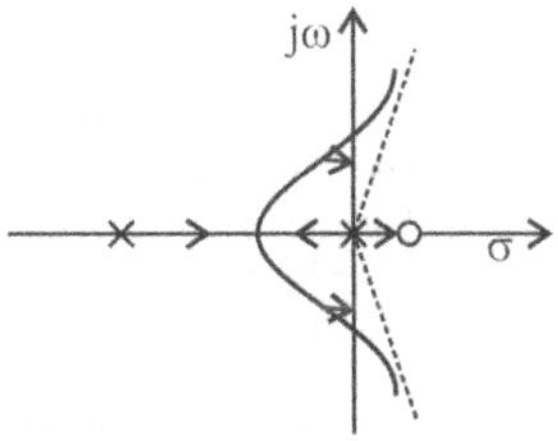

Statement for linked answer questions 9 and 10:

The open loop transfer function of a unity feedback system is given by $G(s)=\dfrac{3e^{-2s}}{s(s+2)}$

9. The gain and phase crossover frequencies in rad/sec are, respectively

(a) 0.632 and 1.26 (b) 0.632 and 0.485

(c) 0.485 and 0.632 (d) 1.26 and 0.632

10. Based on the above results, the gain and phase margins of the system will be

(a) – 7.09 and 87.5° (b) 7.09 and 87.5°

(c) 7.09 dB and – 87.5° (d) – 7.09 dB and – 87.5°

11. A system with zero initial conditions has the closed loop transfer function.

$$T(s)=\frac{s^2+4}{(s+1)(s+4)}$$

The system output is zero at the frequency

(a) 0.5 rad/sec (b) 1 rad/sec

(c) 2 rad/sec (d) 4 rad/sec

12. Figure shows the root locus plot (location of poles not given) of a third order system whose open loop transfer function is

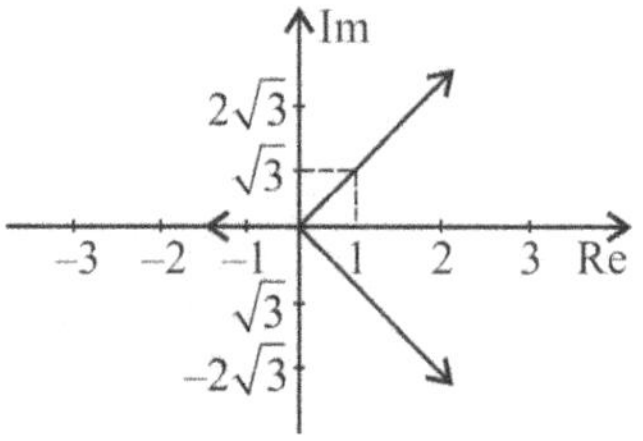

(a) $\dfrac{K}{s^3}$ (b) $\dfrac{K}{s^2(s+1)}$ (c) $\dfrac{K}{s(s^2+1)}$ (d) $\dfrac{K}{s(s^2-1)}$

13. The gain margin of a unity feedback control system with the open loop transfer function $G(s)=\dfrac{(s+1)}{s^2}$ is

(a) 0 (b) $\dfrac{1}{\sqrt{2}}$ (c) $\sqrt{2}$ (d) ∞

14. A unity feedback system, having an open loop gain $G(s)H(s)=\dfrac{K(1-s)}{(1+s)}$, becomes stable when

(a) $|K|>1$ (b) $K>1$ (c) $|K|<1$ (d) $K<-1$

15. When subjected to a unit step input, the closed loop control system shown in the figure will have a steady state error of

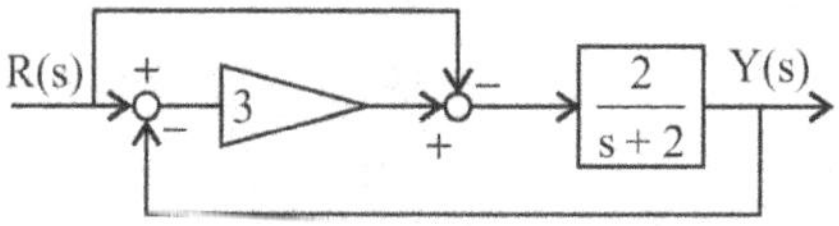

(a) – 1.0 (b) – 0.5 (c) 0 (d) 0.5

16. In the GH(s) plane, the Nyquist plot of the loop transfer function $G(s)\,H(s)=\dfrac{\pi e^{-0.25s}}{s}$ passes through the negative real axis at the point

(a) (– 0.25, j0) (b) (– 0.5, j0) (c) (– 1, j0) (d) (–2, j0)

17. If the compensated system shown in the figure has a phase margin of 60° at the crossover frequency of 1 rad/sec, then value of the gain K is

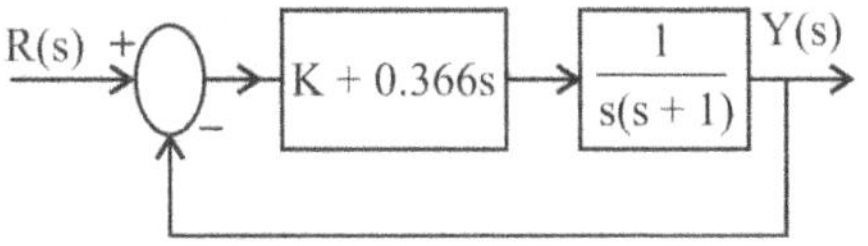

(a) 0.366 (b) 0.732 (c) 1.366 (d) 2.738

Statement for Linked Answer Questions 18 and 19

A state variable system $X(t)=\begin{bmatrix}0 & 1\\ 0 & -3\end{bmatrix}X(t)+\begin{bmatrix}1\\0\end{bmatrix}u(t)$, With the initial condition $X(0)=\begin{bmatrix}-1 & 3\end{bmatrix}^T$ and the unit step input u(t) has

18. The state transition matrix

(a) $\begin{bmatrix}1 & \frac{1}{3}\left(1-e^{-3t}\right)\\ 0 & e^{-3t}\end{bmatrix}$

(b) $\begin{bmatrix}1 & \frac{1}{3}\left(e^{-t}-e^{-3t}\right)\\ 0 & e^{-t}\end{bmatrix}$

(c) $\begin{bmatrix}1 & \frac{1}{3}\left(e^{-t}-e^{-3t}\right)\\ 0 & e^{-3t}\end{bmatrix}$

(d) $\begin{bmatrix}1 & \left(1-e^{-t}\right)\\ 0 & e^{-t}\end{bmatrix}$

19. The state transition equation

(a) $X(t)=\begin{bmatrix}t-e^{-t}\\ 3e^{-3t}\end{bmatrix}$

(b) $X(t)=\begin{bmatrix}t-e^{-t}\\ 3e^{-3t}\end{bmatrix}$

(c) $X(t)=\begin{bmatrix}t-e^{-3t}\\ 3e^{-3t}\end{bmatrix}$

(d) $X(t)=\begin{bmatrix}t-e^{-3t}\\ e^{-3t}\end{bmatrix}$

2006

1. The open-loop transfer function of a unity-gain feedback control system is given by

$$G(s)=\frac{K}{(s+1)(s+2)}$$

The gain margin of the system in dB is given by

(a) 0 (b) 1 (c) 20 (d) ∞

2. In the system shown below, $x(t) = (\sin t)\, u(t)$. In steady-state, the response y(t) will be

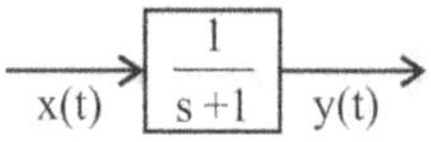

(a) $\frac{1}{\sqrt{2}}\sin\left(t - \frac{\pi}{4}\right)$ (b) $\frac{1}{\sqrt{2}}\sin\left(t + \frac{\pi}{4}\right)$

(c) $\frac{1}{\sqrt{2}}e^{-t}\sin t$ (d) $\sin t - \cos t$

3. The unit-step response of a system starting from rest is given by
$C(t) = 1 - e^{-2t}$ for $t \geq 0$

The transfer function of the system is

(a) $\frac{1}{1+2s}$ (b) $\frac{2}{2+s}$ (c) $\frac{1}{2+s}$ (d) $\frac{2s}{1+2s}$

4. The Nyquist plot of G (jω) H (jω) for a closed loop control system, passes through (–1, j0) point in the GH plane. The gain margin of the system in dB is equal to

(a) Infinite (b) Greater than zero

(c) Less than zero (d) Zero

5. The positive values of "K" and "a" so that the system shown in the figure below oscillates at a frequency of 2 rad/sec respectively are

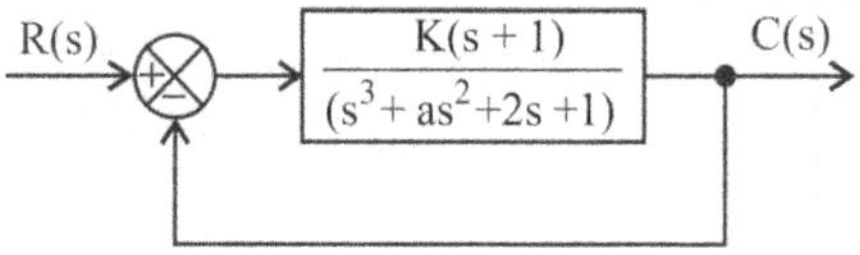

(a) 1, 0.75 (b) 2, 0.75 (c) 1, 1 (d) 2, 2

6. The unit impulse response of a system is $h(t) = e^{-t}$, $t \geq 0$

For this system, the steady-state value of the output for unit step input is equal to

(a) –1 (b) 0 (c) 1 (d) ∞

7. The transfer function of a phase-lead compensator is given by $G_c(s) = \frac{1+3Ts}{1+Ts}$ where $T > 0$. The maximum phase-shift provided by such a compensator is

(a) $\frac{\pi}{2}$ (b) $\frac{\pi}{3}$ (c) $\frac{\pi}{4}$ (d) $\frac{\pi}{6}$

Statement for Linked Answer Questions 8 and 9

Consider a unity-gain feedback control system whose open-loop transfer function is $G(s) = \frac{as+1}{s^2}$

8. The value of 'a' so that system has a phase-margin equal to $\frac{\pi}{4}$ is approximately equal to

(a) 2.40 (b) 1.40 (c) 0.84 (d) 0.74

9. With the value of 'a' set for a phase-margin of $\frac{\pi}{4}$, the value of unit-impulse response of the open-loop system at t = 1 second is equal to

(a) 3.40 (b) 2.40 (c) 1.84 (d) 1.74

10. For a system with the transfer function $H(s) = \frac{3(s-2)}{4s^2 - 2s + 1}$, the matrix A in the state space form $\dot{x} = Ax + Bu$ is equal to

(a) $\begin{bmatrix} 1 & 0 & 0 \\ 0 & 1 & 0 \\ -1 & 2 & -4 \end{bmatrix}$ (b) $\begin{bmatrix} 0 & 1 & 0 \\ 0 & 0 & 1 \\ -1 & 2 & -4 \end{bmatrix}$

(c) $\begin{bmatrix} 0 & 1 & 0 \\ 3 & -2 & 1 \\ 1 & -2 & 4 \end{bmatrix}$ (d) $\begin{bmatrix} 1 & 0 & 0 \\ 0 & 0 & 1 \\ -1 & 2 & -4 \end{bmatrix}$

11. A continuous-time system is described by $y(t) = e^{-|x(t)|}$, where y(t) is the output and x(t) is the input, y(t) is bounded

(a) Only when x(t) is bounded

(b) Only when x(t) is non-negative

(c) Only for $t \geq 0$ if x(t) is bounded for $t \geq 0$

(d) Even when x(t) is not bounded

12. The Bode magnitude plot of

$$H(j\omega) = \frac{10^4(1+j\omega)}{(10+j\omega)(100+j\omega)^2}$$ is

(a)

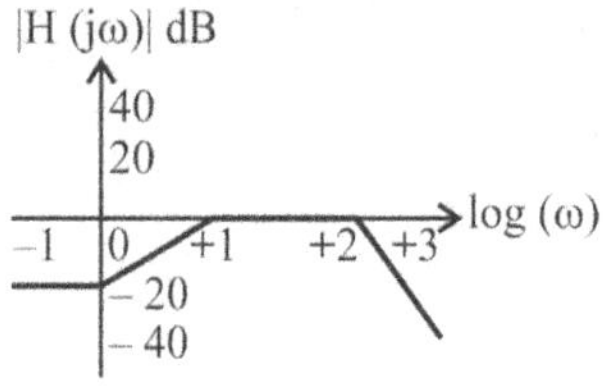

(b)

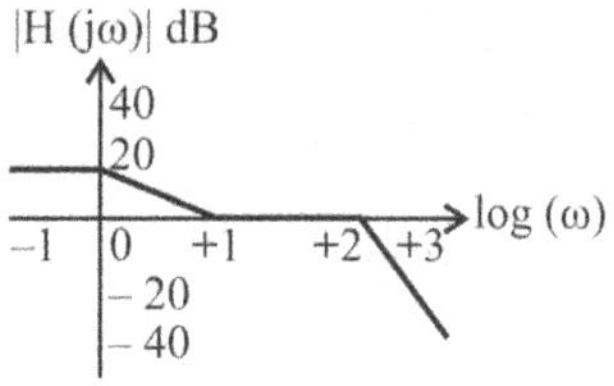

(c)

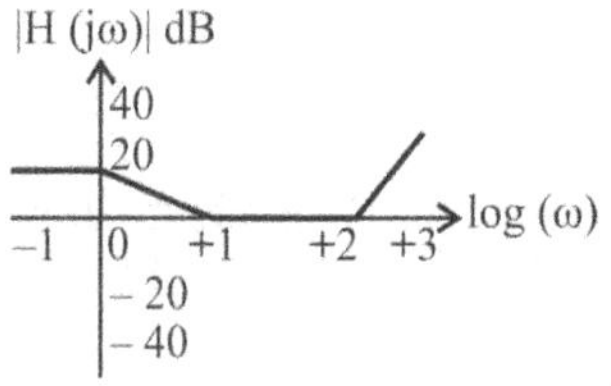

(d)

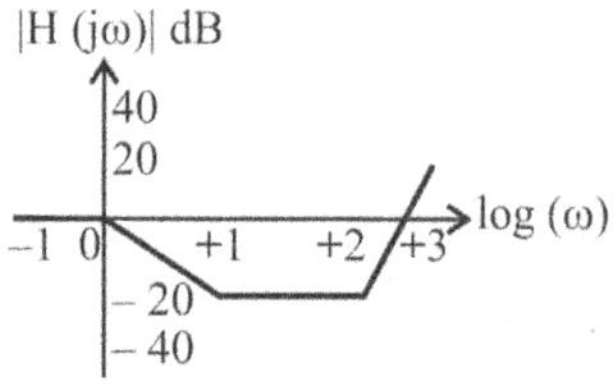

13. The algebraic equation $F(s) = s^5 - 3s^4 + 5s^3 - 7s^2 + 4s + 20$ is given. $F(s) = 0$ has
   (a) A single complex root with the remaining roots being real
   (b) One positive real root and four complex roots, all with positive real parts
   (c) One negative real root, two imaginary roots and two roots with positive real parts
   (d) One positive real root, two imaginary roots, and two roots with negative real parts

14. Consider the following Nyquist plots of loop transfer functions over $\omega = 0$ to $\omega = \infty$. Which of these plots represents a stable closed loop system ?

(1).

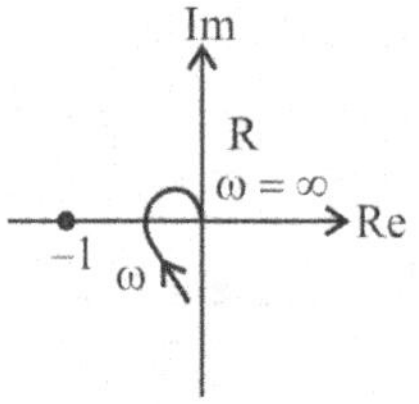

(2).

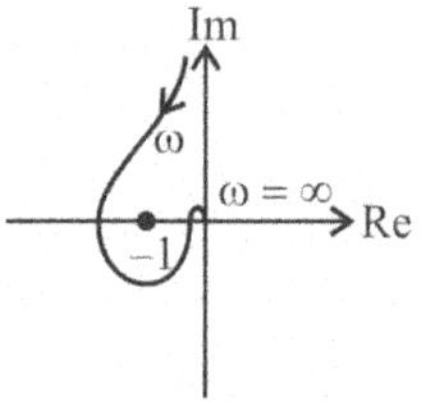

(3).

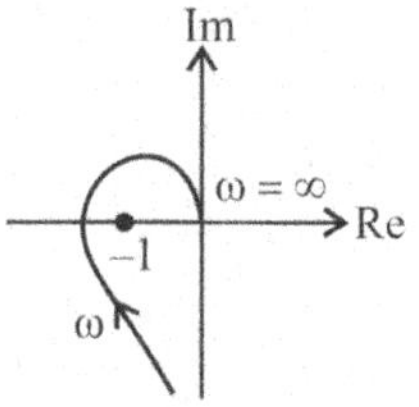

(4).

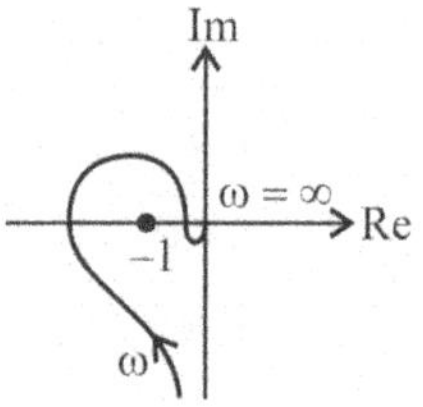

(a) (1) only
(b) All, except (1)
(c) All, except (3)
(d) (1) and (2) only

2007

1. A control system with a PD controller is shown in the figure. If the velocity error constant $K_v = 1000$ and the damping ratio $\zeta = 0.5$, then the values of K_P and K_D are

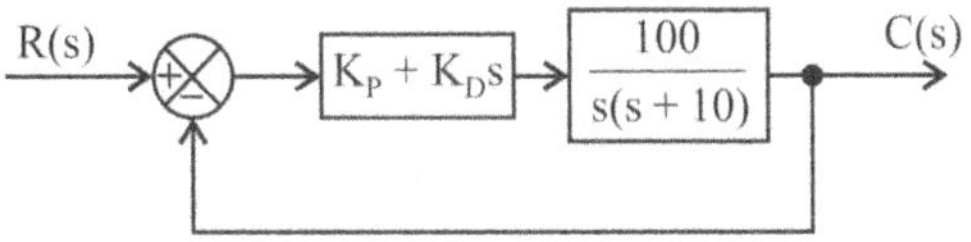

(a) $K_P = 100, K_D = 0.09$ (b) $K_P = 100, K_D = 0.9$
(c) $K_P = 10, K_D = 0.09$ (d) $K_P = 10, K_D = 0.9$

2. The open-loop transfer function of a plant is given as $G(s) = \frac{1}{s^2 - 1}$. If the plant is operated in a unity feedback configuration, then the lead compensator that can stabilize this control system is

(a) $\frac{10(s-1)}{s+2}$ (b) $\frac{10(s+4)}{s+2}$ (c) $\frac{10(s+2)}{s+10}$ (d) $\frac{1(s+2)}{s+10}$

3. A unity feedback control system has an open-loop transfer function $G(s) = \frac{K}{s(s^2 + 7s + 12)}$. The gain K for which $s = -1 + j1$ will lie on the root locus on this system is

(a) 4 (b) 5.5 (c) 6.5 (d) 10

4. The asymptotic Bode plot of a transfer function is as shown in the figure. The transfer function G(s) corresponding to this Bode plot is

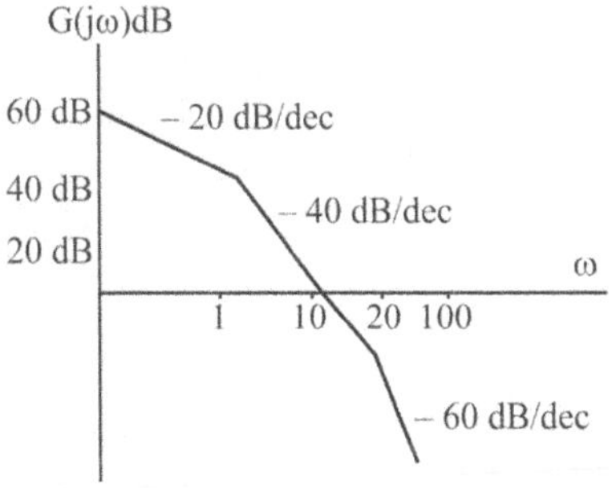

(a) $\frac{1}{(s+1)(s+20)}$ (b) $\frac{1}{s(s+1)(s+20)}$

(c) $\frac{100}{s(s+1)(s+20)}$ (d) $\frac{100}{s(s+1)(1+0.05s)}$

5. The state space representation of a separately excited DC servo motor dynamics is given as

$$\begin{bmatrix} \frac{d\omega}{dt} \\ \frac{di_a}{dt} \end{bmatrix} = \begin{bmatrix} -1 & 1 \\ -1 & -10 \end{bmatrix} \begin{bmatrix} \omega \\ i_a \end{bmatrix} + \begin{bmatrix} 0 \\ 10 \end{bmatrix} u$$

where ω is the speed of the motor, i_a is the armature current and u is the armature voltage. The transfer function $\frac{\omega(s)}{U(s)}$ of the motor is

(a) $\frac{10}{s^2+11s+11}$ (b) $\frac{1}{s^2+11s+11}$

(c) $\frac{10s+10}{s^2+11s+11}$ (d) $\frac{1}{s^2+s+1}$

Statement for linked answer questions 6 and 7

Consider a linear system whose state space representation is $\dot{x}(t) = Ax(t)$. If the initial state vector of the system is $x(0) = \begin{bmatrix} 1 \\ -2 \end{bmatrix}$, then the system response is $x(t) = \begin{bmatrix} e^{-2t} \\ -2e^{-2t} \end{bmatrix}$

If the initial state vector of the system changes to $x(0) = \begin{bmatrix} 1 \\ -1 \end{bmatrix}$, then the system response becomes $x(t) = \begin{bmatrix} e^{-t} \\ -e^{-t} \end{bmatrix}$

6. The eigen value and eigen vector pairs (λ_i, υ_i) for the system are

(a) $\left(-1, \begin{bmatrix} 1 \\ -1 \end{bmatrix}\right)$ and $\left(-2, \begin{bmatrix} 1 \\ -2 \end{bmatrix}\right)$ (b) $\left(-2, \begin{bmatrix} 1 \\ -1 \end{bmatrix}\right)$ and $\left(-1, \begin{bmatrix} 1 \\ -2 \end{bmatrix}\right)$

(c) $\left(-1, \begin{bmatrix} 1 \\ -1 \end{bmatrix}\right)$ and $\left(2, \begin{bmatrix} 1 \\ -2 \end{bmatrix}\right)$ (d) $\left(-2, \begin{bmatrix} 1 \\ -1 \end{bmatrix}\right)$ and $\left(1, \begin{bmatrix} 1 \\ -2 \end{bmatrix}\right)$

7. The system matrix A is

(a) $\begin{bmatrix} 0 & 1 \\ -1 & 1 \end{bmatrix}$ (b) $\begin{bmatrix} 1 & 1 \\ -1 & -2 \end{bmatrix}$

(c) $\begin{bmatrix} 2 & 1 \\ -1 & -1 \end{bmatrix}$ (d) $\begin{bmatrix} 0 & 1 \\ -2 & -3 \end{bmatrix}$

8. Let a signal $a_1 \sin(\omega_1 t + \phi_1)$ be applied to a stable linear time – invariant system. Let the corresponding steady state output be represented as $a_2 F(\omega_2 t + \phi_2)$. Then which of the following statement is true?

   (a) F is not necessarily a "sine" or cosine" function but must be periodic with $\omega_1 = \omega_2$

   (b) F must be a "sine" or "cosine" function with $a_1 = a_2$

   (c) F must be a "sine" function with $\omega_1 = \omega_2$ and $\phi_1 = \phi_2$

   (d) F must be a "sine" or "cosine" function with $\omega_1 = \omega_2$

9. If x = Re G(jω), and y = Im G(jω) then for $\omega \to 0^+$, the Nyquist plot for G(s) = 1/s (s + 1) (s + 2) becomes asymptotic to the line

   (a) $x = 0$ (b) $x = -3/4$

   (c) $x = y - 1/6$ (d) $x = y/\sqrt{3}$

10. The system 900/s(s + 1) (s + 9) is to be compensated such that its gain-crossover frequency becomes same as its uncompensated phase-crossover frequency and provides a 45 phase margin. To achieve this, one may use

    (a) A lag compensator that provides an attenuation of 20 dB and a phase lag of 45 at the frequency of $3\sqrt{3}$ rad/s

    (b) A lead compensator that provides an amplification of 20 dB and a phase lead of 45 at the frequency of 3 rad/s

    (c) A lag-lead compensator that provides an amplification of 20 dB and a phase lag of 45 at the frequency of $\sqrt{3}$ rad/s

    (d) A lag-lead compensator that provides an attenuation of 20 dB and phase lead of 45 at the frequency of 3 rad/s

11. If the loop gain K of a negative feedback system having a loop transfer function $\dfrac{K(s+3)}{(s+8)^2}$ is to be adjusted to induce a sustained oscillation then

    (a) The frequency of this oscillation must be $\dfrac{4}{\sqrt{3}}$ rad/s

    (b) The frequency of this oscillation must be 4 rad/s

    (c) The frequency of this oscillation must be 4 or $\dfrac{4}{\sqrt{3}}$ rad/s

    (d) Such a K does not exist

12. The system shown in figure below

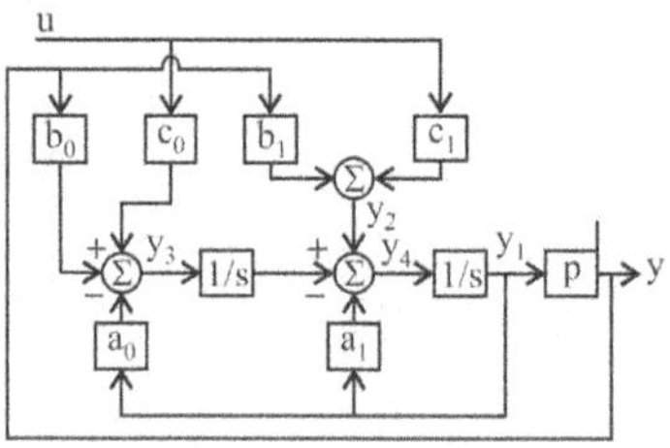

Can be reduced to the form

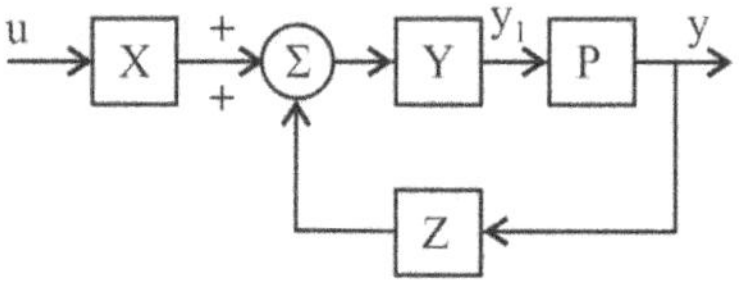

with

(a) $X = c_0 s + c_1, Y = \frac{1}{(s^2 + a_0 s + a_1)}, Z = b_0 s + b_1$

(b) $X = 1, Y = \frac{(c_0 s + c_1)}{(s^2 + a_0 s + a_1)}, Z = b_0 s + b_1$

(c) $X = c_1 s + c_0, Y = \frac{(b_1 s + b_0)}{(s^2 + a_1 s + a_0)}, Z = 1$

(d) $X = c_1 s + C_0, Y = \frac{1}{(s^2 + a_1 s + a_0)}, Z = b_1 s + b_0$

13. Consider the feedback control system shown below which is subjected to a unit step input. The system is stable and has the following parameters $k_p = 4$, $k_i = 10$ $\omega = 500$ and $\zeta = 0.7$.

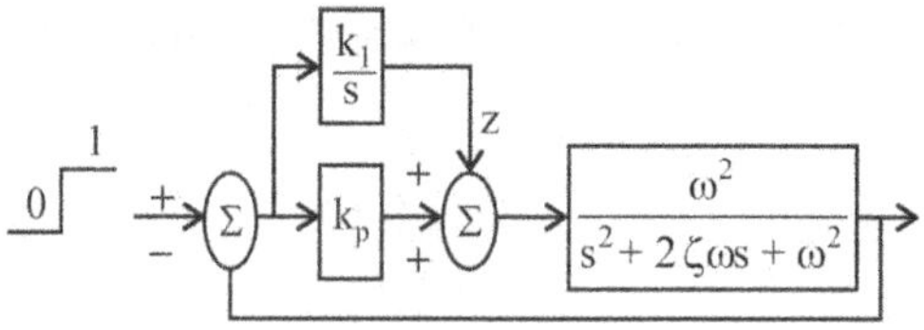

The steady state value of z is

(a) 1 (b) 0.25 (c) 0.1 (d) 0

2008

1. Step responses of a set of three second-order underdamped systems all have the same percentage overshoot. Which of the following diagrams represents the poles of the three systems?

(a)

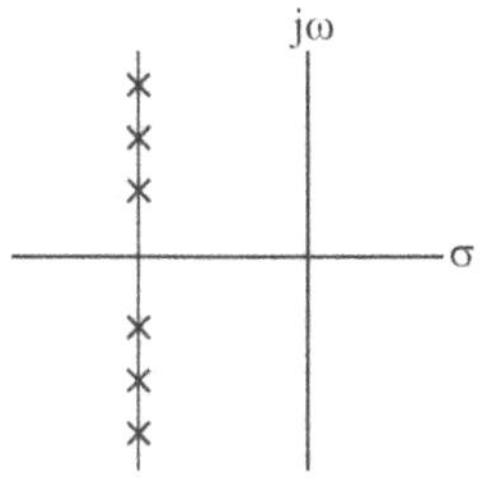

(b)

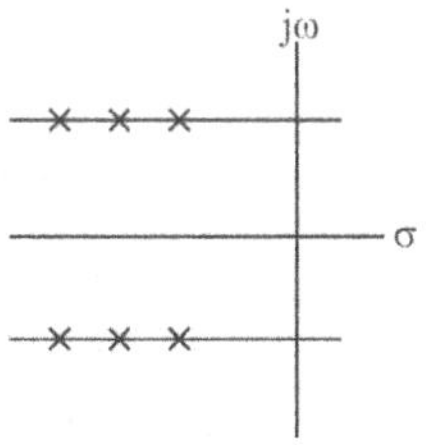

(c)

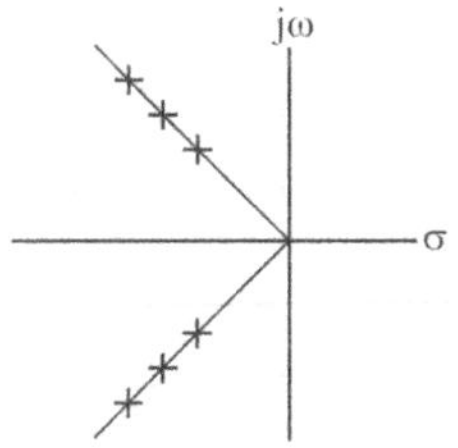

(d)

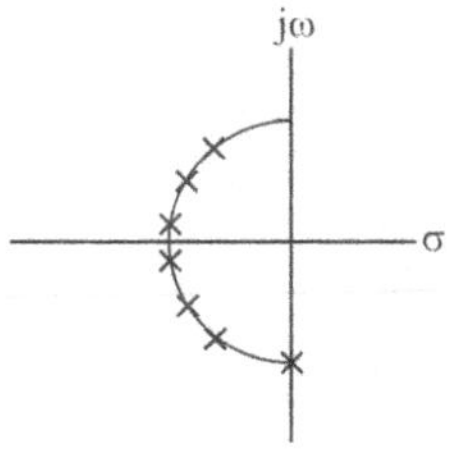

2. A certain system has transfer function $G(s)=\dfrac{s+8}{s^2+\alpha s-4}$, where α is a parameter. Consider the standard negative unity feedback configuration as shown below.

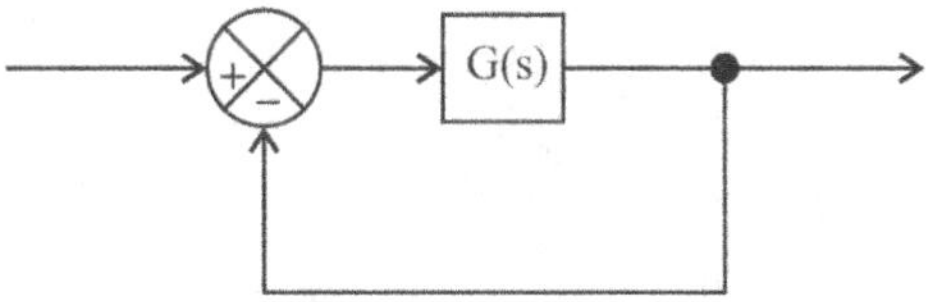

Which of the following statements is true?

(a) The closed loop systems is never stable for any value of α.

(b) For some positive values of α, the closed loop system is stable, but not for all positive values.

(c) For all positive values of α, the closed loop system is stable.

(d) The closed loop systems stable for all values of α, both positive and negative.

3. A signal flow graph of a system is given below.

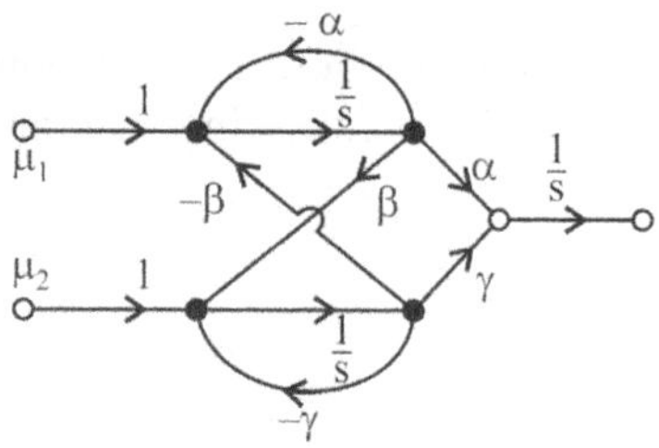

The set of equations that correspond to this signal flow graph is

(a) $\frac{d}{dt}\begin{pmatrix} x_1 \\ x_2 \\ x_3 \end{pmatrix} = \begin{bmatrix} \beta & -\gamma & 0 \\ \gamma & \alpha & 0 \\ -\alpha & \beta & 0 \end{bmatrix}\begin{pmatrix} x_1 \\ x_2 \\ x_3 \end{pmatrix} + \begin{bmatrix} 0 & 0 \\ 0 & 1 \\ 1 & 0 \end{bmatrix}\begin{pmatrix} u_1 \\ u_2 \end{pmatrix}$

(b) $\frac{d}{dt}\begin{pmatrix} x_1 \\ x_2 \\ x_3 \end{pmatrix} = \begin{bmatrix} 0 & \alpha & \gamma \\ 0 & -\alpha & -\gamma \\ 0 & \beta & -\beta \end{bmatrix}\begin{pmatrix} x_1 \\ x_2 \\ x_3 \end{pmatrix} + \begin{bmatrix} 1 & 0 \\ 0 & 1 \\ 0 & 0 \end{bmatrix}\begin{pmatrix} u_1 \\ u_2 \end{pmatrix}$

(c) $\frac{d}{dt}\begin{pmatrix} x_1 \\ x_2 \\ x_3 \end{pmatrix} = \begin{bmatrix} -\alpha & \beta & 0 \\ -\beta & -\gamma & 0 \\ \alpha & \gamma & 0 \end{bmatrix}\begin{pmatrix} x_1 \\ x_2 \\ x_3 \end{pmatrix} + \begin{bmatrix} 1 & 0 \\ 0 & 1 \\ 0 & 0 \end{bmatrix}\begin{pmatrix} u_1 \\ u_2 \end{pmatrix}$

(d) $\frac{d}{dt}\begin{pmatrix} x_1 \\ x_2 \\ x_3 \end{pmatrix} = \begin{bmatrix} -\gamma & 0 & \beta \\ \gamma & 0 & \alpha \\ -\beta & 0 & -\alpha \end{bmatrix}\begin{pmatrix} x_1 \\ x_2 \\ x_3 \end{pmatrix} + \begin{bmatrix} 0 & 1 \\ 0 & 0 \\ 1 & 0 \end{bmatrix}\begin{pmatrix} u_1 \\ u_2 \end{pmatrix}$

4. The number of open right half plane poles of

$G(s) = \frac{10}{s^5 + 2s^4 + 3s^3 + 6s^2 + 5s + 3}$ is

(a) 0 (b) 1 (c) 2 (d) 3

5. The magnitude of frequency response of an underdamped second order system is 5 at 0 rad/sec and peaks to $\frac{10}{\sqrt{3}}$ at $5\sqrt{2}$ rad/sec. The transfer function of the system is

(a) $\frac{500}{s^2 + 10s + 100}$ (b) $\frac{375}{s^2 + 5s + 75}$

(c) $\frac{720}{s^2 + 12s + 144}$ (d) $\frac{1125}{s^2 + 25s + 225}$

Statement for Linked Answer (Q. 6–7):

The impulse response h(t) of a linear time-invariant continuous time system is given by h(t) = exp(– 2t) u(t), where u(t) denotes the unit step function.

6. The frequency response H(ω) of this system in terms of angular frequency ω, is given by H(ω)

(a) $\frac{1}{1+j2\omega}$ (b) $\frac{\sin(\omega)}{\omega}$ (c) $\frac{1}{2+j\omega}$ (d) $\frac{j\omega}{2+j\omega}$

7. The output of this system, to the sinusoidal inuput x(t) = 2cos 2t for all time t, is

(a) 0 (b) $2^{-0.25}\cos(2t-0.125\pi)$

(c) $2^{-0.5}\cos(2t-0.125\pi)$ (d) $2^{-0.5}\cos(2t-0.25\pi)$

8. The transfer function of a linear time invariant system is given as $G(s)=\frac{1}{s^2+3s+2}$

The steady state value of the output of this system for a unit impulse input applied at time instant t = 1 will be

(a) 0 (b) 0.5 (c) 1 (d) 2

9. The transfer functions of two compensators are given below $C_1=\frac{10(s+1)}{(s+10)}$,

$C_2=\frac{s+10}{10(s+1)}$

Which one of the following statements is correct?

(a) C_1 is a lead compensator and C_2 is a lag compensator

(b) C_1 is a lag compensator and C_2 is a lead compensator

(c) Both C_1 and C_2 are lead compensators

(d) Both C_1 and C_2 are lag compensators

10. The asymptotic Bode magnitude plot of a minimum phase transfer function is shown in the figure

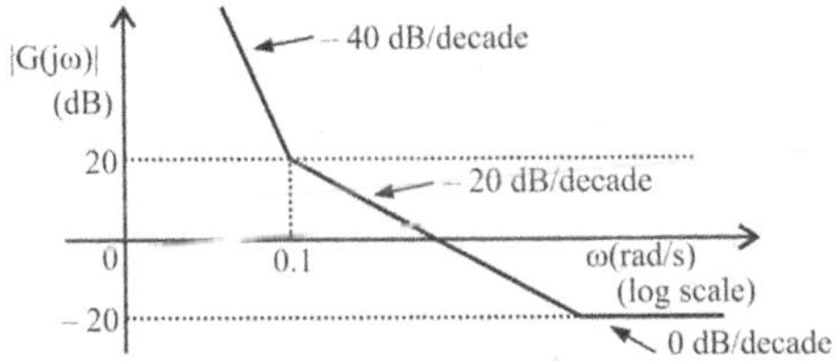

This transfer function has

(a) Three poles and one zero (b) Two poles and one zero

(c) Two poles and two zeros (d) One pole and two zeros

11. Figure shows a feedback system where $K > 0$. The range of K for which the system is stable will be given by

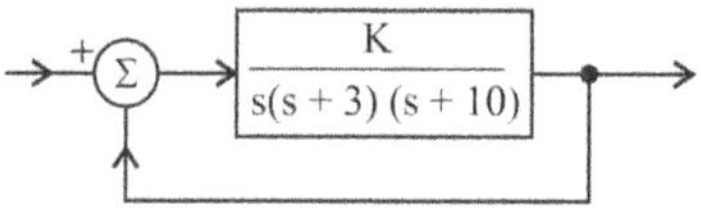

(a) $0 < K < 30$ (b) $0 < K < 39$

(c) $0 < K < 390$ (d) $K > 390$

12. The transfer function of a system is given as $\frac{100}{s^2 + 20s + 100}$. This system is

(a) An overdamped system (b) An underdamped system

(c) A critically damped system (d) An unstable system

Statement for Linked Answer Questions (13-14)

The state space equation of a system is described by

$$x = Ax + Bu$$

$$y = Cx$$

where x is state vector, u is input, y is output and

$$A = \begin{bmatrix} 0 & 1 \\ 0 & -2 \end{bmatrix}, \; B = \begin{bmatrix} 0 \\ 1 \end{bmatrix}, \; C = \begin{bmatrix} 1 & 0 \end{bmatrix}$$

13. The transfer function G(s) of this system will be

(a) $\frac{s}{(s+2)}$ (b) $\frac{s+1}{s(s-2)}$ (c) $\frac{s}{(s-2)}$ (d) $\frac{1}{s(s+2)}$

14. A unity feedback is provided to the above system G(s) to make it a closed loop system as shown in figure.

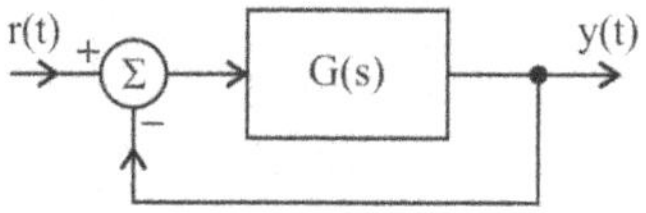

For a unit step input r(t), the steady state error in the output will be

(a) 0 (b) 1 (c) 2 (d) ∞

2009

1. The magnitude plot of a rational transfer function G(s) with real coefficients is shown below. Which of the following compensators has such a magnitude plot?

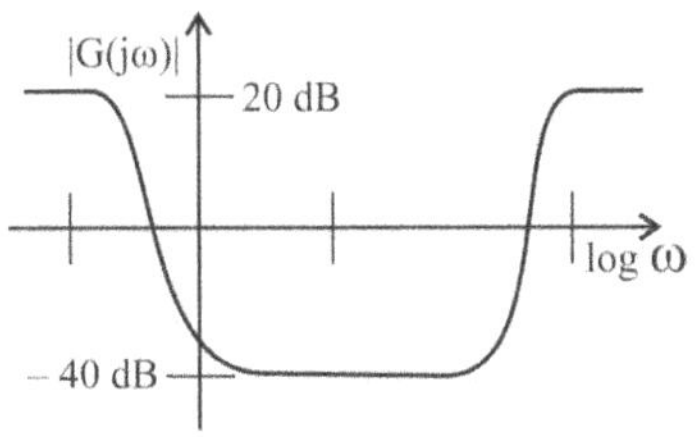

(a) Lead compensator (b) Lag compensator
(c) PID compensator (d) Lead-lag compensator

2. Consider the system $\frac{dx}{dt} = Ax + Bu$ with $A = \begin{bmatrix} 1 & 0 \\ 0 & 1 \end{bmatrix}$, and $B = \begin{bmatrix} p \\ q \end{bmatrix}$

where p and q are arbitrary real numbers. Which of the following statements about the controllability of the system is true?

(a) The system is completely state controllable for any nonzero values of p and q
(b) Only p = 0 and q = 0 result in controllability
(c) The system is uncontrollable for all values of p and q
(d) We cannot conclude about controllability from the given data

3. The feedback configuration and the pole-zero locations of $G(s) = \frac{s^2 - 2s + 2}{s^2 + 2s + 2}$ are shown below. The root locus for negative values of k, i.e., for $-\infty < k < 0$, has breakway/break-in points and angle of departure at pole P (with respect to the positive real axis) equal to

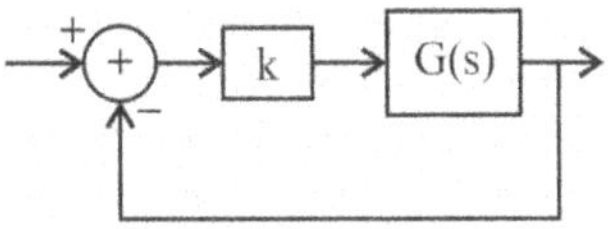

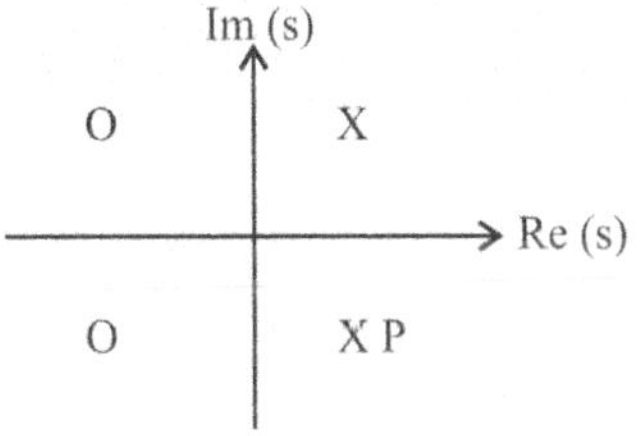

(a) $\mp\sqrt{2}$ and 0° (b) $\pm\sqrt{2}$ and 45°
(c) $\pm\sqrt{3}$ and 0° (d) $\pm\sqrt{3}$ and 45°

4. The unit step response of an under- damped second order system has steady state value of – 2. Which one of the following transfer functions has these properties?

(a) $\dfrac{-2.24}{s^2+2.59s+1.12}$ (b) $\dfrac{-3.82}{s^2+1.91s+1.91}$

(c) $\dfrac{-2.24}{s^2-2.59s+1.12}$ (d) $\dfrac{-3.82}{s^2-1.91s+1.91}$

5. A Linear Time Invariant system with an impulse response h(t) produces output y(t) when input x(t) is applied. When the input $x(t-\tau)$ is applied to a system with impulse response $h(t-\tau)$, the output will be

(a) y(t) (b) $y(2(t-\tau))$

(c) $y(t-\tau)$ (d) $y(t-2\tau)$

6. The polar plot of an open loop stable system is shown below. The closed loop system is

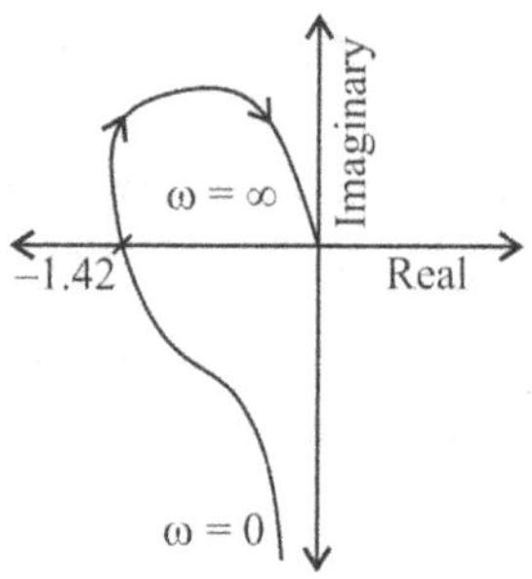

(a) Always stable
(b) Marginally stable
(c) Unstable with one pole on the RH s-plane
(d) Unstable with two poles on the RH s-plane

7. The first two rows of Routh's tabulation of a third order equation are as follows.

S^3 2 2

S^2 4 4

This means there are

(a) Two roots at s = ± j and one root in right half s-plane
(b) Two roots at s = ± j2 and one root in left half s-plane
(c) Two roots at s = ± j2 and one root in right half s-plane
(d) Two roots at s = ± j and one root in left half s-plane

8. The asymptotic approximation of the log-magnitude vs. frequency plot of a system containing only real poles and zeros is shown. Its transfer function is

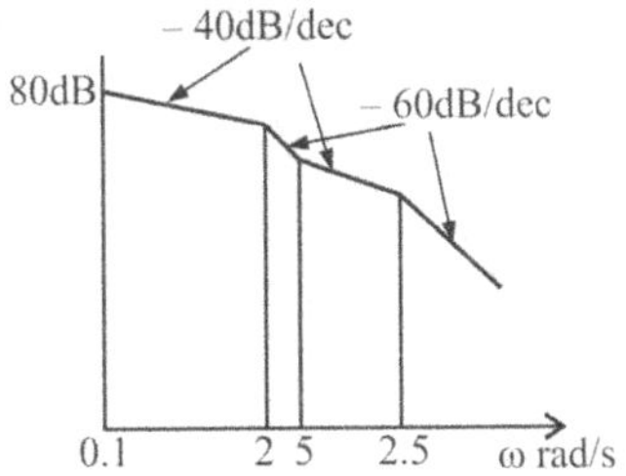

(a) $\dfrac{10(s+5)}{s(s+2)(s+25)}$ (b) $\dfrac{1000(s+5)}{s^2(s+2)(s+25)}$

(c) $\dfrac{100(s+5)}{s(s+2)(s+25)}$ (d) $\dfrac{80(s+5)}{s^2(s+2)(s+25)}$

9. The unit-step response of a unity feedback system with open loop transfer function $G(s)=\dfrac{K}{(s+1)(s+2)}$ is shown in the figure.

The value of K is

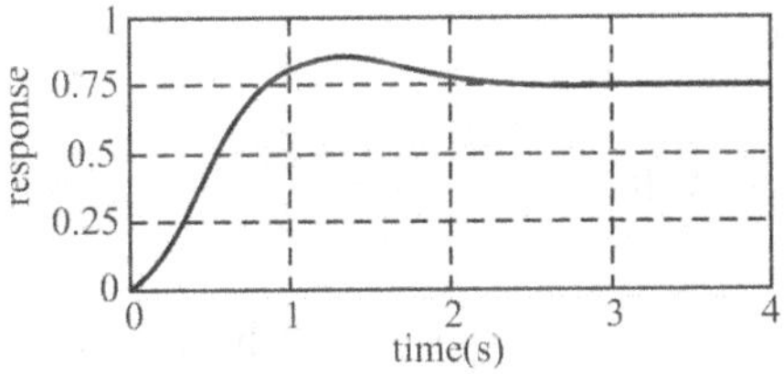

(a) 0.5 (b) 2 (c) 4 (d) 6

10. The open loop transfer function of a unity feedback system is given by $G(s) = (e^{-0.1s})/s$. The gain margin of this system is

(a) 11.95 dB (b) 17.67 dB (c) 21.23 dB (d) 23.9 dB

Common Data for Questions 11 & 12

A system is described by the following state and output equations

$$\frac{dx_1(t)}{dt} = -3x_1(t) + x_2(t) + 2u(t)$$

$$\frac{dx_2(t)}{dt} = -2x_2(t) + u(t)$$

$$y(t) = x_1(t)$$

where u(t) is the input and y(t) is the output

11. The system transfer function is

(a) $\frac{s+2}{s^2+5s-6}$ (b) $\frac{s+3}{s^2+5s+6}$ (c) $\frac{2s+5}{s^2+5s+6}$ (d) $\frac{2s-5}{s^2+5s-6}$

12. The state-transition matrix of the above system is

(a) $\begin{bmatrix} e^{-3t} & 0 \\ e^{-2t}+e^{-3t} & e^{-2t} \end{bmatrix}$ (b) $\begin{bmatrix} e^{3t} & e^{-2t}-e^{3t} \\ 0 & e^{2t} \end{bmatrix}$

(c) $\begin{bmatrix} e^{-3t} & e^{-2t}+e^{3t} \\ 0 & e^{-2t} \end{bmatrix}$ (d) $\begin{bmatrix} e^{-3t} & e^{-2t}-e^{-3t} \\ 0 & e^{-2t} \end{bmatrix}$

2010

1. The transfer function Y(s)/R(s) of the system shown is

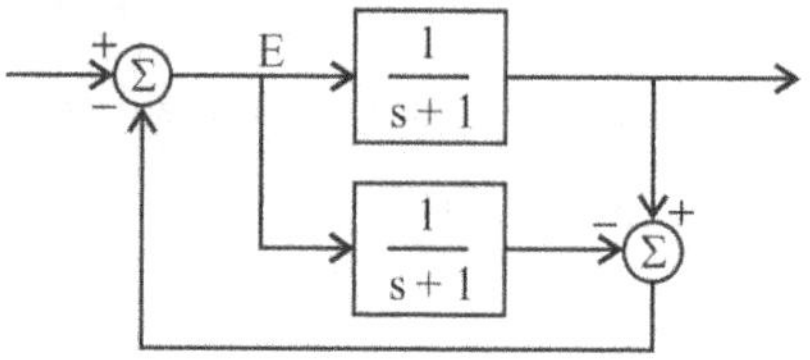

(a) 0 (b) $\frac{1}{s+1}$ (c) $\frac{2}{s+1}$ (d) $\frac{2}{s+3}$

2. A system with the transfer function $\frac{Y(s)}{X(s)} = \frac{s}{s+p}$ has an output $y(t) = \cos\left(2t - \frac{\pi}{3}\right)$ for the input signal $x(t) = p\cos\left(2t - \frac{\pi}{2}\right)$. Then, the system parameter 'p' is

(a) $\sqrt{3}$ (b) $\frac{2}{\sqrt{3}}$ (c) 1 (d) $\frac{\sqrt{3}}{2}$

3. For the asymptotic Bode magnitude plot shown below, the system transfer function can be

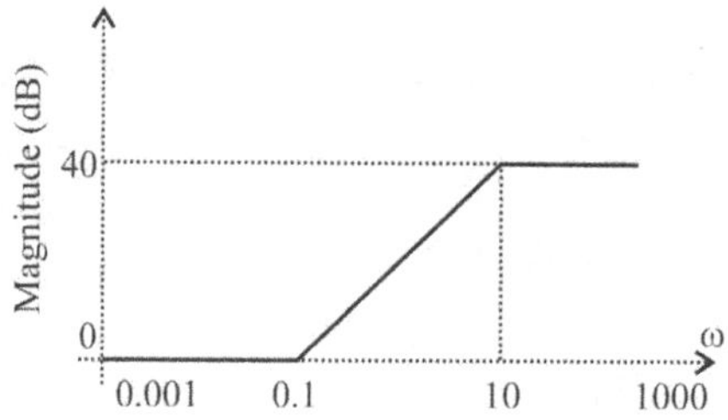

(a) $\frac{10s+1}{0.1s+1}$ (b) $\frac{100s+1}{0.1s+1}$ (c) $\frac{100s}{10s+1}$ (d) $\frac{0.1s+1}{10s+1}$

4. A unity negative feedback closed loop system has a plant with the transfer function $G(s)=\frac{1}{s^2+2s+2}$ and a controller $G_c(s)$ in the feed forward path. For a unit step input, the transfer function of the controller that gives minimum steady state error is

(a) $G_c(s)=\frac{s+1}{s+2}$ (b) $G_c(s)=\frac{s+2}{s+1}$

(c) $G_c(s)=\frac{(s+1)(s+4)}{(s+2)(s+3)}$ (d) $G_c(s)=1+\frac{2}{s}+3s$

Common data for Q. (5 – 6)

The signal flow graph of a system is shown below.

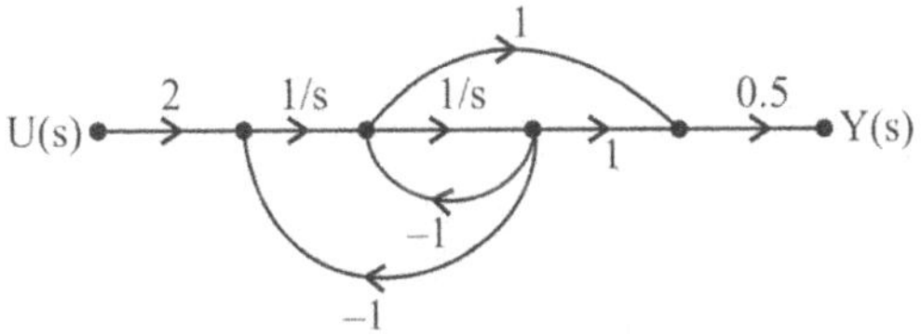

5. The state variable representation of the system can be

(a) $\dot{x}=\begin{bmatrix}1 & 1\\ -1 & 0\end{bmatrix}x+\begin{bmatrix}0\\ 2\end{bmatrix}u,\ y=\begin{bmatrix}0 & 0.5\end{bmatrix}x$

(b) $\dot{x}=\begin{bmatrix}-1 & 1\\ -1 & 0\end{bmatrix}x+\begin{bmatrix}0\\ 2\end{bmatrix}u,\ y=\begin{bmatrix}0 & 0.5\end{bmatrix}x$

(c) $\dot{x}=\begin{bmatrix}1 & 1\\ -1 & 0\end{bmatrix}x+\begin{bmatrix}0\\ 2\end{bmatrix}u,\ y=\begin{bmatrix}0.5 & 0.5\end{bmatrix}x$

(d) $\dot{x}=\begin{bmatrix}-1 & 1\\ -1 & 0\end{bmatrix}x+\begin{bmatrix}0\\ 2\end{bmatrix}u,\ y=\begin{bmatrix}0.5 & 0.5\end{bmatrix}x$

6. The transfer function of the system is

(a) $\frac{s+1}{s^2+1}$ (b) $\frac{s-1}{s^2+1}$ (c) $\frac{s+1}{s^2+s+1}$ (d) $\frac{s-1}{s^2+s+1}$

7. As shown in the figure, a negative feedback system has an amplifier of gain 100 with $\pm 10\%$ tolerance in the forward path, and an attenuator of value $\frac{9}{100}$ in the feedback path. The overall system gain is approximately.

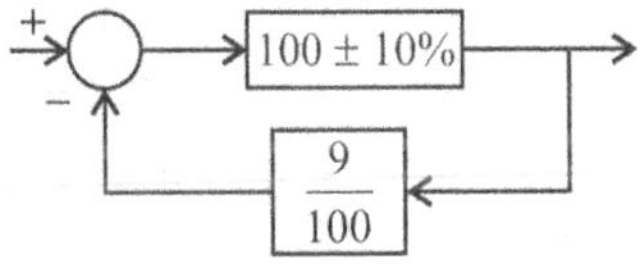

(a) $10 \pm 1\%$ (b) $10 \pm 2\%$ (c) $10 \pm 5\%$ (d) $10 \pm 10\%$

8. For the system 2/(s + 1), the approximate time taken for a step response to reach 98% of the final value is

(a) 1s (b) 2s (c) 4s (d) 8s

9. The frequency response of $G(s) = 1/[s(s + 1)(s + 2)]$ plotted in the complex $G(j\omega)$ plane (for $0 < \omega < \infty$) is

(a)

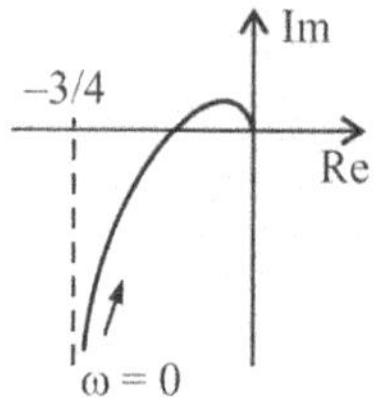

(b)

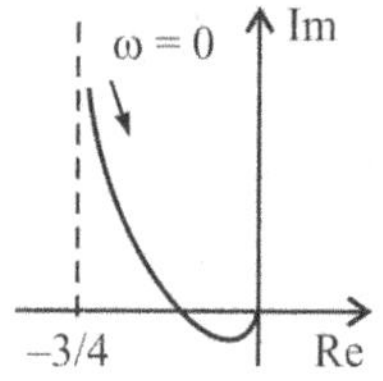

(c)

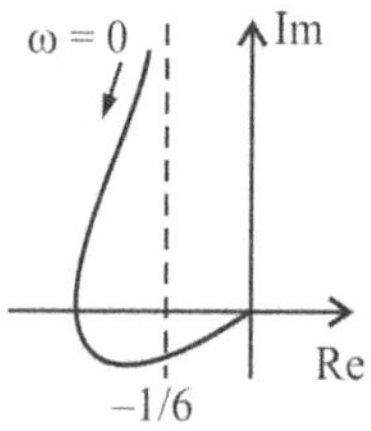

(d)

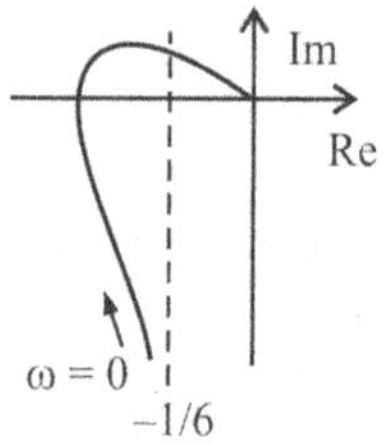

10. The system $\dot{x} = Ax + Bu$ with $A = \begin{bmatrix} -1 & 2 \\ 0 & 2 \end{bmatrix}$, $B = \begin{bmatrix} 0 \\ 1 \end{bmatrix}$ is

(a) Stable and controllable (b) Stable but uncontrollable

(c) Unstable but controllable (d) Unstable and uncontrollable

11. The characteristic equation of a closed-loop system is $s(s + 1)(s + 3) + k(s + 2) = 0$, $k > 0$. Which of the following statements is true?

(a) Its roots are always real

(b) It cannot have a breakaway point in the range $-1 < Re(s) < 0$

(c) Two of its roots tend to infinity along the asymptotes $Rs[s] = -1$

(d) It may have complex roots in the right half plane.

Common data for Q. 12 & Q. 13

Given f(t) and g(t) as shown below.

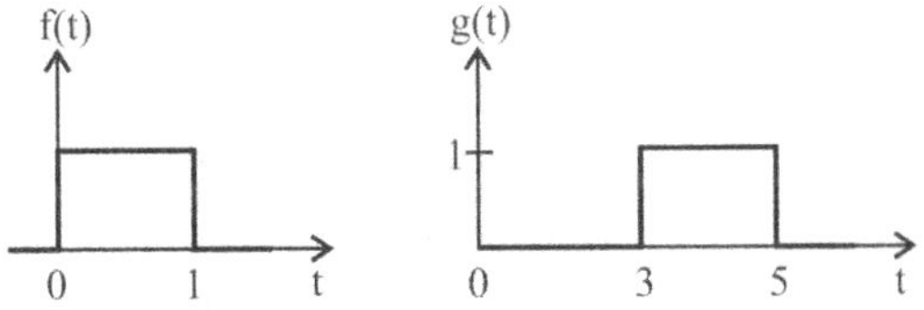

12. g(t) can be expressed as

(a) $g(t) = f(2t - 3)$ (b) $g(t) = f\left(\frac{t}{2} - 3\right)$

(c) $g(t) = f\left(2t - \frac{3}{2}\right)$ (d) $g(t) = f\left(\frac{t}{2} - \frac{3}{2}\right)$

13. The Laplace transform of g(t) is

(a) $\frac{1}{s}\left(e^{3s} - e^{5s}\right)$ (b) $\frac{1}{s}\left(e^{-5s} - e^{-3s}\right)$

(c) $\frac{e^{-3s}}{s}\left(1 - e^{-2s}\right)$ (d) $\frac{1}{s}\left(e^{5s} - e^{3s}\right)$

2011

1. The differential equation $100\frac{d^2y}{dt^2} - 20\frac{dy}{dt} + y = x(t)$ describes a system with an input x(t) and an output y(t). The system, which is initially relaxed, is excited by a unit step input. The output y(t) can be represented by the waveform

(a)

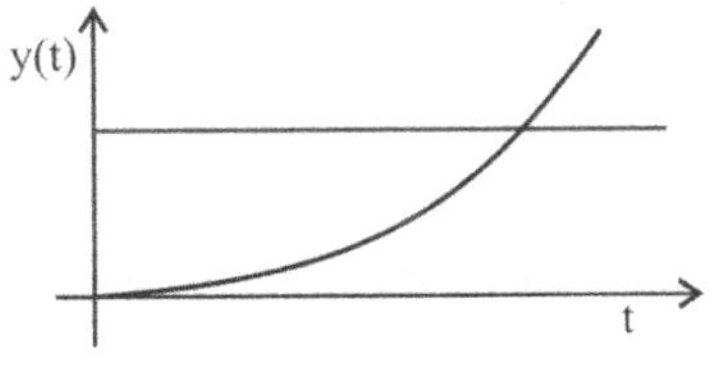

(b)

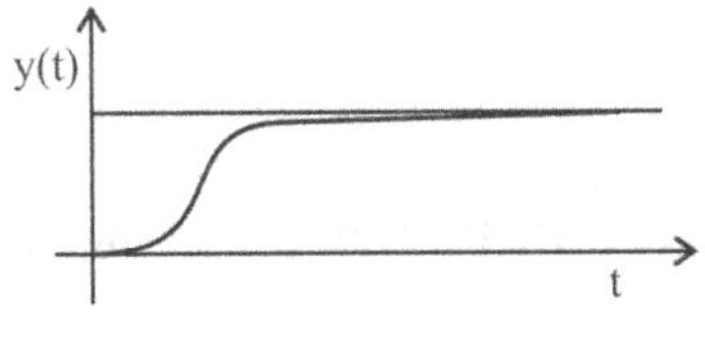

(c)

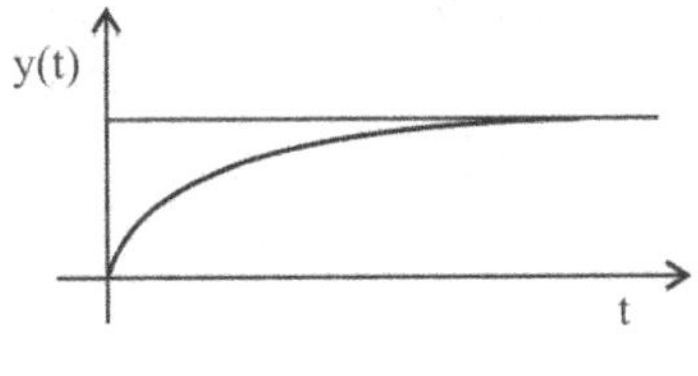

(d)

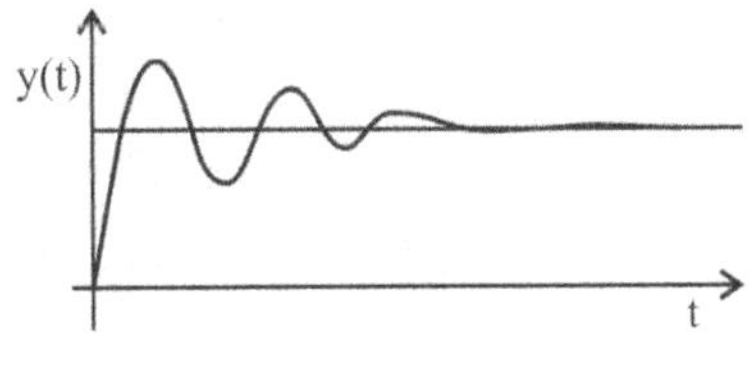

2. For the transfer function G(jω) = 5 + jω, the corresponding Nyquist plot for positive frequency has the form

(a) 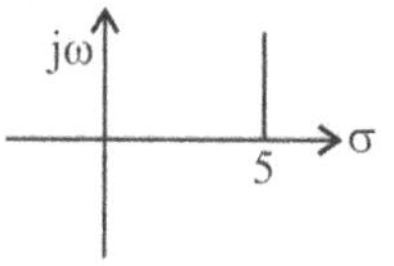

(b) 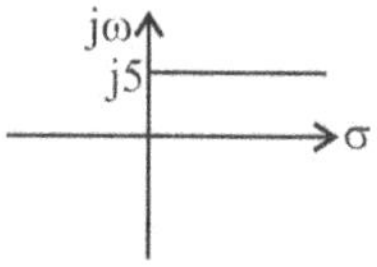

(c) 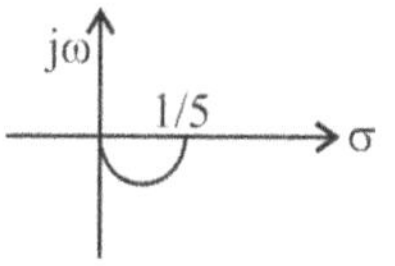

(d) 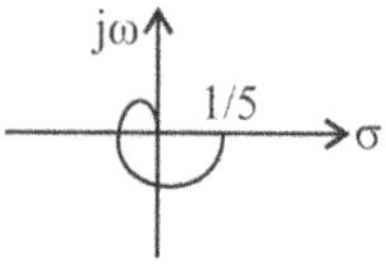

3. The root locus plot for a system is given below. The open loop transfer function corresponding to this plot is given by

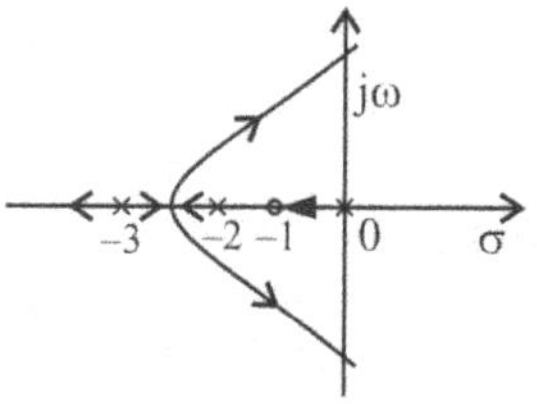

(a) $G(s)H(s) = k\dfrac{s(s+1)}{(s+2)(s+3)}$ (b) $G(s)H(s) = k\dfrac{(s+1)}{s(s+2)(s+3)^2}$

(c) $G(s)H(s) = k\dfrac{1}{s(s-1)(s+2)(s+3)}$ (d) $G(s)H(s) = k\dfrac{(s+1)}{s(s+2)(s+3)}$

4. If the unit step response of a network is $(1 - e^{-at})$, then its unit impulse response is

(a) $a\,e^{-at}$ (b) $a^{-1}\,e^{-at}$ (c) $(1 - a^{-1})\,e^{-at}$ (d) $(1 - a)\,e^{-at}$

5. The block diagram of a system with one input u and two outputs y_1 and y_2 is given below.

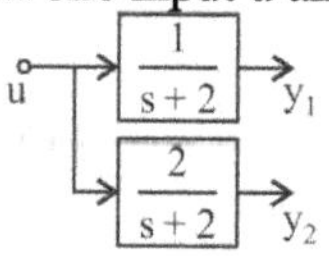

A state space model of the above system in terms of the state vector $\underline{x}$ and the output vector $\underline{y} = [y_1 y_2]^T$ is

(a) $\dot{\underline{x}} = [2]\underline{x} + [1]u;\ \underline{y} = [1\ \ 2]\underline{x}$ (b) $\dot{\underline{x}} = [-2]\underline{x} + [1]u;\ \underline{y} = \begin{bmatrix}1\\2\end{bmatrix}\underline{x}$

(c) $\dot{\underline{x}} = \begin{bmatrix}-2 & 0\\0 & -2\end{bmatrix}\underline{x} + \begin{bmatrix}1\\1\end{bmatrix}u;\ \underline{y} = [1\ \ 2]\underline{x}$ (d) $\dot{\underline{x}} = \begin{bmatrix}2 & 0\\0 & 2\end{bmatrix}\underline{x} + \begin{bmatrix}1\\1\end{bmatrix}u;\ \underline{y} = \begin{bmatrix}1\\2\end{bmatrix}\underline{x}$

Common Data Questions: 6 & 7

The input-output transfer function of a plant $H(s)=\dfrac{100}{s(s+10)^2}$. The plant is placed in a unity negative feedback configuration as shown in the figure below.

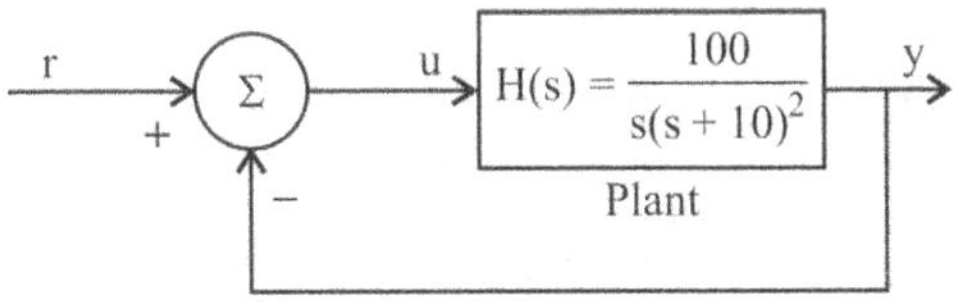

6. The signal flow graph that **DOES NOT** model the plant transfer function H(s) is

(a) 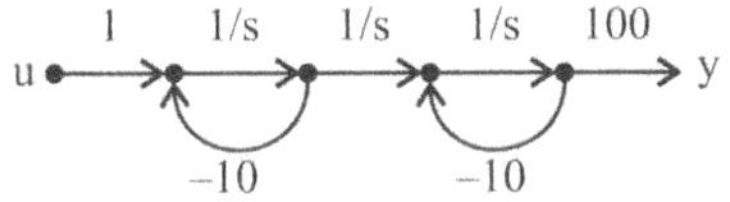

(b) 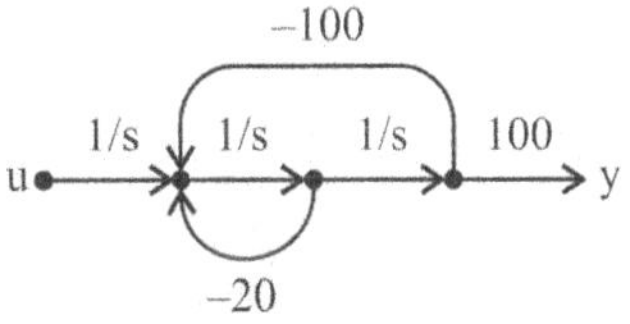

(c) 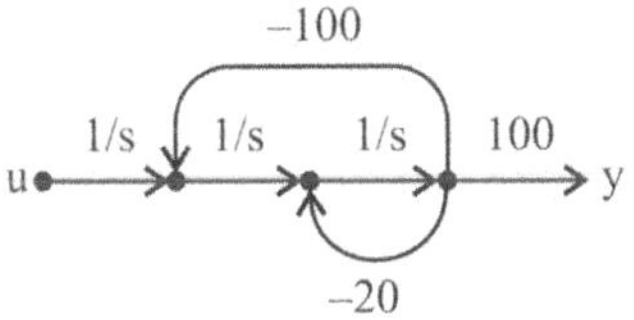

(d) 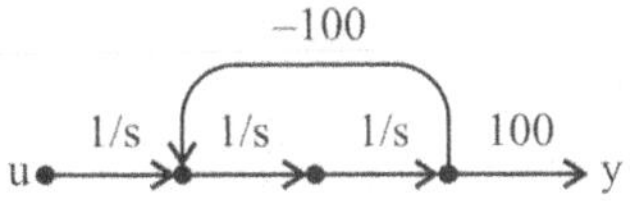

7. The gain margin of the system under closed loop unity negative feedback is

(a) 0 dB (b) 20 dB (c) 26 dB (d) 46 dB

8. The steady state error of a unity feedback linear system for a unit step input is 0.1. The steady state error of the same system, for a pulse input r(t) having a magnitude of 10 and a duration of one second, as shown in the figure is

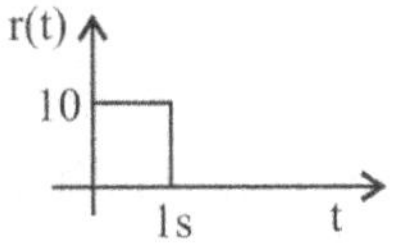

(a) 0 (b) 0.1 (c) 1 (d) 10

9. An open loop system represented by the transfer function $G(s)=\dfrac{(s-1)}{(s+2)(s+3)}$ is

(a) Stable and of the minimum phase type
(b) Stable and of the non-minimum phase type
(c) Unstable and of the minimum phase type
(d) Unstable and of the non-minimum phase type

10. The open loop transfer function G(s) of a unity feedback control system is given as,

$$G(s)=\frac{k\left(s+\frac{2}{3}\right)}{s^2(s+2)}$$

From the root locus, it can be inferred that when k tends to positive infinity,

(a) Three roots with nearly equal real parts exist on the left half of the s-plane

(b) One real root is found on the right half of the s-plane

(c) The root loci cross the $j\omega$ axis for a finite value of k; $k \neq 0$

(d) Three real roots are found on the right half of the s-plane

11. The response h(t) of a linear time invariant system to an impulse $\delta(t)$, under initially relaxed condition is $h(t) = e^{-t} + e^{-2t}$. The response of this system for a unit step input u(t) is

(a) $u(t) + e^{-t} + e^{-2t}$ (b) $(e^{-t} + e^{-2t})\, u(t)$

(c) $(1.5 - e^{-t} - 0.5\, e^{-2t})\, u(t)$ (d) $e^{-t}\, \delta(t) + e^{-2t}\, u(t)$

12. A two-loop position control system is shown below.

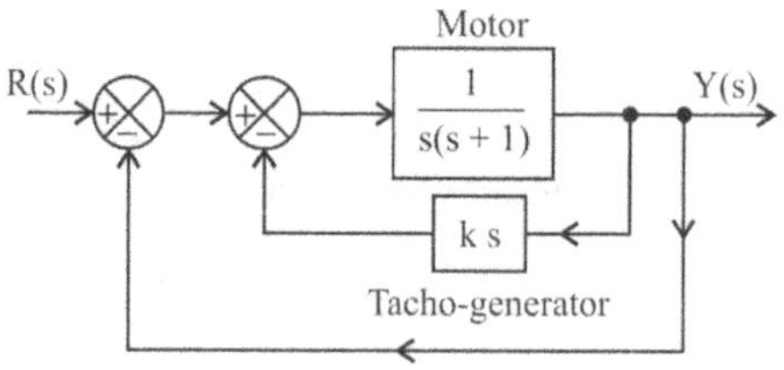

The gain k of the Tacho-generator influences mainly the

(a) Peak overshoot

(b) Natural frequency of oscillation

(c) Phase shift of the closed loop transfer function at very low frequency $(\omega \to 0)$

(d) Phase shift of the closed loop transfer function at very high frequencies $(\omega \to \infty)$

2012

1. A system with transfer function

$$G(s)=\frac{(s^2+9)(s+2)}{(s+1)(s+3)(s+4)}$$

is excited by sin (ωt). The steady-state output of the system is zero at

(a) $\omega = 1$ rad/s (b) $\omega = 2$ rad/s (c) $\omega = 3$ rad/s (d) $\omega = 4$ rad/s

2. The unilateral Laplace transform of f(t) is $\frac{1}{s^2+s+1}$. The unilateral Laplace transform of t f(t) is

(a) $-\frac{s}{\left(s^2+s+1\right)^2}$ (b) $-\frac{2s+1}{\left(s^2+s+1\right)^2}$

(c) $\frac{s}{\left(s^2+s+1\right)^2}$ (d) $\frac{2s+1}{\left(s^2+s+1\right)^2}$

3. The state variable description of an LTI system is given by

$$\begin{pmatrix} \dot{x}_1 \\ \dot{x}_2 \\ \dot{x}_3 \end{pmatrix} = \begin{pmatrix} 0 & a_1 & 0 \\ 0 & 0 & a_2 \\ a_3 & 0 & 0 \end{pmatrix} \begin{pmatrix} x_1 \\ x_2 \\ x_3 \end{pmatrix} + \begin{pmatrix} 0 \\ 0 \\ 1 \end{pmatrix} u$$

$$y = (1\ 0\ 0) \begin{pmatrix} x_1 \\ x_2 \\ x_3 \end{pmatrix}$$

Where y is the output and u is the input. The system is controllable for

(a) $a_1 \neq 0, a_2 = 0, a_3 \neq 0$ (b) $a_1 = 0, a_2 \neq 0, a_3 \neq 0$

(c) $a_1 = 0, a_2 \neq 0, a_3 = 0$ (d) $a_1 \neq 0, a_2 \neq 0, a_3 = 0$

4. The Fourier transform of a signal h(t) is $H(j\omega) = (2\cos\omega)(\sin 2\omega)/\omega$. The value of h(0) is

(a) $\frac{1}{4}$ (b) $\frac{1}{2}$ (c) 1 (d) 2

5. The feedback system shown below oscillates at 2 rad/s when

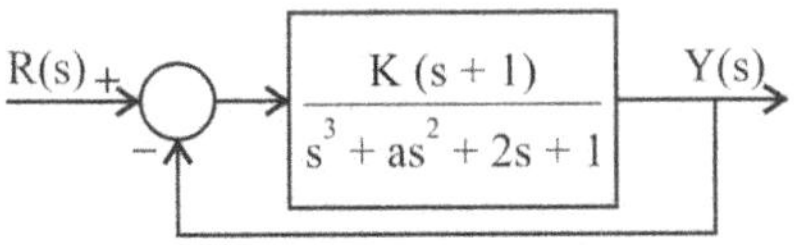

(a) K = 2 and a = 0.75 (b) K = 3 and a = 0.75

(c) K = 4 and a = 0.5 (d) K = 2 and a = 0.5

Common data for Q 6 and Q 7

The transfer function of the compensator is given as $G_c(s) = \frac{s+a}{s+b}$

6. $G_c(s)$ is a lead compensator if

   (a) $a = 1, b = 2$ (b) $a = 3, b = 2$

   (c) $a = -3, b = -1$ (d) $a = 3, b = 1$

7. The phase of the above lead compensator is maximum at

   (a) $\sqrt{2}$ rad/s (b) $\sqrt{3}$ rad/s (c) $\sqrt{6}$ rad/s (d) $\frac{1}{\sqrt{3}}$ rad/s

2013

1. A system is described by the differential equation $\frac{d^2y}{dt^2} + 5\frac{dy}{dt} + 6y(t) = x(t)$.

   Let x(t) be a rectangular pulse given by $x(t) = \begin{cases} 1 & 0 < t < 2 \\ 0 & \text{otherwise} \end{cases}$

   Assuming that y(0) = 0 and $\frac{dy}{dt} = 0$ at $t = 0$, the Laplace transform of y(t) is

   (a) $\frac{e^{-2s}}{s(s+2)(s+3)}$ (b) $\frac{1-e^{-2s}}{s(s+2)(s+3)}$

   (c) $\frac{e^{-2s}}{(s+2)(s+3)}$ (d) $\frac{1-e^{-2s}}{(s+2)(s+3)}$

Statement for linked answer questions 2 and 3

The state diagram of a system is shown below. A system is described by the state-variable equations

$$\dot{X} = AX + Bu;$$

$$y = CX + Du$$

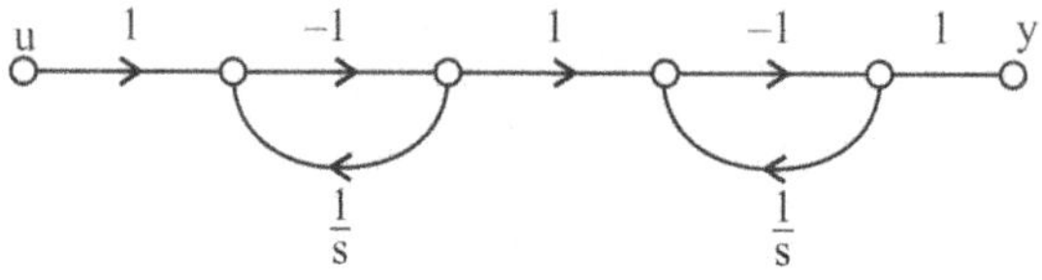

2. The state-variable equations of the system shown in the figure above are

(a) $\dot{X} = \begin{bmatrix} -1 & 0 \\ 1 & -1 \end{bmatrix} X + \begin{bmatrix} -1 \\ 1 \end{bmatrix} u$

$y = \begin{bmatrix} 1 & -1 \end{bmatrix} X + u$

(b) $\dot{X} = \begin{bmatrix} -1 & 0 \\ -1 & -1 \end{bmatrix} X + \begin{bmatrix} -1 \\ 1 \end{bmatrix} u$

$y = \begin{bmatrix} -1 & -1 \end{bmatrix} X + u$

(c) $\dot{X} = \begin{bmatrix} -1 & 0 \\ -1 & -1 \end{bmatrix} X + \begin{bmatrix} -1 \\ 1 \end{bmatrix} u$

$y = \begin{bmatrix} -1 & -1 \end{bmatrix} X - u$

(d) $\dot{X} = \begin{bmatrix} -1 & -1 \\ 0 & -1 \end{bmatrix} X + \begin{bmatrix} -1 \\ 1 \end{bmatrix} u$

$y = \begin{bmatrix} 1 & -1 \end{bmatrix} X - u$

3. The state transition matrix e^{At} of the system shown in the figure above is

(a) $\begin{bmatrix} e^{-t} & 0 \\ te^{-t} & e^{-t} \end{bmatrix}$

(b) $\begin{bmatrix} e^{-t} & 0 \\ -te^{-t} & e^{-t} \end{bmatrix}$

(c) $\begin{bmatrix} e^{-t} & 0 \\ e^{-t} & e^{-t} \end{bmatrix}$

(d) $\begin{bmatrix} e^{-t} & -te^{-t} \\ 0 & e^{-t} \end{bmatrix}$

4. Assuming zero initial condition, the response y(t) of the system given below to a unit step input u(t) is

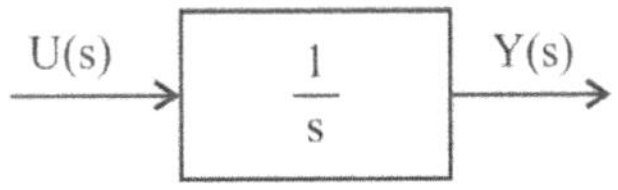

(a) u(t) (b) t u(t) (c) $\frac{t^2}{2}u(t)$ (d) $e^{-t}u(t)$

5. The impulse response of a system is h(t) = t u(t). For an input u(t – 1), the output is

(a) $\frac{t^2}{2}u(t)$

(b) $\frac{t(t-1)}{2}u(t-1)$

(c) $\frac{(t-1)^2}{2}u(t-1)$

(d) $\frac{t^2-1}{2}u(t-1)$

6. Which one of the following statements is NOT TRUE for a continuous time causal and stable LTI system?

(a) All the poles of the system must lie on the left side of the jω axis

(b) Zeros of the system can lie anywhere in the s-plane.

(c) All the poles must lie within $|s| = 1$.

(d) All the roots of the characteristic equation must be located on the left side of the jω axis.

7. Two systems with impulse responses $h_1(t)$ and $h_2(t)$ are connected in cascade. Then the overall impulse response of the cascaded system is given by

(a) Product of $h_1(t)$ and $h_2(t)$

(b) Sum of $h_1(t)$ and $h_2(t)$

(c) Convolution of $h_1(t)$ and $h_2(t)$

(d) Subtraction of $h_2(t)$ from $h_1(t)$

8. The Bode plot of a transfer function G (s) is shown in the figure below.

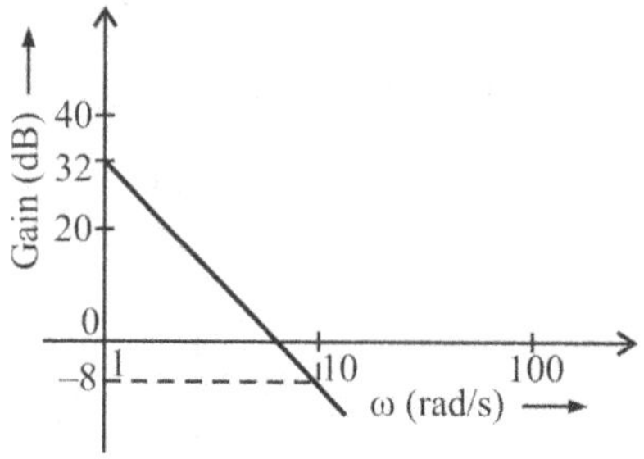

The gain ($20 \log |G(s)|$ is 32 dB and –8 dB at 1 rad/s and 10 rad/s respectively. The phase is negative for all ω. Then G(s) is

(a) $\frac{39.8}{s}$ (b) $\frac{39.8}{s^2}$ (c) $\frac{32}{s}$ (d) $\frac{32}{s^2}$

9. The open-loop transfer function of a dc motor is given as $\frac{\omega(s)}{V_a(s)} = \frac{10}{1+10s}$. When connected in feedback as shown below, the approximate value of K_a that will reduce the time constant of the closed loop system by one hundred times as compared to that of the open-loop system is

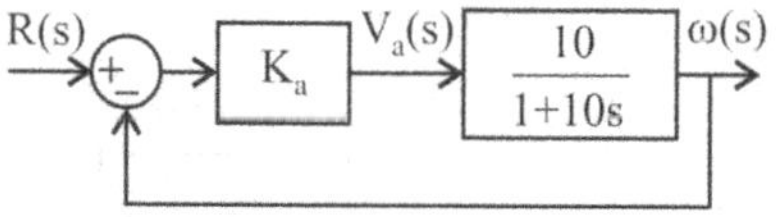

(a) 1 (b) 5 (c) 10 (d) 100

10. The signal flow graph for a system is given below. The transfer function $\frac{Y(s)}{U(s)}$ for this system is

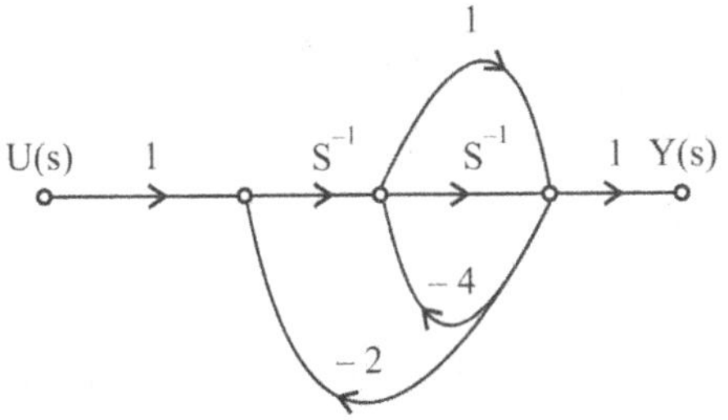

(a) $\dfrac{s+1}{5s^2+6s+2}$ (b) $\dfrac{s+1}{s^2+6s+2}$

(c) $\dfrac{s+1}{s^2+4s+2}$ (d) $\dfrac{1}{5s^2+6s+2}$

11. The impulse response of a continuous time system is given by $h(t) = \delta(t-1) + \delta(t-3)$. The value of the step response at $t = 2$ is

(a) 0 (b) 1 (c) 2 (d) 3

Common Data for Question 12 & 13

The state variable formulation of a system is given as

$$\begin{bmatrix} \dot{x}_1 \\ \dot{x}_2 \end{bmatrix} = \begin{bmatrix} -2 & 0 \\ 0 & -1 \end{bmatrix} \begin{bmatrix} x_1 \\ x_2 \end{bmatrix} + \begin{bmatrix} 1 \\ 1 \end{bmatrix} u,$$

$x_1(0) = 0$

$x_2(0) = 0$

and $y = \begin{bmatrix} 1 & 0 \end{bmatrix} \begin{bmatrix} x_1 \\ x_2 \end{bmatrix}$

12. The system is
    (a) Controllable but not observable
    (b) Not controllable but observable
    (c) Both controllable and observable
    (d) Both not controllable and not observable

13. The response y(t) to a unit step input is

(a) $\frac{1}{2} - \frac{1}{2}e^{-2t}$ (b) $1 - \frac{1}{2}e^{-2t} - \frac{1}{2}e^{-t}$

(c) $e^{-2t} - e^{-t}$ (d) $1 - e^{-t}$

2014

1. A continuous, linear time-invariant filter has an impulse response h(t) described by

$$h(t)=\begin{cases}3 & \text{for } 0\le t\le 3\\ 0 & \text{otherwise}\end{cases}$$

When a constant input of value 5 is applied to this filter, the steady state output is _______.

2. The forward path transfer function of a unity negative feedback system is given by

$$G(s)=\frac{K}{(s+2)(s-1)}$$

The value of K which will place both the poles of the closed-loop system at the same location, is_______.

3. Consider the feedback system shown in the figure. The Nyquist plot of G (s) is also shown. Which one of the following conclusions is correct?

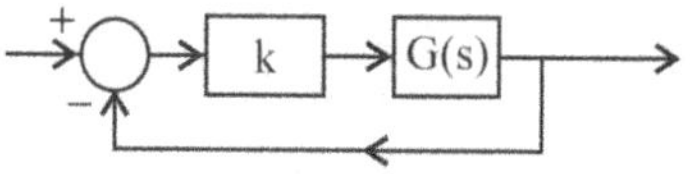

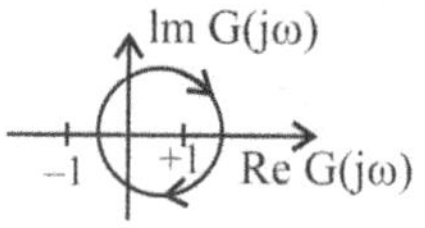

(a) G(s) is an all-pass filter

(b) G(s) is a strictly proper transfer function

(c) G(s) is a stable and minimum-phase transfer function

(d) The closed-loop system is unstable for sufficiently large and positive k

4. A system is described by the following differential equation, where u(t) is the input to the system and y(t) is the output of the system.

$$\dot{y}(t)+5y(t)=u(t)$$

When y(0) = 1 and u(t) is a unit step function, y(t) is

(a) $0.2+0.8e^{-5t}$ (b) $0.2-0.2e^{-5t}$

(c) $0.8+0.2e^{-5t}$ (d) $0.8-0.8e^{-5t}$

5. Consider the state space model of a system, as given below

$$\begin{bmatrix} \dot{x}_1 \\ \dot{x}_2 \\ \dot{x}_3 \end{bmatrix} = \begin{bmatrix} -1 & 1 & 0 \\ 0 & -1 & 0 \\ 0 & 0 & -2 \end{bmatrix} \begin{bmatrix} x_1 \\ x_2 \\ x_3 \end{bmatrix} + \begin{bmatrix} 0 \\ 4 \\ 0 \end{bmatrix} u;$$

$$y = \begin{bmatrix} 1 & 1 & 1 \end{bmatrix} \begin{bmatrix} x_1 \\ x_2 \\ x_3 \end{bmatrix}$$

The system is

(a) Controllable and observable (b) Uncontrollable and observable

(c) Uncontrollable and unobservable (d) Controllable and unobservable

6. The phase margin in degrees of $G(s) = \dfrac{10}{(s+0.1)(s+1)(s+10)}$ calculated using the asymptotic Bode plot is ______.

7. For the following feedback system $G(s) = \dfrac{1}{(s+1)(s+2)}$. The 2%-Settling time of the step response is required to be less than 2 seconds.

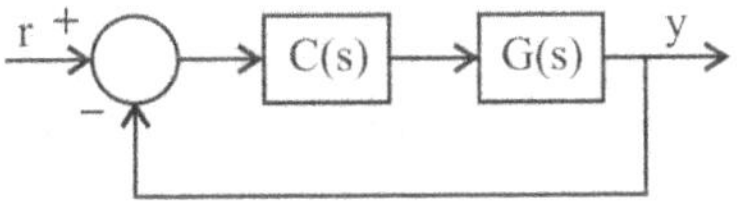

Which one of the following compensators C(s) achieves this?

(a) $3\left(\dfrac{1}{s+5}\right)$ (b) $5\left(\dfrac{0.03}{s}+1\right)$

(c) $2(s+4)$ (d) $4\left(\dfrac{s+8}{s+3}\right)$

8. Consider an LTI system with transfer function $H(s) = \dfrac{1}{s(s+4)}$, If the input to the system is cos (3t) and the steady state output is A sin (3t + a), then the value of A is

(a) 1/30 (b) 1/15 (c) 3/4 (d) 4/3

9. Consider an LTI system with impulse response $h(t) = e^{-5t}\,u(t)$. If the output of the system is $y(t) = (e^{-3t} - e^{-5t})\,u(t)$ then the input, x(t), is given by

(a) $e^{-3t}\,u(t)$ (b) $2\,e^{-3t}\,u(t)$ (c) $e^{-5t}\,u(t)$ (d) $2\,e^{-5t}\,u(t)$

10. The closed-loop transfer function of a system is $T(S)=\dfrac{4}{\left(s^2+0.4s+4\right)}$. The steady state error due to unit step input is ____.

11. The state transition matrix for the system $\begin{bmatrix}\dot{x}_1\\ \dot{x}_2\end{bmatrix}=\begin{bmatrix}1 & 0\\ 1 & 1\end{bmatrix}\begin{bmatrix}x_1\\ x_2\end{bmatrix}+\begin{bmatrix}1\\ 1\end{bmatrix}u$ is

(a) $\begin{bmatrix}e^t & 0\\ e^t & e^t\end{bmatrix}$ (b) $\begin{bmatrix}e^t & 0\\ t^2e^t & e^t\end{bmatrix}$

(c) $\begin{bmatrix}e^t & 0\\ te^t & e^t\end{bmatrix}$ (d) $\begin{bmatrix}e^t & te^t\\ 0 & e^t\end{bmatrix}$

12. A system with the open loop transfer function $G(s)=\dfrac{K}{s(s+2)\left(s^2+2s+2\right)}$ is connected in a negative feedback configuration with a feedback gain of unity. For the closed loop system to be marginally stable, the value of K is ____.

13. For the transfer function $G(s)=\dfrac{5(s+4)}{s(s+0.25)\left(s^2+4s+25\right)}$

The value of the constant gain term and the highest corner frequency of the Bode plot respectively are

(a) 3.2, 5.0 (b) 16.0, 4.0

(c) 3.2, 4.0 (d) 16.0, 5.0

14. The second order dynamic system $\dfrac{dX}{dt}=PX+Qu;\ y=RX$ has the matrices P, Q and R as follows:

$$P=\begin{bmatrix}-1 & 1\\ 0 & -3\end{bmatrix}\quad Q=\begin{bmatrix}0\\ 1\end{bmatrix}\quad R=[0\ 1]$$

The system has the following controllability and observability properties:

(a) Controllable and observable

(b) Not controllable but observable

(c) Controllable but not observable

(d) Not controllable and not observable

Answers

2003

1. (a)	2. (d)	3. (d)	4. (a)	5. (d)	6. (c)
7. (a)	8. (a)	9. (a)	10. (d)	11. (a)	12. (c)
13. (a)	14. (b)	15. (b)			

2004

1. (b)	2. (c)	3. (c)	4. (c)	5. (a)	6. (c)
7. (a)	8. (b)	9. (d)	10. (b)	11. (a)	12. (c)
13. (b)	14. (a)	15. (d)	16. (d)	17. (b)	18. (c)

2005

1. (b)	2. (d)	3. (a)	4. (b)	5. (c)	6. (a)
7. (a)	8. (c)	9. (d)	10. (d)	11. (c)	12. (a)
13. (d)	14. (c)	15. (c)	16. (b)	17. (c)	18. (a)
19. (c)					

2006

1. (d)	2. (a)	3. (b)	4. (d)	5. (b)	6. (c)
7. (d)	8. (c)	9. (c)	10. (b)	11. (d)	12. (a)
13. (c)	14. (d)				

2007

1. (b)	2. (a)	3. (d)	4. (d)	5. (a)	6. (a)
7. (d)	8. (d)	9. (b)	10. (d)	11. (d)	12. (d)
13. (a)					

2008

1. (c)	2. (c)	3. (d)	4. (c)	5. (a)	6. (c)
7. (d)	8. (a)	9. (a)	10. (c)	11. (c)	12. (c)
13. (d)	14. (a)				

2009

1. (d)	2. (c)	3. (b)	4. (b)	5. (d)	6. (d)
7. (d)	8. (b)	9. (d)	10. (d)	11. (c)	12. (d)

2010

1. (b)	2. (b)	3. (a)	4. (d)	5. (d)	6. (c)
7. (a)	8. (c)	9. (a)	10. (c)	11. (c)	12. (d)
13. (c)					

2011

1. (a)	2. (a)	3. (b)	4. (a)	5. (b)	6. (d)
7. (c)	8. (a)	9. (b)	10. (a)	11. (c)	12. (a)

2012

1. (c)	2. (d)	3. (d)	4. (c)	5. (a)	6. (a)
7. (a)					

2013

1. (b)	2. (a)	3. (a)	4. (b)	5. (c)	6. (c)
7. (c)	8. (b)	9. (c)	10. (a)	11. (b)	12. (a)
13. (a)					

2014

1. 4.5	2. 2.25	3. (d)	4. (a)	5. (b)	6. 4.5
7. (c)	8. (b)	9. (b)	10. 0	11. (c)	12. 5
13. (a)	14. (c)				

Bibliography

1. Automatic Control Systems ; B.C Kuo, Prentice Hall, India.
2. Control Systems Engineering ; I.J Nagrath, M. Gopal. New Age International Publishers.
3. Modern Control Engineering ; K Ogata. Prentice Hall, India.
4. Linear System Analysis ; David K. Cheng, Narosa Publishing House.

www.ingramcontent.com/pod-product-compliance
Lightning Source LLC
Chambersburg PA
CBHW061925190326
41458CB00009B/2651